HANDBOOK OF TECHNICAL TEXTILES

HANDBOOK OF TECHNICAL TEXTILES

Edited by
A R Horrocks and S C Anand

The Textile Institute

CRC Press
Boca Raton Boston New York Washington, DC

WOODHEAD PUBLISHING LIMITED
Cambridge England

Published by Woodhead Publishing Limited in association with The Textile Institute
Woodhead Publishing Limited
80 High Street, Sawston, Cambridge CB22 3HJ, UK
www.woodheadpublishing.com

Woodhead Publishing India Private Limited, G-2, Vardaan House, 7/28 Ansari Road, Daryaganj,
New Delhi – 110002, India

Published in North America by CRC Press LLC, 6000 Broken Sound Parkway, NW,
Suite 300, Boca Raton, FL 33487, USA

First published 2000, Woodhead Publishing Limited and CRC Press LLC
© 2000, Woodhead Publishing Limited except Chapter 16 © MOD, 2000
Reprinted 2004, 2007, 2008, 2009, 2011
The authors have asserted their moral rights.

British Library Cataloguing in Publication Data
A catalogue record for this book is available from the British Library.

Library of Congress Cataloging in Publication Data
A catalog record for this book is available from the Library of Congress.

Woodhead Publishing ISBN 978-1-85573-385-5 (print)
Woodhead Publishing ISBN 978-1-85573-896-6 (online)
ISSN 2042-0803 Woodhead Publishing Series in Textiles (print)
ISSN 2042-0811 Woodhead Publishing Series in Textiles (online)
CRC Press ISBN 978-0-8493-1047-8
CRC Press order number WP1047

The publishers' policy is to use permanent paper from mills that operate a sustainable forestry
policy, and which has been manufactured from pulp which is processed using acid-free
and elemental chlorine-free practices. Furthermore, the publishers ensure that the text paper and
cover board used have met acceptable environmental accreditation standards.

Printed by Lightning Source.

To the past and present staff members, support staff and students in Textile Studies at Bolton Institute, for their friendship and support over the years.

Contents

Preface

Technical textiles are reported to be the fastest growing sector of the textile industrial sector and account for almost 19% (10 million tonnes) of the total world fibre consumption for all textile uses, totalling 53 tonnes in 1997. This figure is likely to increase to 14 million tonnes by the year 2005. Technical textiles are estimated to account for well over 40% of the total textile production in many developed countries and, at the year 2000, account for almost 20% of all textile manufacturing in China (Byrne 1997).

The current volume of the market worldwide for technical textiles is more than $60 billion. The average annual growth rate of technical textiles worldwide is expected to be around 3.8% for the period 2000 to 2005.

The uniqueness and challenge of technical textiles lies in the need to understand and apply the principles of textile science and technology to provide solutions, in the main to technological problems but also often to engineering problems as well. With the emphasis on measurable textile performance in a particular field of application, this requires the technologist to have not only an intricate knowledge of fibres and textile science and technology but also an understanding of the application and the scientists, technologists and engineers who service it. Thus the producer of geotextiles requires an intricate knowledge of the world of civil engineering, and the medical textile producer, the requirements of consultant, medical practitioner and nurse. This series attempts to provide a bridge between producer and end-user.

The main principles involved in the selection of raw materials and their conversion into yarns and fabrics followed by dyeing, finishing and coating of technical textiles are explored, followed by the raw materials, processing techniques, finishing, specifications, properties and special technical and commercial features of a wide range of specific areas of application.

Each of the chapters has been specially prepared and edited to cover current developments as well as future trends in both the principles of manufacture and the state-of-the-art constructional specifications, properties, test methods and standards of the major product areas and applications of technical textiles.

A team of internationally famous authors has contributed a great deal of time, effort and above all special and significant expertise and experience to the preparation of this handbook. The editors wish to extend their most sincere thanks to all the authors for their important contribution, patience and cooperation. This book once again confirms that enthusiasm and love of the subject are more important than the financial gains.

Special thanks are also given to Patricia Morrison of Woodhead Publishing Ltd, Cambridge for her consistent interest and effort in keeping this project warm for so long and her continued faith in the editors.

Professor Richard Horrocks
Professor Subhash Anand
Faculty of Technology
Bolton Institute
Deane Road
BOLTON
BL3 5AB
UK

List of contributors

Professor S C Anand
Faculty of Technology, Bolton Institute, Deane Road, Bolton BL3 5AB, UK

Professor P Bajaj
Department of Textile Technology, Indian Institute of Technology, Hauz Khas, New Delhi, India

Mr C Byrne
David Rigby Associates, Peter House, St Peter's Square, Manchester M1 5AQ, UK

Dr X Chen
Department of Textiles, UMIST, P O Box 88, Sackville Street, Manchester M60 1QD, UK

Mr W Fung
Collins and Aikman, P O Box 29, Warley Mills, Walkden, Manchester M28 3WG, UK

Dr R H Gong
Department of Textiles, UMIST, P O Box 88, Sackville Street, Manchester M60 1QD, UK

Dr M Hall
Department of Textiles, Faculty of Technology, Bolton Institute, Deane Road, Bolton BL3 5AB, UK

Mr E Hardman
Madison Filters (formerly Scapa Filtration), Haslingden, Rossendale, Lancashire, UK

Dr I Holme
Department of Textiles, University of Leeds, Leeds LS2 1JT, UK

Dr D Holmes
Department of Textiles, Faculty of Technology, Bolton Institute, Deane Road, Bolton BL3 5AB, UK

Dr M Miraftab
Department of Textiles, Faculty of Technology, Bolton Institute, Deane Road, Bolton BL3 5AB, UK

Dr S Ogin
School of Mechanical and Materials Engineering, University of Surrey, Guildford, Surrey GU2 7HX, UK

Dr M Pritchard
Department of the Built Environment, Faculty of Technology, Bolton Institute, Deane Road, Bolton BL3 5AB, UK

Professor P R Rankilor
9 Blairgowrie Drive, West Tytherington, Macclesfield, Cheshire SK10 2UJ, UK

Mr A J Rigby
Department of Textiles, Faculty of Technology, Bolton Institute, Deane Road, Bolton BL3 5AB, UK

Professor S W Sarsby
Department of the Built Environment, Faculty of Technology, Bolton Institute, Deane Road, Bolton BL3 5AB, UK

Dr R A Scott
MOD, Defence Clothing and Textiles Agency Science and Technology Division, Flagstaff Road, Colchester, Essex CO2 7SS, UK

Professor K Slater
Department of Textiles, University of Guelph, Guelph, Ontario N1G 2W1, Canada

Dr P Smith
26 Newhall Park, Otley, Leeds LS21 2RD, UK

Mr W Sondhelm
10 Bowlacre Road, Hyde, Cheshire SK14 5ES, UK

1

Technical textiles market – an overview

Chris Byrne, Principal Consultant

David Rigby Associates, Peter House, St Peter's Square, Manchester M1 5AQ, UK

1.1 Introduction

Although 'technical' textiles have attracted considerable attention, the use of fibres, yarns and fabrics for applications other than clothing and furnishing is not a new phenomenon. Nor is it exclusively linked to the emergence of modern artificial fibres and textiles. Natural fibres such as cotton, flax, jute and sisal have been used for centuries (and still are used) in applications ranging from tents and tarpaulins to ropes, sailcloth and sacking. There is evidence of woven fabrics and meshes being used in Roman times and before to stabilise marshy ground for road building – early examples of what would now be termed geotextiles and geogrids.

What is relatively new is a growing recognition of the economic and strategic potential of such textiles to the fibre and fabric manufacturing and processing industries of industrial and industrialising countries alike. In some of the most developed markets, technical products (broadly defined) already account for as much as 50% of all textile manufacturing activity and output. The technical textiles supply chain is a long and complex one, stretching from the manufacturers of polymers for technical fibres, coating and speciality membranes through to the converters and fabricators who incorporate technical textiles into finished products or use them as an essential part of their industrial operations. The economic scope and importance of technical textiles extends far beyond the textile industry itself and has an impact upon just about every sphere of human economic and social activity.

And yet this dynamic sector of the textile industry has not proved entirely immune to the effects of economic recession, of product and market maturity, and of growing global competition which are all too well known in the more traditional sectors of clothing and furnishings. There are no easy paths to success and manufacturers and converters still face the challenge of making economic returns commensurate with the risks involved in operating in new and complex markets. If anything, the constant need to develop fresh products and applications, invest in new processes and equipment, and market to an increasingly diverse range of customers, is more demanding and costly than ever.

Technical textiles has never been a single coherent industry sector and market segment. It is developing in many different directions with varying speeds and levels of success. There is continual erosion of the barriers between traditional definitions of textiles and other 'flexible engineering' materials such as paper and plastics, films and membranes, metals, glass and ceramics. What most participants have in common are many of the basic textile skills of manipulating fibres, fabrics and finishing techniques as well as an understanding of how all these interact and perform in different combinations and environments. Beyond that, much of the technology and expertise associated with the industry resides in an understanding of the needs and dynamics of many very different end-use and market sectors. It is here that the new dividing lines within the industry are emerging.

An appreciation of the development and potential of technical textile markets therefore starts with some clarification of the evolving terminology and definitions of scope of the industry and its markets. This chapter goes on to consider some of the factors – technical, commercial and global – which are driving the industry forward.

It also considers how the emergence of new geographical markets in China and other rapidly industrialising regions of the world looks set to be one of the major influences on the growth and location of technical textiles manufacturing in the first 10 years of the 21st century.

1.2   Definition and scope of technical textiles

The definition of technical textiles adopted by the authoritative *Textile Terms and Definitions*, published by the Textile Institute[1], is 'textile materials and products manufactured primarily for their technical and performance properties rather than their aesthetic or decorative characteristics'.

Such a brief description clearly leaves considerable scope for interpretation, especially when an increasing number of textile products are combining both performance and decorative properties and functions in equal measure. Examples are flame retardant furnishings and 'breathable' leisurewear. Indeed, no two published sources, industry bodies or statistical organisations ever seem to adopt precisely the same approach when it comes to describing and categorising specific products and applications as technical textiles.

It is perhaps not surprising that any attempt to define too closely and too rigidly the scope and content of technical textiles and their markets is doomed to failure. In what is one of the most dynamic and broad ranging areas of modern textiles, materials, processes, products and applications are all changing too rapidly to define and document. There are even important linguistic and cultural perceptions of what constitutes a technical textile from geographical region to region in what is now a global industry and marketplace.

1.2.1   Technical or industrial textiles: what's in a name?
For many years, the term 'industrial textiles' was widely used to encompass all textile products other than those intended for apparel, household and furnishing end-uses. It is a description still more widely favoured in the USA than in Europe and elsewhere (see, for example, the Wellington Sears Handbook of Industrial Textiles).[2]

This usage has seemed increasingly inappropriate in the face of developing applications of textiles for medical, hygiene, sporting, transportation, construction, agricultural and many other clearly non-industrial purposes. Industrial textiles are now more often viewed as a subgroup of a wider category of technical textiles, referring specifically to those textile products used in the course of manufacturing operations (such as filters, machine clothing, conveyor belts, abrasive substrates etc.) or which are incorporated into other industrial products (such as electrical components and cables, flexible seals and diaphragms, or acoustic and thermal insulation for domestic and industrial appliances).

If this revised definition of industrial textiles is still far from satisfactory, then the problems of finding a coherent and universally acceptable description and classification of the scope of technical textiles are even greater. Several schemes have been proposed. For example, the leading international trade exhibition for technical textiles, Techtextil (organised biennially since the late 1980s by Messe Frankfurt in Germany and also in Osaka, Japan), defines 12 main application areas (of which textiles for industrial applications represent only one group):

- agrotech: agriculture, aquaculture, horticulture and forestry
- buildtech: building and construction
- clothtech: technical components of footwear and clothing
- geotech: geotextiles and civil engineering
- hometech: technical components of furniture, household textiles and floorcoverings
- indutech: filtration, conveying, cleaning and other industrial uses
- medtech: hygiene and medical
- mobiltech: automobiles, shipping, railways and aerospace
- oekotech: environmental protection
- packtech: packaging
- protech: personal and property protection
- sporttech: sport and leisure.

The search for an all embracing term to describe these textiles is not confined to the words 'technical' and 'industrial'. Terms such as performance textiles, functional textiles, engineered textiles and high-tech textiles are also all used in various contexts, sometimes with a relatively specific meaning (performance textiles are frequently used to describe the fabrics used in activity clothing), but more often with little or no precise significance.

1.2.2 Operating at the boundaries of textiles

If the adjective 'technical' is difficult to define with any precision, then so too is the scope of the term textiles. Figure 1.1 summarises the principal materials, processes and products which are commonly regarded as falling within the scope of technical textiles manufacturing.

However, there remain many grey areas. For example, the manufacture and processing of metallic wires into products such as cables, woven or knitted screens and meshes, and reinforcing carcasses for tyres are not generally regarded as lying within the scope of the textile industry. This is despite the fact that many of the techniques employed and the final products obtained are closely related to conventional textile fibre equivalents.

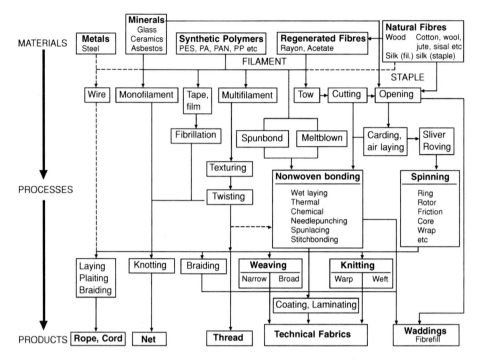

1.1   Technical textile materials, processes and products. PES = polyester, PA = polyamide, PAN = polyacrylonitrile.

Within the composites industry, woven, knitted, braided, nonwoven and wound yarn reinforcements made from glass, carbon fibre and organic polymer materials such as aramids are all now widely accepted as being technical textile products. On the other hand, more loosely structured reinforcing materials such as chopped strand mat, milled glass and pulped organic fibres are often excluded.

The nonwovens industry has developed from several different technology directions, including paper manufacturing. The current definition of a nonwoven promulgated, for example, under the International Standards Organization standard ISO 9092[3] acknowledges a number of borderline areas, including wet-laid products and extruded meshes and nets. Likewise, distinctions between textile fibres and filaments, slit or fibrillated films, monofilaments and extruded plastics inevitably boil down to some fairly arbitrary and artificial criteria. Diameter or width is often used as the defining characteristic, irrespective of the technologies used or the end-uses served. Many of the definitions and categories embodied within existing industry statistics reflect historical divisions of the main manufacturing sectors rather than a functional or market-based view of the products involved.

Polymer membranes, composite materials and extruded grids and nets are other products which challenge traditional notions of the scope of technical textile materials, processes and products. Increasingly, technical textiles are likely to find their place within a broader industry and market for 'flexible engineering materials' (Fig. 1.2). A number of companies and groups have already adopted this outlook and operate across the boundaries of traditional industry sectors, focusing a range

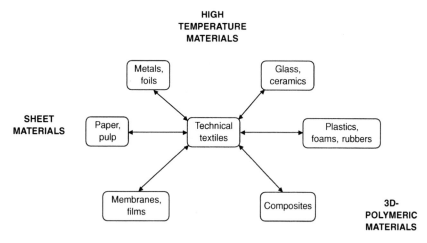

1.2   Scope of flexible engineering materials.

of materials, process technologies and product capabilities upon specific functions and markets such as filtration and health care.

1.2.3   Inconsistent statistical reporting

To add to this complexity, different geographical regions and countries tend to adopt rather different viewpoints and definitions with regard to all of the above. A widely quoted misconception that technical textiles in Japan account for over 40% of all textile output or nearly twice the level in Western Europe can largely be put down to the different statistical bases employed. In Europe, the most authoritative source of fibre consumption (and therefore textile output) data is CIRFS (Comité International de la Rayonne et des Fibres Synthétiques), the European artificial fibre producers association. However, CIRFS' reported statistics (at least until recently) have specifically excluded tape and film yarns (a significant proportion of all polyolefin textiles), coarser monofilaments and all glass products (as well as natural fibres such as jute, flax, sisal, etc.). The merger of CIRFS and EATP, the European Polyolefin Textiles Association, should go some way towards resolving this anomaly.

The Japanese 'Chemical' Fibres Manufacturers Association, JCFA, at the other extreme, includes all these products, including natural fibres, within its definition of technical/industrial textiles while the Fiber Statistics Bureau in the USA includes polyolefin tape and monofilament yarns but excludes glass. Table 1.1 attempts to restate the relative usage of the main technical fibres and yarns on a more consistent basis.

In this new light, Japan still retains a leading position worldwide in terms of the proportion of its total textile manufacturing output devoted to technical textiles. However, this is largely a reflection of the importance of its automotive manufacturing industry (a key user of technical textiles) combined with the relatively smaller size of its apparel and furnishing textile sectors (especially floor coverings). The USA apparently accounts for the lowest proportion of technical

Table 1.1   Comparative levels of technical fibre mill consumption, 1995

	% Total fibre consumption + kg per capita		
	Textile fibres only	Textile fibre, PP tape and monofilaments	Textile fibre, PP tape, monofilaments and glass
W Europe	21% 2.8 kg	28% 4.2 kg	34% 5.6 kg
USA	18% 4.9 kg	22% 6.4 kg	30% 9.7 kg
Japan	30% 3.3 kg	35% 4.3 kg	41% 5.5 kg

Source: CIRFS, Fiber Organon, JCFA and David Rigby Associates estimates.
PP = polypropylene.

textile output of the three major industrial regions but still produces and consumes the largest quantity per capita, especially when all glass textile and technical fibre uses are included.

1.3   Milestones in the development of technical textiles

Although the development of technical and industrial applications for textiles can be traced back many years, a number of more recent milestones have marked the emergence of technical textiles as we know them today. Very largely, these have centred upon new materials, new processes and new applications.

1.3.1   Developments in fibre materials – natural fibres

Until early in the 20th century, the major fibres available for technical and industrial use were cotton and various coarser vegetable fibres such as flax, jute and sisal. They were typically used to manufacture heavy canvas-type products, ropes and twines, and were characterised by relatively heavy weight, limited resistance to water and microbial/fungal attack as well as poor flame retardancy.

Some of the present day regional patterns of technical textiles manufacturing were established even then, for example Dundee, on the east coast of Scotland and located at the centre (then) of an important flax growing area as well as being a whaling port. Following the discovery that whale oil could be used to lubricate the spinning of the relatively coarse jute fibres then becoming available from the Indian subcontinent, jute fabrics were widely used for sacking, furniture and carpet manufacturing, roofing felts, linoleum flooring, twine and a host of other applications.

Although its jute industry was to decline dramatically from a peak at around 1900 owing to competition from other materials as well as from cheaper imports, Dundee and the surrounding industry subsequently become a nucleus for development of the UK polypropylene industry in the 1960s. The then newly available polymer proved not only to be an ideal technical substitute for the natural product but was also much more consistent in terms of its supply and price.

Traditional end-uses for sisal were similarly rapidly substituted throughout the established rope, twine and net making centres of Europe and America.

Wool proved far less versatile and economic for most industrial applications although it is still valued for its insulating and flame retardancy properties and finds use in several high temperature and protective clothing applications. Silk is an even more exotic fibre, rarely used in technical applications other than for highly specialised uses such as surgical suture thread. However, the traces of the early silk industry are still to be seen in the present day location of centres for technical filament weaving such as the Lyons area of France. The traditional silk industry has also contributed to the development of technical textiles in Asia, especially in Japan.

1.3.2 Viscose rayon
The first commercially available synthetic fibre, viscose rayon, was developed around 1910 and by the 1920s had made its mark as reinforcement material for tyres and, subsequently, other mechanical rubber goods such as drive belts, conveyors and hoses. Its relatively high uniformity, tenacity and modulus (at least when kept dry within a rubber casing), combined with good temperature resistance, proved ideal for the fast emerging automotive and industrial equipment markets.

At a much later stage of its lifecycle, other properties of viscose such as its good absorbency and suitability for processing by paper industry-type wet laying techniques contributed to its role as one of the earliest and most successful fibres used for nonwoven processing, especially in disposable cleaning and hygiene end-uses.

1.3.3 Polyamide and polyester
Polyamide (nylon) fibre, first introduced in 1939, provided high strength and abrasion resistance, good elasticity and uniformity as well as resistance to moisture. Its excellent energy absorbing properties proved invaluable in a range of end-uses from climbing ropes to parachute fabrics and spinnaker sails. Polyamide-reinforced tyres are still used much more extensively in developing countries where the quality of road surfaces has traditionally been poor as well as in the emerging market for off-road vehicles worldwide. This contrasts to Western Europe where average road speeds are much greater and the heat-resistant properties of viscose are still valued.

From the 1950s onwards, the huge growth in world production of polyester, initially for apparel and household textile applications, provided the incentive and economies of scale needed to develop and engineer this fibre as a lower cost alternative to both viscose and polyamide in an increasing range of technical applications.

Nowhere is this more true than Japan and the developing industrial economies of Asia, including China, where production capacities for both polyester staple and filament yarn are extremely high and there is an urgent search for new applications. Some high volume applications for technical textiles which would typically use polyolefins in western Europe and North America such as geotextiles, carpet backing and coverstock are more likely to use polyester in Asia largely because of the greater availability and better economics of fibre supplies in those regions.

At a slightly less obvious level, differences in the polyamide supply situation – Western Europe and North America are more strongly oriented towards nylon 66 while Asia and Eastern Europe produce predominantly nylon 6 – are reflected in different manufacturing practices, product types and technical applications for this fibre.

Yet another example is the production and use of Vinylon (PVA, polyvinyl alcohol) fibres in Japan, where they were developed for a variety of industrial and technical applications at a time when that country lacked other raw materials and fibre production capabilities. Use of this fibre for technical textiles is almost non-existent in the West.

1.3.4   Polyolefins

The development of polyolefin (mostly polypropylene but also some polyethylene) fibres as well as tape and film yarns in the 1960s was another milestone in the development of technical textiles. The low cost and easy processability of this fibre, combined with its low density and good abrasion and moisture-resistant properties, have allowed its rapid introduction into a range of applications such as sacks, bags and packaging, carpet backings and furniture linings as well as ropes and netting. Many of these markets were directly taken over from jute and similar fibres but newer end-uses have also been developed, including artificial sports surfaces.

Properties of the polyolefins such as their poor temperature resistance and complete hydrophobicity have been turned to advantage in nonwovens. Initially used in conjunction with viscose to permit thermal bonding, polypropylene has now bene-fited from a growing appreciation of the important role that moisture wicking (as opposed to absorption) can play in hygiene applications such as coverstock for diapers (nappies). Finally, the relatively low extrusion temperatures of the poly-olefins have proved ideally suited to the fast developing technologies of spin laying (spun bonding and melt blowing).

As noted above, the development of the polypropylene industry was initially focused on European and North American markets. However, it is undergoing a major expansion worldwide as new investment in polymer capacity offers more favourable economics to new geographical markets.

1.3.5   High performance fibres

The above 'conventional' fibre types, both chemical and natural, still account for over 95% of all organic fibre technical textiles in use (i.e. excluding glass, mineral and metal fibres). Many of them have been modified and tailored to highly specific end-uses by adjustment of their tenacity, length, decitex, surface profile, finish and even by their combination into hybrid and bicomponent products. However, it is the emergence of the so-called high performance fibres since the early 1980s that has provided some of the most significant and dramatic impulses to the evolution of technical textiles.

First and foremost of these are the aramids, both the highly temperature-resistant *meta*-aramids (widely used in protective clothing and similar applications) and the high strength and modulus *para*-aramids (used in a host of applications ranging from bulletproof vests to reinforcement of tyres, hoses, friction materials, ropes and advanced composites). From their commercial introduction in the 1970s, world demand for *p*-aramids is expected to reach almost 40 000 tonnes per annum by 2000 while for *m*-aramids, consumption will be around 17–18 000 tonnes.

While not huge in overall terms (representing less than 0.5% of total world technical fibre and yarn usage in volume terms but closer to 3–4% in value), the aramids represent a particularly important milestone in the development of the technical textiles industry. Partly practical and partly symbolic, the introduction of the aramids not only led to the injection of large amounts of technical and market support into the industry and for users by leading fibre manufacturers such as DuPont and Akzo, but also concentrated the minds of many developers of new products upon the possibilities (and practicalities) of using similar new generation materials.

The early success of the aramids was a welcome contrast to the development of carbon fibres, which have been commercially available since the 1960s but largely constrained by their high material and processing costs to selected high value markets, particularly aerospace applications. Total world demand for carbon fibres was still only some 8–9000 tonnes per annum as recently as 1995. In fact, their market actually shrank in the early 1990s owing to cutbacks in military spending.

At long last, carbon fibres appear to be emerging from the doldrums, with the appearance not only of important new civil aerospace markets but also of high technology sporting goods and industrial applications such as wind generator turbine blades and reinforced fuel tanks. As new manufacturing methods and greater economies of scale start to bring prices down, the feasibility of even larger scale applications such as the reinforcement of buildings and structures in earthquake zones becomes more attractive. Currently, (2000), consumption is considered to be over 13 000 tonnes per annum, rising to almost 19 000 tonnes by the year 2005.

The introduction of other high performance fibres proliferated, particularly during the late 1980s, and in the wake of the aramids. These included a range of heat and flameproof materials suitable for protective clothing and similar applications (such as phenolic fibres and PBI, polybenzimidazole), ultra-strong high modulus polyethylene (HMPE) for ballistic protection and rope manufacture, and chemically stable polymers such as polytetrafluoroethylene (PTFE), polyphenylene sulphide (PPS) and polyethyletherketone (PEEK) for use in filtration and other chemically aggressive environments.

Individually, none of these other fibres has yet achieved volume sales anywhere near those of the aramids (or even carbon fibres). Indeed, the output of some speciality fibres can still be measured in tens of tonnes per year rather than hundreds or thousands. The widespread industrial recession of the early 1990s caused many fibre manufacturers to review their development strategies and to focus upon narrower ranges of products and markets.

1.3.6 Glass and ceramics
Glass has, for many years, been one of the most underrated technical fibres. Used for many years as a cheap insulating material as well as a reinforcement for relatively low performance plastics (fibre glass) and (especially in the USA) roofing materials, glass is increasingly being recognised as a sophisticated engineering material with excellent fire and heat-resistant properties. It is now widely used in a variety of higher performance composite applications, including sealing materials and rubber reinforcement, as well as filtration, protective clothing and packaging.

The potential adoption of high volume glass-reinforced composite manufacturing techniques by the automotive industry as a replacement for metal body parts and components, as well as by manufacturing industry in general for all sorts of industrial and domestic equipment, promises major new markets. Total world consumption of 'textile' glass in technical applications was some 2.3 million tonnes per annum in 1995 and is considered likely to be over 2.9 million tonnes at 2000, representing over 20% of all technical fibre consumption.

Various higher performance ceramic fibres have been developed but are restricted to relatively specialised applications by their high cost and limited mechanical properties.

1.4   Textile processes

Figure 1.1 summarises the wide range of processes employed in the manufacture of technical textiles. Apart from the use of plaiting and knotting for the manufacture of ropes and nets, weaving was, for many years, the pre-eminent technology employed in the manufacture of 'industrial' textiles. In terms of the total weight of textiles produced, weaving still plays a leading role and developments such as three-dimensional and crimpless weaving have opened up many new product and end-use possibilities.

However, the historical progress of technical textiles has seen the advance of alternative textile forming technologies, most prominently the broad family of non-woven techniques but also warp and weft knitting, stitchbonding and modern braiding methods. The use of loose fibres with sophisticated cross-sectional profiles for insulation, protection and fibrefill applications is another important growth area. Fibres, yarns and textiles of all types also provide the starting point for a diverse and fast expanding range of composite reinforcement and forming technologies.

According to a major study of the world technical textiles industry and its markets projected to 2005 (see Table 1.2), nonwovens are set to overtake weaving (in terms of the total weight of textiles produced) by around 2002/2003. In area terms, nonwovens already far exceed woven and other fabric forming methods because of their lower average weight per unit area. On the other hand, woven and other yarn-based fabrics will remain in the lead in value terms, at least for the fore-seeable future.

There is, therefore, something for every section of the textile industry in the future of technical textiles. Most product areas will see more rapid growth in value

Table 1.2   Worldwide consumption of technical textiles by product type, 2000–2005

	10^3 tonnes			$ million		
	2000	2005	Growth (% pa)	2000	2005	Growth (% pa)
Fabrics	3760	4100	1.7%	26710	29870	2.2%
Nonwovens	3300	4300	5.4%	14640	19250	5.6%
Composites	1970	2580	5.5%	6960	9160	5.6%
Other textiles[a]	2290	2710	3.4%	11950	14060	3.3%
All textile products	11320	13690	3.9%	60260	72340	3.7%

Source: David Rigby Associates/Techtextil.
[a] Includes ropes, twines, thread, fibrefill etc.

Table 1.3   Worldwide consumption of technical textiles by application, 2000–2005

	10³ tonnes			$ million		
	2000	2005	Growth (% pa)	2000	2005	Growth (% pa)
Transport textiles (auto, train, sea, aero)	2 220	2 480	2.2	13 080	14 370	1.9
Industrial products and components	1 880	2 340	4.5	9 290	11 560	4.5
Medical and hygiene textiles	1 380	1 650	3.6	7 820	9 530	4.0
Home textiles, domestic equipment	1 800	2 260	4.7	7 780	9 680	4.5
Clothing components (thread, interlinings)	730	820	2.3	6 800	7 640	2.4
Agriculture, horticulture and fishing	900	1 020	2.5	4 260	4 940	3.0
Construction – building and roofing	1 030	1 270	4.3	3 390	4 320	5.0
Packaging and containment	530	660	4.5	2 320	2 920	4.7
Sport and leisure (excluding apparel)	310	390	4.7	2 030	2 510	4.3
Geotextiles, civil engineering	400	570	7.3	1 860	2 660	7.4
Protective and safety clothing and textiles	160	220	6.6	1 640	2 230	6.3
Total above	**11 340**	**13 680**	**3.9**	**60 270**	**72 360**	**3.7**
Ecological protection textiles[a]	230	310	6.2	1 270	1 610	4.9

Source: David Rigby Associates/Techtextil.
[a] Already counted in several categories above.

than in volume as technical textiles become increasingly sophisticated and employ more specialised and higher value raw materials. On the other hand, the total value of yarns and fibres and of all technical textile products will grow slightly less fast than their volume because of a changing mix of materials and technologies, especially reflecting the growth of nonwovens.

1.5   Applications

The same study identified size and growth trends in each major application area for technical textiles, as defined by the organisers of Techtextil. The results are presented in Table 1.3.

Ecological textiles were identified as a separate and potentially important growth segment but are not consolidated in the total consumption figure because they have already been counted under headings such as industrial textiles (filtration media, oil spill protection and absorption) and geotextiles (geomembrane liners for toxic waste pits, erosion protection textiles, etc.).

Some selected examples of these broad trends which illustrate key aspects of the development and use of technical textiles are discussed in further detail below.

1.5.1   Transport textiles

Transport applications (cars, lorries, buses, trains, ships and aerospace) represent the largest single end-use area for technical textiles, accounting for some 20% of the total. Products range from carpeting and seating (regarded as technical rather than furnishing textiles because of the very stringent performance characteristics which they must fulfil), through tyre, belt and hose reinforcement, safety belts and air bags, to composite reinforcements for automotive bodies, civil and military aircraft bodies, wings and engine components, and many other uses.

The fact that volume and value growth rates in these applications appear to be amongst the lowest of any application area needs to be interpreted with caution. The automotive industry (which accounts for a high proportion of all transport textiles) is certainly one of the most mature in market terms. Growth rates in new end-uses such as air bags and composite materials will continue to outstrip the above averages by a considerable margin for many years to come. However, total technical textile usage is, in many ways, a victim of its own success. Increasing sophistication in the specifications and uses of textile materials has led to the adoption of lighter, stronger, more precisely engineered yarns, woven and knitted fabrics and nonwovens in place of established materials. The decreasing weight per tyre of textile reinforcing cord in modern radial constructions is one example of this. Interior textiles in cars are also making use of lighter weight and lower cost nonwovens.

Modern textiles also last longer. Hoses and belts which used to use substantial quantities of textile reinforcements are now capable of lasting the lifetime of a vehicle, removing much of the large and continuing 'after-market' for textile products.

The automotive industry has led the world in the introduction of tightly organised supply chain structures and textiles are no exception. Technical textile producers have had to learn the language and practice of precision engineering, just-in-time supply relationships and total quality management. The ideas and systems developed to serve the automotive industry have gradually filtered through to other markets and have had a profound effect in many different areas. Meanwhile, the major automotive companies have become increasingly global players in a highly competitive market and have demanded of their suppliers that they follow suit. The supply of textiles to this market is already dominated by a relatively few large companies in each product area. Worldwide manufacturing capabilities and strategic relationships are essential to survival and many smaller players without these resources have already exited from the market. Recessionary cycles in automotive markets as well as in military and civil aerospace applications have dealt some severe blows and only those companies with the long term commitment and strength to survive are likely to benefit from the better times that the market also periodically enjoys.

1.5.2   Industrial products and components

Set to rival transport textiles for first place by the year 2005 or shortly thereafter (in volume terms, although not yet in value) is the diverse field of 'industrial' textiles. As now more precisely defined, this includes textiles used directly in industrial processes or incorporated into industrial products such as filters, conveyor belts and abrasive belts, as well as reinforcements for printed circuit boards, seals and gaskets, and other industrial equipment.

Use of nonwovens already considerably outweighs that of woven and other fabric types here; consumption in 2000 is estimated at 700 000 tonnes and a little over 400 000 tonnes, respectively. However, both are surpassed by the use of technical fibres and textiles for composite reinforcement, over 740 000 tonnes in 2000.

Growth rates are generally well above average in most areas. Because of the universal nature of many industrial requirements, some large companies have emerged with worldwide manufacturing and distribution to dominate markets for industrial textile products. They include companies such as Scapa (UK) and Albany (US), leaders in papermaking felts and related product areas, Milliken (USA) in textiles for rubber reinforcement and other industrial applications and BWF (Germany) in filtration.

1.5.3   Medical and hygiene textiles

The fact that medical and hygiene textiles are expected to show below average growth in volume but above average growth in value reflects the contrasting prospects of at least two main areas of the market.

The largest use of textiles is for hygiene applications such as wipes, babies' diapers (nappies) and adult sanitary and incontinence products. With the possible exception of the last of these, all are relatively mature markets whose volume growth has peaked. Manufacturers and converters now seek to develop them further by adding value to increasingly sophisticated products. Nonwovens dominate these applications which account for over 23% of all nonwoven use, the largest proportion of any of the 12 major markets for technical textiles.

Concern has been expressed at the growth of disposable products and the burden which they place upon landfill and other waste disposal methods. Attempts have been made to develop and introduce more efficient biodegradable fibres for such end-uses but costs remain high. Meanwhile, the fastest areas of growth are in developing and newly industrialised markets where product penetration is still relatively low; Asia is a particular target for many of the big name brand manufacturers who operate in this area.

The other side of the medical and hygiene market is a rather smaller but higher value market for medical and surgical products such as operating gowns and drapes, sterilisation packs, dressings, sutures and orthopaedic pads. At the highest value end of this segment are relatively tiny volumes of extremely sophisticated textiles for uses such as artificial ligaments, veins and arteries, skin replacement, hollow fibres for dialysis machines and so on. Growth prospects in these areas are potentially considerable although the proving and widespread introduction of new life-critical products takes time.

1.5.4   Home textiles

By far the largest area of use for other textiles as defined above, that is other than fabrics, nonwovens and composite reinforcements, over 35% of the total weight of fibres and textiles in that category, lies in the field of household textiles and furnishing and especially in the use of loose fibres in wadding and fibrefill applications. Hollow fibres with excellent insulating properties are widely used in bedding and sleeping bags. Other types of fibre are increasingly being used to replace foams in furniture because of concern over the fire and health hazards posed by such materials.

Woven fabrics are still used to a significant extent as carpet and furniture backings and in some smaller, more specialised areas such as curtain header tapes. However, nonwovens such as spunbondeds have made significant inroads into these larger markets while various drylaid and hydroentangled products are now widely used in household cleaning applications in place of traditional mops and dusters.

1.5.5   Clothing components

This category includes fibres, yarns and textiles used as technical components in the manufacture of clothing such as sewing threads, interlinings, waddings and insulation; it does not include the main outer and lining fabrics of garments, nor does it cover protective clothing which is discussed later.

Although the world's consumption of clothing and therefore of these types of technical textile continues to increase steadily, the major problem faced by established manufacturers is the relocation of garment manufacturing to lower cost countries and therefore the need to develop extended supply lines and marketing channels to these areas, usually in the face of growing local competition.

As for home textile applications, this is a major market for fibrefill products. Some of the latest and most sophisticated developments have seen the incorporation of temperature phase change materials into such insulation products to provide an additional degree of control and resistance to sudden extremes of temperature, be they hot or cold.

1.5.6   Agriculture, horticulture and fishing

Textiles have always been used extensively in the course of food production, most notably by the fishing industry in the form of nets, ropes and lines but also by agriculture and horticulture for a variety of covering, protection and containment applications. Although future volume growth rates appear to be relatively modest, this is partly due to the replacement of heavier weight traditional textiles, including jute and sisal sacking and twine, by lighter, longer lasting synthetic substitutes, especially polypropylene.

However, modern materials are also opening up new applications. Lightweight spunbonded fleeces are now used for shading, thermal insulation and weed suppression. Heavier nonwoven, knitted and woven constructions are employed for wind and hail protection. Fibrillated and extruded nets are replacing traditional baler twine for wrapping modern circular bales. Capillary nonwoven matting is used in horticulture to distribute moisture to growing plants. Seeds themselves can be incorporated into such matting along with any necessary nutrients and pesticides. The bulk storage and transport of fertiliser and agricultural products is increasingly undertaken using woven polypropylene FIBCs (flexible intermediate bulk containers – big bags) in place of jute, paper or plastic sacks.

Agriculture is also an important user of products from other end-use sectors such as geotextiles for drainage and land reclamation, protective clothing for employees who have to handle sprays and hazardous equipment, transport textiles for tractors and lorries, conveyor belts, hoses, filters and composite reinforcements in the construction of silos, tanks and piping.

At sea, fish farming is a growing industry which uses specialised netting and other textile products. High performance fibres such as HMPE (e.g. Dyneema and Spectra) are finding their way into the fishing industry for the manufacture of lightweight, ultra-strong lines and nets.

1.5.7   Construction – building and roofing

Textiles are employed in many ways in the construction of buildings, both permanent and temporary, dams, bridges, tunnels and roads. A closely related but distinct area of use is in geotextiles by the civil engineering sector.

Temporary structures such as tents, marquees and awnings are some of the most obvious and visible applications of textiles. Where these used to be exclusively made from proofed heavy cotton, a variety of lighter, stronger, rot-, sunlight- and weatherproof (also often fireproof) synthetic materials are now increasingly required. A relatively new category of 'architectural membrane' is coming to prominence in the construction of semipermanent structures such as sports stadia, exhibition centres (e.g. the Greenwich Millenium Dome) and other modern buildings.

Nonwoven glass and polyester fabrics are already widely used in roofing applications while other textiles are used as breathable membranes to prevent moisture penetration of walls. Fibres and textiles also have a major role to play in building and equipment insulation. Glass fibres are almost universally used in place of asbestos now. Modern metal-clad roofs and buildings can be lined with special nonwovens to prevent moisture condensation and dripping.

Double wall spacer fabrics can be filled with suitable materials to provide sound and thermal insulation or serve as lightweight cores for composite materials.

Composites generally have a bright future in building and construction. Existing applications of glass-reinforced materials include wall panels, septic tanks and sanitary fittings. Glass, polypropylene and acrylic fibres and textiles are all used to prevent cracking of concrete, plaster and other building materials. More innovative use is now being made of glass in bridge construction. In Japan, carbon fibre is attracting a lot of interest as a possible reinforcement for earthquake-prone buildings although price is still an important constraint upon its more widespread use.

Textiles are also widely employed in the course of construction operations themselves, in uses as diverse as safety netting, lifting and tensioning ropes and flexible shuttering for curing concrete.

The potential uses for textiles in construction are almost limitless. The difficulties for textile manufacturers operating in this market include the strongly cyclical nature of the construction industry and the unevenness of major projects, the long testing and acceptance procedures and, perhaps above all, the task of communicating these developments to a diverse and highly fragmented group of key specifiers, including architects, construction engineers and regulatory bodies. The construction requirements, practices and standards of just about every country and region are different and it has, so far, proved very difficult for any acknowledged global leaders to emerge in this market as they have, for example, in industrial and automotive textiles.

1.5.8   Packaging and containment

Important uses of textiles include the manufacturing of bags and sacks, traditionally from cotton, flax and jute but increasingly from polypropylene. The strength and regularity of this synthetic material, combined with modern materials handling techniques, has allowed the introduction of FIBCs for the more efficient handling, storage and distribution of a variety of powdered and granular materials ranging from fertiliser, sand, cement, sugar and flour to dyestuffs. 'Big bags' with typical carrying capacities from one half to 2 tonnes can be fitted with special liners, carrying straps and filling/discharge arrangements. The ability to re-use these containers in many applications in place of disposable 'one-trip' bags and sacks is another powerful argument for their wider use.

An even faster growing segment of the packaging market uses lighter weight nonwovens and knitted structures for a variety of wrapping and protection applications, especially in the food industry. Tea and coffee bags use wet-laid nonwovens. Meats, vegetables and fruits are now frequently packed with a nonwoven insert to absorb liquids. Other fruits and vegetable products are supplied in knitted net packaging.

Strong, lightweight spunbonded and equivalent nonwoven paper-like materials are particularly useful for courier envelopes while adhesive tapes, often reinforced with fibres, yarns and fabrics, are increasingly used in place of traditional twine. Woven strappings are less dangerous to cut than the metal bands and wires traditionally used with densely packed bales.

A powerful driver of the development and use of textiles in this area is increasing environmental concern over the disposability and recycling of packaging materials. Legislation across the European Union, implemented especially vigorously in countries such as Germany, is now forcing many manufacturers and distributors of products to rethink their packaging practices fundamentally.

1.5.9   Sport and leisure

Even excluding the very considerable use of textiles in performance clothing and footwear, there are plenty of opportunities for the use of technical textiles throughout the sports and leisure market. Applications are diverse and range from artificial turf used in sports surfaces through to advanced carbon fibre composites for racquet frames, fishing rods, golf clubs and cycle frames. Other highly visible uses are balloon fabrics, parachute and paraglider fabrics and sailcloth.

Growth rates are well above average and unit values are often very high. The sports sector is receptive to innovation and developers of new fibres, fabrics and coatings often aim them at this market, at least initially. Many of the products and ideas introduced here eventually diffuse through to the volume leisure market and even the street fashion market.

1.5.10   Geotextiles in civil engineering

Although still a surprisingly small market in volume and value terms, considering the amount of interest and attention it has generated, the geosynthetics market (comprising geotextiles, geogrids and geomembranes) is nevertheless expected to show some of the highest growth rates of any sector over the foreseeable future.

The economic and environmental advantages of using textiles to reinforce, stabilise, separate, drain and filter are already well proven. Geotextiles allow the building of railway and road cuttings and embankments with steeper sides, reducing the land required and disturbance to the local environment. Revegetation of these embankments or of the banks of rivers and waterways can also be promoted using appropriate materials. There has been renewed interest in fibres such as woven jute as a biodegradable temporary stabilising material in such applications.

As in the case of construction textiles, one of the problems faced by manufacturers and suppliers of these materials is the sheer diversity of performance requirements. No two installations are the same in hydrological or geological terms or in the use to which they will subsequently be put. Suppliers to this market need to develop considerable expertise and to work closely with engineers and consultants in order to design and specify suitable products.

Because of the considerable areas (quantities) of fabric that can be required in a single project, cost is always a consideration and it is as essential not to overspecify a product as not to underspecify it. Much of the research and development work undertaken has been to understand better the long term performance characteristics of textiles which may have to remain buried in unpredictable environments (such as landfill and toxic waste sites) for many years and continue to perform to an adequate standard.

Nonwovens already account for up to 80% of geotextile applications. This is partly a question of economics but also of the suitability of such textile structures for many of the filtration and separation duties that they are called upon to perform. Current interest is in 'composite' fabrics which combine the advantages of different textile constructions such as woven, knitted, nonwoven and membrane materials. To supply the diversity of fabrics needed for the many different applications of geotextiles, leading specialist manufacturers are beginning to assemble a wide range of complementary capabilities by acquisition and other means.

1.5.11 Protective and safety clothing and textiles

Textiles for protective clothing and other related applications are another important growth area which has attracted attention and interest somewhat out of proportion to the size and value of the existing market. As in the case of sports textiles, a number of relatively high value and performance critical product areas have proved to be an ideal launch pad for a new generation of high performance fibres, most notably the aramids, but including many other speciality materials.

The variety of protective functions that needs to be provided by different textile products is considerable and diverse. It includes protection against cuts, abrasion, ballistic and other types of severe impact including stab wounds and explosions, fire and extreme heat, hazardous dust and particles, nuclear, biological and chemical hazards, high voltages and static electricity, foul weather, extreme cold and poor visibility.

As well as people, sensitive instruments and processes also need to be protected. Thus, clean room clothing is an important requirement for many industries including electronics and pharmaceuticals.

In Europe and other advanced industrial regions, strict regulations have been placed upon employers through the introduction of legislation such as the Personal Protective Equipment (PPE) at Work Regulations (European Union). Under such legislation, it is not only necessary to ensure that the equipment and clothing provided is adequate to meet the anticipated hazards but also that it is also used effectively, that is that the garments are well designed and comfortable to wear. This has opened up a need for continuing research not only into improved fibres and materials but also into increasingly realistic testing and assessment of how garments perform in practice, including the physiology of protective clothing.

In many developing countries, there has not been the same legislative framework in the past. However, this is rapidly changing and future market growth is likely to concentrate less on the mature industrial markets than upon the newly industrialising countries of Asia and elsewhere. The protective clothing industry is still highly fragmented with much of the innovation and market development being provided by the major fibre and other materials producers. This could change as some global suppliers emerge, perhaps without their own direct manufacturing but relying on contract producers around the world, very much as the mainstream clothing industry does at present.

1.5.12   Ecological protection textiles

The final category of technical textile markets, as defined by Techtextil, is technical textiles for protection of the environment and ecology. This is not a well defined segment yet, although it overlaps with several other areas, including industrial textiles (filtration media), geotextiles (erosion protection and sealing of toxic waste) and agricultural textiles (e.g. minimising water loss from the land and reducing the need for use of herbicides by providing mulch to plants).

Apart from these direct applications, technical textiles can contribute towards the environment in almost every sphere of their use, for example by reducing weight in transport and construction and thereby saving materials and energy. Improved recycleability is becoming an important issue not only for packaging but also for products such as cars.

Composites is an area which potentially presents problems for the recycleability of textile reinforcing materials encased within a thermosetting resin matrix. However, there is considerable interest in and development work being done on thermoplastic composites which should be far simpler to recycle, for example by melting and recasting into lower performance products.

1.6   Globalisation of technical textiles

If North America and Western Europe have the highest levels of per capita consumption of technical textiles at present (see Table 1.1), then they are also relatively mature markets. The emerging countries of Asia, Eastern Europe and the rest of the world are becoming important markets in almost every sphere, from automotive manufacture through to sporting and leisure goods. Technical textiles for food production, construction and geotextiles are likely to be particularly important. In the case of the last of these, geotextiles, consumption up to the year 2005 is expected to grow at over 12% per annum across the whole of Asia compared with less than 6% in Western Europe and the USA. In the case of Eastern Europe and South America, annual growth rates could be as high as 18% and 16% per annum respectively, although from relatively small base levels at present.

In 2000, the major existing users, North America, Western Europe and Japan, are expected to account for less than 65% of total technical textile consumption; by the year 2005, this could be down to 60% and perhaps below 50% by 2010. Consumption of technical textiles in China already exceeds that of Japan, in weight terms at

Table 1.4 Worldwide consumption of technical textiles by geographical region, 2000–2005

	10^3 tonnes			$ million		
	2000	2005	Growth (% pa)	2000	2005	Growth (% pa)
Western Europe	2690	3110	2.9	13770	15730	2.7
Eastern Europe	420	560	5.9	2500	3260	5.5
North America	3450	3890	2.4	16980	18920	2.2
South America	350	430	4.2	1870	2270	3.9
Asia	3560	4510	4.8	20560	25870	4.7
Rest of the world	870	1190	6.5	4590	6280	6.5
Total	**11340**	**13690**	**3.9**	**60270**	**72330**	**3.7**

Source: David Rigby Associates/Techtextil.

least. In 2000, Chinese technical textiles are expected to account for almost 20% of all textile manufacturing in that country and over 12% of total world consumption (see Table 1.4).

But globalisation is not just about increasing internationalisation of markets. It is also about the emergence of companies and supply chains which operate across national and continental boundaries. Such globalisation has already proceeded furthest in the automotive and transport industry, the largest of the 12 market segments defined above. It is a path already being followed within the other major segments, most notably industrial textiles and medical/hygiene textiles and will become increasingly evident in the remainder.

Characteristics of globalisation include higher levels of international trade and increased specialisation of manufacture within individual districts, countries and regions, according to availability of materials, local industry strengths and regional market characteristics.

Relatively low unit value products requiring a significant amount of making-up or other fabrication such as bags and sacks have already seen a significant shift of manufacturing towards the Far East and Eastern Europe. Textiles for tents, luggage and the technical components of footwear and clothing are now increasingly sourced close to where many of these products are manufactured for export, for example China and Indonesia.

Manufacturers in the newly industrialising world are rapidly adopting the latest materials and processing technologies. Taiwan already has an important composites manufacturing sector specialising in sports equipment.

1.7 Future of the technical textiles industry

The future of technical textiles embraces a much wider economic sphere of activity than just the direct manufacturing and processing of textiles. The industry's suppliers include raw materials producers (both natural and artificial), machinery and equipment manufacturers, information and management technology providers, R&D services, testing and certification bodies, consultants, education and training organisations. Its customers and key specifiers include almost every conceivable

Table 1.5 Technical textile functions, markets and end-uses

Markets	Segments	Function						
		Protection	Insulation	Reinforcement	Containment	Filtration	Absorption	Miscellaneous
Industry	Engineering Food, pharmaceuticals Chemicals, plastics Other manufacturing Power, oil, gas Mining, quarrying	High temperature textiles Welders anti-spatter sheets Fire blankets Dustproof fabric Electrostatic shielding Debris, safety nets Solar protection	Acoustic barriers Thermal insulation Seals, joints Packings Pressed felt components Electrical insulation tape Heating elements Electromagnetic shields Electroconductive fabrics Dielectric fabrics Aerials	Composites – FRP and advanced Printed circuit boards Optical fibre/electrical cables Electrical cables Jacquard harness Pressure hoses Drive belts Conveyor belts Bearing materials Abrasive discs	Bags & sacks FIBCs Balewrap Tape Curing tape Hosepipes Nets Webbing Diaphragms Envelopes Floppy disc liners	Dust filtration Air conditioning Process liquid Effluent treatment Papermakers felts Battery separators Tea/coffee bags Cigarette filters Food casing Machine clothing Printers blankets Laundry textiles	Oil spillages Wicks	Thermal stencil paper
Transport	Road Aviation Marine (Military)	Seat belts Air bags Flotation devices Inflatable boats Parachutes Ropes, cables Barriers, nets Camouflage, decoy FR textiles	Sound barriers for roofs, bonnets etc. Tank insulation	Tyre cord Hoses, pipes, drive belts & other MRG Brake/clutch lining Gaskets, seals Composites – FRP & advanced Tow ropes	Containers Tarpaulins Covers Cordage Twine Cargo nets, straps Balloons Sailcloth Gliders Hovercraft skirts	Air filters (engine) Air filters (passenger) Oil filters Fuel filters	Oil booms	Decorative/functional interior textiles (UV, FR)

Category	Sub-sector						
Construction	Buildings Road, rail, tunnel	Tarpaulins Sunblinds, awnings Debris, safety nets Roofing Stadium dome	Silo liners Soundproof panels Swimming pool liners	Inflatable buildings, frames Tape Elevator belts Pipe/sewer linings Ropes Concrete Plaster board	Cordage Ropes Twines Nets Cement shuttering		Wall coverings, blinds
Geotextile	Land Marine	Erosion protection Environmental protection		Road, railway embankment stabilisation Concrete, tarmac	Geomembranes Reservoir lining Waste pit lining	Drainage	
Farming	Agriculture Horticulture Fishing	Wind, storm, frost protection Solar protection Insect/bird nets			Sacks, bags Baler twine Fruit collection Fishing nets Fish farming Other netting Seeding tapes	Drainage	Moisture retention Capillary matts
Medical & hygiene	Hospital Nursing home Domestic	Gauze Plaster Face masks Gowns Tents Bandages Support stockings X-ray machine parts	Wadding	Sutures Adhesive tape Prostheses Support bandages Ligaments Tendons Implants	Body bags Netting Thread Sacks, bags Stretchers Wheelchairs	By-pass filters Dialysis Infusion Stoma Membranes Blood Pads Towels Swabs Napkins Wipes Cotton wool Sponges Mops, brushes Dusters	Veins, arteries Artificial skin
Technical apparel	Industry Offshore oil Forestry Miltary/security	Water/windproof linings Chemical proof Gas tight Anti-radiation NBC suits Bulletproof Diving suits		Sewing thread Binding tape Interlinings Shoe/boot linings Fasteners (Velcro) Shoe/boot laces			Labels

Table 1.5 *Continued*

Markets	Segments	Protection	Insulation	Reinforcement	Containment	Filtration	Absorption	Miscellaneous
					Function			
Technical apparel *continued*		Pressure suits Survival suits Fire-retardant Heat-resistant Dust, asbestos Clean room Chain saw protection Helmets Motor cycle garments Gloves, armguards Aprons Hair nets High visibility Camouflage						
Leisure & environment	Sports Mountaineering Leisure	Tents Climbing ropes Harnesses Safety nets Parachutes Ski fences Muscle supports	Sleeping bags Ground sheets Artificial turf	Rackets Fishing rods Fishing lines Bicycle frames Golf clubs Skis Bow strings Ropes Cords	Marquees Tents Sports nets Bags, rucksacks Luggage Sports balls Litter systems Mosquito nets Book covers Pet leads Equestrian webs			Painters canvasses Cinema screens
Furnishing, decorative		FR fabrics	Carpet underlay Fibre fillings	Webbings Curtain header Carpet backing Linoleum scrim Garden furniture Hammocks	Furniture/ mattress bases Wallcoverings Gift wrapping			Flags/banners

FR = fire retardant, FRP = fibre reinforced plastic, NBC = nuclear biological and chemical.

downstream industry and field of economic activity, including the architects, engineers, designers and other advisors employed by those industries. In between lie many other interested parties, including environmental, health, safety, business and free trade regulators, patent and intellectual property agents and lawyers, investors, bankers, regional investment agencies and providers of development aid.

The task of disseminating and communicating information to all these organisations and individuals is undertaken by a growing number of specialist and generalist publications as well as by international and local trade exhibitions, fairs, seminars and conferences.

The economic importance of technical textiles worldwide therefore undoubtedly far exceeds the $60 billion estimated in Tables 1.2–1.4 just for basic fibres, yarns and textiles.

1.7.1   A changing strategic environment

If the 1980s was a period when the technical textiles industry enjoyed a rapid and increasing awareness of its existence by the outside world (as well as within the mainstream textile industry), then the 1990s was an era of more mature commercial development and consolidation as fibre producers and textile manufacturers alike concentrated on overhauling and refocusing their businesses in the wake of world recession.

The new millennium promises even fiercer international competition which will see manufacturers striving to engineer costs downwards and develop global economies of scale in production and product development. Technical textiles will become better 'value for money' than ever before and this should open the way towards further applications as existing end-uses mature.

Individual companies will become less defined by the technologies and materials they use than by the markets and applications they serve. Table 1.5 summarises some of the key market areas and the functions which technical textiles perform, with examples of individual products in each category. It does not pretend to be an exhaustive list which would run into many thousands of products and would constantly be changing.

References

1. The Textile Institute, *Textile Terms and Definitions, Tenth Edition*, Textile Institute, Manchester, 1994.
2. SABIT ADANUR, *Wellington Sears Handbook of Industrial Textiles*, Technomic, Lancaster PA (USA), 1995.
3. ISO 9092:1988 Definition of nonwovens.

2

Technical fibres

Mohsen Miraftab

Department of Textiles, Faculty of Technology, Bolton Institute, Deane Road, Bolton BL3 5AB, UK

2.1 Introduction

A number of definitions[1-3] have been used to describe the term 'technical textiles' with respect to their intended use, functional ability and their non-aesthetic or decorative requirements. However, none of these carefully chosen words include the fundamental fibre elements, technical or otherwise, which make up the technical textile structures. The omission of the word 'fibre' may indeed be deliberate as most technical textile products are made from conventional fibres that are already well established. In fact over 90% of all fibres used in the technical sector are of the conventional type.[4] Specially developed fibres for use in technical textiles are often expensive to produce and have limited applications.

Historically, utilisation of fibres in technical capacities dates back to the early Egyptians and Chinese who used papyrus mats to reinforce and consolidate the foundations respectively of the pyramids and the Buddhist temples.[5,6] However, their serious use in modern civil engineering projects only began after the floods of 1953 in The Netherlands in which many people lost their lives. The event initiated the famous Delta works project in which for the first time synthetic fibres were written into the vast construction programme.[7] Since then, geotextiles in particular have matured into important and indispensable multifunctional materials.

Use of silk in semitechnical applications also goes back a long way to the lightweight warriors of the Mongolian armies, who did not only wear silk next to their skin for comfort but also to reduce penetration of incoming arrows and enable their subsequent removal with minimal injury. Use of silk in wound dressing and open cuts in web and fabric form also dates back to the early Chinese and Egyptians.

In light of extensive utilization of conventional fibres in the technical sector, this chapter initially attempts to discuss fibres under this category highlighting their importance and the scope of their versatility. The discussion covers concisely an outline of fibre backgrounds, chemical compositions and their salient characteristics. It then introduces other fibres which have been specially developed to perform under extreme stress and/or temperature; ultrafine and novel fibres are also discussed. Finally, the chapter concludes by identifying areas of application and the roles that selected fibres play in fulfilling their intended purpose.

Table 2.1 presents the complete range of fibres available to the end-user and some of their mechanical properties.

2.2   Conventional fibres

2.2.1   Natural fibres

Cotton accounts for half of the world's consumption of fibres and is likely to remain so owing to many of its innate properties and for economical reasons[8] that will not be discussed here. Cotton is made of long chains of natural cellulose containing carbon, hydrogen and oxygen otherwise known as polysaccharides. The length of the chains determines the ultimate strength of the fibre. An average of 10 000 cellulosic repeat or monomeric units make up the individual cellulose chains which are about 2 μm in length. The linear molecules combine into microfibrils and are held together by strong intermolecular forces to form the cotton fibre. The unique physical and aesthetic properties of the fibre, combined with its natural generation and biodegradability, are reasons for its universal appeal and popularity. Chemical treatments such as Proban[9] and Pyrovatex[10] are two examples of the type of durable finishes that can be applied to make cotton fire retardant. High moisture absorbency, high wet modulus and good handle are some of the more important properties of cotton fibre.

Wool, despite its limited availability and high cost, is the second most important natural fibre. It is made of protein: a mixture of chemically linked amino acids which are also the natural constituents of all living organisms. Keratin or the protein in the wool fibre has a helical rather than folded chain structure with strong inter- and intrachain hydrogen bonding which are believed to be responsible for many of its unique characteristics. Geographical location, the breeding habits of the animals, and climatic conditions are some of the additional variables responsible for its properties. The overall high extensibility of wool, its natural waviness and ability to trap air has a coordinated effect of comfort and warmth, which also make it an ideal insulating material. The sophisticated dual morphology of wool produces the characteristic crimp which has also been an inspiration for the development of some highly technical synthetic fibres. Wool is inherently fire retardant, but further improvements can be achieved by a number of fire-retardant treatments. Zirconium- and titanium-treated wool is one such example which is now universally referred to as Zirpro (IWS) wool.[11]

Flax, jute, hemp and ramie, to name but a few of the best fibres, have traditionally taken a secondary role in terms of consumption and functional requirements. They are relatively coarse and durable, and flax has traditionally been used for linen making. Jute, ramie and to a lesser extent other fibres have received attention within the geotextile sector of the fibre markets which seeks to combine the need for temporary to short-term usage with biodegradability, taking into account the regional availability of the fibres.

Silk is another protein-based fibre produced naturally by the silkworm, *Bombyx Mori* or other varieties of moth. Silk is structurally similar to wool with a slightly

Table 2.1 Fibres available to the end-user and associated mechanical properties

Conventional fibres	High strength high modulus organic fibres	High chemical and combustion-resistant organic fibres	High performance inorganic fibres	Ultrafine and novelty fibres
Natural e.g. cotton, wool, silk, jute, etc.	*Para*-aramids e.g. Kevlar (Du Pont) and Twaron (Acordis)	*Meta*-aramids e.g. Nomex (Du Pont) and Conex (Teijin)	Carbon Ceramics Boron Tungsten	Microfibres; (linear density <0.5 dtex) Solar energy absorbing fibres (solar alpha)
Regenerated e.g. viscose, acetates tencel, etc.	Polybenzobisthiazole (PBT) Ultra-high molecular weight polyethylene	Kermel (Rhodia) Kynol (Kynol) Oxidised acrylic fibres, e.g. Panox (SGL)	Alumina (e.g. Saffil) High modulus silicon Carbide & silicon nitride etc.	Heat-sensitive fibres (thermochromics) Scented fibres Antibacterial fibres (aseptic chlorofibres)
Synthetics e.g. polyamide, polyester, polyacrylics, polyurethanes, polyolefins, etc.	e.g. Dyneema (DSM) and Spectra (Allied Signal)	Others: Aromatic polymers; Polyether ether ketone, PEEK (Victrex and Zyex) Polyether ketone, PEK Poly *p*-phenylene sulphide, PPS, e.g. Ryton (Phillips) polytetrafluoroethylene, PTFE, e.g. Teflon (Du Pont) (Inspec formerly Lenzing) P84		Hollow fibres Antistatic fire-retardant fibres; etc.
Tenacity: $0.1–0.5\,N\,tex^{-1}$ Modulus: $2–18\,N\,tex^{-1}$ % Elongation: 2–7	Tenacity: $1.5–3\,N\,tex^{-1}$ Modulus: $25–150\,N\,tex^{-1}$ % Elongation: 1–8 LOI[a]: 0.20–0.40	Tenacity: $1–2\,N\,tex^{-1}$ Modulus: $15–25\,N\,tex^{-1}$ % Elongation: 1–4 LOI: 0.23–0.55	Tenacity: $0.5–2\,N\,tex^{-1}$ Modulus: $70–220\,N\,tex^{-1}$ % Elongation: 0–1.5	Tenacity: $0.1–0.4\,N\,tex^{-1}$ Modulus: $2–15\,N\,tex^{-1}$ % Elongation: 2–17

[a] LOI: limiting oxygen index = minimum fraction of oxygen in nitrogen necessary to sustain burning.

different combination of amino acids which make up the protein or the fibroin, as it is more appropriately known. Silk is the only naturally and commercially produced continuous filament fibre which has high tenacity, high lustre and good dimensional stability. Silk has been and will remain a luxury quality fibre with a special place in the fibre market. However, its properties of biocompatibility and gradual disintegration, as in sutures, have long been recognised in medical textiles.

2.2.2   Regenerated fibres

Viscose rayon was the result of the human race's first attempts to mimic nature in producing silk-like continuous fibres through an orifice. Cellulose from wood pulp is the main constituent of this novel system, started commercially in the early 1920s. Thin sheets of cellulose are treated with sodium hydroxide and aged to allow molecular chain breakage. Further treatment with carbon disulphide, dissolution in dilute sodium hydroxide and ageing produces a viscous liquid, the viscose dope, which is then extruded into an acid bath. The continuous filaments that finally emerge are washed, dried and can be cut to staple lengths. The shorter cellulose molecules in viscose and their partial crystallisation accounts for its rather inferior physical properties relative to cotton. Further development and refinement of the manufacturing technique have created a whole range of fibres with improved properties. High tenacity and high wet modulus viscose compare in all but appearance to cotton in both dry and wet conditions. Chemically altered regenerated cellulose di- and triacetates do not burn like cotton and viscose to leave a fluffy black ash, but melt and drip instead. This characteristic enables them to be shaped or textured to enhance their visual and aesthetic appeal. Hollow viscose modifications give enhanced bulk and moisture absorbency and have an improved cotton-like feel.

Fire-retardant (FR) viscose was first introduced in the 1960s. A major example is produced by Lenzing in Austria by incorporating organophosphorous compounds into the spinning dope prior to extrusion. The additive is reasonably stable and has no chemical interaction with the cellulose molecules. It is also unaffected by bleaching, washing, dry cleaning, dyeing and finishing processes.[10] Early in the 1990s Kemira (now Sateri Fibres) of Finland introduced an alternative version of FR viscose known as Visil in which polysilicic acid is present. The fibre chars upon heating leaving a silica residue.

Lyocell,[12] is the latest addition to this series of fibres, commercially known as Tencel (Acordis), has all the conventional properties of viscose in addition to its much praised environmentally friendly production method. The solvent used is based on non-toxic N-methyl morpholine oxide used in a recyclable closed loop system, which unlike the viscose process avoids discharge of waste. Highly absorbent derivatives of Tencel, known as Hydrocell are establishing a foothold in wound dressing and other medical-related areas of textiles.

2.2.3   Synthetic fibres

All synthetic fibres originate from coal or oil. The first synthetic fibre that appeared on the world market in 1939 was nylon 6.6. It was produced by DuPont and gained rapid public approval. A series of nylons commonly referred to as polyamides now exists in which the amide linkage is the common factor. Nylon 6.6 and nylon 6 are most popular in fibre form. They are melt extruded in a variety of cross-sectional shapes and drawn to achieve the desired tenacity. They are well known for their high extensibility, good recovery, dimensional stability and relatively low moisture absorbency. Nylon 6.6 in particular soon became a popular household carpet fibre and was developed into fibres commonly known as Antron[13] manufactured by DuPont. Antron fibres of various generations use additives, varied cross-sectional shapes and modified surface characteristics to enhance the aesthetic and visual appeal of carpets as well as improving their resilience and dissipating static charges.

Nylon was later surpassed by the even more popular fibre known as polyester, first introduced as Dacron by DuPont in 1951. Polyester is today the second most used fibre after cotton and far ahead of other synthetics both in terms of production and consumption. Polyethylene terephthalate or polyester is made by condensation polymerisation of ethylene glycol and terephthalic acid followed by melt extrusion and drawing. It can be used in either continuous form or as short staple of varying lengths. The popularity of polyester largely stems from its easycare characteristics, durability and compatibility with cotton in blends. Its very low moisture absorbency, resilience and good dimensional stability are additional qualities. Many manufacturers across the world produce polyester under different commercial names with almost tailor-made properties. A high glass transition temperature of around 70 °C with good resistance to heat and chemical degradation also qualifies this polymer for most technical textile applications. These will be discussed in greater detail later. Flame-retardant Trevira CS and Trevira high tenacity, both polyesters developed and marketed by Trevira GmbH in Germany, are examples of the many varieties available today.

Wool-like properties are shown by polyacrylic fibres which are produced by the polymerisation of acrylonitrile using the addition route into polyacrylonitrile. They can then be spun into fibres by dry or wet spinning methods. Orlon[14] was produced by DuPont. It had a distinctive dumbbell shaped cross-section and was extruded by the dry process in which the solvent is evaporated off. Acrilan[15] produced by Monsanto and Courtelle produced by Acordis are spun by the wet extrusion technique and have near circular cross-sections. Acrylic fibres now also appear in bicomponent form with wool-like characteristics. Chemically modified acrylics, principally the modacrylics, include chlorine atoms in their molecular structure which are responsible for their low burning behaviour and, unlike acrylics, have the ability to self-extinguish once the source of ignition has been removed. A selected fibre in terms of fibre chemistry is Oasis,[16] a superabsorbent fibre made by the collaborative efforts of Acordis and Allied Colloids, based on crosslinking copolymers of acrylic acid. This fibre is claimed to absorb moisture many times its own weight and holds it even under pressure. Its application in hygiene and medical care in different forms is being pursued.

Polyolefin fibres include both polyethylene and polypropylene made by addition polymerisation of ethylene and propylene and subsequent melt extrusion, respectively. Polyethylene has moderate physical properties with a low melting temperature of about 110 °C for its low density form and about 140 °C for its high density form which severely restricts its application in low temperature applications. Polypropylene has better mechanical properties and can withstand temperatures of up to 140 °C before melting at about 170 °C. Both polymers have a density less than that of water which allows them to float as ropes, nets and other similar applications. The availability, low cost and good resistance to acid and alkaline environments of polypropylene has greatly influenced its growth and substantial use in geotextile applications.

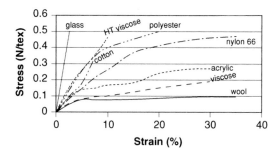

2.1 Stress/strain behaviour of some common fibres.

Finally, elastane fibres[17] are synthetic-based elastomeric polymers with at least 85% segmented polyurethane in their structure. They have rubber-like properties, which means they can be extended up to six or more times their original length. They are used in combination with most natural and synthetic fibres in knitted and woven materials. They were initially produced by DuPont in 1959 under the now well-known trademark of Lycra.

Figure 2.1 presents typical stress/strain behaviour for most conventional fibres. Tenacity, modulus and percentage elongation ranges for most conventional fibres are also given at the bottom of Table 2.1.

2.3 High strength and high modulus organic fibres

Keller's much-published[18] contribution to the understanding of crystal growth in 1957 and confirmation of the tendency of polymers to form folded-chain crystals, provided the insight and inspiration for development of high strength, high modulus organic fibres that would surpass conventional fibres. During crystallisation, long chains of molecules fold back on themselves to form folded-chain crystals which only partly unfold during normal drawing. In the Netherlands in the 1970s DSM developed a super drawing technique known as *gel spinning*, which uses dilute solutions of ultra-high molecular weight polymers such as polyethylene to unfold the chains further and thus increase both tensile strength and fibre modulus. Ultra-high molecular weight polyethylene (UHMWPE) fibres, Dyneema or Spectra, are today the strongest fibres known, with tensile moduli in excess of $70\,GN\,m^{-2}$. Weight for weight this fibre genus is claimed to be 15 times stronger than steel and twice as strong as aromatic polyamides such as Kevlar.[19] It is also low in density, chemically inert and abrasion resistant. It, however, melts at around $150\,°C$ and thermally degrades at $350\,°C$ which restrict its use to low temperature applications.

To achieve even better performance characteristics at higher temperatures, other means of fibre production were explored in the 1960s. One successful approach eventually led to the advent of liquid crystalline polymers. These are based on polymerisation of long stiff molecules such as *para*-phenylene terephthalamide achieving molecular weights averaging to around 20000. The influence of the stiff aromatic rings, together with hydrogen-bonding crosslinks, combines the best features of both the polyamides and the polyesters in an extended-chain configuration.[18–20] Molecular orientation of these fibres is brought about by capillary shear along the flow of the polymer as it exits from the spinneret thus overriding the need for subsequent drawing. Kevlar by DuPont and Twaron by Akzo (now Acordis) were the first of such fibres to appear in the early 1970s. There now exists a series of first, second and third generations of *para*-aramids. Kevlar HT for instance, which has 20% higher tenacity and Kevlar HM which has 40% higher modulus than the original Kevlar 29 are largely utilised in the composite and the aerospace industries. *Para*-aramids generally have high glass transition temperatures nearing 370 °C and do not melt or burn easily, but carbonise at and above 425 °C. All aramid fibres are however prone to photodegradation and need protection against the sun when used out of doors. These will be discussed later.

Other high tenacity and high modulus fibres include the isotropically spun Technora (Teijin) and Supara, based upon *para*-aramid copolymers, with slightly lower maximum strength and modulus values than Kevlar. Several other melt-spinnable liquid crystalline polymers are also available.[21]

2.4   High chemical- and combustion-resistant organic fibres

The fibres discussed in the previous section were developed following earlier observations that aromatic polymer backbones yielded improved tensile and heat resistance compared with conventional fibres. However, if the polymer chains have lower symmetry and order, then polymer tractability and textile fibre characteristics are improved. Solvent-spun Nomex and Conex were the first so-called *meta*-aramids made from poly(meta-phenylene isophthalamide) and were produced by DuPont in 1962 and by Teijin in 1972, respectively. The *meta*-phenylene isophthalamide molecule is identical in all but the position of its —NH— and —CO— groups to *para*-phenylene terephthalamide (or Kevlar) molecules. The different positioning of these groups in *meta*-aramids creates a zig-zag molecular structure that prevents it from full crystallisation thus accounting for its relatively poorer tensile properties. However, Nomex is particularly well known for its resistance to combustion, high decomposition temperature prior to melting and high limited oxygen index (LOI), the minimum amount of oxygen required to induce ignition.

Melt-spinnable aromatic fibres with chains containing paraphenylene rings, like polyether ether ketone (PEEK),[23] polyether ketone (PEK) and poly(*p*-phenylene sulphide) (PPS),[22] also have high melting points but, since their melting points occur prior to their decomposition temperature, they are unsuitable for fire-retardant applications. However, their good chemical resistance renders them suitable for low temperature filtration and other corrosive environments.

The polyheterocyclic fibre, polybenzimidazole or PBI, produced by Hoechst-Celanese has an even higher LOI than the aramids. It has excellent resistance to both heat and chemical agents but remains rather expensive. P84, initially produced by Lenzing and now produced by Inspec Fibres, USA, comprises polyimide groups that yield reasonably high resistance to fire and chemical attack. The acrylic copolymer-based fibre produced by Acordis known as Inidex (although now no longer produced) unlike many aramid fibres has high resistance to UV (ultraviolet) radiation and a fairly high LOI at the expense of much reduced tenacity and rather low long-term exposure resistance to heat.

Oxidised acrylic fibre, best known as Panox (SGL, UK) is another crosslinked, high combustion-resistant material produced by combined oxidation and pyrolysis

Table 2.2   LOI and tenacity range of some high chemical-
and combustion-resistant organic fibres

Fibres brand name	Manufacturer	LOI (%)	Tenacity (GPa)
Nomex	Du Pont	29	0.67
Conex	Conex	29	0.61
Kermel	Rhone-Poulenc	31	0.53
Inidex	Courtaulds	43	0.12
PBI	Hoechst-Celanese	41	0.39
PAN-OX	RK Textiles	55	0.25
PEEK		42	
PEK			
PPS	Phillips	34	0.54

of acrylic fibres at around 300 °C. Although black in appearance, it is not classed as carbon fibre and preserves much of its original non-carbonaceous structure. In addition, it does not have the graphitic or turbostratic structure of carbon fibres. It retains a sufficiently high extension-at-break to be subjected to quite normal textile processes for yarn and fabric formation.[24] Panox has a very high LOI of 0.55. Fully carbonised, or carbon fibres with even higher tensile properties, are achieved by full carbonisation of these oxidised precursor fibres or by the melt spinning of liquid crystalline mesophase pitch followed by their oxidation and pyrolysis. These are discussed again in the section on inorganic fibres.

Table 2.2 shows the LOI and tenacity range of some of the better known high chemical- and combustion-resistant organic fibres.

2.5   High performance inorganic fibres

Any fibre that consists of organic chemical units, where carbon is linked to hydrogen and possibly also to other elements, will decompose below about 500 °C and cease to have long-term stability at considerably lower temperatures. For use at high temperatures it is therefore necessary to turn to inorganic fibres and fibres that consist essentially of carbon.[25]

Glass, asbestos and more recently carbon are three well-known inorganic fibres that have been extensively used for many of their unique characteristics. Use of glass as a fibre apparently dates back to the ancient Syrian and Egyptian civilisations[26] which used them for making clothes and dresses. However, the very high modulus or stiffness displayed by these fibres means that they are quite brittle and can easily be damaged by surface marks and defects. They are, therefore, best utilised by embedding in matrix forms where the fibres are fully protected. Epoxy resins, polyester and other polymers, as well as cement, have commonly been used both to protect and to make use of their contributory strength. Glass-reinforced boat hulls and car bodies, to name but two application areas of such composites, reduce overall weight and cost of fabrication as well as eliminating the traditional problems of rotting wood and rusting metals associated with traditional materials. Their good resistance to heat and very high melting points have also enabled them to be used as effective insulating materials.

Asbestos[27] is a generic name for a variety of crystalline silicates that occur naturally in some rocks. The fibres that are extracted have all the textile-like properties of fineness, strength, flexibility and more importantly, unlike conventional fibres, good resistance to heat with high decomposition temperatures of around 550 °C. Their use as a reinforcement material has been found in clay-rich prehistoric cooking pots discovered in Finland.[28] In more recent times they have been extensively used to reinforce brittle matrices such as cement sheeting, pipes, plastics[29] and also as heat insulators. However, with the discovery of their carcinogenic hazards, their use has gradually declined and alternative fibres have been developed to replace them totally.

High purity, pyrolysed acrylic-based fibres are classified as carbon fibres. The removal of impurities enhances carbon content and prevents the nucleation and growth of graphite crystals which are responsible for loss of strength in these fibres. Carbon fibres with different structures are also made from mesophase pitch. The graphite planes in PAN-based fibres are arranged parallel to the fibre axis rather than perpendicular as is the case with pitch-based carbon fibres.[28] Their high strength and modulus combined with relatively low extensibility means that they are best utilised in association with epoxy or melt-spinnable aromatic resins as composites.

Increasing demand in the defence and aerospace industries for even better performance under extreme conditions led, within the last quarter of the 20th century, to yet another range of new and rather expensive metal oxides, boron and silicon-based fibres. These are often referred to as ceramic fibres and will now be briefly discussed.

Aluminosilicate compounds are mixtures of aluminium oxide (Al_2O) and silicon oxide (SiO_2); their resistance to temperature depends on the mixing ratio of the two oxides. High aluminium oxide content increases their temperature tolerance from a low of 1250 °C to a maximum of 1400 °C. However, despite their high temperature resistance, these fibres are not used in high stress applications owing to their tendency to creep at high temperatures.[29] Their prime applications are in insulation of furnaces and replacement of asbestos fibres in friction materials, gaskets and packings.[30] Both aluminium oxide or alumina fibres and silicon oxide or silica fibres are also produced. Pure boron fibres are too brittle to handle but they can be coated on tungsten or carbon cores. Their complex manufacturing process makes them rather expensive. Their prime application is in lightweight, high strength and high modulus composites such as racket frames and aircraft parts. Boron fibre use is limited by their thickness (about 16 µm), their relatively poor stability in metal matrices and their gradual loss of strength with increasing temperature.[31] Boron nitrides (BN) are primarily used in the electronic industry where they perform both as electrical insulators and thermal conductors.

The most outstanding property of silicon carbide (SiC) is the ability to function in oxidizing conditions up to 1800 °C with little loss of mechanical properties. Silicon carbide exceeds carbon fibre in its greater resistance to oxidation at high temperatures, its higher compressive strength and better electrical resistance. SiC fibres containing carbon, however, lose some tensile properties at the expense of gaining better electrical conductivity.

Finally, many of the inorganic fibres so far referred to are now also produced in microcrystalline or whisker form and not in the more normal textile-fibre form. Whiskers have extremely good mechanical properties and their tensile strength is usually three to four times those of most reinforcing fibres. They are, however, costly to produce and their inclusion into composite structures is both difficult and cumbersome.

2.6   Ultra-fine and novelty fibres

Ultra-fine or microfibres were developed partly because of improved precision in engineering techniques and better production controls, and partly because of the need for lightweight, soft waterproof fabrics that eliminate the more conventional coating or lamination processes.[32] As yet there are no universal definitions of microfibres. *Textile Terms and Definitions*[2] simply describes them as fibres or filaments with linear densities of approximately 1.0 dtex or less. Others[33,34] have used such terms as fine, extra-fine and micro-fine corresponding to linear densities ranging from 3.0 dtex to less than 0.1 dtex. They are usually made from polyester and nylon polymers, but other polymers are now being made into microfibres. The Japanese first introduced microfibres in an attempt to reproduce silk-like properties with the addition of enhanced durability. They are produced by at least three established methods including island-in-sea, split process and melt spinning techniques and appear under brand names such as Mitrelle, Setila, Micrell, Tactel and so on. Once in woven fabric form their fine diameter and tight weave allows up to 30 000 filaments cm^{-2}, making them impermeable to water droplets whilst allowing air and moisture vapour circulation. They can be further processed to enhance other characteristics such as peach-skin and leather-like appearances. The split technique of production imparts sharp-angled edges within the fibre surface, which act as gentle abraders when made into wiping cloths that are used in the optical and precision microelectronic industries. Microfibres are also used to make bacteria barrier fabrics in the medical industries. Their combined effect of low diameter and compact packing also allows efficient and more economical dyeing and finishing.

Finally, constant pressure to achieve and develop even more novel applications of fibres has led to a number of other and, as yet, niche fibrous products. In principle, the new ideas usually strive to combine basic functional properties of a textile material with special needs or attractive effects.

For example, Solar-Aloha, developed by Descente and Unitika in Japan,[35] absorbs light of less than 2 μm wavelength and converts it to heat owing to its zirconium carbide content. Winter sports equipment made from these materials use the cold winter sun to capture more than 90% of this incident energy to keep the wearer warm. Another interesting material gives rise to thermochromic fabrics made by Toray which have a uniform coating of microcapsules containing heat-sensitive dyes that change colour at 5 °C intervals over a temperature range of −40 °C to 80 °C creating 'fun' and special effects.

Cripy 65 is a scented fibre produced by Mitsubishi Rayon (R) who have enclosed a fragrant essence in isolated cavities along the length of hollow polyester fibres. The scent is gradually released to give a consistent and pleasant aroma. Pillows and bed linen made from these materials are claimed[32] to improve sleep and sleeping disorders. The effect can also be achieved by printing or padding microcapsules containing perfumes into fabrics which subsequently burst and release the perfume. With careful handling, garments made from these materials are said to maintain this

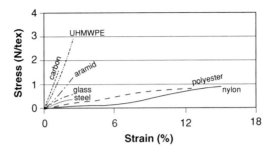

2.2   Stress/strain curves of range of available fibres.

property for up to two years. Infrared-emitting and bacteria-repelling fibres are some of the other emerging novel fibres.

Figure 2.2 shows stress/strain values for a selected range of the fibres discussed.

The remainder of this chapter focuses on the vast application areas of fibres giving examples in each category with respect to their functional requirements, limitations and means of optimising their effectiveness.

Table 2.3 summarises applications into five major areas.

2.7   Civil and agricultural engineering

Natural fibres such as flax, jute and ramie can be used for most temporary applications where, for instance, soil erosion is the problem. The geotextiles made from these natural polymers help to prevent the erosion of soils by allowing vegetative growth and their subsequent root establishment. Once the purpose is served, the geotextile material gradually disintegrates into the soil.

In most medium to long term applications however, where physical and chemical durability and dimensional stabilities are of prime concern, synthetic fibres are preferred. There are currently at least six synthetic polymers considered suitable for this purpose; they include:

- polypropylene
- polyester
- polyethylene
- polyvinyl chloride
- polyamide
- aramids.

Polypropylene is by far the most utilised geotextile, followed by polyester, the other three trailing behind with polyamide as the least used synthetic polymer. *Para*-aramids are only used where very high creep resistance and tolerance to prolonged heating are required.[36]

Generally, geosynthetics must have lifetimes which are defined and relate to their function. Durability, therefore, is mainly determined by the resistance offered by the component fibre and its assembled structure to the degrading species that causes a reduction in the tensile and mechanical properties.[37] At least three main degrading mechanisms have been identified that ultimately determine the durability and life of the contending polymer; they include:

Table 2.3 Applications of technical textiles

Civil & agricultural Engineering	Automotive & aeronautics	Medical & hygiene	Protection & defence	Miscellaneous
Application	Tyre reinforcement	Wound care	Barriers to: chemical	Marine engineering
Geotextiles, geomembranes and	Seat belting	Bandages	compounds, heat,	Electronics
geocomposites	Air bags	Barriers to bacteria	moisture, flame and	Filtration industry
e.g.	Carriage interiors	Sterile wraps	acoustics.	Food processing
Soil reinforcement	Bumpers	Disposable blood	Insulation	Sports & leisure
Separation	Wings	filtration (dialysis)	Workwear	e.g.
Filtration	Engine reinforcement	valves	Body armour	Ropes,
Drainage	Flexible container carriers	Replacement	e.g. Vests, Helmets,	Belts,
Erosion control	Balloons	ligaments	Gloves, etc.	Boat Building
Waterproofing	Parachutes, etc.	Artificial arteries		Sails
Soil stabilization for vegetative		Synthetic sutures, etc.		Cables
growth				Clean room suits
Underground hoses (used for				Gas & liquid filters
irrigation & addition of				Purifiers
fertilisers and pesticides)				Tennis rackets
				Tents, etc.
Form of application				
Nonwoven, woven, warp	Nonwoven, woven, knitted	Nonwoven, woven,	Nonwoven, woven,	Nonwoven, woven,
knitted, grids, composites,	(warp/weft), composites	knitted (warp/weft),	knitted (warp/weft),	knitted (warp/weft),
three-dimensional network	(plastic or metal resins),	braided, etc.	composites, network	braided, composites, etc.
structures, etc.	etc.		structures, etc.	

- physical degradation
- chemical degradation
- biological degradation.

Physical degradation is usually sustained during transport or installation in one form or another. Initiation of cracks is normally followed by their subsequent propagation under environmental or normal stress. In the first instance, polymer susceptibility to physical degradation depends on such factors as the weight of the geotextile; generally lightweight or thin geotextiles suffer larger strength losses than thick and heavy materials. Secondly, woven geotextiles suffer slightly larger strength losses than nonwovens, owing to their greater stiffness and, finally, the generic nature of the polymer itself plays an important role. Polyester, for instance, suffers greater strength losses than polyolefin fibres owing to its high glass transition temperature of around 70 °C. Under normal soil conditions, the polymer is relatively brittle and therefore susceptible to physical damage.[38] Polypropylene with its glass transition averaging at about −10 °C is much more pliable under these conditions but suffers from the rather critical phenomenon of extension with time or creep.

Chemical degradation is the next mechanism by which chemical agencies, often in combination with ultraviolet light or/and heat, attack the polymer whilst in use or being stored. The presence of energy starts off a self-destructing chain of events that ultimately renders the polymer ineffective. Ultraviolet radiation normally attacks the surface of the polymer and initiates chain breakage or scission which leads to embrittlement and eventual failure of the polymer. Generally, chemical degradation is a function of polymer type, thickness and availability of stabilizers. The type, quantity, and distribution of stabilizers[39] also controls the degree of resistance to degradation. Polyolefins are particularly susceptible to ultraviolet degradation and need protection using light-stabilising additives. Oxidation due to heat operates similarly by weakening the polymer thus causing rapid degradation. A range of antioxidants are often included in the polymer during manufacture and processing to minimise this harmful effect.

Biological degradation can result from at least three types of microbiological attack: direct enzymatic attack, chemical production by microorganisms which may react destructively with the polymer, and attack on the additives within the polymer.[40,41] High molecular weight polymers are much more immune to biological attack than low molecular weight polymers because microorganisms cannot easily locate the molecular chain endings. However, some microorganisms can permeate less digestible polymers in order to gain access to food.

2.8   Automotive and aeronautics

Mechanical functionality has increasingly become an almost secondary requirement for travel safety, weight efficiency, comfort and material durability of the transporting medium. From bicycles to spacecraft, fibres in one form or another fulfil these important requirements. Carbon fibre reinforcement of the frame of a bicycle ridden in the 1992 Olympics was the first of its kind to allow a comfortable win ahead of its competitors.[41] The one-piece, light, fibre-reinforced composite structure and design of the bike have since become an earmark of this industry. In the simplest terms, composites utilise unique fibre properties such as strength, stiffness and elasticity whilst incorporating the compression resistance, and torsion and bending characteristics of the matrix used.[42] Glass fibres have been traditionally used for the manufacture of boat hulls and car bodies. UHMWPE, aramids and carbon-reinforced composites in a variety of matrix materials are now routinely produced and utilised in low to very high temperature applications.

Reinforcement of tyres is another area where the transport industry has bene-fited from fibres and where rapid temperature change and changing weather con-ditions demand effective response and durability. High tenacity viscose and polyester yarns built into the internal structure of tyres now address these needs adequately. In addition to all the obvious upholstery materials used in the interior of cars, trains and aeroplanes, fibres are now used in one form or another in such parts as the engine components, fan belts, brake pads, door panels, seat skeletons, seat belts and air bags. Fire-retardant additives or inherently fire-retardant fibres are today standard requirements in the public sector use of all transport in order to improve safety.

2.9 Medical and hygiene applications

Fibres used in relation to health care and surgery may be classified depending on whether they are natural or synthetic, biodegradable or non-biodegradable. All fibres used, however, must be non-toxic, non-allergenic, non-carcinogenic and must be able to be sterilised without imparting any change in their physical or chemical characteristics.[43]

Traditionally, natural fibres such as cotton, silk and later viscose have been exten-sively used in all areas of medical and surgical care. One such area of application is on the wound, where moisture and liquid that exude from the wound are absorbed by the fibrous structure to promote healing in relatively dry conditions. However, upon healing, small fibrous elements protruding from the wound dressing are usually trapped in the pores of the newly formed tissues which make their removal distressing to the patients.

Research work[44] in the early 1960s showed that wounds under 'moist conditions' would in fact heal better and faster, which would also remove the problem of fibres being trapped in the healing wound. The concept of moist healing has since been responsible for the development of many fibres which have vastly improved wound management techniques and patient care. Alginate fibres[45] are one such example where naturally occurring, high molecular weight carbohydrates or polysaccharides obtained from seaweeds have found use in the medical textiles. Chemically, alginate is a copolymer made from α-L-guluronic acid and β-D-mannuronic acid. It is made into fibres by extruding sodium alginate into a calcium chloride bath where calcium alginate filaments precipitate. The filaments are then drawn, washed and dried. Upon contact with wound fluid, these fibres are partially converted to a water-soluble sodium alginate that swells to form a gel around the wound, thus keeping the wound moist during the healing period. They can then be easily removed once the treatment is complete. Artificial polymers based on a crosslinking copolymer of acrylic acid have been developed which are claimed[16] to have superior absorption properties to alginates particularly when used under pressure. Hydrocel, a derivative of Acordis's environmentally friendly Lyocell, is also claimed to be more absorbent than calcium alginate, taking up to 35 times its own weight of water whilst remaining intact.

Chitin[46,47] is another polysaccharide which, after cellulose, is the most abundantly available natural polymer. It is found in the outer shells of shrimps and crabs in combination with protein and minerals. Medicinal and medical use of this polymer has been realised seriously since the early 1970s. High purity chitin, from which protein, heavy metals and pyrogens have been removed, can be used for a range of applications from food additives for controlling cholesterol levels in blood to artificial skins into which tissues can safely grow. A derivative of chitin, chitosan has better processability and is now extensively available in fibre form. The particular appeal of both chitin and chitosan within biomedical applications is due to their being:

- natural and biodegradable
- compatible with most living systems
- versatile in their physical form, i.e. powder, aqueous solutions, films, shaped objects, fibres and sponges, and
- vehicles for transporting and delivering drugs.

Finally, collagen, a protein-based polymer that is collected from bovine skin and has traditionally been used in hydrogel or gelatine form for making jellies and sausage casings, is now available in fibre form. It is very strong and completely biodegradable.

Besides wound care, fibre-based structures in synthetic or natural form are used in extracorporeal devices that may be used to purify blood in kidneys, create artificial livers and function as mechanical lungs,[47] as well as finding use in suture materials, artificial ligaments and cartilages and cardiovascular implants. In general, healthcare and hygiene products, have applications that cover a wide spectrum from disposable items to surgeons uniforms and hospital bedding and all are becoming increasingly important across the world as the need to produce efficient and effective medical care increases.

2.10   Protection and defence

In textile terms, protection and defence can be a passive response in which the textile product receives and absorbs the impending impact or energy in order to protect the underlying structure. Ballistic garments are obvious examples, where the assembled fibrous material is deliberately designed to slow down and reduce the penetration of an incoming projectile. But they may also have a more active role, where the fibres show positive response by generating char or protective gases, shrinking or expanding to prevent penetration of moisture or vapour and so on. In each scenario, the protection of the underlying structure is the common objective.

In the early days, leather and metal mesh garments were used to protect the body against sword and spear attacks, but with the passage of time development of new materials occurred and, for example, early in the 1940s, nylon-based flack jackets[48] were introduced. They were a considerable improvement on the leather and metal garments but were still rather heavy and uncomfortable to wear. With the advent of *para*-aramids in the early 1970s, advanced fibres were for the first time used to make much more acceptable protective gear. In these garments, Kevlar or Twaron continuous filament yarns are woven into tight structures and assembled in a multilayer form to provide maximum protection. Their high tenacity and good energy absorption combined with high thermal stability enables these garments to receive and neutralise a range of projectiles from low calibre handguns, that is 0.22 to 0.44 inch (5.6–11.2 mm) to military bullets within the 5.56–7.62 mm range.[49] In the latter case, the fabric will require facing with ceramic tiles or other hard materials to blunt the tips of metal-jacket spitzer-pointed bullets.

To strike a better balance between garment weight, comfort and protection an even stronger and lighter fibre based on ultra-high molecular weight polyethylene (see Section 2.3) was developed and used in the early 1990s, which immediately reduced average garment weight by 15%.[50,51] UHMWPE, commercially known as Dyneema (DSM) and in composite form as Spectra Shield by Allied Signal, is now used to make cut-resistant gloves and helmets, as well as a wide range of protective garments. However, unlike Kevlar, with a fairly low melting temperature of 150 °C, it is best suited to low temperature applications.

Fire is another means by which fatal and non-fatal injuries can be sustained. No textile-based material can withstand the power and ferocity of a fire for sustained periods of greater than 10 min or so. However, if the fire ignition point could be increased or the organic structure converted to a carbonaceous char replica and its spreading rate delayed, then there may just be enough time gained to save an otherwise lost life. This is in essence the nature and objective of all fire-retardant compounds. Such treatments give fibres the positive role of forming char, reducing the emission of combustible volatile gases and effectively starving the fire of oxygen or providing a barrier to underlying surfaces.

In contrast, water-repellent fabrics may shed water by preventing water droplets from physically passing through them owing to their fine fibre dimensions and tight weaves, as is the case with ventile fabrics[52] or perforated barrier inlays such as Goretex,[53] whose tiny holes allow water vapour to pass through them but not water droplets. Wax and chemically treated fabrics reduce water/fabric surface tension by allowing water to roll off but do not allow permeation of moisture and air, so compromising comfort.

2.11   Miscellaneous

It is not possible to categorise fully all disciplines within which textile-based materials are increasingly being applied. The spectrum of fibre utilisation has already grown to include anything from conventional ropes and industrial belts to sophisticated two- and three-dimensional composite structures. The industries they cover, other than those already discussed, include wet/dry filtration, sports and leisure gear, inland water and marine applications, food processing, purifiers, electronic kit, clean room suits and many more. It will not be long before each area develops into its own major category.

2.12   Conclusions

Conventional fibres dominate the technical fibre market and are likely to do so for a long time to come. The rate of growth in consumption of fibres destined for technical applications, however, is now faster than those going to the traditional clothing and furnishing sectors. It was estimated that the world's technical fibres share of the market would have reached the 40% mark by the year 2000.[1] Much of this growth will depend on greater realisation of the technological and financial benefits that fibre-based structures could bring to the traditional and as yet unyielding sectors of engineering. Issues such as environment, recycling and biodegradability, which are increasingly subjects of concern for the public, will further encourage and benefit the growth and use of technical textiles.

References

1. T NOBOLT, Keeping a competitive edge in technical textiles, *Textile Month*, Union des Industries, February p16, 1992.
2. M C TUBBS (ed.), *Textile Terms and Definitions*, 9th edition, The Textile Institute, Manchester, 1991.
3. C BYRNE, 'Technical textiles', *Textiles Mag.*, issue 2.95, pp. 12–16, 1995.
4. J E MCINTYRE, *Speciality Fibres for Technical Textiles*, Department of Textile Industries, University of Leeds, notes.
5. A R HORROCKS, *The Durability of Geotextiles*, Comet Eurotex, UK 1992.
6. P R RANKILOR, *Membranes in Ground Engineering*, Wiley Interscience, London, 1981.
7. V van ZANTEN, *Geotextiles and Geomembranes in Civil Engineering*, Balkema, Rotterdam, 1986.
8. P W HARRISON (ed.), *Cotton in a Competitive World*, The Textile Institute, 1979.
9. B K KANDOLA and A R HORROCKS, 'Complex char formation in flame-retarded fibre-intumescent combinations-II. Thermal analytical studies', *Polymer Degradation and Stability*, **54**, 289–303, 1996.
10. A R HORROCKS, 'Flame-retardant finishing of textiles', *Rev. Prog. Coloration*, **16**, 62–101, 1986.
11. L BENISEK and W A PHILLIPS, *Textile Res. J.*, 51, 1981.
12. D J COLE, 'A new cellulosic fibre – Tencel', *Advances in Fibre Science*, The Textile Institute, 1992.
13. C M LEGAULT, 'Antron – a special nylon carpet fibre', *Can. Textile J.*, April, 1982.
14. C B CHAPMAN, *Fibres*, Butterworths, 1974.
15. C Z PORCZYNSKI, *Manual of Man-made Fibres*, Astex, 1960.
16. P AKERS and R HEATH, Superabsorbent Fibres, The Key to the Next Generation of Medical Products, Technical Absorbents Ltd., UK, 1996.
17. D H MORTON-JONES, *Polymer Processing*, Chapman and Hall, London, 1989.
18. J W S HEARLE, 'Understanding and control of fibre structure, parts 1 to 3', *Textile Horizon*, 1997.
19. F HAPPEY, *Appl. Fibre Sci.*, **3**.
20. A SHARPLES, *Introduction to Polymer Crystallisation*, Arnold, 1966.
21. D S VARMA, 'High performance organic fibres for industrial applications', *Man-made Textiles in India*, 380–386, August, 1986.
22. L MILES (ed.), 'Flame-resistant fabrics – Some like it hot', *Textile Month*, 37–41, April, 1996.
23. P HEIDEL, A G HOECHST, Frankfurt, Germany, 'New opportunities with aromatics', *Textile Month*, 43–44, July, 1989.
24. L BAKER, 'Polyphenylene sulphide fibres for high temperature filtration', *Tech. Textiles Int.*, 11–16, September, 1992.
25. E MCINTYRE, 'High-performance fibres for industrial textiles', *Textile Horizon*, 43–45, October, 1988.
26. F O ANDEREGG, 'Strength of glass fibres', *Ind. Eng. Cem.*, **31**, 290–298, 1939.
27. L MICHAELS and S S CHISSICK (eds.), *Asbestos Properties, Applications and Hazards, Vol. 1*, Wiley, New York 1979.
28. I N ERMOLENKO *et al*, *Chemically Modified Fibres*, VCH, 1990.
29. T F COOKE, 'Inorganic fibres, A literature review', *J. Am. Ceramics Soc.*, p. 2959.
30. F D BIRCHALL, 'Oxide inorganic fibres', *Concise Encyclopedia of Composite Materials*, Oxford, UK, 1989.
31. A K DHINGRA, 'Advances in inorganic fibre development', *Contemporary Topics in Polymer Science*, New York, 1982.
32. K NOSSE, 'New developments in Japanese fibres', *Textile Month*, 44–46, June, 1990.
33. D and K CONSULTING, 'The European potential of microfibres', *Textile Month*, 23–25, January, 1992.
34. M GOLDING, 'The microfibre business in Japan', *Tech. Textiles Int.*, 18–23, May, 1992.
35. T HONGU and G O PHILLIPS, *New Fibres*, Ellis Horwood, UK, 1990.
36. A R HORROCKS and J A D'SOUZA, 'Degradation of polymers in geomembranes and geotextiles', *Handbook of Polymer Degradation*, S H Hamid, M B Amin and A G Maadhah (eds), Marcel Dekker, New York, 1992, 433–501.
37. T M ALLEN, 'Determination of long-term tensile strength of geosynthetics: A state of the art review', *Geosynthetics'91 Conference*, Atlanta, USA, 351–379.

38. T M ALLEN, J R BELL and T S VINSON, 'Properties of geotextiles in cold region applications', *Transportation Research Report*, Oregon State University, 1983.
39. R M KOERNER *et al*, 'Long-term durability and aging of geotextiles,' *Geotextiles Geomembranes*, 1988.
40. G R KOERNER *et al*, 'Biological clogging in leachate collection systems', *Durability Aging Geosynthetics*, 1989.
41. S K MUKHOPADHYAY, 'High performance fibres', *Textile Prog.*, **25**(3/4), 1993.
42. P BRACKE *et al*, *Inorganic Fibres and Composite Materials*, Pergamon Press, Oxford 1984.
43. A J RIGBY, S C ANAND and M MIRAFTAB, 'Textile materials in medicine and surgery', *Textile Horizons*, December, 1993 and Knitting Int. February, 1994, vol. 101.
44. G D WINTER, *Nature*, 1962, 193, 293.
45. Y QIN, C AGBOH, X WANG and D K GILDING, 'Novel polysaccharide fibres for advanced wound dressing', *Medical Textiles 96*, conference proceedings, Woodhead, 15–20, 1997.
46. P A SANDFORD and A STEINNES, 'Biomedical applications of high purity chitosan', *American Chemical Society*, chapter 28, 1991.
47. A S HOFFMAN, 'Medical applications of polymeric fibres', *Appl Polym Symp* **31**, 313–334 J. Wiley and Sons, New York, 1977.
48. A ELDIB and R WALTHER, 'Advanced fibres for personal protective clothing', *Tech. Textiles Int.*, June, 1992.
49. S A YOUNG, 'The use of Kevlar *para*-aramid fibre in ballistic protection garments', *Tech. Textiles Int.*, 26–27, July/August, 1992.
50. S A YOUNG, 'Dyneema – the strongest fibre in the world', *Tech. Textiles Int.*, May, 1992.
51. The use of Kevlar *para*-aramid fibre in ballistic protection garments, *Tech. Textiles Int.*, 24–27, July/August, 1992.
52. Ventile fabrics commercial publications, Courtaulds, Chorley, Lancashire.
53. W L GORE and ASSOCIATES, Inc, 'Porous materials derived from tetrafluoroethylene and processes for their production', US Patent no. 1355373, USA, 1974.

3

Technical yarns

R H Gong and X Chen

Department of Textiles, UMIST, PO Box 88, Sackville Street, Manchester M60 1QD, UK

3.1 Introduction

Technical yarns are produced for the manufacture of technical textiles. They have to meet the specific functional requirements of the intended end-use. This may be achieved through special yarn production techniques or through the selection of special fibre blends or a combination of both. This chapter describes the yarn production technologies that are applicable to technical yarns and discusses the structures and properties of the yarns that may be produced using these technologies.

3.2 Staple fibre yarns

3.2.1 Ring spinning

Ring spinning is currently the most widely used yarn production method. Initially developed in America in the 1830s, its popularity has survived the emergence of much faster spinning technologies. In addition to the superior yarn quality, ring spinning is extremely versatile. It is capable of producing yarns with wide ranges of linear density and twist from a great variety of fibre materials. It is also used for doubling and twisting multifold and cabled yarns.

Fibre materials must be properly prepared before they can be used on the ring spinning machine. The preparation processes are dependent on the fibre material. Figures 3.1 and 3.2 illustrate the typical process routes for cotton and wool. The ultimate objectives of the many preparation processes are to produce a feed material for the final spinning process that is clean, even, homogeneous and free from fibre entanglement. The fibres must also be in the preferred orientation.

On the ring spinning machine, the feed material is attenuated to the required linear density by a drafting system, typically a roller drafting system with three lines of rollers. The drafted fibre strand is then twisted by the ring spindle illustrated in Fig. 3.3.

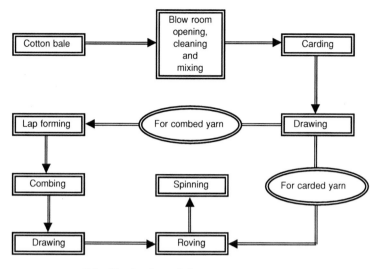

3.1 Production of ring-spun cotton yarn.

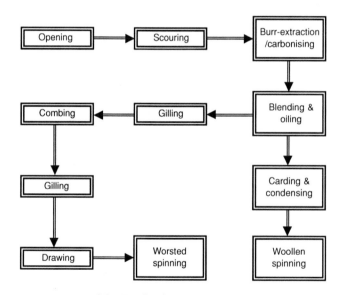

3.2 Production of wool yarn.

The yarn leaving the front rollers is threaded through a yarn guide (the lappet), which is located directly above the spindle axis. The yarn then passes under the C-shaped traveller onto the bobbin. The bobbin is mounted on the spindle and rotates with the spindle. When the bobbin rotates, the tension of the yarn pulls the traveller around the ring. The traveller rotational speed, the spindle rotational speed and the yarn delivery speed follow Equation 3.1:

$$N_t = N_s - \frac{V_d}{\pi D_b} \qquad\qquad (3.1)$$

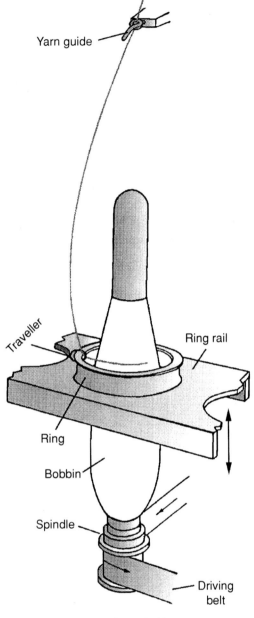

Yarn guide

Traveller

Ring rail

Ring

Bobbin

Spindle

Driving
belt

3.3   Ring spindle.

where N_t is the traveller rotational speed (rpm), N_s is the spindle rotational speed (rpm), V_d is the yarn delivery speed (m min^{-1}) and D_b is the bobbin diameter (m).

During production, the bobbin rail moves up and down to spread the yarn along the length of the bobbin so that a proper package can be built. The movement of the ring rail is quite complicated, but the aim is to build a package that is stable, easy to unwind and contains the maximum amount of yarn. As the yarn is wound on the bobbin, the bobbin diameter increases steadily during production. The spindle speed and the yarn delivery speed are normally kept constant, it is therefore obvious from Equation 3.1 that the traveller speed increases during production.

Each rotation of the traveller inserts one full turn in the yarn, so the twist inserted in a unit length of yarn can be calculated by Equation 3.2:

$$t = \frac{N_t}{V_d} \tag{3.2}$$

where t is the yarn twist (turns m^{-1}).

Because the traveller speed is not constant, the twist in the yarn also varies. However, because this variation is usually very small, it is commonly ignored and the twist is simply calculated from the spindle speed, Equation 3.3:

$$t = \frac{N_s}{V_d} \tag{3.3}$$

As can be seen from Equation 3.3, for a given yarn twist, the higher the spindle speed the higher the yarn delivery speed. A spindle speed of up to 25 000 rpm is possible, although speeds of between 15 000 and 20 000 rpm are more usually used. The spindle speed is restricted by the traveller speed, which has an upper limit of around 40 m s^{-1}. When the traveller speed is too high, the friction between the traveller and the ring will generate too much heat, which accelerates the wear on the traveller and the ring and may also cause yarn damage. The yarn between the yarn guide and the traveller rotates with the traveller and balloons out owing to centrifugal force. The tension in the yarn increases with the rotational speed of the yarn balloon. When the spindle speed is too high, the high yarn tension will increase the yarn breakage. The traveller speed and the yarn tension are the two most critical factors that restrict the productivity of the ring spinning system. The increasing power cost incurred by rotating the yarn package at higher speeds can also limit the economic viability of high spindle speeds. For the same traveller linear speed, using a smaller ring allows a higher traveller rotational speed and increases delivery speed. A smaller ring also reduces yarn tension, as the yarn balloon is also smaller. However, a smaller ring leads to a smaller bobbin which results in more frequent doffing.

Ring-spun yarns have a regular twist structure and, because of the good fibre control during roller drafting, the fibres in the yarn are well straightened and aligned. Ring spun yarns therefore have excellent tensile properties, which are often important for technical applications.

The ring spinning system can be used for spinning cover yarns where a core yarn, spun or filament, is covered by staple fibres. This can provide yarns with a combination of technical properties. For example, a high strength yarn with good comfort characteristics may be spun from a high strength filament core with natural fibre covering. Other technical yarns, such as flame-retardant and antistatic yarns can also be made by incorporating flame-retardant and electricity conductive fibres.

The main limitation of the ring spinning system is the low productivity. The other limitations are the high drafting and spinning tensions involved. These high tensions can become a serious problem for spinning from fibres such as alginate fibres that have low strength.

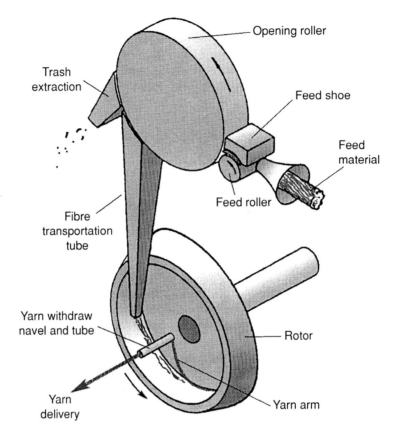

3.4   Rotor spinning principle.

3.2.2   Rotor spinning

The productivity limitation of the ring spinning system was recognised long before the commercial introduction of rotor spinning in 1967. In ring spinning, the twist insertion rate is dependent on the rotational speed of the yarn package. This is so because of the continuity of the fibre flow during spinning. Numerous attempts have been made since before the end of the 19th century, particularly since the 1950s, to introduce a break into the fibre flow so that only the yarn end needs to be rotated to insert twist. Very high twisting speeds can thus be achieved. In addition, by separating twisting from package winding, there will be much more flexibility in the form and size of the yarn package built on the spinning machine. This increases the efficiency of both the spinning machine and of subsequent processes. Rotor spinning was the first such new technology to become commercially successful and it is the second most widely used yarn production method after ring spinning.

The principles of rotor spinning are illustrated in Fig. 3.4. The fibre material is fed into an opening unit by a feed roller in conjunction with a feed shoe. The feed material is usually a drawn sliver. An opening roller is located inside the opening unit and is covered with carding wire, usually saw-tooth type metallic wire. The surface speed of the opening roller is in the region of 25–30 m s^{-1}, approximately 2000 times the feed roller surface speed. This high speed-ratio enables the fibres to be opened up into a very thin and open fibre flow. The fibres are taken off the opening roller by an air stream with a speed about twice that of the opening roller. The fibres are carried by the air stream, through the fibre transportation tube, into the spinning rotor. The air speed in the transportation tube increases, owing to the narrowing cross-section of the tube, as the air reaches the exit point inside the rotor. This ensures that the fibres are kept aligned along the airflow direction and as straight as possible. The exit angle of the fibres from the transportation tube is at a tangent to the rotor wall and the surface speed of the rotor is faster than the exit air speed, so the fibres emerging from the transportation tube are pulled into the rotor, keeping the fibres aligned in the direction of the fibre flow. The centrifugal force generated by the rotor forces the fibres into the rotor groove. Because of the high surface speed of the rotor, only a very thin layer of fibres, usually one or two fibres in the cross-section, is deposited in the rotor when the rotor passes the fibre exit point of the transportation tube. Many such layers of fibres are needed to make up the yarn. This doubling up of the fibres in the rotor is called back doubling.

The tail of the yarn arm inside the rotor is thrown against the rotor groove because of the centrifugal force. The yarn arm rotates with the rotor and each rotation of the yarn arm inserts one full turn in the yarn. As the yarn is withdrawn continuously through the navel and tube, the contact point of the yarn arm with the rotor groove must move around the rotor. Because the yarn arm is rotating axially, the fibres in the rotor groove are twisted into the yarn. The machine twist of the yarn can be calculated by Equation 3.4:

$$t = \frac{N_y}{V_d} \qquad (3.4)$$

where t is the yarn twist (turns m^{-1}), N_y is the rotational speed of the yarn arm (rpm) and V_d is the yarn delivery speed (m min^{-1}).

The following relationship exists between the yarn arm speed, the rotor speed and the yarn delivery speed, Equation 3.5:

$$(N_y - N_r)\pi D = V_d \qquad (3.5)$$

where D is the diameter of rotor groove.

The relative speed between the yarn arm and the rotor is normally very small in comparison with the rotor speed and the machine twist of the yarn is commonly calculated by Equation 3.6:

$$t = \frac{N_r}{V_d} \qquad (3.6)$$

The back-doubling ratio β is equal to the ratio of the rotor speed to the relative speed between the yarn arm and the rotor, Equation 3.7:

$$\beta = \frac{N_r}{N_y - N_r} = \frac{N_r}{V_d}\pi D = t\pi D \qquad (3.7)$$

Because there is no need to rotate the yarn package for the insertion of twist, rotor spinning can attain much higher twisting speeds than ring spinning. The rotor speed can reach 150 000 rpm. The roving process needed in ring spinning is eliminated in rotor spinning, further reducing the production cost. The package can be much larger, with fewer knots in the product and with a more suitable form for subsequent processes.

Because the yarn is formed in an enclosed space inside the rotor, trash particles remaining in the fibres can accumulate in the rotor groove. This leads to a gradual deterioration of yarn quality and in severe cases yarn breakage. The cleanliness of fibres is more critical for rotor spinning than for ring spinning. In order to improve the cleanliness of the fibres, a trash extraction device is used at the opening roller.

As the twist in the yarn runs into the fibre band in the rotor groove, inner layers of the yarn tend to have higher levels of twist than outer layers. Fibres landing on the rotating fibre band close to the yarn tail, or directly on the rotating yarn arm when the yarn arm passes the exit of the transportation tube, tend to wrap around the yarn instead of being twisted into the yarn. These wrapping fibres are characteristic of rotor-spun yarns.

Rotor-spun yarns usually have lower strength than corresponding ring-spun yarns because of the poorer fibre disposition in the yarn. This is the result of using an opening roller to open up the fibres, of transporting the fibres by airflow, and of the low yarn tension during yarn formation. The wrapping fibres also lead to a rougher yarn surface. Rotor yarns have better short term evenness than ring-spun yarns because of the back-doubling action.

The main advantage of rotor spinning is the high production rate. However, because of the lower yarn strength, rotor spinning is limited to medium to course yarn linear densities. It is also limited to the spinning of short staple fibre yarns.

3.2.3   Friction spinning

Friction spinning is an open end spinning technique. Instead of using a rotor, two friction rollers are used to collect the opened-up fibres and twist them into the yarn. The principle is shown in Fig. 3.5.

The fibres are fed in sliver form and opened by a carding roller. The opened fibres are blown off the carding roller by an air current and transported to the nip area of two perforated friction drums. The fibres are drawn onto the surfaces of the friction drums by air suction. The two friction drums rotate in the same direction and because of the friction between the fibre strand and the two drum surfaces, twist is inserted into the fibre strand. The yarn is withdrawn in the direction parallel to the friction drum axis and delivered to a package forming unit. The friction drum diameter is much larger than the yarn diameter. The diameter ratio can be as high as 200. A high twisting speed can thus be achieved by using a relatively low speed for the friction drums. Owing to the slippage between the drum surface and the yarn end, the yarn takes up only 15–40% of the drum rotation. Nevertheless, a high production speed, up to $300\,\mathrm{m\,min^{-1}}$, can be achieved. For a finer yarn the twist insertion rate is higher with the same drum speed so the delivery speed can be practically independent of yarn linear density.

Because the yarn is withdrawn from the side of the machine, fibres fed from the machine end away from the yarn delivery tend to form the yarn core while fibres fed from the machine end closer to the yarn delivery tend to form the sheath. This characteristic can be conveniently exploited to produce core–sheath yarn structures. Extra core components, filaments or drafted staple fibres, can be fed from the side of the machine while the fibres fed from the top of the machine, the normal input, form the sheath.

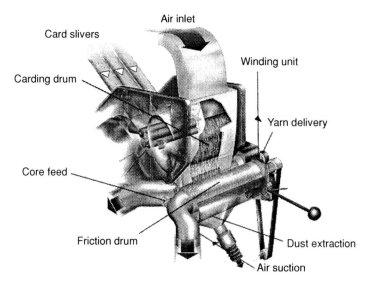

3.5   DREF 2 friction spinner.

Unlike ring or rotor spinning machines that are produced by many manufacturers around the world, friction spinning machines are only currently made by Dr. Ernst Fehrer AG of Austria. The diagram shown in Fig. 3.5 is the DREF 2 machine that has recently been upgraded to DREF 2000. The company also produces the DREF 3 machine that has an extra drafting unit on the side of the machine for feeding drafted staple fibres as a core component.

The fibre configuration in friction-spun yarns is very poor. When the fibres come to the friction drum surface, they have to decelerate sharply from a high velocity to almost stationary. This causes fibre bending and disorientation. Because of the very low tension in the yarn formation zone, fibre binding in the yarn is also poor. As a result, the yarn has a very low tensile strength and only coarse yarns, 100 tex and above, are usually produced.

The main application of friction spinning is for the production of industrial yarns and for spinning from recycled fibres. It can be used to produce yarns from aramid and glass fibres and with various core components including wires. The yarns can be used for tents, protective fabrics, backing material, belts, insulation and filter materials.

3.2.4   Wrap spinning

Wrap spinning is a yarn formation process in which a twistless staple fibre strand is wrapped by a continuous binder. The process is carried out on a hollow spindle machine as illustrated in Fig. 3.6. The hollow spindle was invented by DSO 'Textil' in Bulgaria. The first wrap spinning machine was introduced in the 1979 ITMA.

The staple roving is drafted on a roller drafting system similar to those used on ring frames and is passed through a rotating hollow spindle that carries a binder bobbin. The rotation of the hollow spindle and the bobbin wraps the binder around the staple strand.

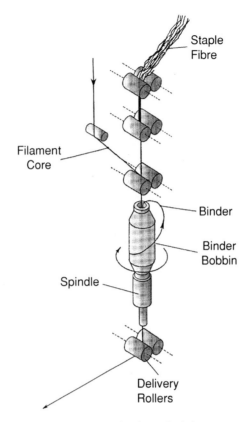

3.6 Wrap spinning principle.

To prevent the drafted staple strand falling apart before it is wrapped by the binder, false twist is usually generated in the staple strand by the spindle. To introduce the false twist, the staple strand is not threaded through the hollow spindle directly. It is wrapped around either a twist regulator at the bottom of the spindle or part of the spindle top. The false twist also allows the staple strand to be compacted before the binder is wrapped around it. This improves the yarn strength.

Two hollow spindles can be arranged one above the other to wrap the staple strand with two binders in opposite directions. This is used to produce special effect yarns with a more stable structure. Real twist may also be added to the yarn by passing the wrapped yarn onto a ring spindle, usually arranged directly underneath the hollow spindle.

Core yarns, mostly filaments, can be added to the feed. This can be used to provide extra yarn strength or other special yarn features. An example is to use this method to spin alginate yarns. Alginate fibres are very weak and cause excessive breakages during spinning without the extra support of core filaments.

A variety of binders can be used to complement the staple core or to introduce special yarn features. For example, a carbon-coated nylon filament yarn can be used to produce yarns for antistatic fabrics. Soluble binders can be used for making yarns for medical applications.

Wrap spinning is highly productive and suitable for a wide range of yarn linear densities. Yarn delivery speeds of up to 300 m min^{-1} are possible. Because the binder is normally very fine, each binder bobbin can last many hours, enabling the pro-

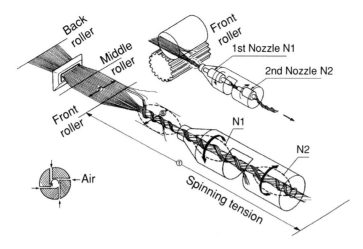

3.7   Murata jet spinner.

duction of large yarn packages without piecing. Because the staple core is composed of parallel fibres with no twist, the yarn has a high bulk, good cover and very low hairiness. The main limitation of wrap spinning is that it is only suitable for the production of multicomponent yarns. The binder can be expensive, increasing the yarn cost.

3.2.5   Air-jet spinning

Air-jet spinning technology was first introduced by Du Pont in 1963, but it has only been made commercially successful by Murata since 1980. Du Pont used only one jet, which produced a low strength yarn. The Murata system has two opposing air jets, which improves the yarn strength. The twin-jet Murata Jet Spinner is illustrated in Fig. 3.7. Staple fibres are drafted using a roller drafting system with three or four pairs of rollers. The fibres are then threaded through the twin-jet assembly. The second jet N_2 has a higher twisting torque than the first jet N_1. Immediately after leaving the front drafting rollers, the fibres in the core of the yarn are twisted in the twist direction of N_2. The fibres on the edges of the drafted ribbon are twisted by the weaker N_1 and wrap around the core fibres in the opposite direction. Because the jet system is located between the front drafting rollers and the yarn delivery rollers, neither of which rotates around the axis of the yarn, the twist inserted by the jets is not real twist and after the yarn passes through the jet system, the core fibres become twistless. The yarn strength is imparted by the wrapping of the edge fibres. Because of the small jet dimensions, very high rotational jet speeds are possible. Although twist efficiency is only 6–12% because of the twist resistance of the yarn, delivery speeds of up to $300\,m\,min^{-1}$ are possible. In a further development by Murata the second jet is replaced with a pair of roller twisters. The principle of yarn formation is similar to the twin-jet system. The new machine, the roller-jet spinner, is capable of delivery speeds of up to $400\,m\,min^{-1}$. However, the yarn has a harsher handle.

Air-jet spinning is used mainly for spinning from short staple fibres, especially cotton and polyester blends. The vortex spinner, the latest addition to the Murata jet spinner range, was shown in ITMA 99 for spinning from 100% cotton.

The air-jet system can be used to produce core–sheath yarn structures by feeding the core and sheath fibres at different stages of the drafting system. Fibres fed in from the back of the drafting system tend to spread wider under the roller pressure and form the sheath of the yarn while fibres fed in nearer to the front tend to form the yarn core. Filament core can also be introduced at the front drafting roller. Two spinning positions can be combined to produce a two-strand yarn that is then twisted using the usual twisting machinery.

Air-jet yarns have no real twist, therefore they tend to have higher bulk than ring and rotor yarns and better absorbency. They are more resistant to pilling and have little untwisting tendency. Because the yarn strength is imparted by the wrapping fibres, not the twisting of the complete fibre strand, air-jet yarns have lower tensile strength than ring and rotor yarns. The system is only suitable for medium to fine yarn linear densities as the effectiveness of wrapping decreases with the yarn thickness. The rigid yarn core of parallel fibres makes the yarn stiffer than the ring and rotor yarns.

3.2.6   Twistless spinning

Numerous techniques have been developed to produce staple yarns without twisting so that the limitations imposed by twisting devices, notably the ring traveller system, can be avoided and production speed can be increased. Because of the unconventional yarn characteristics, these techniques have not gained widespread acceptance commercially, but they do offer an alternative and could be exploited to produce special products economically.

Most of these twistless methods use adhesives to hold the drafted staple fibre strand together. They can produce low linear density yarns at a high speed. The adhesives may later be removed after the fabric is made and the fibres are then bound by the interfibre forces imposed by fabric constraints. This type of yarn has high covering power due to the untwisted yarn structure. However, these processes mostly involve additional chemicals and require high power consumption. The yarns can only be used for fabrics that offer good interfibre forces.

As an example, the TNO twistless spinning method is shown in Fig. 3.8. In the system shown here, the roving is drafted under wet conditions, which gives better fibre control. An inactive starch is then applied to the drafted roving, which is also false twisted to give it temporary strength. The starch is activated by steaming the package that is then dried. The later version of TNO twistless system replaces the starch with PVA (polyvinyl alcohol) fibres (5–11%), which melt at above 80 °C to bind the staple fibres. This is also known as the Twilo system.

Another twistless spinning method is the Bobtex process, which can produce high strength yarns for industrial/leisure fabrics, such as tents, workwear and sacks. In this process, staple fibres (30–60%) are bonded to a filament core (10–50%) by a layer of molten polymer (20–50%). Production speeds of up to $1000\,\mathrm{m\,min^{-1}}$ can be achieved. The process can use all types of staple fibres including waste fibres.

3.2.7   Ply yarn

Single yarns are used in the majority of fabrics for normal textile and clothing applications, but in order to obtain special yarn features, particularly high strength

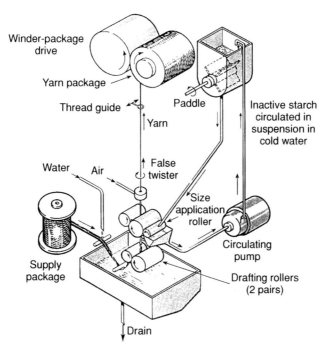

3.8   TNO twistless yarn production method.

and modulus for technical and industrial applications, ply yarns are often needed. A folded yarn is produced by twisting two or more single yarns together in one operation, and a cabled yarn is formed by twisting together two or more folded yarns or a combination of folded and single yarns.

The twisting together of several single yarns increases the tenacity of the yarn by improving the binding-in of the fibres on the outer layers of the component single yarns. This extra binding-in increases the contribution of these surface fibres to the yarn strength. Ply yarns are also more regular, smoother and more hard wearing. By using the appropriate single yarn and folding twists, a perfectly balanced ply yarn can be produced for applications that require high strength and low stretch, for example, for tyre cords.

A typical process route of a ply yarn involves the following production stages:

1.   single yarn production
2.   single yarn winding and clearing
3.   assembly winding: to wind the required number of single yarns as one (doubling) on a package suitable for folding twisting
4.   twisting
5.   winding

Twisting can be carried out in a two-stage process or with a two-for-one twister. In the two-stage process, the ring frame is used to insert a low folding twist in the first stage and an up-twister to insert the final folding twist in the second

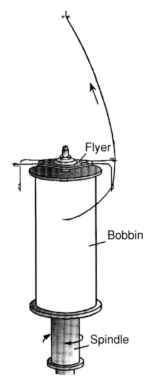

3.9   The up-twister.

stage. The ring frame uses a low twist to enable higher delivery speeds. A suitable package is formed on the ring frame for the over-end withdrawal of yarn on the up-twister. Figure 3.9 shows the principle of the up-twister. The supply package rotates with the spindle while the yarn is withdrawn over the package end from the top. The free-rotating flyer is pulled around by the yarn and inserts twist in the yarn.

The two-for-one twister is illustrated in Fig. 3.10. The supply yarn package is stationery. After withdrawal from the package, the yarn is threaded through the centre of the spindle and rotates with the spindle. Each rotation of the spindle inserts one full turn in the yarn section inside the spindle centre and also one turn in the yarn section outside the yarn package (the main yarn balloon). The yarn therefore gets two turns for each spindle rotation. If the supply package is rotated in the opposite direction of the spindle, then the twisting rate will increase by the package rotational speed. The Saurer Tritec Twister is based on this principle. In the Tritec Twister, the package rotates at the same speed as the spindle, but in the opposite direction, so each spindle rotation inserts three twists in the yarn. The package is magnetically driven.

The production of a ply yarn is much more expensive than the production of a single yarn of equivalent linear density. Not only does the ply yarn production require the extra assembly winding and twisting processes, but the production of the finer component single yarn is also much more expensive.

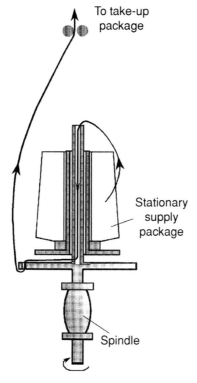

To take-up
package

Stationary
supply
package

Spindle

3.10 The two-for-one twister.

3.3 Filament yarns

3.3.1 Definitions

A filament yarn is made from one or more continuous strands called filaments where each component filament runs the whole length of the yarn. Those yarns composed of one filament are called monofilament yarns, and those containing more filaments are known as multifilament yarns. For apparel applications, a multifilament yarn may contain as few as two or three filaments or as many as 50 filaments. In carpeting, for example, a filament yarn could consist of hundreds of filaments. Most manufactured fibres have been produced in the form of a filament yarn. Silk is the only major natural filament yarn.

According to the shape of the filaments in the yarn, filament yarns are classified into two types, flat and bulk. The filaments in a flat yarn lie straight and neat, and are parallel to the yarn axis. Thus, flat filament yarns are usually closely packed and have a smooth surface. The bulked yarns, in which the filaments are either crimped or entangled with each other, have a greater volume than the flat yarns of the same linear density.

Texturing is the main method used to produce the bulked filament yarns. A textured yarn is made by introducing durable crimps, coils, and loops along the length of the filaments. As textured yarns have an increased volume, the air and vapour permeability of fabrics made from them is greater than that from flat yarns.

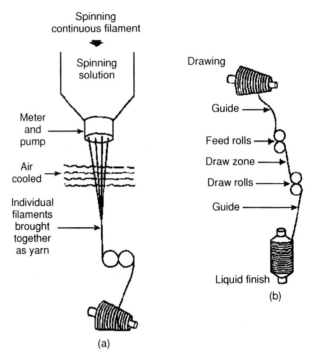

3.11   (a) Melt spinning process, (b) drawing process.

However, for applications where low air permeability is required, such as the fabrics for air bags, flat yarns may be a better choice.

3.3.2   Manufacture of filament yarns

Most manufactured fibres are extruded using either melt spinning, dry spinning, or wet spinning, although reaction spinning, gel spinning and dispersion spinning are used in particular situations. After extrusion, the molecular chains in the filaments are basically unoriented and therefore provide no practical strength. The next step is to draw the extruded filaments in order to orient the molecular chains. This is conventionally carried out by using two pairs of rollers, the second of which forwards the filaments at approximately four times the speed of the first. The drawn filaments are then wound with or without twist onto a package. The tow of filaments at this stage becomes the flat filament yarn. Figure 3.11 shows the melt spinning process and the subsequent drawing process.

For many applications, flat filament yarns are textured in order to gain increased bulkiness, porosity, softness and elasticity in some situations. Thermoplastic filament yarns are used in most texturing processes. The interfibre bonds break and reform during the texturing process. A filament yarn is generally textured through three steps. The first step is to distort the filament in the yarn so that the interfibre bond is broken. Twisting or other means are used to distort the filaments within a yarn. The second step is to heat the yarn, which breaks bonds between polymers, allowing the filaments to stay crimped. The last step is to cool the yarn in the distorted state to enable new bonds to form between the polymers. When the yarn is untwisted

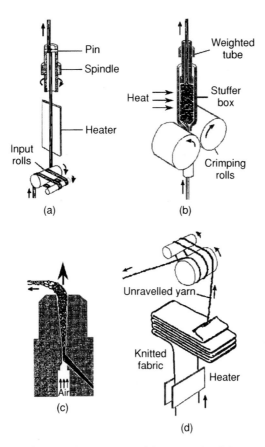

3.12   Principles of main texturing methods: (a) false twist, (b) stuffer box, (c) air-jet, (d) knit-de-knit.

or otherwise released from its distorted state, the filaments remain in a coiled or crimped condition.

There are many methods for yarn texturing, including false-twist, air-texturing, knit-de-knit, stuffer box and gear crimp. Among these, the false-twist is the most popular method. Figure 3.12 shows the principles of the main methods of yarn texturing.

3.3.3   Filament technical yarns

There have been many types of filament yarns developed for technical applications, such as reinforcing and protecting. The reinforcing technical yarns have either high modulus, high strength, or both. Yarns for protecting applications can be resistant to safety hazards such as heat and fire, chemical and mechanical damage. There are many types of technical filament yarns used in various applications, it is only possible, therefore, to list just a few yarns here that are popularly used in the development of some technical textile products.

3.3.3.1   Aramid filament yarns

Aramid fibre is a chemical fibre in which the fibre-forming substance is a long chain synthetic polyamide where at least 85% of the amide linkages are attached directly to two aromatic rings. Nomex and Kevlar are two well-known trade names of the aramid fibre, owned by Du Pont. Aramid fibres have high tenacity and high resistance to stretch, to most chemicals and to high temperature. The Kevlar aramid is well known for its relatively light weight and for its fatigue and damage resistance. Because of these properties, Kevlar 29 is widely used and accepted for making body armour. Kevlar 49, on the other hand, has high tenacity and is used as reinforcing material for many composite uses, including materials for making boat and aircraft parts. The Nomex aramid, on the other hand, is heat resistant and is used in making fire fighters' apparel and similar applications.

Aramid yarns are more flexible than their other high performance counterparts such as glass and Kevlar, and thus are easier to use in subsequent fabric making processes, be it weaving, knitting, or braiding. Care should be taken, though, as aramid yarns are much stronger and much more extensible than the conventional textile yarns, which could make the fabric formation process more difficult.

3.3.3.2   Glass filament yarns

Glass is an incombustible textile fibre and has high tenacity too. It has been used for fire-retardant applications and also is commonly used in insulation of buildings. Because of its properties and low cost, glass fibre is widely used in the manufacture of reinforcement for composites. There are different types of glass fibres, such as E-glass, C-glass, and S-glass. E-glass has very high resistance to attack by moisture and has high electrical and heat resistance. It is commonly used in glass-reinforced plastics in the form of woven fabrics. C-glass is known for its chemical resistance to both acids and alkalis. It is widely used for applications where such resistance is required, such as in chemical filtration. The S-glass is a high strength glass fibre and is used in composite manufacturing.

Glass filament yarns are brittle compared with the conventional textile yarns. It has been shown that the specific flexural rigidity of glass fibre is $0.89\,mN\,mm^2\,tex^{-2}$, about 4.5 times more rigid than wool. As a result, glass yarns are easy to break in textile processing. Therefore, it is important to apply suitable size to the glass yarn to minimise the interfibre friction and to hold the individual fibres together in the strand. Dextrinised starch gum, gelatine, polyvinyl alcohol, hydrogenerated vegetable oils and non-ionic detergents are commonly used sizes.

When handling glass fibres, protective clothing and a mask should be worn to prevent skin irritation and inhalation of glass fibres.

3.3.3.3   Carbon filament yarns

Carbon fibres are commonly made from precursor fibres such as rayon and acrylic. When converting acrylic fibre to carbon, a three-stage heating process is used. The initial stage is oxidative stabilisation, which heats the acrylic fibre at $200-300\,^\circ C$ under oxidising conditions. This is followed by the carbonisation stage, when the oxidised fibre is heated in an inert atmosphere to temperatures around $1000\,^\circ C$. Consequently, hydrogen and nitrogen atoms are expelled from the oxidised fibre, leaving the carbon atoms in the form of hexagonal rings that are arranged in oriented fibrils. The final stage of the process is graphitisation, when the carbonised filaments are heated to a temperature up to $3000\,^\circ C$, again in an inert atmosphere. Graphitisation increases the orderly arrangement of the carbon atoms, which are organised into a crystalline structure of layers. These layers are well oriented in the direction of fibre axis, which is an important factor in producing high modulus fibres.

Like the glass yarns, most carbon fibres are brittle. Sizes are used to adhere the filaments together to improve the processability. In addition to protecting operatives against skin irritation and short fibre inhalation, protecting the processing machinery and auxiliary electric and electronic devices needs to be considered too, as carbon fibre is conductive.

3.3.3.4   HDPE filament yarns

HDPE refers to high density polyethylene. Although the basic theory for making super strong polyethylene fibres was available in the 1930s, commercial high performance polyethylene fibre was not manufactured until recently. Spectra, Dyneema, and Tekmilon are among the most well-known HDPE fibres. The gel spinning process is used to produce the HDPE fibre. Polyethylene with an extra high molecular weight is used as the starting material. In the gel spinning process, the molecules are dissolved in a solvent and spun through a spinneret. In solution, the molecules which form clusters in the solid state become disentangled and remain in this state after the solution is cooled to give filaments. The drawing process after spinning results in a very high level of macromolecular orientation in the filaments, leading to a fibre with very high tenacity and modulus. Dyneema, for example, is characterised by a parallel orientation of greater than 95% and a high level of crystallinity of up to 85%. This gives unique properties to the HDPE fibres. The most attractive properties of this type of fibre are: (1) very high tenacity, (2) very high specific modulus, (3) low elongation and (4) low fibre density, that is lighter than water.

HDPE fibres are made into different grades for different applications. Dyneema, for example, is made into SK60, SK65 and SK66. Dyneema SK60 is the multi-purpose grade. It is used, for example, for ropes and cordage, for protective clothing and for reinforcement of impact-resistant composites. Dyneema SK65 has a higher tenacity and modulus than SK60. This fibre is used where high performance is needed and maximum weight savings are to be attained. Dyneema SK66 is specially designed for ballistic protection. This fibre provides the highest energy absorption at ultrasonic speeds.

Table 3.1 compares the properties of the above mentioned filament yarns to steel.

3.3.3.5   Other technical yarns

There have been many other high performance fibres developed for technical applications, among which are PTFE, PBI, and PBO fibres.

PTFE (polytetrafluoroethylene) fibres offer a unique blend of chemical and temperature resistance, coupled with a low fraction coefficient. Since PTFE is virtually chemically inert, it can withstand exposure to extremely harsh temperature and chemical environments. The friction coefficient, claimed to be the lowest of all fibres, makes it suitable for applications such as heavy-duty bearings where low relative speeds are involved.

PBI (polybenzimidazole) is a manufactured fibre in which the fibre-forming substance is a long chain aromatic polymer. It has excellent thermal resistance and a good hand, coupled with a very high moisture regain. Because of these, the PBI

Table 3.1   Comparison of filament yarn properties

Yarns	Density (g cm^{-3})	Strength (GPa)	Modulus (GPa)	Elongation (%)
Aramid – regular	1.44	2.9	60	3.6
Aramid – composite	1.45	2.9	120	1.9
Aramid – ballistic	1.44	3.3	75	3.6
E Glass	2.60	3.5	72	4.8
S Glass	2.50	4.6	86	5.2
Carbon HS	1.78	3.4	240	1.4
Carbon HM	1.85	2.3	390	0.5
Dyneema SK60	0.97	2.7	89	3.5
Dyneema SK65	0.97	3.0	95	3.6
Dyneema SK66	0.97	3.2	99	3.7
Steel	7.86	1.77	200	1.1

Table 3.2   High performance fibres

Fibre	Density (g cm^{-3})	Tenacity (mN tex^{-1})	Elongation (%)	Regain (%)
PTFE	2.1	0.9–2.0	19–140	0
PBI	1.43	2.6–3.0	25–30	15
PBO	1.54	42	2.5–3.5	0.6–2.0

fibre is ideal for use in heat-resistant apparel for fire fighters, fuel handlers, welders, astronauts, and racing car drivers.

PBO (polyphenylenebenzobisoxazole) is another new entrant in the high performance organic fibres market. Zylon, made by Toyobo, is the only PBO fibre in production. PBO fibre has outstanding thermal properties and almost twice the strength of conventional *para*-aramid fibres. Its high modulus makes it an excellent material for composite reinforcement. Its low LOI gives PBO more than twice the flame-retardant properties of *meta*-aramid fibres. It can also be used for ballistic vests and helmets.

Table 3.2 lists some properties of these fibres.

Bibliography

A BREARLEY and J IREDALE, *The Worsted Industry*, 2nd edition, WIRA, Leeds, 1980.

J G COOK, Handbook of Textile Fibres, Manmade Fibres, 5th edition, Merrow, 1984.

Dyneema – the Top in High Performance Fibres, Properties and Applications, DSM High Performance Fibres B.V., Eisterweg 3, 6422 PN Heerlen, Netherlands.

E T GRIFFITHS, 'Wrap Spun Plain and Fancy Yarns', *Tomorrow's Yarns*, Conference Proceedings, UMIST, 1984.

S J KADOLPH and A L LANGFORD, *Textiles*, 8th edition, Prentice-Hall, 1998.

H KATO, 'Development of MJS Yarns', *J. Text. Machinery Soc. Jpn*, **32**(4) 1986.

W KLEIN, *New Spinning Systems*, The Textile Institute, Manchester, 1993.

W KLEIN, *Man-made Fibres and Their Processing*, The Textile Institute, Manchester, 1994.

W KLEIN, *The Technology of Short Staple Spinning*, 2nd edition, The Textile Institute, Manchester, 1998.

W E MORTON and G R WRAY, *An Introduction to the Study of Spinning*, 3rd edition, Longmans, London, 1962.

W E MORTON and J W S HEARLE, *Physical Properties of Textile Fibres*, 3rd edition, Textile Institute, Heinemann, London, 1993.

E OXTOBY, *Spun Yarn Technology*, Butterworth, London, 1987.

H J SELLING, *Twistless Yarns*, Merrow, 1971.

W C SMITH, High Performance Fibres Protect, Improve Lives, http://www.textileworld.com/categories/9905/fibres.html.

Textile Terms and Definitions, 10th edition, Textile Institute, Manchester, 1995.

4

Technical fabric structures – 1. Woven fabrics

Walter S Sondhelm

10 Bowlacre Road, Hyde, Cheshire SK14 5ES, UK

4.1 Introduction

Technical textiles[1] are textile materials and products manufactured primarily for their technical performance and functional properties rather than their aesthetic or decorative characteristics. Most technical textiles consist of a manufactured assembly of fibres, yarns and/or strips of material which have a substantial surface area in relation to their thickness and have sufficient cohesion to give the assembly useful mechanical strength.

Textile fabrics are most commonly woven but may also be produced by knitting, felting, lace making, net making, nonwoven processes and tufting or a combination of these processes. Most fabrics are two-dimensional but an increasing number of three-dimensional woven technical textile structures are being developed and produced.

Woven fabrics generally consist of two sets of yarns that are interlaced and lie at right angles to each other. The threads that run along the length of the fabric are known as warp ends whilst the threads that run from selvedge to selvedge, that is from one side to the other side of the fabric, are weft picks. Frequently they are simply referred to as ends and picks. In triaxial and in three-dimensional fabrics yarns are arranged differently.

Woven technical textiles are designed to meet the requirements of their end use. Their strength, thickness, extensibility, porosity and durability can be varied and depend on the weave used, the thread spacing, that is the number of threads per centimetre, and the raw materials, structure (filament or staple), linear density (or count) and twist factors of the warp and weft yarns. From woven fabrics higher strengths and greater stability can be obtained than from any other fabric structure using interlaced yarns. Structures can also be varied to produce fabrics with widely different properties in the warp and weft directions.

4.2   Weave structures

The number of weave structures that can be produced is practically unlimited. In this section basic structures, from which all other weave structures are developed, are discussed. Also briefly referred to are lenos, because of their importance in selvedge constructions, and triaxial fabrics, because they show simple structural changes which can affect the physical properties of fabrics. Most two-dimensional woven technical fabrics are constructed from simple weaves and of these at least 90% use plain weave. Simple cloth constructions are discussed in greater detail by Robinson and Marks[2] whilst Watson[3,4] describes a large variety of simple and complicated structures in great detail.

4.2.1   Plain weave

4.2.1.1   Construction of a plain weave
Plain weave is the simplest interlacing pattern which can be produced. It is formed by alternatively lifting and lowering one warp thread across one weft thread. Figure 4.1 shows 16 repeats (four in the warp and four in the weft direction) of a plain

Plan View

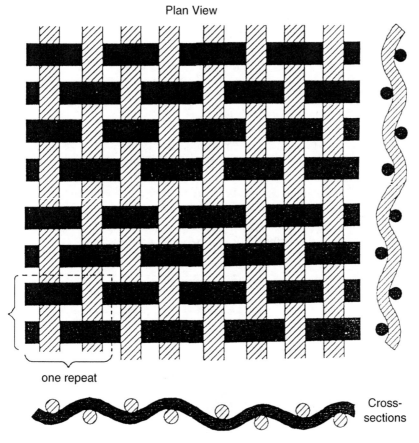

one repeat

Cross-sections

4.1   Fabric woven with plain weave – plan view (4 × 4 repeats) and warp and weft cross-sections.

weave fabric in plan view and warp way and weft way cross-sections through the same fabric. The diagrams are idealized because yarns are seldom perfectly regular and the pressure between the ends and picks tends to distort the shape of the yarn cross-sections unless the fabrics are woven from monofilament yarns or strips of film. The yarns also do not lie straight in the fabric because the warp and weft have to bend round each other when they are interlaced. The wave form assumed by the yarn is called crimp and is referred to in greater detail in Section 4.4.3.

4.2.1.2   Constructing a point paper diagram
To illustrate a weave either in plan view and/or in cross-section, as in Fig. 4.1, takes a lot of time, especially for more complicated weaves. A type of shorthand for depicting weave structures has therefore been evolved and the paper used for producing designs is referred to as squared paper, design paper or point paper. Generally the spaces between two vertical lines represent one warp end and the spaces between two horizontal lines one pick. If a square is filled in it represents an end passing over a pick whilst a blank square represents a pick passing over an end. If ends and picks have to be numbered to make it easier to describe the weave, ends are counted from left to right and picks from the bottom of the point paper design to the top. The point paper design shown in Fig. 4.2(a) is the design for a plain weave fabric. To get a better impression of how a number of repeats would look, four repeats of a design (two vertically and two horizontally) are sometimes shown. When four repeats are shown the first repeat is drawn in the standard way but for the remaining three repeats crossing diagonal lines may be placed into the squares, which in the firsts repeat, are filled in. This method is shown for a plain weave in Fig. 4.2(b).

4.2.1.3   Diversity of plain weave fabrics
The characteristics of the cloths woven will depend on the type of fibre used for producing the yarn and whether it is a monofilament yarn, a flat, twisted or textured (multi-)filament yarn or whether it has been spun from natural or manufactured staple fibres. The stiffness of the fabrics and its weavability will also be affected by the stiffness of the raw materials used and by the twist factor of the yarn, that is the number of turns inserted in relation to its linear density. Very highly twisted yarns are sometimes used to produce special features in plain weave yarns. The resulting fabrics may have high extensibility or can be semiopaque.

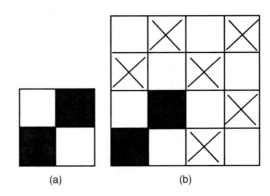

(a)   (b)

4.2   Point paper diagram of a plain weave fabric. (a) One repeat, (b) four repeats.

The area density of the fabric can be varied by changing the linear density or count of the yarns used and by altering the thread spacing, which affects the area covered by the yarns in relation to the total area. The relation between the thread spacing and yarn linear density is called the cover factor and is discussed in Section 4.4. Changing the area density and/or the cover factors may affect the strength, thickness, stiffness, stability, porosity, filtering quality and abrasion resistance of fabrics.

Square sett plain fabrics, that is fabrics with roughly the same number of ends and picks per unit area and warp and weft yarns of the same linear densities are produced in the whole range of cloth area densities and cloth cover factors. Low area density fabrics of open construction include bandages and cheese cloths, light area density high cover factor fabrics include typewriter ribbons and medical filter fabrics, heavy open cloths include geotextile stabilization fabrics and heavy closely woven fabrics include cotton awnings.

Warp-faced plain fabrics generally have a much higher warp cover factor than weft cover factor. If warp and weft yarns of similar linear density are used, a typical warp faced plain may have twice as many warp ends as picks. In such fabrics the warp crimp will be high and the weft crimp extremely low. The plan view and cross-sections of such a fabric are shown in Fig. 4.3. By the use of suitable cover factors and choice of yarns most of the abrasion on such a fabric can be concentrated on the warp yarns and the weft will be protected.

Weft-faced plains are produced by using much higher weft cover factors than warp cover factors and will have higher weft than warp crimp. Because of the difference in weaving tension the crimp difference will be slightly lower than in warp-faced plain fabrics. Weft-faced plains are little used because they are more difficult and expensive to weave.

Plan View

Cross-sections

4.3 Plan view and cross-sections of warp-faced plain weave fabric with a substantially higher warp than weft cover factor.

4.2.2   Rib fabrics and matt weave fabrics

These are the simplest modifications of plain weave fabrics. They are produced by lifting two or more adjoining warp threads and/or two or more adjoining picks at the same time. It results in larger warp and/or weft covered surface areas than in a plain weave fabric. As there are fewer yarn intersections it is possible to insert more threads into a given space, that is to obtain a higher cover factor, without jamming the weave.

4.2.2.1   Rib fabrics

In warp rib fabrics there are generally more ends than picks per unit length with a high warp crimp and a low weft crimp and vice versa for weft rib fabrics. The simplest rib fabrics are the 2/2 warp rib and the 2/2 weft rib shown respectively in Fig. 4.4(a) and (b). In the 2/2 warp rib one warp end passes over two picks whilst in the 2/2 weft rib one pick passes over two adjoining ends. The length of the floats can be extended to create 4/4, 6/6, 3/1 or any similar combination in either the warp or weft direction. 3/1 and 4/4 warp ribs are shown in Fig. 4.4(c) and (d).

In rib weaves with long floats it is often difficult to prevent adjoining yarns from overlapping. Weft ribs also tend to be expensive to weave because of their relatively high picks per unit length which reduces the production of the weaving machine unless two picks can be inserted at the same time.

4.2.2.2   Matt fabrics (or hopsack)

Simple matt (or hopsack) fabrics have a similar appearance to plain weave. The simplest of the matt weaves is a 2/2 matt shown in Fig. 4.5(a), where two warp ends are lifted over two picks, in other words it is like a plain weave fabric with two ends and two picks weaving in parallel. The number of threads lifting alike can be increased to obtain 3/3 or 4/4 matt structures. Special matt weaves, like a 4/2 matt, shown in Fig. 4.5(b), are produced to obtain special technical effects. Larger matt structures give the appearance of squares but are little used because they tend to become unstable, with long floats and threads in either direction riding over each other. If large matt weaves are wanted to obtain a special effect or appearance they can be stabilized by using fancy matt weaves containing a binding or stitching lift securing a proportion of the floats.

Matt weave fabrics can be woven with higher cover factors and have fewer intersections. In close constructions they may have better abrasion and better filtration properties and greater resistance to water penetration. In more open constructions matt fabrics have a greater tear resistance and bursting strength. Weaving costs may also be reduced if two or more picks can be inserted at the same time.

4.2.3   Twill fabrics

A twill is a weave that repeats on three or more ends and picks and produces diagonal lines on the face of a fabric. Such lines generally run from selvedge to selvedge. The direction of the diagonal lines on the surface of the cloth are generally described as a fabric is viewed along the warp direction. When the diagonal lines are running upwards to the right they are 'Z twill' or 'twill right' and when they run in the opposite direction they are 'S twill' or 'twill left'. Their angle and definition can be varied

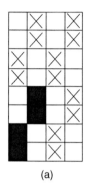

(a)

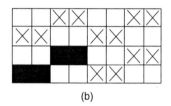

(b)

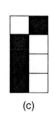

(c)

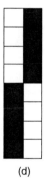

(d)

4.4 Rib fabrics. (a) 2/2 Warp rib (four repeats), (b) 2/2 weft rib (four repeats), (c) 3/1 warp rib (one repeat), (d) 4/4 warp rib (one repeat).

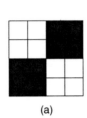

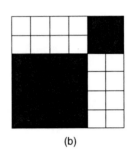

(a)   (b)

4.5 Matt fabrics. (a) 2/2 Matt, (b) 4/2 matt.

by changing the thread spacing and/or the linear density of the warp and weft yarns. For any construction twills will have longer floats, fewer intersections and a more open construction than a plain weave fabric with the same cloth particulars.

Industrial uses of twill fabrics are mainly restricted to simple twills and only simple twills are discussed here. Broken twills, waved twills, herringbone twills and elongated twills are extensively used for suiting and dress fabrics. For details of such twills see Robinson and Marks[2] or Watson.[3,4]

The smallest repeat of a twill weave consists of 3 ends × 3 picks. There is no theoretical upper limit to the size of twill weaves but the need to produce stable fabrics with floats of reasonable length imposes practical limits.

The twill is produced by commencing the lift sequence of adjacent ends on one pick higher or one pick lower. The lift is the number of picks which an end passes over and under. In a 2/1 twill, an end will pass over two picks and under one, whilst in a 1/2 twill the end will pass over one pick and then under two picks. Either weave can be produced as a Z or S twill. There are, therefore, four combinations of this simplest of all twills and they are illustrated in Fig. 4.6(a) to (d). To show how pronounced the twill line is even in a 3 × 3 twill, four repeats (2 × 2) of Fig. 4.6(a) are shown in Fig. 4.6(e) with all lifted ends shown in solid on the point paper. 2/1 twills are warp faced twills, that is fabrics where most of the warp yarn is on the surface, whilst 1/2 twills have a weft face. Weft-faced twills impose less strain on the weaving machine than warp-faced twills because fewer ends have to be lifted to allow picks to pass under them. For this reason, warp-faced twills are sometimes woven upside down, that is as weft-faced twills. The disadvantage of weaving twills upside down is that it is more difficult to inspect the warp yarns during weaving.

Twills repeating on 4 ends × 4 picks may be of 3/1, 2/2 or 1/3 construction and may have 'Z' or 'S' directions of twill. Weaves showing 4 × 4 twills with a Z twill line are shown in Fig. 4.7(a) to (c). As the size of the twill repeat is increased further, the number of possible variations also increases. In the case of a 5 × 5 the possible combinations are 4/1, 3/2, 2/3 and 1/4 and each of these can be woven with the twill

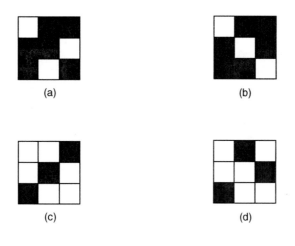

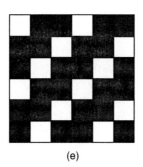

4.6   3 × 3 Twill weaves. (a) 2/1 Twill with Z twill line, (b) 2/1 twill with S twill line, (c) 1/2 twill with Z twill line, (d) 1/2 twill with S twill line, (e) four repeats of (a) (2/1 twill with a Z twist line).

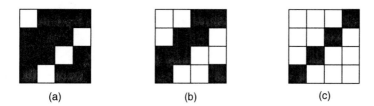

4.7   4 × 4 Twill weaves. (a) 3/1 Twill with Z twill line, (b) 2/2 twill with Z twill line, (c) 1/3 twill with Z twill line.

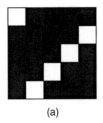

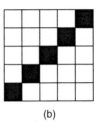

(a)   (b)

4.8   5 × 5 Twill weaves. (a) 4/1 Twill with Z twill line, (b) 1/4 twill with Z twill line.

line in either direction. In Fig. 4.8(a) and (b) 4/1 and 1/4 twills with a Z twill line are shown.

4.2.4   Satins and sateens

In Britain a satin is a warp-faced weave in which the binding places are arranged to produce a smooth fabric surface free from twill lines. Satins normally have a much greater number of ends than picks per centimetre. To avoid confusion a satin is frequently described as a 'warp satin'. A sateen, frequently referred to as a 'weft sateen', is a weft-faced weave similar to a satin with binding places arranged to produce a smooth fabric free of twill lines. Sateens are generally woven with a much higher number of picks than ends. Satins tend to be more popular than sateens because it is cheaper to weave a cloth with a lower number of picks than ends. Warp satins may be woven upside down, that is as a sateen but with a satin construction, to reduce the tension on the harness mechanism that has to lift the warp ends.

To avoid twill lines, satins and sateens have to be constructed in a systematic manner. To construct a regular satin or sateen weave (for irregular or special ones see Robinson and Marks[2] or Watson[3]) without a twill effect a number of rules have to be observed. The distribution of interlacing must be as random as possible and there has only to be one interlacing of each warp and weft thread per repeat, that is per weave number. The intersections must be arranged in an orderly manner, uniformly separated from each other and never adjacent. The weaves are developed from a 1/x twill and the twill intersections are displaced by a fixed number of steps. The steps that must be avoided are:

(i)   one or one less than the repeat (because this is a twill),
(ii)   the same number as the repeat or having a common factor with the repeat (because some of the yarns would fail to interlace).

The smallest weave number for either weave is 5 and regular satins or sateens also cannot be constructed with weave numbers of 6, 9, 11, 13, 14 or 15. The most popular weave numbers are 5 and 8 and weave numbers above 16 are impracticable because of the length of floats. Weave numbers of 2 or 3 are possible for five-end weaves and 3 or 5 for 8-end weaves.

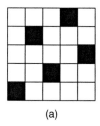   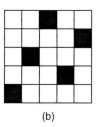

(a)   (b)

4.9   5-End weft sateen. (a) 5-end and two step sateen, (b) 5-end three step sateen.

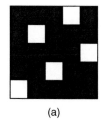   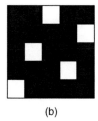

(a)   (b)

4.10   5-End warp satin. (a) 5-end two step satin, (b) 5-end three step satin.

Figure 4.9(a) and (b) shows 5-end weft sateens with two and three steps, respectively. They have been developed from the 1/4 twill shown in Fig. 4.8(b). Mirror images of these two weaves can be produced. Five-end warp satins with two and three steps are shown in Fig. 4.10(a) and (b). It is the cloth particulars, rather than the weave pattern, which generally decides on which fabric is commercially described as a sateen or a satin woven upside down. Five-end satins and sateens are most frequently used because with moderate cover factors they give firm fabrics. Figure 4.11(a) and (b) shows 8-end weft sateens with three and five step repeats and Fig. 4.12 shows an 8-end warp satin with a 5 step repeat.

Satins and sateens are widely used in uniforms, industrial and protective clothing. They are also used for special fabrics such as downproofs.

In North America a satin is a smooth, generally lustrous, fabric with a thick close texture made in silk or other fibres in a satin weave for a warp-face fabric or sateen weave for a filling (weft) face effect. A sateen is a strong lustrous cotton fabric generally made with a five-harness satin weave in either warp or filling-face effect.

4.2.5   Lenos

In lenos adjoining warp ends do not remain parallel when they are interlaced with the weft but are crossed over each other. In the simplest leno one standard end and

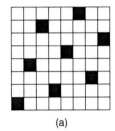
(a)

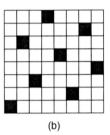
(b)

4.11   8-End weft sateen. (a) 8-end three step sateen, (b) 8-end five step sateen.

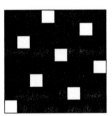

4.12   8-End five step warp satin.

one crossing end are passed across each other during consecutive picks. Two variations of this structure are shown in plan view and cross-section in Fig. 4.13(a) and (b).[4] Whenever the warp threads cross over each other, with the weft passing between them, they lock the weft into position and prevent weft movement. Leno weaves are therefore used in very open structures, such as gauzes, to prevent thread movement and fabric distortion. When the selvedge construction of a fabric does not bind its edge threads into position, leno ends are used to prevent the warp threads at each side of a length of cloth from slipping out of the body of the fabric. They are also used in the body of fabrics when empty dents are left in weaving because the fabrics are to be slit into narrower widths at a later stage of processing.

Leno and gauze fabrics may consist of standard and crossing ends only or pairs or multiples of such threads may be introduced according to pattern to obtain the required design. For larger effects standard and crossing ends may also be in pairs or groups of three. Two or more weft threads may be introduced into one shed and

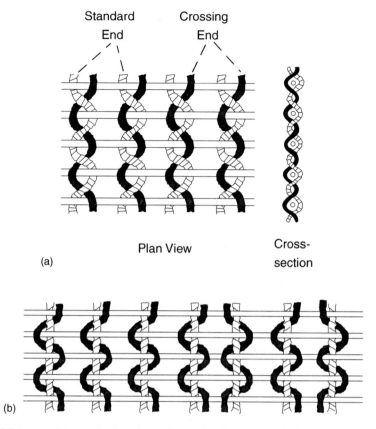

Standard End   Crossing End

Plan View   Cross-section

(a)

(b)

4.13 (a) Leno with standard and crossing ends of same length (woven from one beam), (b) leno with standard and crossing ends of different length (woven from two beams).

areas of plain fabric may be woven in the weft direction between picks where warp ends are crossed over to give the leno effect. Gauze fabrics used for filtration generally use simple leno weaves.

Standard and crossing ends frequently come from separate warp beams. If both the standard and crossing ends are warped on to one beam the same length of warp is available for both and they will have to do the same amount of bending, that is they will have the same crimp. Such a leno fabric is shown in plan view and cross-section in Fig. 4.13(a). If the two series of ends are brought from separate beams the standard ends and the crossing ends can be tensioned differently and their crimp can be adjusted separately. In such a case it is possible for the standard ends to lie straight and the crossing ends to do all the bending. Such a fabric is shown in Fig. 4.13(b). This figure also shows that crossing threads can be moved either from the right to the left or from the left to the right on the same pick and adjoining leno pairs may either cross in the same or opposite directions. The direction of crossing can affect locking, especially with smooth monofilament yarns. When using two beams it is also possible to use different types or counts of yarn for standard and crossing ends for design or technical applications. The actual method of weaving the leno with doups or similar mechanisms is not discussed in this chapter.

When lenos are used for selvedge construction only one to four pairs of threads are generally used at each side and the leno selvedge is produced by a special leno mechanism that is independent of the shedding mechanism of the loom. The leno threads required for the selvedge then come from cones in a small separate creel rather than from the warp beam. The choice of selvedge yarns and tensions is particularly important to prevent tight or curly selvedges. The crimps selected have to take into account the cloth shrinkage in finishing.

4.2.6   Triaxial weaves

Nearly all two-dimensional woven structures have been developed from plain weave fabrics and warp and weft yarns are interlaced at right angles or at nearly right angles. This also applies in principle to leno fabrics (see Section 4.2.5 above) and to lappet fabrics where a proportion of the warp yarns, that is the yarns forming the design, are moved across a number of ground warp yarns by the lappet mechanism. The only exceptions are triaxial fabrics, where two sets of warp yarns are generally inserted at 60° to the weft, and tetra-axial fabrics where four sets of yarns are inclined at 45° to each other. So far only weaving machines for triaxial fabrics are in commercial production. Triaxial weaving machines were first built by the Barber-Colman Co. under licence from Dow Weave and have been further developed by Howa Machinery Ltd., Japan.

Triaxial fabrics are defined as cloths where the three sets of threads form a multitude of equilateral triangles. Two sets of warp yarn are interlaced at 60° with each other and with the weft. In the basic triaxial fabric, shown in Fig. 4.14(a), the warp travels from selvedge to selvedge at an angle of 30° from the vertical. When a warp yarn reaches the selvedge it is turned through an angle of 120° and then travels to the opposite selvedge thus forming a firm selvedge. Weft yarn is inserted at right angles (90°) to the selvedge. The basic triaxial fabric is fairly open with a diamond-shaped centre. The standard weaves can be modified by having biplane, stuffed or basic basket weaves, the latter being shown in Fig. 4.14(b). These modified weaves form closer structures with different characteristics. Interlacing angles of 75° to the selvedge can be produced. At present thread spacing in the basic weave fabric is limited to 3.6 or 7.4 threads per centimetre.

The tear resistance and bursting resistance of triaxial fabrics is greatly superior to that of standard fabrics because strain is always taken in two directions. Their shear resistance is also excellent because intersections are locked. They have a wide range of technical applications including sailcloths, tyre fabrics, balloon fabrics, pressure receptacles and laminated structures.

4.3   Selvedge

The selvedges form the longitudinal edges of a fabric and are generally formed during weaving. The weave used to construct the selvedge may be the same, or may differ from, the weave used in the body of the cloth. Most selvedges are fairly narrow but they can be up to 20-mm wide. Descriptions may be woven into the selvedge, using special selvedge jacquards, or coloured or fancy threads may be incorporated for identification purposes. For some end-uses selvedges have to be discarded but,

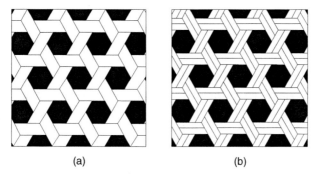

4.14   Triaxial fabrics (weaving machine: Howa *TRI-AX* Model TWM). (a) Basic weave, (b) basic basket weave.

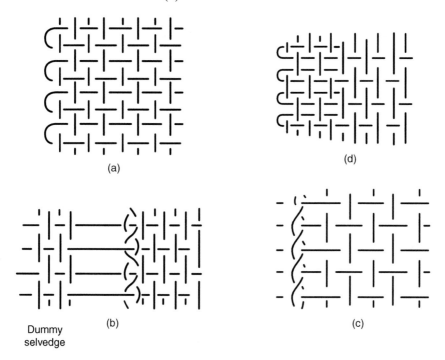

4.15   (a) Hairpin selvedge – shuttle loom, (b) leno selvedge with dummy selvedge, (c) helical selvedge, (d) tucked-in or tuck selvedge.

whenever possible, selvedges should be constructed so that they can be incorporated into the final product so as to reduce the cost of waste.

In cloths woven on weaving machines with shuttles selvedges are formed by the weft turning at the edges after the insertion of each pick. The weft passes continuously across the width of the fabric from side to side. In cloths woven on shuttleless weaving machines the weft is cut at the end of each pick or after every second pick. To prevent the outside ends from fraying, various selvedge motions are used to bind the warp into the body of the cloth or edges may be sealed. The essential requirements are that the selvedges should protect the edge of the fabric during weaving and subsequent processing, that they should not detract from the appearance of the cloth and that they should not interfere with finishing or cause waviness, contraction or creasing. Four types of woven selvedges are shown in Fig. 4.15.

4.3.1   Hairpin selvedge – shuttle weaving machine

Figure 4.15(a) shows a typical hairpin selvedge which is formed when the weft package is carried in a reciprocating media, for example a shuttle. With most weaves it gives a good edge and requires no special mechanism. Frequently strong two-ply yarns are incorporated into the selvedge whilst single yarns are used in the body of the cloth because the edge threads are subjected to special strains during beat-up. To ensure a flat edge a different weave may be used in the selvedge from the body of the cloth. For twills, satins and fancy weaves this may also be necessary to ensure that all warp threads are properly bound into the edge. If special selvedge yarns are used it is important to ensure that they are not, by mistake, woven into the body of the fabric because they are likely to show up after finishing or cause a reduction in the tear and/or bursting resistance of industrial fabrics.

When two or more different weft yarns are used in a fabric only one weft yarn is being inserted at a time and the other yarn(s) are inactive until the weft is changed. In shuttle looms the weft being inserted at any one time will form a normal selvedge whilst the weft yarn(s) not in use will float along the selvedge. If long floats are formed in weaving, because one weft is not required for a considerable period, the floats may have to be trimmed off after weaving to prevent them from causing problems during subsequent processing. Whenever a pirn is changed, or a broken pick is repaired, a short length of yarn will also protrude from the selvedge and this has to be trimmed off after weaving.

Industrial fabrics with coarse weft yarn are sometimes woven with loop selvedges to ensure that the thread spacing and cloth thickness remains the same right to the edge of the cloth. To produce a loop selvedge a wire or coarse monofilament yarn is placed 3 or 4 mm outside the edge warp end and the pick reverses round the wire to form a loop. As the wire is considerably stiffer than a yarn it will prevent the weft pulling the end threads together during beat-up. The wire extends to the fell of the cloth and is automatically withdrawn during weaving. Great care is required to ensure that no broken-off ends of wire or monofilament remain in the fabric because they can cause serious damage to equipment and to the cloth during subsequent processing.

4.3.2   Leno and helical selvedges

In most shuttleless weaving machines a length of weft has to be cut for every pick. For looms not fitted with tucking motions the warp threads at the edges of the fabric have to be locked into position to prevent fraying. With some weft insertion systems the weft has to be cut only after every second pick and it is possible to have a hairpin selvedge on one side of the fabric and a locked selvedge on the second side. Leno and helical selvedges are widely used to lock warp yarns into position and these are shown in Fig. 4.15(b) and (c), respectively. When shuttleless weaving machines were first introduced there was considerable customer resistance to fringe selvedges but, in the meantime, it has been found that they meet most requirements.

With leno selvedges (see also Section 4.2.5) a set of threads at the edge of the fabric is interlaced with a gauze weave which locks round the weft thread and prevents ravelling of the warp. As the precut length of the picks always varies slightly, a dummy selvedge is sometimes used at the edge of the cloth as shown in Fig. 4.15(b). This makes it possible to cut the weft close to the body of the cloth resulting in a narrow fringe which has a better appearance than the selvedge from which weft threads of varying lengths protrude. It also has the advantage in finishing and coating that there are no long lengths of loose weft that can untwist and shed fly. Because of the tails of weft and because of the warp in dummy selvedges, more waste is generally made in narrow shuttleless woven fabrics than in fabrics woven on shuttle looms. For wide cloth the reverse frequently applies because there is no shuttle waste. Leno selvedges, sometimes referred to as 'centre selvedges', may be introduced into the body of the cloth if it is intended to slit the width of cloth produced on the loom into two or more widths either on the loom or after finishing.

Helical selvedges consist of a set of threads which make a half or complete revolution around one another between picks. They can be used instead of leno selvedges and tend to have a neater appearance.

4.3.3   Tucked-in selvedges

A tuck or tucked-in selvedge is shown in Fig. 4.15(d). It gives a very neat and strong selvedge and was first developed by Sulzer[5] Brothers (now Sulzer-Textil), Winterthur, Switzerland, for use on their projectile weaving machines. Its appearance is close to that of a hairpin selvedge and it is particularly useful when cloths with fringe selvedges would have to be hemmed. High-speed tucking motions are now available and it is possible to produce tucked-in selvedges even on the fastest weaving machines.

Tucked-in selvedges, even when produced with a reduced number of warp threads, are generally slightly thicker than the body of the cloth. When large batches of cloth with such selvedges are produced it may be necessary to traverse the cloth on the cloth roller to build a level roll. The extra thickness will also have to be allowed for when fabrics are coated. When tucked-in selvedges can be incorporated into the finished product no yarn waste is made because no dummy selvedges are produced but the reduced cost of waste may be counterbalanced by the relatively high cost of the tucking units.

4.3.4   Sealed selvedges

When fabrics are produced from yarns with thermoplastic properties the edge of a fabric may be cut and sealed by heat. The edge ends of fringe selvedges are frequently cut off in the loom and the edge threads with the fringe are drawn away into a waste container. Heat cutting may also be used to slit a cloth, in or off the loom, into a number of narrower fabrics or tapes.

For special purposes ultrasonic sealing devices are available. These devices are fairly expensive and can cut more cloth than can be woven on one loom. Whilst they can be mounted on the loom they are, because of their cost, more frequently mounted on a separate cutting or inspection table.

4.4   Fabric specifications and fabric geometry

Fabric specifications[6] describe a cloth but frequently need experience for correct interpretation. The most important elements are cloth width, threads per centimetre in the warp and weft directions, linear density (count) and type of warp and weft yarns (raw material, filament or staple, construction, direction of twist and twist factors), weave structure (see Section 4.2 above) and finish. From these, if the weaving machine particulars and finishing instructions are known, the cloth area density can be calculated.

It has been assumed that other cloth particulars, such as warp and weft cover factors, crimp, cloth thickness, porosity and drape, have either to be estimated or measured by various test methods. Peirce[7] showed that standard physical and engineering principles can be applied to textiles and cloth specifications can be forecast if the effect of interlacing the warp and weft threads and the distortion of the yarns caused thereby is allowed for.

4.4.1   Fabric width
The width of fabrics is generally expressed in centimetres and has to be measured under standard conditions to allow for variations which are caused by moisture and tension. It is necessary to know whether the cloth width required is from edge to edge or whether it excludes the selvedges. Before deciding on the weaving specifications of a fabric, allowance has to be made not only for shrinkage from reed width to grey width during weaving but also for contraction (if any) during finishing. If the cloth width changes, other parameters such as cloth area density and threads per centimetre, will be effected.

4.4.2   Fabric area density
The fabric area density is generally expressed in grams per square metre although sometimes it is reported as grams per linear metre. It is essential to specify whether the area density required is loomstate or finished. The loomstate area density depends on the weaving specification, that is, yarns, thread spacing and weave, and on any additives, such as size, which are used to improve the weaving process. During finishing the cloth area density is frequently altered by tension and chemical treatments or compressive shrinkage which affect cloth width and length, by the removal of additives needed in weaving and by substances added during finishing.

4.4.3   Crimp
The waviness or crimp[1] of a yarn in a fabric, shown in the cross-sections of Fig. 4.1, is caused during weaving and may be modified in finishing. It is due to the yarns being forced to bend around each other during beat-up. It depends on the warp and weft cover factors, which are described in Section 4.4.4, the characteristics of the yarns and the weaving and finishing tensions.

The crimp is measured by the relation between the length of the fabric sample and the corresponding length of yarn when it has been straightened after being removed from the cloth, as shown in Fig. 4.16, which, as drawn, is idealized and only applies to monofilament yarns. Multifilament and staple yarn will be deformed to some extent at intersections by the pressure exerted by the warp on the weft or vice versa depending on yarn particulars and cloth construction. It is most convenient to express crimp as a percentage, which is 100 divided by the fabric length and multiplied by the difference between the yarn length and the fabric length.

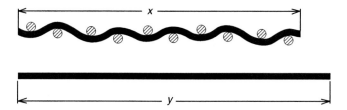

4.16 Crimp – relation of length of yarn in cloth to length of cloth. x is the width of sample, y is the length of yarn extracted from sample and percentage crimp = $(y-x)/x \times 100$.

Warp crimp is used to decide the length of yarn which has to be placed on to a warper's beam to weave a specified length of fabric. Allowance has to be made for stretching of the yarns during weaving which is generally low for heavily sized yarns but can be of importance when unsized or lightly sized warps are used. For fabrics woven on shuttleless weaving machines, the estimated weft crimp has to be adjusted for the length of yarn used to produce a selvedge and, sometimes, a dummy selvedge.

4.4.4   Cover factors

The cover factor[1] indicates the extent to which the area of a fabric is covered by one set of threads. For any fabric there are two cover factors: the warp cover factor and the weft cover factor. The cloth cover factor is obtained by adding the weft cover factor to the warp cover factor. The cover factors can be adjusted to allow for yarns of different relative densities – either because of the yarn structure or because of the raw material used.

The cover factors in SI units[8] are calculated by multiplying the threads per centimetre by the square root of the linear density of the yarn (in tex) and dividing by 10. The resultant cover factor differs by less than 5% from the cotton cover factor pioneered by Peirce[7] and expressed as the number of threads per inch divided by the square root of the cotton yarn count.

For any given thread spacing plain weave has the largest number of intersections per unit area. All other weaves have fewer intersections than plain weave. The likely weavability of all fabrics woven with the same weave and from similar yarns can be forecast from their cover factors. Plain weave fabrics with warp and weft cover factors of 12 in each direction are easy to weave. Thereafter weaving becomes more difficult and for cover factors of 14 + 14 fairly strong weaving machines are required. At a cover factor of 16 + 16 the plain structure jams and a very strong loom with heavy beat-up is needed to deform the yarns sufficiently to obtain a satisfactory beat-up of the weft. Some duck and canvas fabrics woven on special looms achieve much greater cover factors. Three cloths with cover factors of 12 + 12 are shown in Table 4.1. It shows how thread spacing and linear density have to be adjusted to maintain the required cover factor and how cloth area density and thickness are affected.

When widely varying cover factors are used for warp and weft, a high cover factor in one direction can generally be compensated by a low cover factor in the other direction. In Fig. 4.3, which shows a poplin-type weave, the warp does all the bending and the weft lies straight. With this construction warp yarns can touch because they bend around the weft and cloth cover factors of well above 32 can be woven fairly easily. Peirce calculated the original adjustment factors for a number of weave

Table 4.1 Comparison of fabrics with identical warp and weft cover factors woven with yarns of different linear densities (in SI units)

Cloth	Threads per cm		Linear density		Cover factor		Cloth weight[a] (g/m²)	Thickness[b] (mm)
	n_1	n_2	N_1	N_2	K_1	K_2		
A	24	24	25	25	12.0	12.0	130	0.28
B	12	12	100	100	12.0	12.0	260	0.56
C	6	6	400	400	12.0	12.0	520	1.12

n_1 are warp threads, n_2 are weft threads, N_1 is the linear density of warp, N_2 is the linear density of weft, K_1 is the warp cover factor, K_2 is the weft cover factor.
[a] Allowing for 9% of crimp.
[b] Allowing 25% for flattening and displacement of yarns.

Table 4.2 Cover factor adjustment factors for weave structure

Weave	Adjustment factor[a]
Plain weave	1.0
2/2 weft rib	0.92
1/2 and 2/1 twills	0.87
2/2 matt	0.82
1/3, 3/1 and 2/2 twills	0.77
5 end satin and 5 end sateen	0.69

Source: Rüti[10].
[a] Multiply actual cover factor by adjustment factor to obtain equivalent plain weave cover factor.

structures. Rüti[9] published an adjustment factor that they found useful to establish whether fabrics with various weave structures can be woven in their machines and these are shown in Table 4.2. Sulzer have prepared graphs showing how easy or difficult it is to weave fabrics with different weave structures, thread spacings and linear densities in various types of weaving machines.

4.4.5   Thickness
Yarn properties are as important as cloth particulars when forecasting cloth thickness. It is difficult to calculate the cloth thickness because it is greatly influenced by yarn distortion during weaving and by pressures exerted on the cloth during finishing. It is also difficult to measure thickness because the results are influenced by the size of the presser feet used in the test instrument, the pressure applied and the time that has elapsed before the reading is taken. It is therefore essential to specify the method of measurement of thickness carefully for many industrial fabrics.

4.5   Weaving – machines (looms) and operations

Whilst the principle of weft interlacing in weaving has not changed for thousands of years, the methods used and the way that weaving machines are activated and controlled has been modified. During the last few decades of the 20th century the rate of change increased continuously and weft insertion rates, which control machine productivity, increased ten-fold from around 1950–2000 and they are likely to double again during the first five to ten years of the 21st century. The productivity of the weaver has increased even more and is likely soon to have increased a hundred-fold. Weaving, which used to be a labour-intensive industry, is now capital intensive using the most modern technology.

The essential operations in the weaving of a cloth are:

1. Shedding, i.e. the separation of the warp threads into two (or more) sheets according to a pattern to allow for weft insertion
2. Weft insertion (picking) and
3. Beating-up, i.e. forcing the pick, which has been inserted into the shed, up to the fell of the cloth (line where the cloth terminates after the previous pick has been inserted).

Provision has also to be made for the supply of warp and weft warp yarns and for the cloth, after weaving, to be collected. The warp yarn is usually supplied on warp beam(s) and the weft on pirn (shuttle looms only) or cones. A typical cross-section through a shuttle weaving machine[10] showing the main motions, together with the arrangement of some ancillary motions, is shown schematically in Fig. 4.17. Most single phase machines, irrespective of their weft insertion system, use similar motions and a nearly horizontal warp sheet between the back rest(2) and the front rest(9). Although this is the most common layout, other successful layouts have been developed. One of these, for an air jet,[11] is shown in Fig. 4.18 and an interesting double-sided air jet with two vertical warp lines back to back is being developed by Somet.[12]

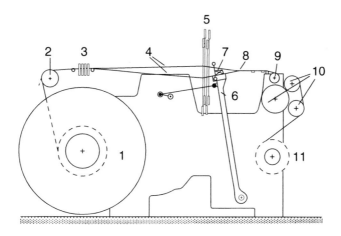

4.17   Schematic section through Rüti shuttle weaving machine (machine motions and parts shown in *italics*). 1, *Weaver's beam* holding warp yarn (controlled by *let-off*); 2, *back rest* guiding yarn and controlling warp tension; 3, *drop wires* of *warp stop motion*; 4, shed formed by top and bottom warp sheet; 5, *healds* controlled by *shedding motion (crank, cam or dobby)* (Jacquard shedding motions do not use healds); 6, *sley* carrying *reed* and *shuttle*. It reciprocates for beat-up. (In many shuttleless machines the weft insertion system and the reed have been separated); 7, *shuttle* carrying the *pirn*; 8, fell where cloth is formed; 9, *front rest* for guiding cloth; 10, *take-up motion* controlling the pick spacing and cloth wind-up; 11, cloth being wound on to *cloth roller*.

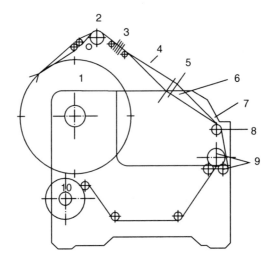

4.18 Schematic section through Elitex air jet weaving machine (machine motions and parts shown in *italics*). 1, *Weaver's beam* holding warp yarn (controlled by *let-off*); 2, *back rest* guiding warp yarn; 3, *drop wires* of *warp stop motion*; 4, shed formed by top and bottom warp sheet; 5, *healds* controlled by *cam shedding motion*; 6, *weft insertion area* (the reed for beating-up located in this area is not shown); 7, fell where cloth is formed; 8, *front rest* for guiding cloth; 9, *take-up motion* controlling the pick spacing and cloth wind-up; 10, cloth being wound on to *cloth roller*.

Weaving machines can be subdivided into single phase machines, where one weft thread is laid across the full width of the warp sheet followed by the beat-up and the formation of the next shed in preparation for the insertion of the next pick, and multiphase machines, in which several phases of the working cycle take place at any instant so that several picks are being inserted simultaneously. Single phase machines are further subdivided according to their weft insertion system, whilst multiphase machines are classified according to their method of shed formation. The classification of weaving machines is summarized in Table 4.3.

In this chapter the details of the different types of weaving machines and their functions can only be outlined and the most important points highlighted. Weaving machines and their operation are described and discussed by Ormerod and Sondhelm.[13]

For the successful operation of a weaving machine good warps are essential. Warp preparation is, therefore, discussed in Section 4.5.1. This is followed by shedding in Section 4.5.2, weft insertion and beating-up in Section 4.5.3 and other motions used on machines in Section 4.5.4. The importance of machine width is considered in Section 4.5.5.

4.5.1 Warp preparation
The success of the weaving operation depends on the quality of the weaver's beam presented to the weaving machine because each fault in the warp will either stop the machine and require rectification, or cause a fault in the cloth, which is woven from it.

Before most fabrics can be woven a weaver's beam (or beams) holding the warp yarn has to be prepared. For very coarse warp yarns or when very long lengths of

Table 4.3 Classification of weaving machines

Single-phase weaving machines
Machines with shuttles (looms):
   Hand operated (hand looms)
   Non-automatic power looms (weft supply in shuttle changed by hand)
   Automatic weaving machines
   • shuttle changing
      rotary batteries, stack batteries, box loaders or
      pirn winder mounted on machine (Unifil)
   • pirn changing
Shuttleless weaving machines:
   Projectile
   Rapier
   • rigid rapier(s)
      single rapier, single rapier working bilaterally
      or two rapiers operating from opposite sides of machine
   • telescopic
   • flexible
   Jet machines
   • air (with or without relay nozzles)
   • liquid (generally water)

Multiphase weaving machines
Wave shed machines:
   Weft carriers move in straight path
   Circular weaving machines (weft carriers travel in circular path)
Parallel shed machines (rapier or air jet)

---

filament can be woven without modification of the warping particulars, a separate cone creel placed behind each weaving machine can be used economically. It improves the weaving efficiency by avoiding frequent beam changes but requires a large amount of space. For most yarns, especially sized yarns, it is more economical to prepare a weaver's beam and use it in the weaving machine.

The object of most warp preparation systems is to assemble all ends needed in the weaving machine on one beam and to present the warp with all ends continuously present and the integrity and elasticity of the yarns as wound fully preserved. Before this can be done the yarns have first to be wound on to a cone, then rewound on to a warper's beam and finally sized before the weaver's beam is made. The purpose of warp sizing is to apply a protective coating to the yarn to enable it to withstand the complex stresses to which it is subjected in the weaving machine. Some coarse ply yarns and some strong filament yarns can be woven without being sized.

Details of the various warp preparation processes are described by Ormerod and Sondhelm.[14]

4.5.2 Shedding

Irrespective of whether cloth is woven on an ancient handloom or produced on the most modern high speed multiphase weaving machine a shed has to be formed before a pick can be inserted prior to beat-up and cloth formation. The shed must be clean, that is slack warp threads or hairy or taped ends must not obstruct the passage of the weft or of the weft carrier. If the weft cannot pass without obstruction either the machine will stop for rectification, a warp end may be cut or damaged or a faulty weave pattern may be produced.

4.5.2.1 Shedding in single phase machines

In most single phase weaving machines, before pick insertion commences, an upper and a lower warp sheet is formed and the lifting pattern is not changed until the weft thread has been inserted across the full width of the warp. The shedding mechanism is used to move individual ends up or down in a prearranged order governed by the weave pattern. To maintain a good separation of the ends during weaving and avoid adjoining ends from interfering with each other the ends in each warp sheet can be staggered but in the area of weft insertion an unobstructed gap through which the weft can pass has to be maintained. The shedding mechanism chosen for a given weaving machine depends on the patterns that it is intended to weave on it. Most shedding mechanisms are expensive and the more versatile the mechanism the greater its cost. On some weaving machines there are also technical limitations governing the shedding mechanisms that can be fitted.

When crank, cam (or tappet) or dobby shedding is used, ends are threaded through eyes in healds that are placed into and lifted and lowered by heald frames. All healds in one heald frame are lifted together and all ends controlled by it will therefore lift alike. The weave pattern therefore controls the minimum number of healds required. To prevent overcrowding of healds in a heald frame or to even out the number of ends on a heald frame, more than one frame may lift to the same pattern. To weave a plain fabric, for example, 2, 4, 6, or 8 heald frames may be used with equal numbers of frames lifting and lowering the warp on each pick. Crank motions are generally limited to 8 shafts, cams to 10 or 12 and dobbies to 18 or 24. When the necessary lifting pattern cannot be obtained by the use of 24 shafts a Jacquard shedding mechanism, in which individual ends can be controlled separately, has to be used.

The crank shedding mechanism is the simplest and most positive available. It can only be used to weave plain weave. It is cheap, simple to maintain and in many high speed machines increases weft insertion rates by up to 10%. This mechanism has not been used as much as would be expected because of its lack of versatility. It is however particularly useful for many industrial fabrics that, like the majority of all fabrics, are woven with plain weave.

Cam or tappet shedding motions on modern high speed machines use grooved or conjugate cams because they give a positive control of the heald shafts. Negative profile cams are, however, still widely used especially for the weaving of fairly open fabrics of light and medium area density. The cam profile is designed to give the necessary lifting pattern to healds in the sequence required by the lifting plan which is constructed from the weave structure.

The third method of controlling heald shafts is by dobby and its main advantage is that there is practically no limit to the size of pattern repeat that can be woven, whilst with cam motions it is difficult and expensive to construct a repeat of more than eight or ten picks. It is also easier to build dobbies for a large number of heald shafts. Dobbies were controlled by pattern chains containing rollers or pegs which accentuated the lifting mechanism of the heald shafts. Punched rolls of paper or plastic were used instead of the heavier wooden pegs or metal chains for long repeats. During the 1990s electronic dobbies have replaced mechanical ones enabling weaving machines to operate at much greater speeds, reducing the cost of preparing a pattern and the time required for changing to a new design. Following the development of electronic dobbies, cam machines are becoming less popular because cams for high speed machines are expensive. If large numbers are required, because of the weave structure or because of frequent pattern changes, it may be cheaper to buy machines fitted with dobbies.

Shedding mechanisms are still undergoing intensive development and electronic control of individual shafts may soon reduce the difference in cost between crank, cam and dobby machines. Developments are likely to simplify the shedding units, reduce the cost of shedding mechanisms and their maintenance and make weaving machines more versatile.

When the patterning capacity of dobbies is insufficient to weave the required designs, Jacquards have to be used. Modern electronic Jacquards can operate at very high speeds and impose practically no limitation on the design. Every end across the full width of the weaving machine can be controlled individually and the weft repeat can be of any desired length. Jacquards are expensive and, if a very large number of ends have to be controlled individually, rather than in groups, the Jacquard may cost as much as the basic weaving machine above which it is mounted.

4.5.2.2 *Shedding in multiphase weaving machines*

In all multiphase weaving machines[1] several phases of the working cycle take place at any instant so that several picks are being inserted simultaneously. In wave shed machines different parts of the of the warp sheet are in different parts of the weaving cycle at any one moment. This makes it possible for a series of shuttles or weft carriers to move along in successive sheds in the same plane. In parallel shed machines several sheds are formed simultaneously with each shed extending across the full width of the warp and with the shed moving in the warp direction. The difference between the methods of shed formation are shown schematically in Fig. 4.19.[15]

Weaving machines in which shuttles travel a circular path through the wave shed are generally referred to as 'circular weaving machines'. They have been widely used for producing circular woven polypropylene fabrics for large bags for heavy loads.

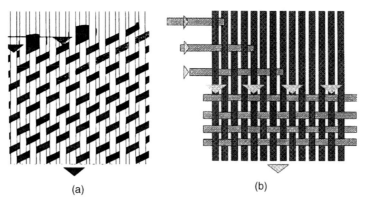

(a)   (b)

4.19 Weft insertion on multiphase weaving machines (plain weave). (a) Wave shed principle, (b) multiphase parallel shed principle. Reproduced by kind permission of Sulzer Textil.

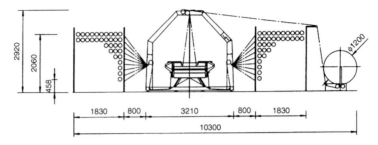

4.20   Schematic drawing of Starlinger circular loom SL4. Reproduced by kind permission of Starlinger & Co. G.m.b.H. (all dimensions in mm).

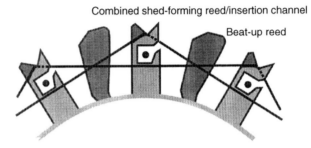

4.21   Shed formation of Sulzer Textil multiphase parallel shed M8300. Shed-forming reed with weft insertion channel and beat-up reed mounted on drum. Reproduced by kind permission of Sulzer Textil.

The layout of a Starlinger circular weaving machine[16] is shown in Fig. 4.20. The creels containing cheeses of polypropylene tape are located on either side of the weaving unit that is in the centre. The warp yarns are fed into the weaving elements from the bottom, interlaced with weft carried in either four or six shuttles through the wave shed in the weaving area, and the cloth is drawn off from the middle top and wound on to a large cloth roll by the batching motion located on the right.

No wave shed machine using weft carriers moving in a straight path has proved commercially successful because the shedding elements have no upper constraint, making it impossible to guarantee a correct cloth structure, and because maintaining a clean shed has proved very difficult. Their attraction was a weft insertion rate of up to 2200 m min⁻¹ but they became obsolescent when simpler single phase air jet machines begun to exceed this speed.

The first parallel shed machine to win limited commercial acceptance was the Orbit 112,[13] which on a two head machine weaving used rigid rapiers for weft insertion and reached weft insertion rates of up to 3600 m min⁻¹ when weaving bandage fabrics. Sulzer Textil released their new multiphase linear air jet machine, which is already operating at up to 5000 m min⁻¹ and is only at the beginning of its operational development. This machine, which needs no healds, has a rotating drum on which the shed-forming reed and beat-up reed are mounted (Fig. 4.21).[17] On this machine shedding is gentle and individual picks are transported across the warp sheet at relatively low speeds. Weft yarns do not have to be accelerated and decelerated continuously thus reducing the stresses imposed on the yarns. This reduces demands on yarn and, for the first time, we have a machine with a high insertion rate which should be able to weave relatively weak yarns at low stop rate and high efficiency.

4.5.3   Weft insertion and beat-up (single phase machines)

All single phase weaving machines are classified in accordance with their weft insertion system. The different types have been summarised in Table 4.3. The main methods of single phase weft insertion are by shuttle (Fig. 4.22), projectile (Fig. 4.23), rapier and air or water jet (Fig. 4.24).

4.5.3.1   Weft insertion by shuttle

Looms using shuttles for carrying the weft through the warp sheet dominated cloth production until the 1980s even in high wage countries like the USA. They are now obsolete, except for use in weaving a few highly specialised fabrics. In spite of this, large numbers of automatic bobbin changing looms are still in use but they are being rapidly replaced by shuttleless weaving machines. Shuttleless machines produce more regular fabrics with fewer faults and need less labour for weaving and maintenance. Millions of handlooms are still in operation in south east Asia being protected by legislation.

Figure 4.22 shows schematically the production of cloth on a shuttle loom. The shuttle carrying the pirn, on which the weft is wound, is reciprocated through the warp by a picking motion (not shown) on each side of the machine. For each pick the shuttle has to be accelerated very rapidly and propelled along the race board. Whilst crossing through the shed, one pick of weft is released and when the shuttle reaches the second shuttle box, the shuttle has to be stopped very rapidly. After each pick has been inserted it has to be beaten-up, that is moved to the fell of the cloth. The reed and the race board are mounted on the sley and during the weaving cycle are reciprocated backward and forward. Whilst the shuttle passes through the shed the sley is close to the healds to enable the shuttle to pass without damaging the warp yarns. The sley is then moved forward for beat-up. The need to have an open shed for weft insertion during a considerable part of the picking cycle, and the weight of the sley carrying the race board and the reed, impose restrictions on the picking speed, i.e. the number of revolutions at which the loom can operate.

The basic weakness of fly shuttle weaving machines is the unsatisfactory ratio existing between the large projected mass of the shuttle and the weft bobbin in relation to the small variable mass of the weft yarn carried in the shuttle.[13] Only about

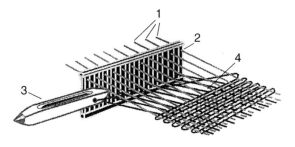

4.22   Weft insertion by shuttle (schematic). 1, Warp yarns; 2, reed; 3, shuttle carrying pirn (entering shed); 4, fell of cloth. Picking motions and race board not shown. Reproduced by kind permission of Sulzer Textil.

3% of the energy imparted to the shuttle is used for the actual weft insertion. Further limitations on machine speed are imposed by the need to reciprocate the heavy sley. Whilst theoretically it is possible to attain weft insertion rates of up to 450 m min^{-1} for wide machines, few shuttle machines in commercial use exceed 250 m min^{-1}.

In a non-automatic power loom each time a pirn is nearly empty the weaver has to stop the loom and replace it. Pirns should be replaced when there is still a little weft left on them to prevent the weft running out in the middle of the shed and creating a broken pick which has to be repaired. In industrialized countries most power looms have been replaced by automatic pirn changing weaving machines that, in turn, are now being replaced by shuttleless weaving machines. In automatic weaving machines the pirns are changed without attention from the weaver and without the loom stopping. The replacement pirns are periodically placed into a magazine by an operative so that the machine can activate the pirn replacement whenever necessary. The magazine fillers can be replaced by a 'box loader' attachment when the pirns are brought to the loom in special boxes from where they are transferred automatically to the change mechanism. Shuttle change looms, where the shuttle rather than the pirn is changed whenever the pirn empties, are available for very weak yarns. All these methods require pirns to be wound prior to being supplied to the loom. Alternatively a Unifil attachment can be fitted to wind the pirns on the weaving machine and feed them into the change mechanism.

There are practically no restrictions on the widths or area density of fabrics which can be woven on shuttle machines. Automatic looms can be fitted with extra shuttle boxes and special magazines so that more than one weft yarn can be inserted in accordance with a prearranged pattern. Compared with similar equipment for shuttleless machines this equipment is clumsy and labour intensive.

4.5.3.2 Projectile machines

Projectile machines can use either a single projectile, which is fired from each side of the machine alternately and requires a bilateral weft supply, or use a unilateral weft supply and a number of projectiles which are always fired from the same side and are returned to the picking position on a conveyor belt. Since Sulzer commenced series production of their unilateral picking multiple projectile machines in the 1950s they have dominated the market and have sold more shuttleless machines than any other manufacturer. Sulzer Textil have continuously developed their machines, improved weft insertion rates and machine efficiency and extended the range of fabrics that can be woven on them. They are now used not only for weaving a vast range of standard fabrics but also for heavy industrial fabrics of up to 8 m wide, for sailcloth, conveyor belts, tyre cord fabrics, awnings, geotextiles, airbags and a wide range of filter fabrics of varying area density and porosity.

One of the major advantages of all shuttleless machines is that weft on cone does not have to be rewound before it is used. This eliminates one process and reduces the danger from mixed yarn and ensures that weft is used in the order in which it is spun. On shuttle looms weft is split into relatively short lengths, each one of which is reversed during weaving which can show up long periodic faults in a yarn.

The standard projectile is 90 mm long and weighs only 40 g, a fraction of the mass of a shuttle. For pick insertion on a Sulzer machine (see Fig. 4.23) weft is withdrawn from the package through a weft brake and a weft tensioner to the shuttle feeder,

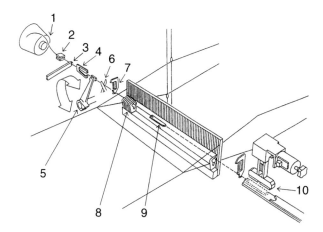

4.23 Weft insertion by projectile (Sulzer system) (schematic). 1, Weft (on cone); 2, yarn brake (adjustable); 3, weft tensioner; 4, weft presenter; 5, torsion rod; 6, weft cutter (scissors); 7, gripper (to hold cut end); 8, guide teeth; 9, projectile; 10, projectile brake (receiving side). Reproduced with kind permission of Sulzer Textil.

which places it into the gripper of the projectile. A torsion bar system is used for picking which transfers the maximum possible strain-energy to the projectile before it separates from the picker shoe. The torsion bar can be adjusted to deliver the energy required to propel the projectile through the guide teeth to the shuttle brake. Sulzer redesigned the reed and the beat-up mechanism so as to obtain a stronger and more rapid beat-up thus making a higher proportion of the picking cycle available for weft insertion.

Narrow machines can operate at weft insertion rates of up to 1000 m min^{-1} whilst 3600 mm wide machines can insert weft at up to 1300 m min^{-1}. Models are available for weaving heavy fabrics, for weaving coarse and fancy yarns and for up to six weft colours. The machines can be fitted with a variety of shedding motions and are equipped with microprocessors to monitor and adjust machine performance. Because of the increase in weft insertion rates with increases in reed width and because of the decrease in capital cost per unit width for wider projectile machines, it is often attractive to weave a number of widths of fabric side by side in one projectile machine.

For even wider and heavier fabrics Jürgens[18] build a machine using the Sulzer Rüti system of pick insertion. They can propel a heavier projectile carrying a weft yarn of up to 0.7 mm diameter across a reed width of up to 12 m. Their machines take warp from one, two or three sets of warp beams and maintain weaving tensions of up to 30000 N m^{-1}, accommodate up to 24 shafts controlled by an extra heavy dobby, and deliver the cloth on to a large batching motion. Jäger have developed a projectile weaving machine for fabrics of medium area density and up to 12 m wide using a hydraulically propelled projectile.

4.5.3.3 Rapier machines
At ITMA Paris 1999, out of 26 machinery manufacturers showing weaving machines no fewer than 17 offered rapier machines and some offered machines of more than one type. Rapier machines were the first shuttleless machines to become available

but, at first, they were not commercially successful because of their slow speed. With the introduction of precision engineering and microprocessor controls, the separation of the weft insertion from the beat-up and improved rapier drives and heads, their weft insertion rates have increased rapidly. For machines of up to 2500 mm reed space they equal those of projectile machines with which they are now in direct competition.

Machines may operate with single or double rapiers. Single rapier machines generally use rigid rapiers and resemble refined versions of ancient stick looms. They have proved attractive for weaving fairly narrow cloths from coarse yarns. Wide single rapier machines are too slow for most applications. In single rapier machines the rapier traverses the full width of the shed and generally picks up the weft and draws it through the shed on its return. A variation of the single rigid rapier is the single rapier working bilaterally, sometimes referred to as a two-phase rapier. As it has not been used to any extent for industrial fabrics it is not considered here but details can be found in the book by Ormerod and Sondhelm.[13]

Most rapier machines use double rapiers, one rapier entering the shed from each side. They meet in the middle and transfer the yarn. With the Gabler system weft is inserted alternately from both sides of the machine and yarn is cut every second pick with hairpin selvedges being formed alternately on both selvedges. The Gabler system has now been largely superseded by the Dewas system where weft is inserted from one side only and is cut after every pick.

Double rapiers machines use either rigid or flexible rapiers. Rigid rapier machines need more space than machines fitted with any other weft insertion system. Rigid rapier machines, of which the Dornier HTV and P[19] series are prime examples, are capable of weaving most types of industrial fabrics with weft linear densities of up to 3000 tex, in widths of up to 4000 mm and at weft insertion rates of up to 1000 m min^{-1}. Typical fabrics being produced on them range from open-coated geotextile mesh to heavy conveyor belting. A variation of the rigid rapier is the telescopic rapier.

By far the largest number of rapier machines use double flexible rapiers which are available in widths up to 4600 mm with even wider machines being custom built for industrial applications. Standard machines have a relatively low capital cost and can be used to weave a wide range of low and medium area density fabrics. They are ideal for weaving short runs and for fabrics woven with more than one weft because their weft change mechanism for up to eight colours is simple and cheap. They are widely used for furnishing and fashion fabrics, often with Jacquards. They are also used for some industrial cloths.

4.5.3.4   Fluid jet weaving machines
Fluid jet machines use either air or water to propel the yarn through the shed. They do not need a weft carrier or a rapier for weft insertion and therefore have fewer moving parts and less mass to move. Water jets are only suitable for hydrophobic yarns whilst most yarns can be woven on air jets. Water jets generally use a single nozzle at the picking side to propel the yarn through the full width of the shed and this limits their width to about 2 m. As the flow of air is more difficult to control than the flow of water under pressure, air jets with single nozzles have only been commercially successful in widths of up to 1700 mm. For wider machines, booster or relay nozzles are placed along the reed to ensure the smooth movement of the weft across the full reed width. Although theoretically wide air jets can be built, com-

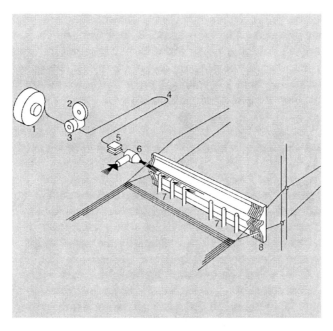

4.24   Weft insertion by airjet (Sulzer Rüti L5000). 1, Supply package; 2, measuring disc; 3, rollers; 4, storage tube; 5, clamp; 6, main nozzle; 7, relay nozzles; 8, reed with tunnel.
Reproduced by kind permission of Sulzer Textil.

mercially single width machines are most attractive and machinery makers limit their ranges to 3600 or 4000 mm reed width.

Compressed air is expensive to produce and its flow is difficult to control. The air flow in the shed has, therefore, to be restricted either by special air guides or 'confusers' or by passing the weft through a channel in a special 'profile' reed. The former method was pioneered by Elitex and is used in most of their 'P' type weaving machines (Fig. 4.18) of which a large number are in use for weaving light and medium weight fabrics up to 150 cm wide. Sulzer Rüti developed the 'te strake' profile reeds with relay nozzles and this system is shown schematically in Fig. 4.24. There is one main nozzle per colour and one set of relay nozzles spaced at regular intervals along the reed. The weft is measured to length in the weft feeder and then carried by the air stream of the main nozzle into the weft duct, accelerated and transported further by air discharged from the relay nozzles. After insertion, a stretch nozzle at the receiving side holds the pick under tension until it is bound into the cloth.

Since air jets came into large scale commercial use in the 1970s they have been developed rapidly. They can now weave the majority of fabrics and are dominant for the mass production of fairly simple cloths. They have reached weft insertion rates of up to $3000 \, \text{m min}^{-1}$, twice that of any other single phase weft insertion system and are still under intensive development. Their capital cost per metre of weft inserted is highly competitive. Their operating costs depend largely on the local cost of electricity and whether low grade waste heat from the compressors can be used for other operations in the plant.

Air jet machines fitted with an automatic weft fault repair system can correct the majority of weft faults, including part picks, which occur between the main nozzle and the selvedge on the receiving side. The unit removes the broken thread from the shed without disturbing the warp ends and then restarts the machine. Only if the machine cannot locate and repair the fault will it signal for attention. As weft stops represent the majority of stops on an air jet, the system greatly reduces the weaver's work load, frequently by more than 50%. It also reduces machine interference and improves the quality of many fabrics.

4.5.4   Other motions and accessories for single phase weaving machines

4.5.4.1   Warp supply and let-off motion

Warp yarns are generally supplied to the weaving machine on one or more weaver's beams. In special cases, as already mentioned above, cone creels can be used. The warp yarns assembled on the weaver's beam should be evenly spaced and under standard tension to ensure that all are of exactly the same length when they are unwound for weaving. The larger the diameter of the beams the longer the length of warp which can be wound on to it and the fewer the warp changes required but the greater the tension variations which have to be compensated. Different weaving machines accommodate beams of different maximum diameter but most modern machines can take beams of up to 1000 mm diameter. If even larger diameter beams are needed for very coarse warp yarns, for example for industrial fabrics or denims, the beams can be placed into a separate beam creel outside the loom frame. Such units cater for beams of up to 1600 mm diameter.

The width of yarn on the weaver's beam has to be at least as wide as the yarn in the reed. When the warp width required exceeds 2800 mm, more than one beam is frequently used to simplify sizing and warp transport. If yarns from more than one beam are used in one cloth it is essential that both beams are prepared under identical conditions to prevent variations which can cause cloth faults after finishing. The let-off tensions of different beams have to be carefully controlled and this has become simpler with the introduction of electronic sensors. When the cloth design requires the use of warp yarns of widely differing linear densities or results in different warp yarns weaving with widely differing crimps, two or more warp beams may have to be used in parallel. They can be placed either above each other or behind each other in the weaving machine. Comparison of Figs. 4.17 and 4.18 shows how the layout of machines can be modified without impairing their efficiency.

During weaving the let-off motion will release the required amount of yarn for each pick cycle. It must also hold the warp yarns under even tension so that they separate easily into two or more sheets during shedding prior to weft insertion and so that the required tension is maintained during beat-up when the newly inserted weft is moved by the reed to the fell of the cloth. Let-off motions used to be mechanically controlled with tension being measured by the deflection of the back rest but now electronic sensors are used for tension measurements and the let-off is frequently controlled by separate servo motors.

4.5.4.2   Take-up motion

Take-up motions are required to withdraw the cloth at a uniform rate from the fell. The speed of withdrawal controls the pick spacing, which has to be regular to prevent weft bars and other faults. In most weaving machines the take-up motion also controls the winding of the cloth on to the cloth roller. If large rolls of large diameter have to be made, as is common for heavy fabrics, a separate batching motion is placed outside the loom frame. Batching motions can be on a different level from the weaving machine, either below or above, to reduce the area required for the weaving shed. To prevent the weaving of long lengths of reject fabric, batching motions can incorporate cloth inspection facilities, sometimes with an intermediate cloth storage unit.

4.5.4.3 Automatic stop motions

The first group, warp protector motions, are only necessary on machines which use a free flying shuttle or projectile. They are designed to prevent the forward movement of the reed if the shuttle fails to reach the receiving side. This prevents damage to the machine and the breakage of large numbers of ends if the shuttle is trapped.

Warp stop motions stop the machine if an end breaks. They are activated when a drop wire, through which an end has been threaded, drops because a broken end will no longer support it. Drop wires can be connected to mechanical or electrical stop motions. Yarns have to be properly sized to prevent them being damaged by the drop wires. Electronic warp stop motions, which do not require physical contact with the warp, are now being introduced especially for fine filament yarns.

Weft stop motions are used to activate weft changes in automatic shuttle looms and to stop weaving machines if the weft breaks during weft insertion. Electronic motions are available that will stop the machine even if a broken end catches on again before it reaches the receiving side. In air jet machines fitted with automatic repair facilities the weft stop motion also starts the weft repair cycle.

4.5.4.4 Quick style change

QSC (quick style change) equipment, first shown by Picanol in 1991 and now available from most manufacturers, greatly reduces the time a weaving machine has to be stopped for a warp change. The warp beam, back rest, warp stop motion, heald frames and reed are located on a module which separates from the main frame of the weaving machine and which is transferred by a special transport unit to and from the entering and knotting department where the module becomes the preparation frame for a replacement warp.[14] Thus 90% of the work load, which is normally carried out on the stopped weaving machine, is eliminated from the warp replenishment cycle and the weaving machine efficiency is improved. It also results in cleaner reeds and healds resulting in better machine performance and improved cloth quality.

4.5.5 Machine width

The reed width of the machine must be equal to or greater than the width in reed of the fabric to be woven. Width in reed must allow for the width of selvedge and dummy ends. If the machine is narrower the cloth cannot be woven in the machine. Generally it is impossible to increase the available reed width of a machine.

Whilst the machine width cannot be exceeded, it is generally possible to weave narrower fabrics. In Sulzer projectile machines weaving down by up to 50% is possible. Different manufacturers and models have different arrangements for weaving down – some allow for only 200 mm which is often insufficient considering likely changes in materials and styles. It is most economic to use a high proportion of the reed width because weaving down is likely to reduce the weft insertion rate. Wider machines also tend to have higher capital and running costs. On some occasions it is economic to weave a number of fabrics side by side in one machine. Five, six or seven roller towels, each with its own tucked selvedges, can be produced in a wide Sulzer projectile machine.

4.6   The future

Because of rapid technological advances modern weaving machines have become highly automated with most functions electronically controlled. Machine settings can be adjusted and transferred and many faults can be repaired without requiring attention by an operative. The frequency of machine stops and their duration has been reduced. The cost of labour has been desensitized and the cost of production has been reduced whilst product quality has improved. Because of the cost of modern machines it becomes ever more important to use the right equipment for the job and operate machines at high efficiency.

At present projectile weaving machines are the most versatile of conventional machines and by bolting the right equipment to the machine there is hardly a fabric they cannot weave well. Most simple fabrics can be woven at much lower capital cost on air jets fitted with simple shedding systems. Their economic range is being extended continuously. In between there are fabrics which can be woven most cheaply on rapiers. Multiphase machines, like Sulzer Textil's linear air jet M8300, are likely to become cheap producers of simple fabrics in the not too distant future.

References

1. The Textile Institute Terms and Definitions Committee, *Textile Terms and Definitions*, 10th edition, The Textile Institute, Manchester, 1995.
2. A T ROBINSON and R MARKS, *Woven Cloth Construction*, The Textile Institute, Manchester, 1973.
3. W WATSON, *Textile Design and Colour–Elementary Weaves and Figured Fabrics*, Longmans, London, 1912 (later editions revised by Grosicki and published by Butterworth).
4. W WATSON, *Advanced Textile Design*, Longmans, London, 1912 (later editions revised by Grosicki and published by Butterworth).
5. Sulzer Brothers (now Sulzer-Textil), Winterthur, Switzerland.
6. R MARKS, *An Intoduction to Textiles – Workbook: Fabric Production – Weaving*, ed. W S Sondhelm, Courtaulds Textiles plc, Manchester, 1989.
7. F T PEIRCE, *The Geometry of Cloth Structure*, Manchester, Journal of the Textile Institute, T45, 1937.
8. The Textile Institute Terms and Definitions Committee, *Textile Terms and Definitions*, 10th edition, *Table of SI Units and Conversion Factors*, The Textile Institute, Manchester, 1995.
9. Rüti Machinery Works Ltd, Verkauf, Rüti, Switzerland, 1977.
10. Rüti Machinery Works Ltd, Switzerland, 'C' Model shuttle weaving machine.
11. Investa, Czechoslovakia, Elitex air jet, 'P' type.
12. Somet, Colzate (BG), Italy, *'Mach 3' air jet*.
13. A ORMEROD and W S SONDHELM, *Weaving–Technology and Operations*, Chapter 5: 'Weaving,' The Textile Institute, Manchester, 1995 (reprinted 1999).
14. A ORMEROD and W S SONDHELM, *Weaving–Technology and Operations*, Chapter 2: 'Warping' and Chapter 3: 'Sizing,' The Textile Institute, Manchester, 1995 (reprinted 1999).
15. Sulzer Rüti, Wilmslow, *Fabric Structure*, 3085/06.10.95/Legler/sei, 1995.
16. Maschinenfabrik Starlinger, Vienna, Austria, Circular Loom SL4, Drwg. 0304A3, 1997.
17. Sulzer Rüti, Wilmslow, *Shed Formation*, 3138/10.10.95/Legler/kaa, 1995.
18. Jürgens GmbH, Emsdetten, Germany, *Projectile weaving machine JP-2000*, 1996.
19. Lindauer Dornier GmbH, Germany, 1999.

5

Technical fabric structures – 2. Knitted fabrics

Subhash C Anand

Faculty of Technology, Bolton Institute, Deane Road, Bolton BL3 5AB, UK

5.1 Terms and definitions

- *Warp knitting* is a method of making a fabric by normal knitting means, in which the loops made from each warp are formed substantially along the length of the fabric. It is characterised by the fact that each warp thread is fed more or less in line with the direction in which the fabric is produced. Each needle within the knitting width must be fed with at least one separate and individual thread at each course. It is the fastest method of converting yarn into fabric, when compared with weaving and weft knitting (Fig. 5.1).
- *Weft knitting* is a method of making a fabric by normal knitting means, in which the loops made by each weft thread are formed substantially across the width of the fabric. It is characterised by the fact that each weft thread is fed more or less at right angles to the direction in which the fabric is produced. It is possible to knit with one thread only, but up to 144 threads can be used on one machine. This method is the more versatile of the two in terms of the range of products produced as well as the type of yarns utilised (Fig. 5.2).
- *Single-jersey fabric* is a weft-knitted fabric made on one set of needles.
- *Double-jersey fabric* is a weft-knitted fabric made on two sets of needles, usually based on rib or interlock gaiting, in a manner that reduces the natural extensibility of the knitted structure. These fabrics can be non-Jacquard or Jacquard.
- *Course* is a row of loops across the width of the fabric. Courses determine the length of the fabric, and are measured as courses per centimetre.
- *Wale* is a column of loops along the length of the fabric. Wales determine the width of the fabric, and are measured as wales per centimetre.
- *Stitch density* is the number of stitches per unit area of a knitted fabric (loops cm^{-2}). It determines the area of the fabric.
- *Stitch length* is the length of yarn in a knitted loop. It is the dominating factor for all knitted structures. In weft knitting, it is usually determined as the average length of yarn per needle, while in warp knitting, it is normally determined as the average length of yarn per course.

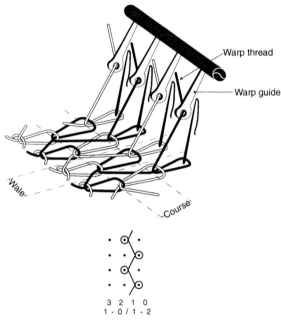

5.1   Warp knitting.

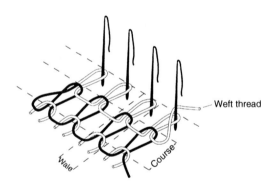

5.2   Weft knitting.

- *Yarn linear density* indicates the thickness of the yarn and is normally determined in tex, which is defined as the mass in grams of 1 km of the material. The higher the tex number, the thicker is the yarn and vice-versa.
- *Overlap* is the lateral movement of the guide bars on the beard or hook side of the needle. This movement is normally restricted to one needle space. In the fabric a loop or stitch is also termed the overlap.
- *Underlap* is the lateral movement of the guide bars on the side of the needle remote from the hook or beard. This movement is limited only by the mechani-

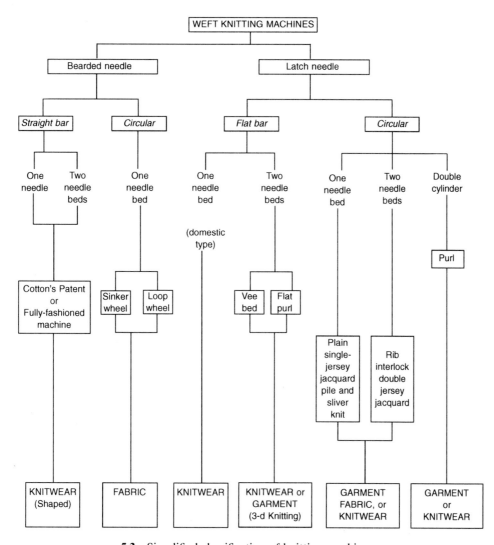

5.3   Simplified classification of knitting machinery.

cal considerations. It is the connection between stitches in consecutive courses in a warp knitted fabric.

- *Tightness factor K* is a number that indicates the extent to which the area of a knitted fabric is covered by the yarn. It is also an indication of the relative looseness or tightness of the knitting. ($K = \text{tex}^{1/2} l^{-1}$), where l is the stitch length.
- *Area density* is a measure of the mass per unit area of the fabric (g m^{-2}).

5.2   Weft knitting machines

Figure 5.3 shows a simplified classification of weft knitting equipment. It will be noticed from Fig. 5.3 that the latch needle is the most widely used needle in weft knitting, because it is self-acting or loop controlled. It is also regarded as more ver-

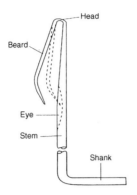

5.4   Bearded needle.

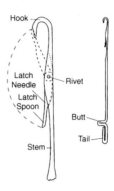

5.5   Latch needle.

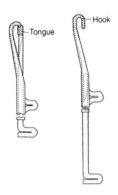

5.6   Compound needle (fly needle frame).

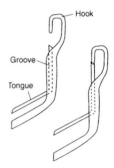

5.7   Compound needle (Kokett needle).

satile in terms of the range of materials that can be processed on latch needle machines. Bearded needles are less expensive to manufacture, can be produced in finer gauges and supposedly knit tighter and more uniform stitches compared with latch needles, but have limitations with regard to the types of material that can be processed as well as the range of structures that can be knitted on them. Bearded needle machines are faster than the equivalent latch needle machines. The compound needle has a short, smooth and simple action, and because it requires a very small displacement to form a stitch in both warp and weft knitting, its production rate is the highest of the three main types of needle. Compound needles are now the most widely used needles in warp knitting and a number of manufacturers also offer circular machines equipped with compound needles. The operation speeds of these machines are up to twice those of the equivalent latch needle machines.

The main parts of the bearded, latch, compound needle (fly needle frame) and compound needle (Kokett) are shown in Figs. 5.4, 5.5, 5.6 and 5.7, respectively. Variations of latch needles include rib loop transfer needles and double-ended purl needles, which can slide through the old loop in order to knit from an opposing bed and thus draw a loop from the opposite direction.

5.2.1   Loop formation with latch needles

Figure 5.8(a) illustrates the needle at tuck height, that is high enough to receive a new yarn, but not high enough to clear the old loop below the latch. The needle is kept at this position because the loop formed at the previous course (A) lies on the open latch and stops the latch from closing. Note that once a latch is closed, it can only be opened by hand, after stopping the machine.

In Fig. 5.8(b), the needle has been lifted to the clearing position and the new yarn (B) is presented. The old loops (A) are below the latch and the new yarn (B) is fed into the needle hook. A latch guard normally prevents the latch from closing at this point.

In Fig. 5.8(c), the needle has now moved to its lowest position or knitting point and has drawn the new loop (B) through the old loop (A) which is now cast-off or knocked over. The needle now rises and the sequence of movements is repeated for the next course.

The loop formation on a latch needle machine is also illustrated in Fig. 5.9(a) and a typical cam system on such a machine is shown in Fig. 5.9(b).

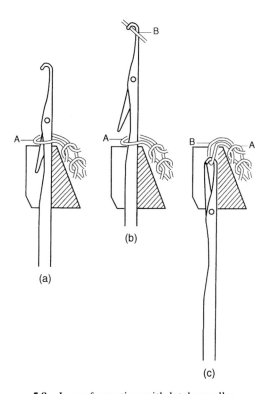

5.8   Loop formation with latch needles.

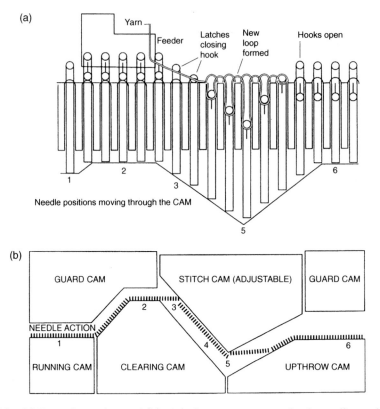

5.9 (a) Loop formation and (b) typical cam system on a latch needle machine.

5.2.2 Single-jersey latch needle machines

This type of machinery is employed throughout the world, either as a basic machine or with certain refinements and modifications, to produce fabric ranging from stockings to single-jersey fabric for dresses and outerwear, as well as a wide range of fabrics and products for technical applications. The machine sizes vary from 1 feeder 1 inch diameter to 144 feeds 30 inches diameter. Most single-jersey machines are rotating cylinder type, although a few rotating cam box machines are still used for specialised fabrics. These machines are referred to as sinker top machines and use web holding sinkers. Comprehensive reviews of single- and double-jersey knitting machinery and accessories exhibited at ITMA'95 and ITMA'99 were published that illustrate the versatility and scope of modern weft knitting equipment.[1,7]

5.2.2.1 Knitting action of a sinker top machine
Figures 5.10(a) to (d) show the knitting action of a sinker top machine during the production of a course of plain fabric.

Figure 5.10(a) illustrates the rest position. This shows the relative position of the knitting elements in-between the feeders with the needle at tuck height and the fabric loop held on the needle latch by the forward movement of the sinker towards the centre of the machine.

Figure 5.10(b) illustrates the clearing position. The needle has been raised to its

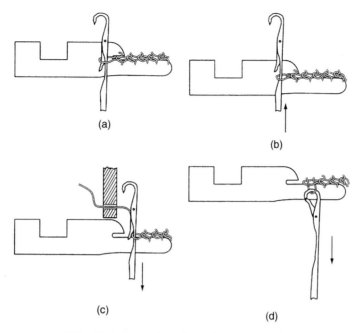

5.10   Knitting action of a sinker top machine.

highest position by the clearing cam acting on the needle butt; the old loop slides down from the open latch on to the needle stem.

Figure 5.10(c) illustrates the yarn feeding position. The sinker is partially withdrawn allowing the feeder to present its yarn to the descending needle hook, at the same time freeing the old loop so that it can slide up the needle stem and under the open latch spoon.

Figure 5.10(d) illustrates the knock-over position. The needle has now reached its lowest position and has drawn the new loop through the old loop which is now knocked over the sinker belly.

Stitch length may be controlled in a number of ways. On machines without positive feed mechanism, it is controlled mainly by the distance the needle descends below the sinker belly. Other factors such as input tension (T_i), yarn to metal coefficient of friction (μ) and take-down tension also influence the final stitch length in the fabric. When a positive feed device is used, the length of yarn fed to the needles at a particular feed is the factor that decides the stitch length. Other factors such as input tension T_i, μ, stitch cam setting and take-down tension influence the yarn or fabric tension during knitting, and hence determine the quality of the fabric. Stitch length is fixed by the positive feed device setting.

The sinker has two main functions and these are:

• to hold the fabric loop in a given position whenever the needles rise and
• to provide a surface over which the needles draw the loops.

Other advantages of using sinkers include:

• The control exerted by the sinker allows minimum tension on the fabric thus producing a good quality fabric with even loops.

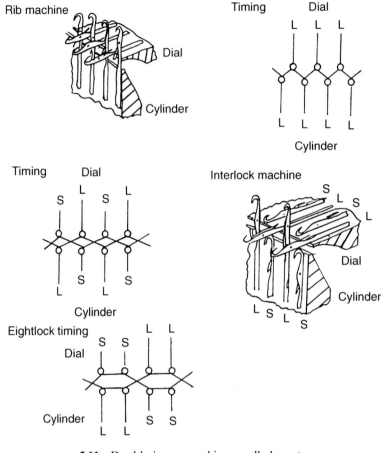

5.11 Double-jersey machine needle layouts.

- Fine adjustments in quality and those required in the knitting of certain difficult yarns and structures are possible.
- The sinker facilitates the setting-up of the machine after a partial or full press-off (after the latches have been opened manually).

5.2.3 Double-jersey machines

Figure 5.11 shows the needle layout of rib and interlock machines. Both types of machine are available as circular or flat machines, whereas straight bar or fully fashioned machines are available in rib-type only.

Rib and interlock double-jersey machines are used either as garment length machines or for producing rolls of fabric. They can be either plain or equipped with a wide range of mechanical patterning mechanisms at each feed in the cylinder. Both types can also be equipped with electronic needle mechanisms to produce large area Jacquard patterns at high speeds.

5.2.3.1 Rib machines

Rib machines use two sets of needles and can be flat, circular or fully fashioned. The needles in the two beds are staggered or have spaces between them (rib gaiting). Most machines have revolving needle cylinders but in some cases the cams rotate past stationary needles. Patterning is obtained by altering cam and needle set-out or by using various needle selection mechanisms including individual needle electronic selection with computer aided design system. Machine diameters range from $7\frac{1}{2}$–20 inches for garment length and from 30–33 inches for fabric machines. An example of a modern double-jersey machine is the Monarch V-7E20, a 30 inch diameter, E20, 72 feeders 8-lock machine with RDS on dial and ACT II motorised automatic friction take-down system. The machine has 2×2 cam tracks and can be converted from rib to interlock or 8-lock timing in minutes. All basic non-jacquard double-jersey structures can be knitted at a speed factor of 900 (machine diameter (inches) $\times$ machine rpm).

5.2.3.2 Interlock machines

These are latch needle circular machines of the rib type, provided with a cylinder and dial. Unlike rib machines where the tricks of the dial alternate with the tricks of the cylinder (rib gaiting) the needle tricks of the cylinder are arranged exactly opposite those of the dial (interlock gaiting). Long-and short-stemmed needles are used that are arranged alternately, one long, one short in both cylinder and dial as shown in Fig. 5.11. An example of a modern high-speed interlock machine is Sulzer Morat Type 1L 144, which is a 30 inches diameter, 144 feeds, 28 or 32 gauge (npi, needles per inch), 28 rpm, producing at 100% efficiency $86.4\,\mathrm{m\,h^{-1}}$ ($15.55\,\mathrm{kg\,h^{-1}}$) of 76 dtex polyester with 14 cpc (courses per centimetre) and with an area density of $180\,\mathrm{g\,m^{-1}}$ (60 inches wide) finished fabric.

To accommodate the long- and short-stemmed needles, the cam system is provided with a double cam track. The long dial needles knit with the long cylinder needles at feeder 1 and the short cylinder needles knit with the short dial needles at feeder 2. Thus two feeders are required to make one complete course of loop.

5.2.3.3 Needle timing

Two different timings can be employed on 1×1 rib and 1×1 interlock machines.

- *Synchronised timing* is the timing of a machine that has two sets of needles where the point of knock-over of one set is aligned with the point of knock-over of the other set.
- *Delayed timing* is the setting of the point of knock-over of one set of needles on a two-bed knitting machine out of alignment with that of the other set so as to permit the formation of a tighter stitch. Broad ribs (i.e. 2×2, 3×3 etc.), and rib Jacquard fabrics cannot be produced in delayed timing because there will not always be cylinder needles knitting either side of the dial needles from which to draw yarn. Up to nine needles delay is possible, but 4–5 needles delay is normal.

5.2.3.4 Knitting action of V-bed flat machine

Figure 5.12 illustrates the different stages of loop formation on a V-bed flat knitting machine, and Fig. 5.13 shows the cam system used on a simple single system flat

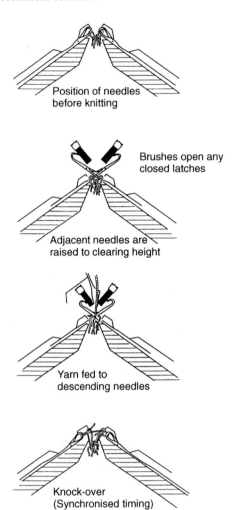

Position of needles
before knitting

Brushes open any
closed latches

Adjacent needles are
raised to clearing height

Yarn fed to
descending needles

Knock-over
(Synchronised timing)

5.12   Loop formation on a V-bed flat knitting machine.

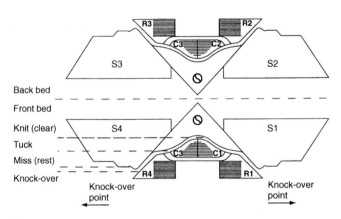

5.13   V-bed single-system cams. S are stitch cams, R are raising cams and C
are clearing cams.

machine. Power V-bed flat machines are used mainly for the production of knitwear for children, women and men. They range from simple machines through mechanical jacquard machines to fully electronic and computerised flat machines, even equipped with presser foot. The developments in the automation of fabric designing, pattern preparation, and electronic needle selection, as well as in the range of structures and effects which can be produced, have been tremendous and flat machines and their products are now regarded as extremely sophisticated. High quality garments can now be produced at competitive prices owing to revolutionary garment production systems feasible with presser foot. Two- and three-dimensional structures as well as complete garments without any seams or joins can be produced on the latest electronic flat knitting machines and the associated design systems.

5.3   Weft-knitted structures

The basic weft-knitted structures and stitches are illustrated in Fig. 5.14 (A to G), and the appearance, properties and end-use applications of plain, 1×1 rib, 1×1 purl and interlock structures are summarised in Table 5.1. These basic stitches are often combined together in one fabric to produce an enormous range of single- and double-jersey fabrics or garments. Weft-knitted fabrics are produced commercially for apparel, household and technical products and they are used for an extremely large array of products, ranging from stockings and tights to imitation furs and rugs.

The importance and diversity of warp- and weft-knitted fabrics used for various technical applications has been discussed by Anand,[2] who highlighted the fact that knitted fabrics are being increasingly designed and developed for technical products ranging from scouring pads (metallic) to fully fashioned nose cones for supersonic aircrafts. Warp- and weft-knitted products are becoming popular for a wide spectrum of medical and surgical products.[3]

5.4   Process control in weft knitting

5.4.1   Main factors affecting the dimensional properties of knitted fabrics or garments

- *Fabric structure:* different structures relax differently.
- *Fibre(s) type:* fabrics or garments made from different fibre(s) relax differently.
- *Stitch length:* the length of yarn in a knitted loop is the dominating factor for all structures.
- *Relaxation/finishing route:* the fabric dimensions vary according to relaxation/finishing sequence.
- *Yarn linear density:* affects the dimensions slightly, but affects fabric tightness, area density ($g\,m^{-2}$) and other physical properties.

5.4.2   Laboratory stages of relaxation

- *On machine – Strained state:* this is predominantly length strain.
- *Off machine – Dry relaxed state:* the fabric moves to this state with time. The dry

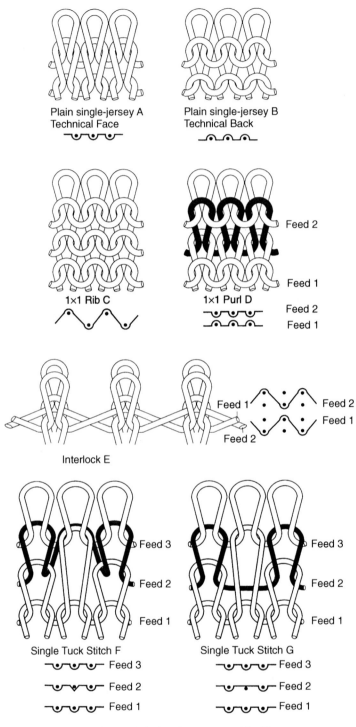

5.14 Weft-knitted structures (A–G).

Table 5.1 Comparison of appearance and properties

Property	Plain	1 × 1 Rib	1 × 1 Purl	Interlock
Appearance	Different on face and back; V-shapes on face, arcs on back	Same on both sides, like face of plain	Same on both sides, like back of plain	Same on both sides, like face of plain
Extensibility				
Lengthwise	Moderate (10–20%)	Moderate	Very high	Moderate
Widthwise	High (30–50%)	Very high (50–100%)	High	Moderate
Area	Moderate–high	High	Very high	Moderate
Thickness and warmth	Thicker and warmer than plain woven made from same yarn	Much thicker and warmer than plain woven	Very much thicker and warmer than plain woven	Very much thicker and warmer than plain woven
Unroving	Either end	Only from end knitted last	Either end	Only from end knitted last
Curling	Tendency to curl	No tendency to curl	No tendency to curl	No tendency to curl
End-uses	Ladies' stockings	Socks	Children's clothing	Underwear
	Fine cardigans	Cuffs	Knitwear	Shirts
	Men's and ladies' shirts	Waist bands	Thick and heavy	Suits
	Dresses	Collars	Outerwear	Trouser suits
	Base fabric for coating	Men's outerwear		Sportswear
		Knitwear		Dresses
		Underwear		

relaxed state is restricted by fabric structure and fibre type. Only *wool* can attain this state.
- *Static soak in water and dry flat – Wet relaxed state:* tight structures do not always reach a 'true' relaxed state. Only *wool* and *silk* can attain this state.
- *Soak in water with agitation, or Agitation in steam, or Static soak at selected temperatures (>90°C) plus, dry flat – Finished relaxed state:* the agitation and/or temperature induces a further degree of relaxation, producing a denser fabric. *Wool, silk, textured yarn fabrics, acrylics.*
- *Soak in water and Tumble dry at 70°C for 1 hour – Fully relaxed state:* three-dimensional agitation during drying. *All* fibres and structures.

5.4.3 Fabric geometry of plain single-jersey structures

1. Courses per cm (cpc) $\alpha\, 1/l = \dfrac{k_c}{l}$

2. Wales per cm (wpc) $\alpha\, 1/l = \dfrac{k_w}{l}$

3. $s = (\text{cpc} \times \text{wpc})\ \alpha\, 1/l^2 = \dfrac{k_s}{l^2}$

4. $\dfrac{\text{cpc}}{\text{wpc}}\ \alpha\, c = \dfrac{k_c}{k_w}$ (shape factor)

k_C, k_W, k_S are dimensionless constants, l is the stitch length and s is the stitch density.

5.4.4 Practical implications of fabric geometry studies

- Relationship between yarn tex and machine gauge is given by Equation (5.1):

$$\text{Optimum tex} = \frac{\text{constant}}{(\text{gauge})^2} \tag{5.1}$$

For single-jersey machines, the optimum tex $= 1650/G^2$, and for double-jersey machines, the optimum tex $= 1400/G^2$, where G is measured in needles per centimetre (npc).
- Tightness factor is given by Equation (5.2):

$$K = \sqrt{\frac{\text{tex}}{l}} \tag{5.2}$$

where l is the stitch length, measured in millimetres. For single-jersey fabrics: $1.29 \le K \le 1.64$. Mean $K = 1.47$. For most weft-knitted structures (including single- and double-jersey structures and a wide range of yarns): $1 \le K \le 2$. Mean $K = 1.5$. The tightness factor is very useful in setting up knitting machines. At mean tightness factor, the strain on yarn, machine, and fabric is constant for a wide range of conditions.
- Fabric area density is given by Equation (5.3):

$$\text{Area density} = \frac{s \times l \times T}{100}\, \text{g m}^{-2} \tag{5.3}$$

Table 5.2 k-Constant values for wool plain single jersey[a]

	k_c	k_w	k_s	k_c/k_w
Dry relaxed	50	38	1900	1.31
Wet relaxed	53	41	2160	1.29
Finished relaxed	56	42.2	2360	1.32
Fully relaxed	55 ± 2	42 ± 1	2310 ± 10	1.3 ± 0.05

[a] Courses and wales are measured per centimetre and l is measured in millimetres.
For a relaxed fabric cpc/wpc = 1.3. cpc/wpc > 1.3 indicates widthwise stretching. cpc/wpc < 1.3 indicates lengthwise stretching. k_s > 2500 indicates felting or washing shrinkage. Relaxation shrinkage is the change in loop shape. Felting/washing shrinkage is the change in loop length.

where s is the stitch density/cm^2; l is the stitch length (mm) and T is the yarn tex, or, Equation (5.4):

$$\frac{k_s}{l} \times \frac{T}{100} \, \text{g m}^{-2} \tag{5.4}$$

where k_s is a constant and its value depends upon the state of relaxation, that is, dry, wet, finished or fully relaxed. The area density can also be given by Equations (5.5) and (5.6)

$$\text{Area density} = \frac{n \times l \times \text{cpc} \times T}{10\,000} \, \text{g m}^{-1} \tag{5.5}$$

where n is the total number of needles, l is the stitch length (mm) and T is the yarn tex, or

$$\frac{n \times k_c \times T}{10\,000} \, \text{g m}^{-1} \tag{5.6}$$

where k_c is a constant and its value depends upon the state of relaxation, that is, dry, wet, finished or fully relaxed.
• Fabric width is given by Equation (5.7)

$$\text{Fabric width} = \frac{n \times l}{k_w} \, \text{cm} \tag{5.7}$$

where k_w is a constant, and its value depends upon the state of relaxation, that is, dry, wet, finished or fully relaxed. It can also be given by Equation (5.8)

$$n \times l = L \, (\text{course length}) \quad \therefore \text{Fabric width} = \frac{L}{k_w} \, cm \tag{5.8}$$

Fabric width depends upon course length and not upon the number of needles knitting.
• Fabric thickness. In the dry and wet relaxed states, fabric thickness (t) is dependent upon fabric tightness, but in the fully relaxed state, it is more or less independent of the fabric tightness factor. In the fully relaxed state $t \approx 4d$ where d is the yarn diameter.

5.4.5   Quality control in weft knitting

The dimensions of a weft-knitted fabric are determined by the number of stitches and their size, which in turn is determined by stitch length. Most knitting quality control therefore reduces to the control of stitch length; differences in mean stitch length give pieces of different size; variation of stitch length within the piece gives appearance defects, by far the most common one being the occurrence of widthwise bars or streaks owing to variation in stitch length between adjacent courses.

5.4.5.1   Measurement of stitch length, l

* Off machine (in the fabric):
  – HATRA course length tester
  – Shirley crimp tester.
* On machine (during knitting):
  – Yarn speed meter (revolving cylinder only)
  – Yarn length counter (both revolving cylinder and cambox machines).

5.4.5.2   Control of stitch length, l

* Positive feed devices:
  – Capstan feed: cylindrical or tapered
  – Nip feed: garment length machines
  – Tape feed: (Rosen feed) circular machines producing plain structures
  – Ultrapositive feed: IPF or MPF
* Constant tension device: Storage feed device: flat, half-hose, hose and circular machines producing either plain or jacquard structures.
* Specialised positive feed devices:
  – Positive Jacquard Feeder MPF 20 KIF
  – Striper Feeder ITF
  – IROSOX Unit (half-hose machines)
  – Elastane Feed MER2
  – Air controlled feeds for flat and fully fashioned machines. Figure 5.15 shows a tape feed. A modern ultrapositive feed and a yarn storage feed device are illustrated in Figs. 5.16 and 5.17, respectively.

5.5   End-use applications of weft-knitted fabrics

Weft-knitted fabrics are used for apparel, household and technical products. The main outlets for the different types of weft-knitted fabrics are as shown below. The knitting equipment used to produce these fabrics is also given.

5.5.1   Flat bar machines

* Machine gauge: normally needles per inch, 3–18 npi
* Machine width: up to 78.7 inches
* Needle type: latch (compound needle machines are being developed)
* Needle bed type: single (hand machines), but mainly rib type

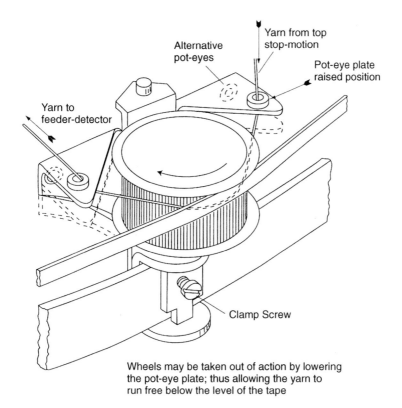

Alternative pot-eyes

Yarn from top stop-motion

Pot-eye plate raised position

Yarn to feeder-detector

Clamp Screw

Wheels may be taken out of action by lowering the pot-eye plate; thus allowing the yarn to run free below the level of the tape

5.15 Tape positive feed device.

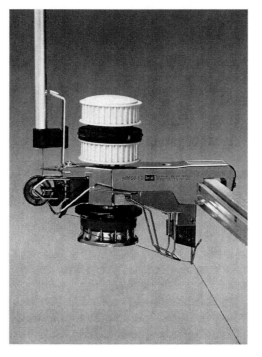

5.16 Ultrapositive feed device.

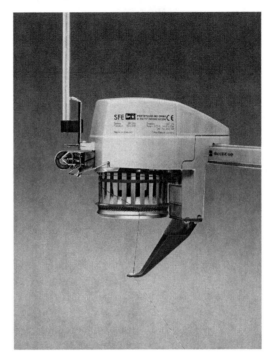

5.17 Yarn storage feed device.

- Products: jumpers, pullovers, cardigans, dresses, suits, trouser suits, trimmings, hats, scarves, accessories, ribs for straight-bar machines (fully fashioned machines). Cleaning clothes, three-dimensional and fashioned products for technical applications, multiaxial machines are under development.

5.5.2 Circular machines

- Machine gauge: normally needles per inch, 5–40 npi
- Machine diameter: up to 30 inches. Up to 60 inch diameter machines are now available
- Needle type: latch, (bearded on sinker wheel and loop wheel, some compound needle machines)
- Needle bed type: single, rib, interlock, double cylinder
- Products
  - Hose machines: seamfree hose, tights, industrial use dye bags, knit-de-knit yarns, industrial fabrics
  - Half-hose machines: men's and boy's half-hose, ladies' stockings, children's tights, sports socks
  - Garment blank machines: underwear, T-shirts, jumpers, pullovers, cardigans, dresses, suits, trouser suits, vests, briefs, thermal wear, cleaning cloths, technical fabrics
  - Fabric machines: rolls of fabric with the following end-uses: jackets, ladies'

tops, sports and T-shirts, casual wear, suits, dresses, swimwear, bath robes, dressing gowns, track suits, jogging suits, furnishing, upholstery, automotive and technical fabrics, household fabrics.

5.5.3   Straight-bar machines (fully fashioned machines)

- Machine gauge: normally needles per $1^1/_2$ inch, 9–33 (up to 60 gauge machines have been produced)
- Machine width: from 2–16 section machines – each section up to 36 inches wide (up to 40 section machines have been produced)
- Needle type: bearded or bearded and latch
- Needle bed type: single and rib
- Products: jumpers, pullovers, cardigans, dresses, suits, trouser suits, fully fashioned hose, sports shirts, underwear, thermal wear.

5.6   Warp-knitting machines

5.6.1   Introduction

The first weft-knitting machine was built by William Lee in 1589. In 1775, just under 200 years later, the first warp-knitting machine was invented by Crane, an Englishman. It was a single guide bar machine to make blue and white zig-zag striped silk hosiery and these fabrics were named after Van Dyck, the painter. With the advent of acetate continuous-filament yarns after World Ward I, the first bulk production of tricot fabrics was commenced by British Celanese on German Saupe 2-guide bar, 28 gauge machines. Locknits replaced the single guide bar atlas fabrics for lingerie, the latter being difficult to finish and laddered easily.

From 1950 to 1970, the growth of the warp-knitting industries in the UK and other western countries was phenomenal. The main reasons for this colossal expansion are summarised below (although developments in the various fields mentioned here were taking place concurrently). The state of the art and current developments in tricot and raschel machinery have been summarised below.

Anand also published a review of warp-knitting equipment exhibited by Karl Mayer at ITMA in 1995[4] and ITMA'99.[8]

5.6.1.1   Yarn developments

- The discovery of thermoplastic yarns and their suitability, even in very low linear densities (deniers) and in flat or low-twist form, to be knitted with very low yarn breakage rates on modern high speed tricot and raschel machines
- The extra design scope offered by differential dye yarns
- Improved cover-comfort attained through textured and producer-bulked yarns
- Elastomeric yarns, which have given a tremendous fillip to the raschel power-net industry.

5.6.1.2   Machinery developments

- Higher machine speeds, (up to 3500 cpm)
- Finer gauges (up to 40 needles per inch)

- Wider machines (up to 260 inches)
- Increased number of guide bars (up to 78 guide bars)
- Special attachments such as cut presser, fallplate, swanwarp, etc.
- Some speciality raschel machines such as Co-we-nit and Jacquard machines and, more recently, redesigned full-width weft insertion raschel and tricot machines
- High speed direct-warping machines and electronic yarn inspection equipment during warping
- Electronic stop motions for the knitting machines
- Larger knitting beams and cloth batches
- Modern heat-setting and beam-dyeing machinery
- Electronic warp let-off, electronic patterning, electronic jacquard and electronic fabric take-up mechanisms
- Loop-raised fabrics
- Stable constructions, such as sharkskins, queenscord, etc.
- Various net constructions utilising synthetic yarns
- Mono-, bi-, tri- and multiaxial structures for technical applications
- Three-dimensional and shaped (fashioned) structures for medical and other high technology products.

It is well known that the warp-knitting sector, particularly tricot knitting has grown in step with the expansion of manufactured fibres. In 1956, 17.8 million lbs of regenerated cellulosic and synthetic fibre yarns were warp knitted; the figure reached a staggering 70.6 million lbs in 1968.

In the mid-1970s, the tricot industry suffered a major setback, mainly because of a significant drop in the sale of nylon shirts and sheets, which had been the major products of this sector. It is also true that the boom period of textured polyester double-jersey was also a contributing factor in the sudden and major decline in the sales of tricot products. A change in fashion and the growth in the demand for polyester/cotton woven fabrics for shirting and sheeting was another cause of this decline. The two major manufacturers of warp-knitting equipment, Karl Mayer and Liba, both in West Germany, have been actively engaged in redesigning their machinery in order to recapture some of the lost trade. The compound needle is the major needle used on both tricot and raschel machines, and many specialised versions of warp-knitting machines are now available for producing household and technical products. One of the major developments in warp knitting has been the commercial feasibility of using staple-fibre yarns for a wide range of products. It is also significant to note that the warp-knitting sector has broadened its market base and has expanded into household and technical fabric markets, such as lace, geotextiles, automotive, sportswear and a wide spectrum of surgical and healthcare products. The current and future potential of warp-knitted structures in engineering composite materials has been discussed by Anand.[5]

5.6.2 Tricot and raschel machines

The principal differences between tricot and raschel machines are listed here:

1. Latch needles are generally used in raschel machines, while bearded or compound needle machines are referred to as tricot machines. Compound needle

raschel machines are also now fairly common. The compound needle is the most commonly used needle on warp knitting equipment.

2.   Raschel machines are normally provided with a trick plate, whereas tricot machines use a sinker bar.
3.   In raschel machines the fabric is taken up parallel to the needle stems; in the tricot machines, however, it is taken up at approximately right angles to the needles.
4.   Raschel machines are normally in a coarser gauge; they are also slower compared with tricot machines, because more guide bars are frequently used and they also require a longer and slower needle movement.
5.   Raschel machines are much more versatile in terms of their ability to knit most types of yarns such as staple yarns, and split films, etc. Only continuous-filament yarns can be successfully knitted on most tricot machines.
6.   Generally, warp beams are on the top of the machine on raschel machines; on tricot machines, they are generally at the back of the machine.

A simplified classification of warp-knitting equipment is given in Fig. 5.18; it will be noticed that apparel, household and technical fabrics are produced on modern warp-knitting machinery. It is in fact in the technical applications that the full potential of warp knitting is being exploited. It is virtually possible to produce any product on warp-knitting equipment, but not always most economical.

The simplest warp-knitted structures are illustrated in Fig. 5.19. It can be seen that both closed- and open-loop structures can be produced and there is normally very little difference in the appearance and properties between the two types of loops.

5.6.3   Knitting action of compound needle warp-knitting machine

In Fig. 5.20(a) the sinkers move forward holding the fabric down at the correct level in their throats. The needles and tongues rise with the needle rising faster until the hook of the needle is at its highest position and is open. In Fig. 5.20(b) the guides then swing through to the back of the machine and Fig. 5.20(c) shows the guides shog for the overlap and swing back to the front of the machine.

Figure 5.20(d) shows the needles and the tongues starting to descend, with the tongues descending more slowly thus closing the hooks. The sinkers start to withdraw as the needles descend so that the old loop is landed onto the closed hook and the new loops are secured inside the closed hook.

In Fig. 5.20(e) the needle descends below the sinker belly and the old loop is knocked-over. At this point, the underlap occurs and in Fig. 5.20(f) the sinkers move forward to hold down the fabric before the needles commence their upward rise to form a fresh course.

5.6.4   Knitting action of standard raschel machine

In Fig. 5.21(a) the guide bars are at the front of the machine completing their underlap shog. The web holders move forward to hold the fabric down at the correct level, whilst the needle bar starts to rise from knock-over to form a fresh course.

Figure 5.21(b) shows that the needle bar has risen to its full height and the old

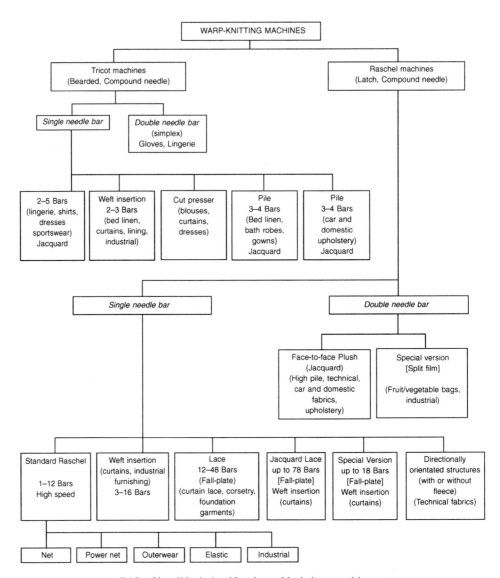

5.18 Simplified classification of knitting machinery.

loops slip down from the latches onto the stems after opening the latches. The latches are prevented from closing by the latch guard. The web holders then start to withdraw to allow the guide bars to form the overlap movement.

In Fig. 5.21(c) the guide bars swing to the back of the machine and then shog for the overlap and in Fig. 5.21(d) the guide bars swing back to the front and the warp threads are laid into the needle hooks. Note: only the front guide bar threads have formed the overlap movement, the middle and back guide bar threads return through the same pair of needles as when they swung towards the back of the machine. This type of movement is called laying-in motion.

In Fig. 5.21(e) the needle bar descends so that the old loops contact and

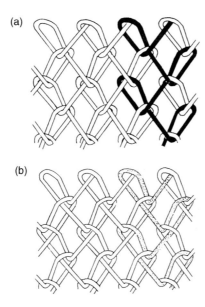

5.19   Single-guide bar warp-knitted fabrics. (a) Closed lap fabric, (b) open lap fabric.

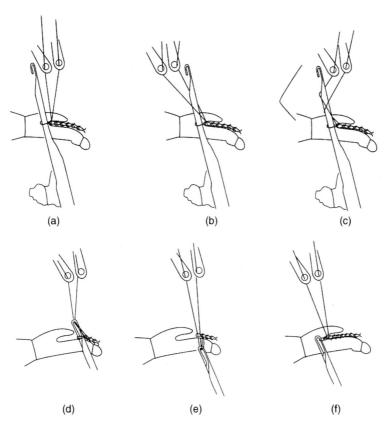

5.20   Knitting action of compound needle warp-knitting machine.

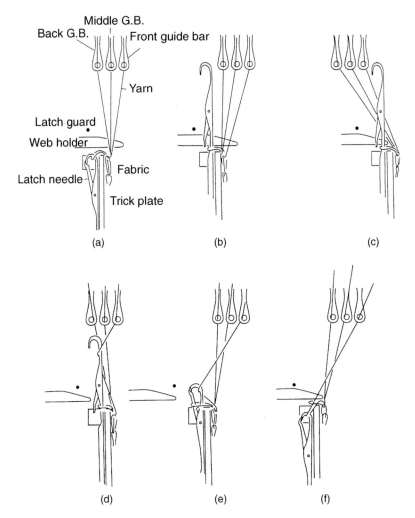

5.21 Knitting action of a standard raschel machine. (a) Start of new course, (b) start of overlap, (c) guide bar swinging motion, (d) return swing after overlap, (e) old loop closing latch, (f) knock-over and underlap movements.

close the latches, trapping the new loops inside. The web holders now start to move forward.

Figure 5.21(f) shows the needle bar continuing to descend, its head passing below the surface of the trick-plate, drawing the new loops through the old loops, which are cast-off, and as the web holders advance over the trick-plate, the underlap shog of the guide bar is commenced.

The knitting action of bearded needle warp-knitting machines has not been given here because in the main the machines likely to be used for technical textile products would use either latch or compound needles. Also the proportion of new bearded needle machines sold has decreased steadily over the years. This is mainly due to the lack of versatility of these machines in terms of the variety of yarns that can be processed and the range of structures that can be normally knitted on them. The displacement curves for the three main types of needle are shown in Fig. 5.22.

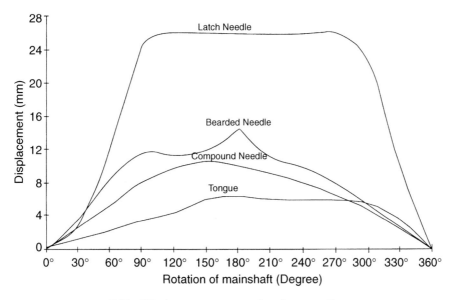

5.22 Displacement curves of various needles.

It is obvious that compound needle machines would operate at faster rates, provided all other factors are similar.

5.7   Warp-knitted structures

5.7.1   Stitch notation

Some of the more popular stitches used in the production of warp-knitted fabrics are given in Fig. 5.23. These stitches, together with the number of guide bars used, a comprehensive range of types and linear densities of yarns available, fancy threading, controlling individual run-ins and run-in ratios, and various finishing techniques are combined and modified to construct an endless variety of fabrics. The lapping movements of the individual guide bars throughout one repeat of the pattern are normally indicated on special paper, called point paper. Each horizontal row of equally spaced dots represents the same needles at successive courses. The spaces between the dots, or needles, are numbered 0, 1, 2, 3, 4, and so on, and show the number of needles transversed by each guide bar. Although three links per course are normally employed, only two are actually required; the third (last link) is only used to effect a smoother movement of the guide bar during the underlap. The first link determines the position of the guide bars at the start of the new course. The second link determines the direction in which the overlap is made. The links, therefore, are grouped together in pairs and the lapping movements at each course are separated by a comma. For instance, the lapping movements shown in Fig. 5.23(c) are interpreted as follows:

- (1-0) is the overlap at the first course
- (0,1) is the underlap at the same course, but made in the opposite direction to the overlap
- (1-2) is the overlap at the second course, and

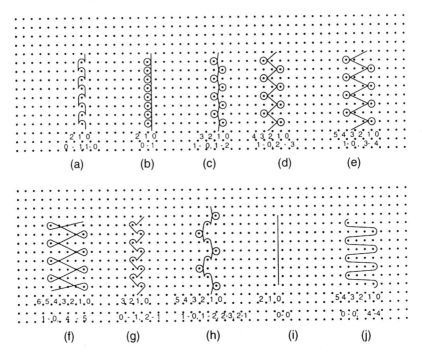

5.23 Stitch notation in tricot knitting. (a) Open pillar, (b) closed pillar, (c) tricot stitch,
(d) 2 × 1 closed lap, (e) 3 × 1 closed lap, (f) 4 × 1 closed lap, (g) open tricot stitch,
(h) two-course atlas, (i) misslapping, (j) laying-in.

- (2,1) is the underlap at the second course, but made in the opposite direction
to the previous underlap.

It will also be observed from Fig. 5.23 that when the underlap is made in the opposite direction to the immediately preceding overlap, a closed loop is formed, but
when the underlap is made in the same direction as the immediately preceding
overlap, or no underlap is made, then an open loop will result.

It is vital to ensure when placing a pattern chain around the drum that the correct
link is placed in contact with the guide bar connecting rod, otherwise the underlap
will occur on the wrong side of the needles, or open loops may be formed instead
of the intended closed loops.

5.7.2   Single-guide bar structures

Although it is possible to knit fabrics using a single fully threaded guide bar, such
fabrics are now almost extinct owing to their poor strength, low cover, lack of
stability and pronounced loop inclination on the face of the fabric. Three examples
of single guide bar structures are shown in Fig. 5.24.

5.7.3   Two-guide bar full-set structures

The use of two guide bars gives a much wider pattern scope than is possible
when using only one, and a large proportion of the fabrics produced in industry

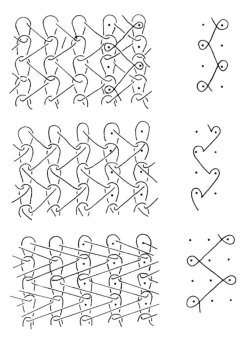

5.24   Single-guide bar structures.

are made with two guide bars. The first group of fabrics to consider are those made with fully threaded guide bars, as many different effects may be obtained by altering the lapping movements and these effects may be increased still further by the use of colour, mixing different yarn, linear densities or using different yarn types, such as yarns with different dyeing characteristics, textured yarns, and so on.

5.7.3.1   Loop plating
With two fully threaded guide bars, each loop in the fabric will contain two threads, one supplied by each bar. The underlaps made by the front guide bar are plated on the back of the fabric and the loops from this bar are plated on the face of the fabric, whereas the loops and the underlaps formed by the back guide bar are sandwiched between those from the front guide bar (see Fig. 5.25). It will be observed from Fig. 5.20(c) that when the guide bars swing through the needles to form the overlap, the ends will be crossed on the needle hook (normally the two bars form overlaps in opposite directions). As the guide bars return to the front of the machine, the threads of the front guide bar are first to strike the needles and are wrapped around the needle hook first, whereas the back guide bar threads are placed later and above those from the front guide bar. If the tensions of the two warp sheets are similar and the heights of the guide bars are correctly adjusted, the front bar loops will always be plated on the face of the fabric. Any coloured thread in the front guide bar will thus appear prominent on both fabric surfaces, an important factor to be remembered in warp-knit fabric designing (see Fig. 5.25 for loop plaiting).

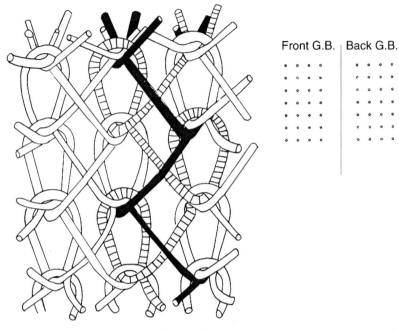

Front G.B. | Back G.B.

5.25   Full tricot.

5.7.3.2   Different structures

The two guide bars may make their underlaps in the same or opposite directions. If made in the same direction, the fabric will show distortion similar to the single bar fabric (see Fig. 5.29(a)) as the loops will be inclined. If, however, the underlaps are made in opposite directions, an equal tension will be imposed in both directions, and loops will be upright.

The structure of the simplest fabric made with two guide bars is shown in Fig. 5.25 and is known as full tricot. The appearance of full tricot may be varied by threading the guide bars with different coloured threads to give vertical stripes of colour.

The most common fabric of all is locknit and its structure and the lapping movements are shown in Fig. 5.26. When correctly knitted, the fabric shows even rows of upright loops on the face of the fabric, and the two needle underlaps on the back of the fabric give a smooth sheen. It has a soft handle and is very suitable for lingerie. If the lapping movements for the bars are reversed to give reverse locknit, the fabric properties are completely changed (Fig. 5.29(e)). The short underlaps will now appear on the back of the fabric and will trap in the longer ones to give a more stable and stiff structure, with far less width shrinkage from the needles than ordinary locknit. The underlaps of the back guide bar may be increased to give even greater stability and opacity with practically no width shrinkage from the needles. An example of this is sharkskin, whose structure and lapping movements are shown in Fig. 5.27. Another stable structure is shown in Fig. 5.28, and is know as queens cord. The long back guide bar underlaps are locked firmly in the body of the fabric by the chain stitches of the front guide bar. Both sharkskin and queenscord structures can be made more stable, heavier and stronger by increasing the back guide underlaps to four or five needle spaces. The vertical chains of loops from the front

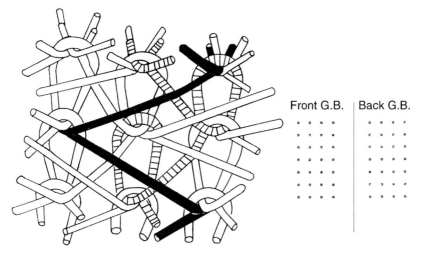

Front G.B.	Back G.B.

5.26 Locknit.

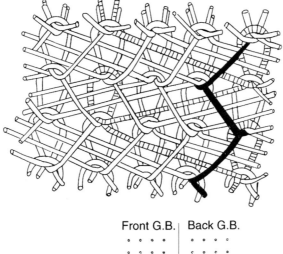

Front G.B.	Back G.B.

5.27 Three-needle sharkskin.

guide bar may be used to give single wale vertical stripes of colour, such as pin stripes in men's suiting.

   If the guide bars making a sharkskin are reversed, that is, if the front bar makes the longer underlaps, the resultant fabric is known as satin which is a lustrous soft fabric similar to the woven satin. Because of the long floats on the back of the fabric,

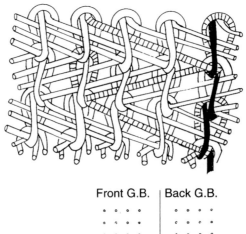

5.28   Three-needle queenscord.

satin laps are used to make loop-raised fabrics. The raising machine is set so that the underlaps are raised into loops without actually breaking any filaments. In order to achieve the maximum raising effect, the two guide bars in a loop-raised fabric are normally made to traverse in the same direction, and open loops may also be used. The lapping movements of three-needle satin are shown in Fig. 5.29(b) and those for a three-needle loop-raised fabric are shown in Fig. 5.29(a). The density and height of pile can be increased by increasing the front guide bar underlaps to four, five or six needle spaces.

Yarns may be introduced into the fabric without actually knitting. Figure 5.30 shows the structure lapping movements and pattern chains of a laid-in fabric. The laid-in thread is trapped between the loop and the subsequent underlap of the guide bar which must be situated in front of the laying-in bar. In order to lay-in a yarn, therefore, that yarn must be threaded in a guide bar to the rear of the guide bar (knitting bar), and it must make no overlaps. Laying-in is a useful device because a laid-in thread never goes round the needle, and therefore very thick or fancy yarns may be introduced into the fabric, such as heavy worsted yarn or metallic threads. Figure 5.31 shows the laid-in thread being trapped in the fabric by the front guide bar threads knitting an open tricot stitch (0-1, 2-1).

5.7.4   Grey specification of a warp-knitted fabric

A complete grey specification of a warp-knitted fabric should include the following details:

1.   gauge of machine in needles per inch
2.   number of guide bars in use

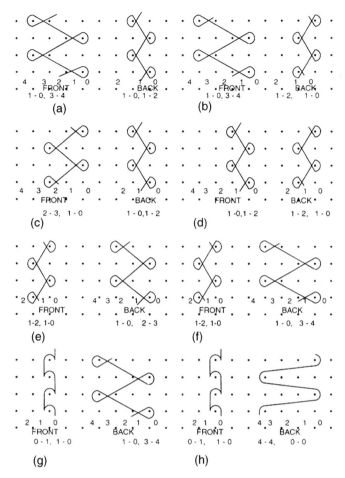

5.29 Some two-guide bar full set structures. (a) Loop raised, (b) satin, (c) locknit, (d) full tricot, (e) reverse locknit, (f) sharkskin, (g) queenscord, (h) laid-in fabric.

3.   number of ends in each warp
4.   types and linear densities of yarns used
5.   run-in per rack for each warp
6.   knitted quality of the fabric in courses per centimetre
7.   order of threading in each guide bar
8.   lapping movements of each guide bar during one repeat of the pattern or details of the pattern wheels or pattern chains
9.   relative lateral positions of the guide bars at a given point in the lapping movements
10. any special knitting instructions.

5.7.5   Fabric quality

The main parameter controlling the quality and properties of a given structure is the run-in per rack, or the amount of yarn fed into the loop. Run-in per rack is

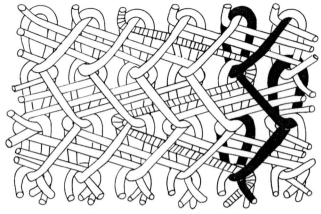

Front G.B. | Back G.B.

5.30 Laid-in structure.

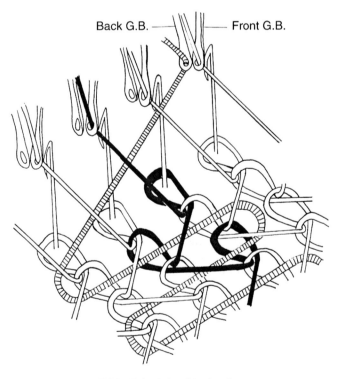

Back G.B. ——— Front G.B.

5.31 Principle of laying-in.

defined as the length of warp fed into the fabric over 480 courses (1 rack = 480 courses). In two-guide bar fabrics, the run-in per rack for each guide bar may be the same or different, depending upon the fabric structure. For example, in full tricot structures (front: 1-2, 1-0 and back: 1-0, 1-2), it is normal to use the same run-in per rack from both beams or 1:1, whereas in three-needle sharkskin fabrics (front: 1-2, 1-0 and back: 1-0, 3-4), the run-in per rack required from the back beam would be more than the front beam say 1:1.66.

The run-in may be altered in two different ways, first by altering the total run-in of the bars, and second by altering the ratio or difference between the bars. Altering the total run-in will affect the finished number of courses per centimetre and hence the area density of the fabric, the stability and the cover, but not the general shape of the loop. Altering the difference between the guide bars will change the balance of the fabric, affect the inclination of the loops and, because it puts more or less strain on the individual yarns, change the strength.

Fabric take-up on the machine is adjusted to attain trouble-free knitting and also to effect ease of finishing.

5.7.6 Tightness factor

The tightness factor K of a knitted fabric is defined as the ratio of the fabric area covered by the yarn to the total fabric area. It is regarded as a measure of looseness or tightness of the structure, and influences dimensions such as the length, width, and thickness and many other fabric characteristics such as area density, opacity, abrasion resistance, modulus, strength and shrinkage.

If the tightness factor of a single-guide bar fabric is defined as in Equation 5.2

$$K = \sqrt{\frac{\text{tex}}{l}} \qquad (5.9)$$

where l is the stitch length, measured in millimetres, and tex is the yarn linear density, then the tightness factor of a two-guide bar, full-set fabric is given by Equation (5.10)

$$K = \frac{\sqrt{\text{tex}_f}}{l_f} + \frac{\sqrt{\text{tex}_b}}{l_b} \qquad (5.10)$$

where suffixes f and b refer to front and back guide bars, and l is the stitch length equal to (run-in/rack)/480 and is measured in millimetres. If the same tex is employed in both bars, then

$$K = \sqrt{\text{tex}}\left(\frac{1}{l_f} + \frac{1}{l_b}\right) \qquad (5.11)$$

For most commercial two-guide bar full-set fabrics $1 \le K \le 2$ with a mean tightness factor value of 1.5.[6]

5.7.7 Area density

The area density of a single-guide bar fabric can be determined from Equation (5.12)

$$\text{Mass of the fabric} = \text{cpc} \times \text{wpc} \times l \times T \times 10^{-2}\,\text{g\,m}^{-2}$$
$$= s \times l \times T \times 10^{-2}\,\text{g\,m}^{-2} \qquad (5.12)$$

where s is the stitch density (cm^{-2}) or (cpc × wpc), l is the stitch length (mm) and T is the yarn tex.

Similarly, the area density of a two-guide bar full-set fabric would be Equation (5.13):

$$\text{Mass of the fabric} = s[(l_f \times T_f) + (l_b \times T_b)] \times 10^{-2} \, \text{gm}^{-2} \tag{5.13}$$

where suffixes f and b refer to the front and back guide bars. If the same tex is used in both guide bars, then the above equation can be written as Equation (5.14):

$$\text{Mass of the fabric} = s \times T \times 10^{-2}(l_f + l_b) \text{gm}^{-2}$$

or

$$= s \times T \times 10^{-2}\left(\frac{\text{Total run-in}}{480}\right) \text{gm}^{-2} \tag{5.14}$$

If the stitch density, that is, the number of loops cm^{-2}, stitch length in millimetres of the individual guide bars and tex of yarns employed in individual beams are known, the fabric area density can be readily obtained using the above equation in any fabric state, that is, on the machine, dry relaxed or fully relaxed.

The geometry and dimensional properties of warp-knitted structures have been studied by a number of researchers including Anand and Burnip.[6]

5.7.8 End-use applications of warp-knitted fabrics

Specification for tricot machines is:

- *Type of needle*: compound or bearded
- *Machine gauge*: from 18 to 40 needles per inch (E18–E40)
- *Machine width*: from 213 to 533 cm (84–210 inches)
- *Machine speed*: from 2000 to 3500 courses per minute (HKS 2 tricot machine operates at 3500 cpm)
- *Number of needle bars*: one or two
- *Number of guide bars*: from two to eight
- *Products*: lingerie, shirts, ladies' and gents' outerwear, leisurewear, sportswear, swimwear, car seat covers, upholstery, technical fabrics, bed linen, towelling, lining, nets, footwear fabrics, medical textiles.

Specification for raschel machines is:

- *Type of needle*: latch or compound
- *Machine gauge*: from 12 to 32 needles per inch (E12–E32).
- *Machine width*: from 191 to 533 cm (75–210 inches)
- *Machine speed*: from 500 to 2000 courses per minute
- *Number of needle bars*: one or two
- *Number of guide bars*: from two to seventy-eight
- *Products*: marquisettes, curtains, foundation garments, nets, fishing nets, sports nets, technical fabrics, curtain lace, power nets, tablecloths, bed covers, elastic bandages, cleaning cloths, upholstery, drapes, velvets, carpets, ladies' underwear, fruit and vegetable bags, geotextiles, medical textiles.

References

1. s c ANAND, 'ITMA '95 review of circular knitting machines and accessories', *Asian Textile J.*, February 1996, 49.
2. s c ANAND, '*Contributions of knitting to current and future developments in technical textiles*', Inaugural Conference of Technical Textiles Group, The Textile Institute, Manchester, 23–24 May, 1988.
3. s c ANAND, 'Knitting's contribution to developments in medical textiles', *Textile Technol. Int.*, 1994, 219.
4. s c ANAND, 'Fine array of warp knitting machines from Karl Mayer', *Asian Textile J.*, April 1996, 51.
5. s c ANAND, 'Warp knitted structures in composites', *ECCM-7 Proceedings*, Volume 2, Woodhead Publishing, Cambridge, UK, 14–16 May 1996, p. 407.
6. s c ANAND and m s BURNIP, 'Warp–knit construction', *Textile Asia*, September 1981, 65.
7. s c ANAND, 'Speciality knitting equipment at ITMA'99', *Asian Textile J.*, December 1999, 49.
8. s c ANAND, 'Karl Mayer warp knitting equipment at ITMA'99', *Asian Textile J.*, September 1999, 49.

6

Technical fabric structures – 3. Nonwoven fabrics

Philip A Smith

Department of Textile Industries, University of Leeds, Leeds LS2 9JT, UK

6.1 Introduction

It is an unfortunate fact that there is no internationally agreed definition of non-wovens, in spite of the fact that the International Standards Organization published a definition in 1988 (ISO 9092:1988). Many countries, particularly those that have played an active part in the development of nonwovens, still prefer their own national definition, which is generally wider in its scope than the very narrow definition of ISO 9092.

As it is essential to be clear on the subject matter to be included in this chapter, I have decided to use the definition of the American Society for Testing Materials (ASTM D 1117-80). This definition is as follows: 'A nonwoven is a textile structure produced by the bonding or interlocking of fibres, or both, accomplished by mechanical, chemical, thermal or solvent means and combinations thereof. The term does not include paper or fabrics that are woven, knitted or tufted.' It has to be admitted that this definition is not very precise, but it has been chosen because it includes many important fabrics which most people regard as nonwovens, but which are excluded by ISO 9092. Nonwovens are still increasing in importance; production is increasing at the rate of 11% per annum.

One of the major advantages of nonwoven manufacture is that it is generally done in one continuous process directly from the raw material to the finished fabric, although there are some exceptions to this. This naturally means that the labour cost of manufacture is low, because there is no need for material handling as there is in older textile processes. In spite of this mass-production approach, the nonwovens industry can produce a very wide range of fabric properties from open waddings suitable for insulation containing only 2–3% fibres by volume to stiff reinforcing fabrics where the fibre content may be over 80% by volume. How is this wide range of properties produced? All nonwoven processes can be divided into two stages, the preparation of the fibres into a form suitable for bonding and the bonding process itself. There are a number of different ways of fibre processing, each producing its own particular characteristic in the final fabric. Equally there are a number of different bonding methods which have an even bigger effect on the finished fabric properties. Almost all the fibre processing methods can be combined with all the bonding methods, so that the range of different possible manufacturing lines is enormous, allowing a great range of final properties.

However, this does raise a difficulty in describing the nonwoven process. We know that the process is essentially a continuous one in which the fibre processing and bonding take place in two machines tightly linked together, but it is impossible to describe the combined machines together owing to the wide number of machine combinations that are possible. Instead it is necessary to explain the methods of fibre processing and the methods of bonding separately.

In fibre processing it is common to make first a thin layer of fibre called a web and then to lay several webs on top of each other to form a batt, which goes directly to bonding. The words web and batt are explained by the previous sentence, but there are cases where it is difficult to decide if a fibre layer is a web or a batt. Nevertheless the first stage of nonwoven processing is normally called batt production.

6.2   Methods of batt production using carding machines

The principles of carding and the types of carding machines have already been discussed in Chapter 3. The machines in the nonwoven industry use identical principles and are quite similar but there are some differences. In particular in the process of yarn manufacture there are opportunities for further opening and for improving the levelness of the product after the carding stage, but in nonwoven manufacture there is no further opening at all and very limited improvements in levelness are possible. It therefore follows that the opening and blending stages before carding must be carried out more intensively in a nonwoven plant and the card should be designed to achieve more opening, for instance by including one more cylinder, though it must be admitted that many nonwoven manufacturers do not follow this maxim.

Theoretically either short-staple revolving flat cards or long-staple roller cards could be used, the short-staple cards having the advantages of high production and high opening power, especially if this is expressed per unit of floor space occupied. However, the short-staple cards are very narrow, whereas long-staple cards can be many times wider, making them much more suitable for nonwoven manufacture, particularly since nonwoven fabrics are required to be wider and wider for many end-uses. Hence a nonwoven installation of this type will usually consist of automatic fibre blending and opening feeding automatically to one or more wide long-staple cards. The cards will usually have some form of autoleveller to control the mass per unit area of the output web.

6.2.1   Parallel laying

The mass per unit area of card web is normally too low to be used directly in a nonwoven. Additionally the uniformity can be increased by laying several card webs over each other to form the batt. The simplest and cheapest way of doing this is by parallel laying. Figure 6.1 shows three cards raised slightly above the floor to allow a long conveyor lattice to pass underneath. The webs from each card fall onto the

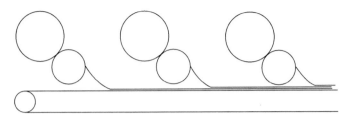

6.1   Parallel laying.

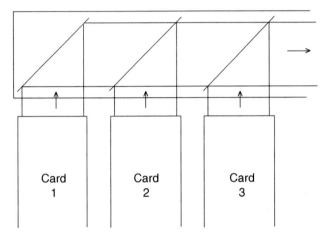

6.2   Alternative layout for parallel laying.

lattice forming a batt with three times the mass per unit area. If the cards are longer this method becomes unwieldy and instead the cards are placed side-by-side as in Figure 6.2.

The card webs are turned through a right angle by a guide at 45°, but the batt produced by this method is identical in all respects to the previous method. It is important to recognise that it is not cross laid, as in Section 6.2.2, in spite of the similarities between the layouts.

In any card web there is a marked tendency for the fibres to lie along the web rather than across it. Since in parallel laying all the card webs are parallel to each other (and to the batt), it follows that most of the fibres will lie along the batt and very few across it. At this stage it is important to introduce two terms used widely in nonwovens. The direction along the batt is called the 'machine direction' and the direction at right angles is called the 'cross direction'. Whatever method of bonding is used it is found that the bonds are weaker than the fibres. A tensile test on a bonded parallel-laid material in the machine direction will depend mainly on the fibre strength, whereas in the cross direction it will depend more on the bond strength. The effect of these facts on the relative fibre frequency and on the directional strength of a typical parallel-laid fabric in various directions is shown in Figure 6.3.

The weakness of the fabric in the cross direction has a profound effect on possible uses of the fabric. Briefly it can be used when strength is not required in either

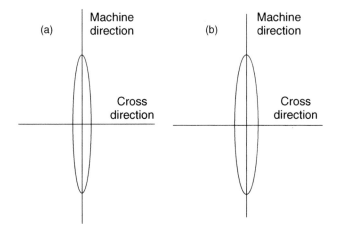

6.3 Polar diagrams showing (a) the relative frequency of fibres lying in various directions and (b) the relative strength in various directions for a parallel-laid fabric.

direction, for example, a filter fabric which is completely supported or as a wiping cloth. It is especially useful when high strength is required in one direction but strength in the other direction is not important, but examples of this are rare; the one usually quoted is a narrow tape cut in the machine direction and used mainly for medical purposes.

This situation has been altered by the advent of cards designed especially for the nonwoven industry. These cards are given a randomising doffer, which as its name implies, makes the fibre directions more random, together with 'scrambling rollers' that condense the card web in length, having the effect of buckling those fibres lying in the machine direction and forming segments lying in the cross direction. By using these two techniques together it is possible to bring the strength ratio of parallel-laid fabric down from the normal 10 or 20:1 to 1.5:1, which is about as isotropic as any nonwoven. It may also be worth mentioning that similar carding techniques have been expanded even further to make a card that produces a really thick web, making laying unnecessary. However, all parallel-lay processes suffer from a further fundamental problem; the width of the final fabric cannot be wider than the card web, while current trends demand wider and wider fabrics.

6.2.2 Cross laying

When cross laying, the card (or cards) are placed at right angles to the main conveyor just as in Fig. 6.2, but in this case the card web is traversed backwards and forwards across the main conveyor, which itself is moving. The result is a zig-zag as shown in Fig. 6.4.

Usually the conveyor B is moving only slowly so that many layers of card web are built up, as shown in the diagram. The thickness of the card web is very small in comparison with the completed batt, so that the zig-zag marks, which appear so prominent in the figure, do not usually show much. There are two major problems with cross layers; one is that they tend to lay the batt heavier at the edges than in the middle. This fault can be corrected by running the traversing mechanism rather slower in the centre and more rapidly at the edges, with a very rapid change

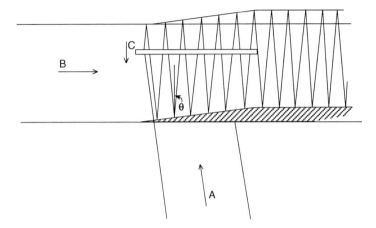

6.4   Cross laying. A, card web; B, main conveyor; C, traverse mechanism; θ, angle of cross laying.

of direction at the edge. The other problem is trying to match the input speed of the cross layer with the card web speed. For various reasons the input speed of the cross layer is limited and the speed of the card web has to be reduced to match it. Because for economic reasons the card is forced to run at maximum production, the card web at the lower speed is thicker and the cross-laying marks discussed above will tend to show more. In spite of these problems, cross layers are used much more frequently than parallel layers.

The diagram in Fig. 6.4 showing webs crossing at an angle seems to imply that the batt will be fairly isotropic. However, this is not so because the cross-laying angle (θ in Fig. 6.4) is normally less than 10°, so that the great majority of fibres lie in or near the cross direction. Cross-laid fabrics are consequently very strong in the cross direction and weak in the machine direction. In many cases this may not matter, because cross-laid fabrics are often quite heavy and may not require much strength, but in many other cases a more isotropic batt is required. The obvious solution is to combine parallel laying and cross laying together; this is done very occasionally but it is uncommon because it combines the limitations of both systems, that is, the relatively narrow width of parallel laying and the slow output speed of the cross laying. The common solution is to stretch the batt in the machine direction as it exits from the cross layer. Various machines are available for doing this; the important criterion is that the stretching should be even, otherwise it could create thick and thin places in the batt. Cross laying, with or without stretching, is much more popular than parallel laying and is probably the most widely used system of all.

6.3   Air laying

The air-laying method produces the final batt in one stage without first making a lighter weight web. It is also capable of running at high production speeds but is similar to the parallel-lay method in that the width of the final batt is the same as the width of the air-laying machine, usually in the range of 3–4 m.

The degree of fibre opening available in an air-lay machine varies from one manufacturer to another, but in all cases it is very much lower than in a card. As a consequence of this more fibre opening should be used prior to air laying and the fibres used should be capable of being more easily opened, otherwise the final batt would show clumps of inadequately opened fibre. In the past the desire for really good fibre opening (which is needed for lightweight batts) led to a process consisting of carding, cross laying, then feeding the cross-laid batt to an air-laying machine. The only purpose of the air-lay machine in this example was to obtain the desired machine:cross-direction strength ratio, but it is a very expensive way compared to the stretching device discussed above and such a process would not be installed today.

The diagram in Fig. 6.5 shows the principle of an air-lay machine, although the actual machines may vary considerably from this outline. Opened fibre from the opening/blending section is fed into the back of hopper A, which delivers a uniform sheet of fibres to the feed rollers. The fibre is then taken by the toothed roller B, which is revolving at high speed. There may or may not be worker and stripper rollers set to the roller B to improve the opening power. A strong air stream C dislodges the fibres from the surface of roller B and carries them onto the permeable conveyor on which the batt is formed. The stripping rail E prevents fibre from recirculating round the cylinder B. The air flow at D helps the fibre to stabilise in the formation zone.

Figure 6.6 shows the formation zone in more detail. This shows that the fibres fall onto an inclined plane, and that the angle between this plane and the plane of the fabric depends on the width of the formation zone, w, and the thickness of the

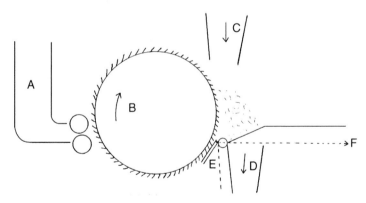

6.5   Principle of air-lay machine. A, Hopper/chute feed; B, opening roller; C, air flow from fan; D, suction to fan; E, stripping rail; F, batt conveyor.

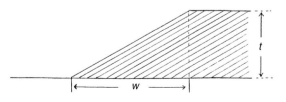

6.6   Close-up of the formation zone.

batt, t. It is possible when making thick fabrics to reduce the width, w, so that the fibres lie at a substantial angle to the plane of the batt. It is claimed that for this reason air-laid fabrics show better recovery from compression compared with cross-laid fabrics. It is often thought because the fibres are deposited without control from an air stream that they will be random in the plane of the batt. This belief is so widespread that air-laid fabrics are frequently called 'random-laid' fabrics. The use of this term should be strongly discouraged, because it gives a false impression. In fact air-laid fabrics can have strength ratios as high as 2.5:1, which is far from random. It is thought that the increase in the machine direction strength is caused by the movement of the conveyor; a fibre in the air-stream approaching the conveyor will tend to be pulled into the machine direction as it lands on the moving conveyor.

The parallel-laid, cross-laid and air-laid methods discussed so far, are collectively known as dry-laid processes. They are the most similar to traditional textile processing and currently account for slightly less than half the total nonwoven production.

6.4   Wet laying

The wet-laid process is a development from papermaking that was undertaken because the production speeds of papermaking are very high compared with textile production. Textile fibres are cut very short by textile standards (6–20 mm), but at the same time these are very long in comparison with wood pulp, the usual raw material for paper. The fibres are dispersed into water; the rate of dilution has to be great enough to prevent the fibres aggregating. The required dilution rate turns out to be roughly ten times that required for paper, which means that only specialised forms of paper machines can be used, known as inclined-wire machines. In fact most frequently a blend of textile fibres together with wood pulp is used as the raw material, not only reducing the necessary dilution rate but also leading to a big reduction in the cost of the raw material. It is now possible to appreciate one of the problems of defining 'nonwoven'. It has been agreed that a material containing 50% textile fibre and 50% wood pulp is a nonwoven, but any further increase in the wood pulp content results in a fibre-reinforced paper. A great many products use exactly 50% wood pulp. Wet-laid nonwovens represent about 10% of the total market, but this percentage is tending to decline. They are used widely in disposable products, for example in hospitals as drapes, gowns, sometimes as sheets, as one-use filters, and as coverstock in disposable nappies.

6.5   Dry laying wood pulp

The paper industry has attempted for many years to develop a dry paper process because of the problems associated with the normal wet process, that is, the removal of very large volumes of water by suction, by squeezing and finally by evaporation. Now a dry process has been developed using wood pulp and either latex binders or thermoplastic fibres or powders to bond the wood pulp together to replace hydrogen bonding which is no longer possible. Owing to the similarity of both the bonding methods and some of the properties to those of nonwovens, these products are being referred to as nonwovens in some areas, although it is clear from the definition above they do not pass the percentage wood pulp criterion. Hence although the dry-laid paper process cannot be regarded as a nonwoven process at present, it is very likely that the process will be modified to accept textile fibres and will become very important in nonwovens in the future. It is for this reason that it has been included here.

6.6   Spun laying

Spun laying includes extrusion of the filaments from the polymer raw material, drawing the filaments and laying them into a batt. As laying and bonding are normally continuous, this process represents the shortest possible textile route from polymer to fabric in one stage. In addition to this the spun-laid process has been made more versatile. When first introduced only large, very expensive machines with large production capabilities were available, but much smaller and relatively inexpensive machines have been developed, permitting the smaller nonwoven producers to use the spun-laid route. Further developments have made it possible to produce microfibres on spun-laid machines giving the advantages of better filament distribution, smaller pores between the fibres for better filtration, softer feel and also the possibility of making lighter-weight fabrics. For these reasons spun-laid production is increasing more rapidly than any other nonwoven process.

Spun laying starts with extrusion. Virtually all commercial machines use thermoplastic polymers and melt extrusion. Polyester and polypropylene are by far the most common, but polyamide and polyethylene can also be used. The polymer chips are fed continuously to a screw extruder which delivers the liquid polymer to a metering pump and thence to a bank of spinnerets, or alternatively to a rectangular spinneret plate, each containing a very large number of holes. The liquid polymer pumped through each hole is cooled rapidly to the solid state but some stretching or drawing in the liquid state will inevitably take place. Up to this stage the technology is similar to the fibre or filament extrusion described in Chapter 2, except that the speeds and throughputs are higher, but here the technologies tend to divide. In spun laying the most common form of drawing the filaments to obtain the correct modulus is air drawing, in which a high velocity air stream is blown past the filaments moving down a long tube, the conditions of air velocity and tube length being chosen so that sufficient tension is developed in the filaments to cause drawing to take place. In some cases air drawing is not adequate and roller drawing has to be used as in normal textile extrusion, but roller drawing is more complex and tends to slow the process, so that air drawing is preferred.

The laying of the drawn filaments must satisfy two criteria; the batt must be as even as possible in mass per unit area at all levels of area, and the distribution of filament orientations must be as desired, which may not be isotropic. Taking the regularity criterion first, the air tubes must direct the filaments onto the conveyor belt in such a way that an even distribution is possible. However, this in itself is not sufficient because the filaments can form agglomerations that make 'strings' or 'ropes' which can be clearly seen in the final fabric. A number of methods have been suggested to prevent this, for instance, charging the spinneret so that the filaments become charged and repel one another or blowing the filaments from the air tubes against a baffle plate, which tends to break up any agglomerations. With regard to

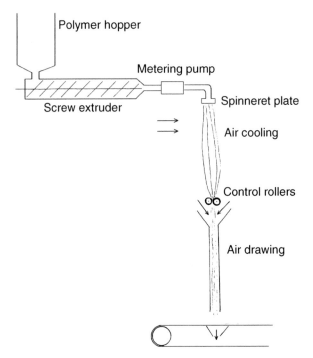

6.7 Diagram of a spun-laid process.

the filament orientation, in the absence of any positive traversing of the filaments in the cross direction, only very small random movements would take place. However, the movement of the conveyor makes a very strong machine direction orientation; thus the fabric would have a very strong machine-direction orientation. Cross-direction movement can most easily be applied by oscillating the air tubes backwards and forwards. By controlling the speed and amplitude of this oscillation it is possible to control the degree of cross-direction orientation. A simplified diagram of a spun-laid plant is shown in Fig. 6.7.

6.7   Flash spinning

Flash spinning is a specialised technique for producing very fine fibres without the need to make very fine spinneret holes. It is used in making only two fabric types. Flash spinning depends on the fact that the intramolecular bonds in polyethylene and polypropylene are much weaker than similar bonds in polyester and polyamide. When a thin sheet of polyolefin is very highly drawn (10 or 12 to 1) the molecules align in the direction of drawing giving high strength in that direction, but across the drawing direction the strength is based only on the intramolecular bonds and so is very low, so low that any mechanical action such as bending, twisting or abrasion causes the sheet to split. Depending on the original thickness of the sheet and the amount of splitting, the result may be fibre-like but with a rectangular cross-section. The process is known as fibrillation.

In flash spinning the polymer is dissolved in a solvent and is extruded as a sheet at a suitable temperature so that when the pressure falls on leaving the extruder the solvent boils suddenly. This blows the polymer sheet into a mass of bubbles with a large surface area and consequently with very low wall thickness. Subsequent drawing of this sheet, followed by mechanical fibrillation, results in a fibre network of very fine fibres joined together at intervals according to the method of production. This material is then laid to obtain the desired mass per unit area and directional strength.

Flash-spun material is only bonded in two ways, so that it seems sensible to discuss both the bonding and the products at this juncture. One method (discussed later in Section 6.10.3) involves melting the fibres under high pressure so that virtually all fibres adhere along the whole of their length and the fabric is almost solid with very little air space. This method of construction makes a very stiff material with high tensile and tear strengths. It is mainly used in competition with paper for making tough waterproof envelopes for mailing important documents and for making banknotes. The fact that the original fibres were very fine means that the material is very smooth and can be used for handwriting or printing. The alternative method of bonding is the same, that is, heat and pressure (see Section 6.10.4) but is only applied to small areas, say 1 mm square, leaving larger areas, say 4 mm square, completely unbonded. The bonded areas normally form a square or diagonal pattern. This material, known as Tyvek, has a lower tensile strength than the fully bonded fabric, but has good strength and is flexible enough to be used for clothing. Because the fine fibres leave very small pores in the fabric, it is not only waterproof but is also resistant to many other liquids with surface tensions lower than water. The presence of the pores means that the fabric is permeable to water vapour and so is comfortable to wear. Tyvek is used principally for protective clothing in the chemical, nuclear and oil industries, probably as protection for the armed forces and certainly in many industries not requiring such good protection but where it is found to be convenient. The garments can be produced so cheaply that they are usually regarded as disposable.

6.8   Melt blown

The process of melt blowing is another method of producing very fine fibres at high production rates without the use of fine spinnerets. Figure 6.8 shows that the polymer is melted and extruded in the normal way but through relatively large holes.

As the polymer leaves the extrusion holes it is hit by a high speed stream of hot air at or even above its melting point, which breaks up the flow and stretches the many filaments until they are very fine. At a later stage, cold air mixes with the hot and the polymer solidifies. Depending on the air velocity after this point a certain amount of aerodynamic drawing may take place but this is by no means as satisfactory as in spun laying and the fibres do not develop much tenacity. At some point the filaments break into staple fibres, but it seems likely that this happens while the polymer is still liquid because if it happened later this would imply that a high tension had been applied to the solid fibre, which would have caused drawing before breaking. The fine staple fibres produced in this way are collected into a batt on a permeable conveyor as in air laying (Section 6.3) and spun laying (Section 6.6). The big difference is that in melt blowing the fibres are extremely fine so that there are many more fibre-to-fibre contacts and the batt has greater integrity.

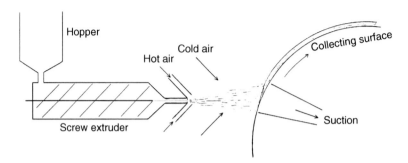

6.8   Diagram of melt-blown equipment.

For many end-uses no form of bonding is used and the material is not non-woven, but simply a batt of loose fibres. Such uses include ultrafine filters for air conditioning and personal face masks, oil-spill absorbents and personal hygiene products. In other cases the melt-blown batt may be laminated to another nonwoven, especially a spun-laid one or the melt-blown batt itself may be bonded but the method must be chosen carefully to avoid spoiling the openness of the very fine fibres. In the bonded or laminated form the fabric can be used for breathable protective clothing in hospitals, agriculture and industry, as battery separators, industrial wipes and clothing interlinings with good insulation properties. Melt blowing started to develop rapidly in about 1975, although the process was known before 1950. It is continuing to grow at about 10% per annum.

6.9   Chemical bonding

Chemical bonding involves treating either the complete batt or alternatively isolated portions of the batt with a bonding agent with the intention of sticking the fibres together. Although many different bonding agents could be used, the modern industry uses only synthetic lattices, of which acrylic latex represents at least half and styrene–butadiene latex and vinyl acetate latex roughly a quarter each. When the bonding agent is applied it is essential that it wets the fibres, otherwise poor adhesion will be achieved. Most lattices already contain a surfactant to disperse the polymer particles, but in some cases additional surfactant may be needed to aid wetting. The next stage is to dry the latex by evaporating the aqueous component and leaving the polymer particles together with any additives on and between the fibres. During this stage the surface tension of the water pulls the binder particles together forming a film over the fibres and a rather thicker film over the fibre intersections. Smaller binder particles will form a more effective film than larger particles, other things being equal. The final stage is curing and in this phase the batt is brought up to a higher temperature than for the drying. The purpose of curing is to develop crosslinks both inside and between the polymer particles and so to develop good cohesive strength in the binder film. Typical curing conditions are 120–140 °C for 2–4 min.

6.9.1   Saturation bonding
Although the principles discussed above apply to all forms of chemical bonding whatever the method of binder application, this latter factor has a profound

influence on the properties of the nonwoven material. As implied by the name, saturation bonding wets the whole batt with bonding agents, so that all fibres are covered in a film of binder. It is clear that saturation bonding would not be suitable for flash-spun or melt-blown batts, since it would impair the fine fibres, but it is inherently suitable for all other batts, although other methods of bonding are generally preferred for spun-laid batts.

To saturate the batt, it is carried under the surface of the bonding agent. In most cases the batt is very open and weak and care is needed to avoid distortion. The action of the liquid greatly reduces the thickness of the batt and the thickness is further reduced by the squeeze rollers which follow. Hence saturation bonded fabrics are generally compact and relatively thin. However, as in most technologies, there is an exception to this rule; if coarse fibres containing a lot of crimp are used they can spring back after being crushed by the squeeze rollers and produce a thick open fabric, but this is very rare.

The drying is often done in a dryer designed basically for woven fabric, modified by having a permeable conveyor to support and transport the nonwoven through the machine. Hot air is blown against the top and bottom surfaces to cause drying, the top air pressure being slightly greater than the bottom pressure to press and control the nonwoven against the conveyor. Because the air only penetrates the immediate surfaces of the nonwoven, drying is confined to these areas and the central layers of the nonwoven remain wet. The result is that the liquid wicks from the wet areas to the dry ones, unfortunately carrying the suspended binder particles with the water. This is the cause of the problem known as binder migration; under adverse conditions virtually all the binder can be found in the surface layers and the central layers contain no binder at all, leaving them very weak. Such a fabric can easily split into two layers, a problem known as delamination. Fortunately a number of ways have been found to control binder migration, several of them being rather complicated. Only one method will be mentioned here, through-drying. In this case only one air stream is used to blow down and through the fabric. Drying conditions are then almost the same in all parts of the fabric and no binder migration takes place. However, the air pressure may exert a significant force on the nonwoven, pressing it against the conveyor so hard that it may be imprinted with the pattern of the conveyor.

In contrast the curing stage is simpler; it should be done in a separate compartment in order to achieve the correct temperature. However, it is quite common for curing to be done in the final part of the dryer in order to keep down machinery costs.

Many of the physical properties of saturation-bonded fabric derive from the fact that all fibres are covered with a film of binder. First, the fabric feels like the binder and not the fibres it is made from. However, in some cases this can be an advantage, because by using either a hydrophobic or a hydrophilic binder the reaction of the fabric to water can be changed regardless of which fibres are used.

The mechanical properties can be explained from a model of a network of fibres bonded together at close intervals. The fabric cannot stretch without the fibres also stretching by a similar amount. Hence the fabric modulus is of the order of the fibre modulus, that is, extremely high. A high modulus in a spatially uniform material means that it will be stiff, which explains why saturation-bonded fabrics are very stiff relative to conventional textiles. At the same time tensile strength is low, because the bonds tend to break before most fibres break. Efforts to make the fabrics more flexible include using a more flexible binder and using a lower percentage of binder, but both of these reduce the tensile strength; in fact the ratio of fabric modulus to tensile strength remains remarkably constant.

One of the major uses of saturation bonded fabric turns this apparent disadvantage into an advantage. Interlining fabric for textile clothing is required to be stiff and to have a high modulus. Other uses are as some types of filter fabric, in some coverstock and in wiping cloths.

6.9.2   Foam bonding
It can be inferred above that one of the problems of saturation bonding is that too much water is used. This not only increases the cost of drying but also increases the risk of binder migration. Application of chemicals as a foam was developed not only for nonwovens but also for the dyeing and finishing industry as a means of using less water during application. The binder solution and a measured volume of air are passed continuously through a driven turbine which beats the two components into a consistent foam. The foam is then delivered to the horizontal nip of the impregnating roller, as shown in Fig. 6.9. The foam delivery has to be traversed because the foam does not flow easily; end plates, which cannot be shown in the cross-section diagram prevent the foam from running out of the gap at the end of the rollers.

The rollers serve the dual purpose of metering the amount of foam applied and also of squeezing the foam into the batt. If the set-up shown in Fig. 6.9 does not give adequate foam penetration, then the batt can be entered vertically and the foam can be applied from both sides.

Foam application should only be thought of as an alternative method of saturation bonding. The properties and uses of the fabrics are identical.

6.9.3   Print bonding
Print bonding involves applying the same types of binder to the batt but the application is to limited areas and in a set pattern. The binder does not penetrate well into the dry batt so the batt is first saturated with water and then printed with either a printing roller or a rotary screen printer. The final properties of the fabric depend

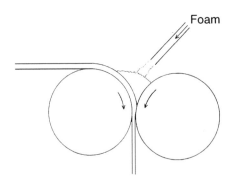

6.9   Foam impregnation.

vitally on the printed/unprinted area ratio, which could be changed significantly if the binder migrated sideways from the printed area. To prevent this the binder formulation must contain some thickener.

Print-bonded fabrics are much softer in feel and also much more flexible owing to the strong effect of the free fibres in the unbonded areas. They are also significantly weaker than saturation-bonded fabrics owing to the fibres slipping in unbonded areas, but knowing the fibre length and the fibre orientation distribution it is possible to design a print pattern which will minimise the strength loss.

Print-bonded fabrics tend to be used in applications where the textile-like handle is an advantage. Examples are disposable/protective clothing, coverstock and wiping cloths, particularly domestic ones (washing-up and dusting).

6.9.4 Spray bonding

Similar latex binders may also be applied by spraying, using spray guns similar to those used in painting, which may be either operated by compressed air or be airless. On the first passage the spray penetrates about 5 mm into the top surface, then the batt is turned over for a spray application on the lower surface. Each spray application reduces the thickness of the batt slightly, but it is still left substantially lofty; the drying and curing stage also causes some small dimensional changes. The final product is a thick, open and lofty fabric used widely as the filling in quilted fabrics, for duvets, for some upholstery and also for some types of filter media.

6.10   Thermal bonding

Thermal bonding is increasingly used at the expense of chemical bonding for a number of reasons. Thermal bonding can be run at high speed, whereas the speed of chemical bonding is limited by the drying and curing stage. Thermal bonding takes up little space compared with drying and curing ovens. Also thermal bonding requires less heat compared with the heat required to evaporate water from the binder, so it is more energy efficient.

Thermal bonding can use three types of fibrous raw material, each of which may be suitable in some applications but not in others. First, the fibres may be all of the same type, with the same melting point. This is satisfactory if the heat is applied at localised spots, but if overall bonding is used it is possible that all the fibres will melt into a plastic sheet with little or no value. Second, a blend of fusible fibre with either a fibre with a higher melting point or a non-thermoplastic fibre can be used. This is satisfactory in most conditions except where the fusible fibre melts completely, losing its fibrous nature and causing the batt to collapse in thickness. Finally, the fusible fibre may be a bicomponent fibre, that is, a fibre extruded with a core of high melting point polymer surrounded by a sheath of lower melting point polymer. This is an ideal material to process because the core of the fibre does not melt but supports the sheath in its fibrous state. Thermal bonding is used with all the methods of batt production except the wet-laid method, but it is worth pointing out that the spun-laid process and point bonding (see Section 6.10.4) complement each other so well that they are often thought of as the standard process.

6.10.1   Thermal bonding without pressure

The batt may be processed through a hot air oven with just sufficient air movement to cause the fusible portion to melt. This method is used to produce high loft fabrics; the products are similar to spray-bonded materials except that in this case the bonding is uniform all the way through and there is virtually no limit to the thickness of fabric made. The uses of the thermal-bonded fabric are basically the same as those of a spray-bonded fabric but they would be used in situations where a higher specification is required. One interesting example is in the fuel tanks of Formula 1 cars to limit the rate of fuel loss in the event of a crash.

6.10.2   Thermal bonding with some pressure

This method is basically the same as the previous one, except that as the batt leaves the thermobonding oven it is calendered by two heavy rollers to bring it to the required thickness. The products could be used as insulation, as rather expensive packaging or for filtration.

6.10.3   Thermal bonding with high pressure

The batt is taken between two large heated rollers (calender rollers) that both melt the fusible fibre and compress the batt at the same time. Provided the batt is not too heavy in mass per unit area the heating is very rapid and the process can be carried out at high speed ($300\,m\,min^{-1}$). The design of calender rollers for this purpose has become highly developed; they can be 4–5 m wide and can be heated to give less than 1 °C temperature variation across the rollers. Also the rollers have to be specially designed to produce the same pressure all the way across the rollers, because rollers of this width can bend quite significantly.

The products tend to be dense and heavily bonded, although of course the amount of bonding can be adjusted by varying the percentage of fusible fibre in the blend. Typical properties are high strength, very high modulus and stiffness but good recovery from bending. The main uses are in some geotextiles, stiffeners in some clothing and in shoes, some filtration media and in roofing membranes.

6.10.4   Thermal bonding with point contact

Although it is very strong, the fabric produced with bonding all over the batt (area bonding) is too stiff and non-textile-like for many uses. It is far more common to use point bonding, in which one of the calender rollers is engraved with a pattern that limits the degree of contact between the rollers to roughly 5% of the total area. The bonding is confined to those points where the rollers touch and leaves roughly 95% of the batt unbonded. The area, shape and location of the bonding points are of great importance.

Fabrics made in this way are flexible and relatively soft owing to the unbonded areas. At the same time they maintain reasonable strength, especially in the case of the spun-laid fabrics. These fabrics have many uses, for example, as a substrate for tufted carpets, in geotextiles, as a filtration medium, in protective/disposable clothing, as a substrate for coating, in furniture and home furnishings and as coverstock.

6.10.5   Powder bonding

Thermoplastic powders may be used as an alternative to thermoplastic fibres for bonding in all the methods of thermobonding except for point bonding, where powder in the unbonded areas would be wasted and would drop out in use. Products made by powder bonding seem to be characterised by softness and flexibility but in general they have relatively low strength. Again there is a very wide range of uses covering particularly the high bulk applications, protective apparel and coverstock areas.

6.11   Solvent bonding

This form of bonding is only rarely used but it is interesting from two points of view; first, the solvent can be recycled, so the process is ecologically sound, although whether or not recycling is practical depends on the economics of recovering the solvent. Second, some of the concepts in solvent bonding are both interesting and unique. In one application of the method, a spun-laid polyamide batt is carried through an enclosure containing the solvent gas, NO_2 which softens the skin of the filaments. On leaving the enclosure bonding is completed by cold calender rolls and the solvent is washed from the fabric with water using traditional textile equipment. This is a suitable method of bonding to follow a spun-laid line because the speeds of production can be matched. The other application uses a so-called latent solvent, by which is meant one that is not a solvent at room temperature but becomes a solvent at higher temperatures. This latent solvent is used in conjunction with carding and cross laying and is applied as a liquid before carding. The action of carding spreads the solvent and at the same time the solvent lubricates the fibres during carding. The batt is passed to a hot air oven which first activates the solvent and later evaporates it. The product will normally be high loft, but if less loft is required a compression roller could be used.

6.12   Needlefelting

All the methods of bonding discussed so far have involved adhesion between the fibres; hence they can be referred to collectively as adhesive bonding. The final three methods, needlefelting, hydroentanglement and stitch bonding rely on frictional forces and fibre entanglements, and are known collectively as mechanical bonding.

The basic concept of needlefelting is apparently simple; the batt is led between two stationary plates, the bed and stripper plates, as shown in Fig. 6.10. While between the plates the batt is penetrated by a large number of needles, up to about $4000 \, m^{-1}$ width of the loom. The needles are usually made triangular and have barbs cut into the three edges as shown in Fig. 6.11.

When the needles descend into the batt the barbs catch some fibres and pull them through the other fibres. When the needles return upwards, the loops of fibre formed on the downstroke tend to remain in position, because they are released by the barbs. This downward pressure repeated many times makes the batt much denser, that is, into a needlefelt.

The above description illustrates how simple the concept seems to be. Without going into too much detail it may be interesting to look at some of the complica-

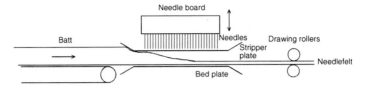

6.10   Diagram of a needleloom.

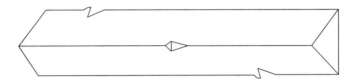

6.11   Diagram of three needle barbs on a section of a needle.

tions. First, the needles can only form vertical loops or 'pegs' of fibre and increase the density of the batt. This alone does not form a strong fabric unless the vertical pegs pass through loops already present in the horizontal plane of the batt. It follows from this that parallel-laid fabric is not very suitable for needling since there are few fibre loops present, so most needling processes are carried out with cross-laid, air-laid and spun-laid batts. Second, the amount of needling is determined partly by the distance the drawing rollers move between each movement of the needleboard, the 'advance', and partly by the number of needles per metre across the loom. If the chosen advance happens to be equal to, or even near the distance between needle rows as shown in Fig. 6.10, then the next row of needles will come down in exactly the same position as the previous row, and so on for all the rows of needles. The result will be a severe needle patterning; to avoid this the distance between each row of needles must be different. Computer programs have been written to calculate the best set of row spacings. Third, if it is necessary to obtain a higher production from a needleloom, is it better to increase the number of needles in the board or to increase the speed of the needleboard? Finally, in trying to decide how to make a particular felt it is necessary to choose how many needle penetrations there will be per unit area, how deep the needles will penetrate and what type of needles should be used from a total of roughly 5000 different types. The variations possible seem to be infinite, making an optimisation process very difficult.

There are several different types of needleloom. Figure 6.10 is called a 'down-punch' because it is pushing the fibres downwards. Similarly an 'up-punch' pushes the fibres upwards. There is some advantage in combining an up-punch with a down-punch when trying to make a dense felt, rather than to punch continually in the same direction. For this reason some looms are made 'double-punch', that is, one board punching down and one board punching up. Although this type of double punch loom is better than unidirectional punching, it has the disadvantage that all the down-punching is completed before the up-punching even starts. The up-punch and the down-punch are far from symmetrical. In order to improve this situation the modern form of double-punch loom has been developed in which both boards penetrate through the same bed and stripper plates. The normal operation is for the boards to penetrate the fabric alternately, one from above and the other from below. To all intents and purposes the two boards are now symmetrical and this appears to give optimum needling. This type of loom can often be adjusted so that both sets of needles enter and leave the batt simultaneously. (Alternate rows of needles have to be taken out of both boards to prevent serious needle damage.) This is not found to be advantageous with long fibres (50 mm upwards) because there is quite a high chance that the same fibre may be pulled upwards by one needle and pushed downwards by another, leading to serious fibre breakage. However, the very short mineral fibres used for high temperature insulation are found to consolidate better using this form of simultaneous needling.

Needlefelts have a high breaking tenacity and also a high tear strength but the modulus is low and the recovery from extension is poor. For these last two reasons any needlefelt which is likely to be subjected to a load has to have some form of reinforcement to control the extension. Needled carpets, for instance, may be impregnated with a chemical binder that gives better dimensional stability and increases the resistance to wear. In other cases thermal bonding may be used for the same object. For heavy duty applications such as filter media and papermakers' felts, yarns are used, either spun yarns or filament yarns. In some cases with a cross-laid batt the threads may run only in the machine direction, but it is also common to use a woven fabric as the foundation and to build up the nonwoven on one or both sides. In both applications the presence of the woven fabric or base fabric reduces the efficiency by restricting the liquid flow; in a few cases 'baseless' fabrics, that is nonwovens without woven supports have been made, but there are still very many situations where a base fabric is essential.

Needlefelts are used widely in gas filtration media and in some wet filtration. The principal advantage is that the nonwoven is practically homogeneous in comparison with a woven fabric so that the whole area of a nonwoven filter can be used for filtration, whereas in a woven fabric the yarns effectively stop the flow, leaving only the spaces between the yarns for filtration. In papermakers' felts the same considerations apply, but in some cases the design of the felt has also to allow for the fact that the felt is acting as an enormous drive belt to drive some parts of the machine. In both gas filtration and papermakers' felts needlefelts hold virtually the whole market. Needlefelts are used in geotextiles, but in view of the low modulus their application is mainly in removing water rather than as reinforcement. If the water is flowing through the fabric this is termed filtration. For instance a simple drain may be formed by lining a trench with a geotextile, partly filling it with graded gravel, drawing the edges of the geotextile together and back-filling the trench as shown in Fig. 6.12. The geotextile will keep the drain open for many years by filtering out fine particles. Less frequently the geotextile is required for drainage, which means water movement in the plane of the fabric. Because of the inevitable contamination of the surface layers, together with the compression of the geotextile caused by the soil pressure, only a limited part of the fabric cross-section is actually available for drainage and more complicated compound structures are frequently used in preference.

Many makers of synthetic leather have taken the view that the structure should be similar to natural leather. In these cases the backing or foundation of the synthetic leather is a needlefelt. The method of production is to include heat-shrinkable fibres which are made to shrink after intense needling in order to make the felt even more dense. After shrinking the felt is impregnated using a binder

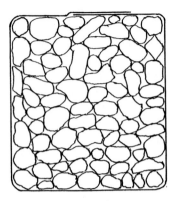

6.12   Simple drain using geotextiles.

that fills the voids in the felt but does not adhere to the fibres, otherwise the felt would become very stiff. The leather is then finished by coating the face side with one or two layers of suitable polymer. Needlefelts are also widely used in home and commercial carpeting; in many cases the carpet may have either a velour or a loop pile surface to improve the appearance. Both velour and loop pile are produced in a separate needling operation using special needles and a special bedplate in the needleloom. Similar materials in a lighter weight are used for car carpeting, head-liners and other decorative features on cars.

6.13   Stitch bonding

The idea of stitch bonding was developed almost exclusively in Czechoslovakia, in the former East Germany and to some extent in Russia, though there was a brief development in Britain. The machines have a number of variants which are best dis-cussed separately; many possible variants have been produced but only a limited number are discussed here for simplicity.

6.13.1   Batt bonded by threads
Stitch bonding uses mainly cross-laid and air-laid batts. The batt is taken into a modification of a warp knitting machine (see Chapter 5) and passes between the needles and the guide bar(s). The needles are strengthened and are specially designed to penetrate the batt on each cycle of the machine. The needles are of the compound type having a tongue controlled by a separate bar. After the needles pass through the batt the needle hooks open and the guide bar laps thread into the hooks of the needles. As the needles withdraw again the needle hooks are closed by the tongues, the old loops knock over the needles and new loops are formed. In this way a form of warp knitting action is carried out with the overlaps on one side of the batt and the underlaps on the other. Generally, as in most warp knitting, con-tinuous filament yarns are used to avoid yarn breakages and stoppages on the machine. Two structures are normally knitted on these machines, pillar (or chain) stitch, or tricot. When knitting chain stitch the same guide laps round the same needle continuously, producing a large number of isolated chains of loops. When

knitting tricot structure, the guide bar shogs one needle space to the left, then one to the right. Single-guide bar structure is called tricot, whereas the two-guide bar structure is often referred to as full tricot.

The nature of this fabric is very textile-like, soft and flexible. At one time it was widely used for curtaining but is now used as a backing fabric for lamination, as covering material for mattresses and beds and as the fabric in training shoes. In deciding whether to use pillar or tricot stitch, both have a similar strength in the machine direction, but in the cross direction the tricot stitch fabric is stronger, owing to the underlaps lying in that direction. A cross-laid web is already stronger in that direction so the advantage is relatively small. The abrasion resistance is the same on the loop or overlap side, but on the underlap side the tricot fabric has significantly better resistance owing to the longer underlaps. However, continuous filament yarn is very expensive relative to the price of the batt, so tricot fabric costs significantly more. The decision then becomes a purely financial one; is it worth paying more for the greater abrasion resistance?

6.13.2 Stitch bonding without threads

In this case the machine is basically the same as in the previous section, but the guide bar(s) are not used. The needle bar moves backwards and forwards as before, pushing the needles through the batt. The main difference is that the timing of the hook closing by the tongues is somewhat delayed, so that the hook of the needle picks up some of the fibre from the batt. These fibres are formed into a loop on the first cycle and on subsequent cycles the newly formed loops are pulled through the previous loops, just as in normal knitting. The final structure is felt-like on one side and like a knitted fabric on the other. The fabric can be used for insulation and as a decorative fabric.

6.13.3 Stitch bonding to produce a pile fabric

To form a pile fabric two guide bars are usually used, two types of warp yarns (pile yarn and sewing yarn) and also a set of pile sinkers, which are narrow strips of metal over which the pile yarn is passed and whose height determines the height of the pile. The pile yarn is not fed into the needle's hook so does not form a loop; it is held in place between the underlap of the sewing yarn and the batt itself. It is clear that this is the most efficient way to treat the pile yarn, since any pile yarn in a loop is effectively wasted. This structure has been used for making towelling with single-sided pile and also for making loop-pile carpeting in Eastern Europe. The structure has not been popular in the West owing to competition with double-sided terry towelling and tufted carpets. Equally it has not been used in technical textiles, but it could be a solution in waiting for the correct problem. Strangely a suitable problem has been proposed. Car seating usually has a polyester face with a foam backing, but this material cannot be recycled, because of the laminated foam. It has been suggested that a polyester nonwoven pile fabric could replace the foam and would be 100% recyclable.

6.13.4 Batt looped through a supporting structure

In this technique the needles pass through the supporting fabric and pick up as much fibre from the batt as possible. Special sinkers are used to push fibre into the needle's hook to increase this pick-up. The fibre pulled through the fabric forms a chain of loops, with loose fibre from the batt on the other surface of the fabric. The fabric is finished by raising, not as one might expect on the loose side of the fabric but instead the loops are raised because this gives a thicker pile. This structure was widely used in Eastern Europe, particularly for artificial fur, but in the West it never broke the competition from silver knitting, which gives a fabric with similar properties. The method could be used for making good quality insulating fabrics.

6.13.5   Laid yarns sewn together with binding threads

Two distinct types of fabric can be made using the same principle. The first is a simulated woven fabric in which the cross-direction yarns are laid many at a time in a process a bit like cross laying. The machine direction yarns, if any are used, are simply unwound into the machine. These two sets of yarns are sewn together using chain stitch if there are only cross-direction threads and tricot stitch if machine-direction threads are present, the underlaps holding the threads down. Although fabric can be made rapidly by this system this turns out to be a situation in which speed is not everything and in fact the system is not usually economically competitive with normal weaving. However, it has one great technical advantage; the machine and cross threads do not interlace but lie straight in the fabric. Consequently the initial modulus of this fabric is very high compared with a woven fabric, which can first extend by straightening out the crimp in the yarn. These fabrics are in demand for making fibre-reinforced plastic using continuous filament glass and similar high modulus fibres or filaments.

The alternative system makes a multidirectional fabric. Again sets of yarns are cross-laid but in this case not in the cross direction but at, say, 45° or 60° to the cross direction. Two sets of yarns at, say, +45° and −45° to the cross direction plus another layer of yarns in the machine direction can be sewn together in the usual way. Again high modulus yarns are used, with the advantage that the directional properties of the fabric can be designed to satisfy the stresses in the component being made.

6.14   Hydroentanglement

The process of hydroentanglement was invented as a means of producing an entanglement similar to that made by a needleloom, but using a lighter weight batt. A successful process was developed during the 1960s by Du Pont and was patented. However, Du Pont decided in the mid-1970s to dedicate the patents to the public domain, which resulted in a rush of new development work in the major industrial countries, Japan, USA, France, Germany and Britain.

As the name implies the process depends on jets of water working at very high pressures through jet orifices with very small diameters. A very fine jet of this sort is liable to break up into droplets, particularly if there is any turbulence in the water passing through the orifice. If droplets are formed the energy in the jet stream will still be roughly the same, but it will spread over a much larger area of batt so that the energy per unit area will be much less. Consequently the design of the jet to avoid turbulence and to produce a needle-like stream of water is critical. The jets are arranged in banks and the batt is passed continuously under the jets held up by a perforated screen which removes most of the water. Exactly what happens to the batt underneath the jets is not known, but it is clear that fibre ends become twisted together or entangled by the turbulence in the water after it has hit the batt. It is also known that the supporting screen is vital to the process; changing the screen with all other variables remaining constant will profoundly alter the fabric produced.

Although the machines have higher throughputs compared with most bonding systems, and particularly compared with a needleloom, they are still very expensive and require a lot of power, which is also expensive. The other considerable problem lies in supplying clean water to the jets at the correct pH and temperature. Large quantities of water are needed, so recycling is necessary, but the water picks up air bubbles, bits of fibre and fibre lubricant/fibre finish in passing through the process and it is necessary to remove everything before recycling. It is said that this filtration process is more difficult than running the rest of the machine.

Fabric uses include wipes, surgeons' gowns, disposable protective clothing and backing fabrics for coating. The wipes produced by hydroentanglement are guaranteed lint free, because it is argued that if a fibre is loose it will be washed away by the jetting process. It is interesting to note that the hydroentanglement process came into being as a process for entangling batts too light for a needleloom, but that the most recent developments are to use higher water pressures (400 bar) and to process heavier fabrics at the lower end of the needleloom range.

Bibliography

P LENNOX-KERR (ed.), *Needlefelted Fabrics*, Textile Trade Press, Manchester, 1972.

G E CUSICK (ed.), Nonwoven Conference, Manchester, University of Manchester Institute of Science and Technology, 1983.

A NEWTON and J E FORD, 'Production and properties of nonwoven fabrics', *Textile Progr.*, 1973 **5**(3) 1–93.

P J COTTERILL, 'Production and properties of stitch-bonded fabrics', *Textile Progr.*, 1975 **7**(2) 101–135.

A T PURDY, 'Developments in non-woven fabrics', *Textile Progr.*, 1980 **12**(4) 1–97.

K PFLIEGEL, 'Nonwovens for use in the geotextile area', *Textile Institute Ind.*, 1981 **19** 178–181.

A KRAUTER and P EHLER, 'Aspects of fibre to binder adhesion in nonwovens', *Textil Praxis Int.*, 1980, No. **10**, 1206–1212; No. **11**, 1325–1328.

ANON, 'Foam padder and foam production for the nonwoven industry', *Chemiefaser/Textilindustrie* (English edition), 1981 4, E.35–36 in English (pp. 336–338 in German).

G E CUSICK and A NEWTON, (eds), 'Nonwovens Conference', Manchester, University of Manchester Institute of Science and Technology, 1988.

G E CUSICK and K L GHANDI, (eds), 'Conference on nonwovens', Huddersfield, University of Huddersfield, 1992.

A J RIGBY and S C ANAND, 'Nonwovens in medical and healthcare products', *Tech. Textiles Int.*, 1996, Part I, Sept. pp. 22–28 Part II, Oct. pp. 24–29.

Index 1999 Congress, Geneva, EDANA (European Disposables and Nonwovens Association), Brussels, 1999.

7

Finishing of technical textiles

Michael E Hall

Department of Textiles, Faculty of Technology, Bolton Institute, Deane Road, Bolton BL3 5AB, UK

7.1 Introduction

The name textile finishing covers an extremely wide range of activities, which are performed on textiles before they reach the final customer. They may be temporary, for example the way bed sheets are pressed before packing, or they may be permanent, as in the case of a flame-retardant tenting fabric. However, all finishing processes are designed to increase the attractiveness or serviceability of the textile product. This could involve such techniques as putting a glaze on an upholstery fabric, which gives it a more attractive appearance, or the production of water-repellent finishes, which improve the in-service performance of a tenting fabric. Thus a further aim of textile finishing may be described as improvement in customer satisfaction, which finishing can bring about. This improvement in the perceived value of a product to the consumer forms the basis of modern ideas on product marketing. Technical textiles are defined as those materials with non-clothing applications. Thus the fashion aspects of textiles will be ignored, although aesthetic aspects of say upholstery and drapes will be covered.

7.2 Finishing processes

The finishing processes that are available can be divided into four main groups, which are:

- **Mechanical processes:** these involve the passage of the material through machines whose mechanical action achieves the desired effects. A heating process, the purpose of which is usually to enhance these desired effects, frequently accompanies this.
  These mechanical finishes, which will be discussed in detail later in this section, are:
  - *Calendering:* compression of the fabric between two heavy rolls to give a flattened, smooth appearance to the surface of the fabric.

- – *Raising:* plucking the fibres from a woven or knitted fabric to give a nap effect on the surface.
- – *Cropping:* cutting the surface hairs from the a fabric to give a smooth appearance, often used on woollen goods where the removal of surface hair by a singeing process is not possible.
- – *Compressive shrinkage:* the mechanical shrinking of the warp fibres in woven fabrics so that shrinkage on washing is reduced to the desired level.
- • **Heat setting:** this is a process for the stabilisation of synthetic fibres so that they do not shrink on heating.
- • **Chemical processes:** these may be described as those processes that involve the application of chemicals to the fabric. The chemicals may perform various functions such as water repellency or flame retardancy, or may be used to modify the handle of a fabric. Chemical finishes are normally applied in the form of an aqueous solution or emulsion and may be applied via a variety of techniques, the main one being the pad mangle, which is illustrated in Fig. 7.1.

  After the padding or the application stage of the chemical finishing the fabric is usually dried to remove the water from the fabric and some form of fixation of the finish is then performed. This commonly takes the form of a baking process, where the fabric is subjected to a high temperature for a short period, which enables the applied chemicals to form a more durable finish on the fabric than would otherwise be achieved.

Surface coating is a most important part of the finishing of technical textiles and as such deserves a separate chapter (see Chapter 8).

7.3   Mechanical finishes

7.3.1   Calendering

Calendering may be defined as the modification of the surface of a fabric by the action of heat and pressure.[1] The finish is obtained by passing the fabric between heated rotating rollers (or bowls as they are frequently called), when both speed of rotation and pressure applied are variable. The surfaces of the rollers can be either smooth or engraved to provide the appropriate finish to the fabric, while the actual construction of the rollers may be varied from hardened chromium-plated steel to elastic thermoplastic rollers.

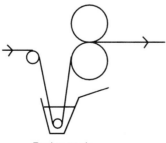

Pad mangle

7.1 Simple pad mangle.

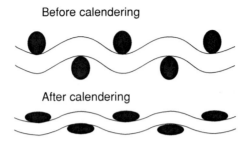

7.2   Flattening effect on fabric of calendering.

7.3.1.1   Effects which may be achieved by calendering

Calendering is done for many purposes but the main effects are:

- smoothing the surface of the fabric
- increasing the fabric lustre
- closing the threads of a woven fabric
- decreasing the air permeability
- increasing the fabric opacity
- improving the handle of a fabric, i.e. softening
- flattening slubs
- obtaining silk-like to high gloss finishes
- surface patterning by embossing
- consolidation of nonwovens.

The flattening of the fabric is illustrated in Fig. 7.2, which shows the effect of the flattened yarn structure. This gives a greater light reflectance than the original rounded yarn structure.

7.3.1.2   Types of calender

In general calenders usually have between two and seven rollers, with the most common being the three-bowl calender. Perhaps the most important factor in calender design is the composition of the rollers and the surface characteristics of these.[2] Textile calenders are made up from alternate hard steel and elastic bowls. The elastic bowls are made from either compressed paper or compressed cotton, however, a lot of modern calenders are made with a thermoplastic thick covering, which is usually nylon 6. The latter have the advantage that they are less liable to damage from knots, seams and creases than cotton and paper bowls, damage that would then mark off onto the calendered fabric. In fact the development of the 'in situ' polymerisation technique for nylon 6 has enabled the simple production of elastic rollers from this material. Because of this improved performance, nylon 6 covered rollers often enable the required effects to be achieved in a single nip thus reducing the overall number of bowls.

In two-bowl calenders (Fig. 7.3), it is normal to have the steel bowl on top so that the operator can see any finish. This type of arrangement is often used with the nylon bottom bowl mentioned previously, especially where the calender is used for glazing or the embossed type of finishes.

The arrangement where two steel bowls are used together only occurs in exceptional circumstances, for example, in the compaction of nonwovens. Here both bowls

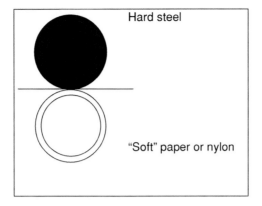

7.3   Two-bowl calender.

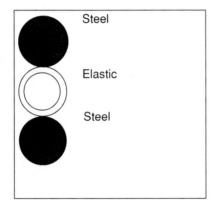

7.4   Three-bowl calender.

are usually oil heated so that some form of permanent setting occurs. Finally, the arrangement with two elastic bowls is not common but is sometimes used on cotton knitgoods to obtain a soft handle.

The three-bowl calender (Fig. 7.4) was developed from the two-bowl calender and with this type of calender it is normal to use only the top nip, with the bowls arranged steel–elastic–steel. The bottom bowl is used to keep the central elastic bowl smooth and thus assist in the finishing. The same arrangement also serves the same purpose on embossing calenders, where there is the possibility of permanent indentation from the top roller.

Pressure used in all of the above calenders can be varied between 10 and 40 tonnes, with running speeds up to 60 m min^{-1}. However, these are very much average figures with figures as low as 6 tonnes for a 1 m wide calender to as high as 120 tonnes for a 3 m wide calender. In addition, running speeds of 20 m min^{-1} are used on an embossing calender, while on a glazing calender speeds of over 150 m min^{-1} have been quoted.

The temperatures which are used in calender rollers can, of course, vary from room temperature to 250 °C. However, it must be stressed that temperature control is of vital importance, with a tolerance of ±2 °C being commonly quoted. Some generalisations can be made as follows:

- Cold bowls give a soft handle without much lustre; warm bowls (40–80 °C) give a slight lustre.
- Hot bowls (150–250 °C) give greatly improved lustre, which can be further improved by the action of friction and waxes.

7.3.1.3 Types of finish

- **Swissing or normal gloss:** a cold calender produces a smooth flat fabric. However, if the steel bowl of the calender is heated then in addition to smoothness the calender produces a lustrous surface. If a seven-bowl multipurpose calender is used then a smooth fabric with surface gloss on both sides is produced.
- **Chintz or glazing:** this gives the highly polished surface which is associated with glazed chintz. The effect is obtained by heating the top bowl on a three-bowl calender and rotating this at a greater speed than that of the fabric. The speed of this top bowl can vary between 0 and 300% of the speed of the fabric. In certain cases where a very high gloss is required, the fabric is often preimpregnated with a wax emulsion, which further enhances the polished effect. This type of calendering is often called friction calendering.
- **Embossing:** in this process the heated top bowl of a two-bowl calender is engraved with an appropriate pattern which is then transferred to the fabric passing through the bowls. The effect can be made permanent by the use of thermoplastic fibres or in the case of cellulosics by the use of an appropriate crosslinking resin.
- **Schreiner or silk finishing:** this is a silk-like finish on one side of the fabric. It is produced (see Fig. 7.5) by embossing the surface of the fabric with a series of fine lines on the surface of the bowls. These lines are usually at an angle of about 30° to the warp threads. The effect can be made permanent by the use of thermoplastic fabric or, in the case of cotton, by the use of a resin finish. This finish is particularly popular on curtains and drapes because of the silk-like appearance this type of finish gives to the product.
- **Delustering:** this is commonly achieved by passing the fabric through the bottom two bowls of a three-bowl calender, where these are elastic. However, steel bowls with a special matt finish have been manufactured that are very effective for this purpose.
- **Chasing:** the fabric is threaded through the calender in such a way as to press the fabric against itself several times. It is common to use a five- or seven-bowl calender, the fabric passing through each nip of the calender in two or three layers.

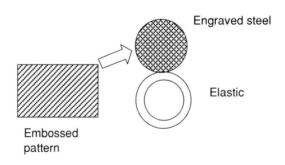

7.5   Principle of Schreiner calender.

- **Palmer finishing:** in this type of finish the damp fabric is carried on the inside of a felt blanket round a steam-heated cylinder, often called a Palmer drier. The face of the fabric, which is run on the surface of the heated cylinder, takes a light polish from the cylinder, while the back of the fabric, which is in contact with the felt blanket, takes a roughened effect from the cylinder. This finish is particularly popular for cotton drill fabric.

These descriptions of calendering should not be regarded in any way as complete and for a more complete description the reader is recommended to contact the calender manufacturers. Some of these are Küsters in Germany, Parex Mather in the UK, Kleinewefers in Germany and Metallmeccanica in Italy.

7.3.2   Raising

Raising is the technique whereby a surface effect is produced on the fabric that gives the fabric a brushed or napped appearance. It is achieved by teasing out the individual fibres from the yarns so that they stand proud of the surface.[3] The way this was done originally was to use the seedpod of the thistle, which was known as a teasel. These teasels were nailed to a wooden board and the fabric was drawn over them to produce a fabric with a hairy surface, which had improved insulating properties. This method has largely been superseded by the use of rotating wire brushes, although where a very gentle raising action is required, such as in the case of mohair shawls, teasels are still used.

7.3.2.1   Modern raising machines

All modern raising machines (see Fig. 7.6) use a hooked or bent steel wire to tease the fibres from the surface of fabric. The most important factor in the raising operation is the relationship between the point and the relative speed of the cloth. The raising wires or 'card' wires are mounted on a flexible base, which is then wrapped around a series of rollers, which are themselves mounted on a large cylindrical frame, illustrated in Fig. 7.6.

The raising action is brought about by the fabric passing over the rotating rollers and the wire points teasing out the individual fibres in the yarn. Because there are a large number of points acting on the fabric at any one time, the individual fibres must be sufficiently free to be raised from the fibre surface. This is a combination of the intrafibre friction and the degree of twist in the raised yarns. Thus for 'ideal' raising, the yarns should be of low twist and be lubricated. One further point to note is that because the fabric runs in the warp direction over the machine, only the weft threads are at right angles to the rotating raising wires and therefore only the weft threads take part in the raising process.

7.3.2.2   Raising action

From Fig. 7.6 it can be seen that both the card wire rollers and the cylinder to which these are attached may be rotated; it is the relative speed of these in relation to that of the fabric that produces the various raising effects that may be achieved. There are two basic actions in raising and these are governed by the direction in which the card wires point and the relative speed of rotation of these in relation to the fabric. These actions are called the pile and the counterpile actions and are shown in Fig. 7.7.

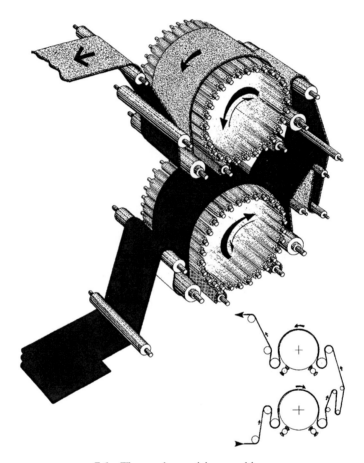

7.6   The modern raising machine.

In the counterpile action, the working roller rotates in the opposite direction to that of the cylinder with the points of the wire moving in the direction of rotation. This action pulls the individual fibres free from the surface.

In the pile action, the points of the wire are pointing away from the direction of movement of the fabric. This results in an action where the raised fibres are subject to a combing action which tends to tuck back the fibres into the body of the fabric.

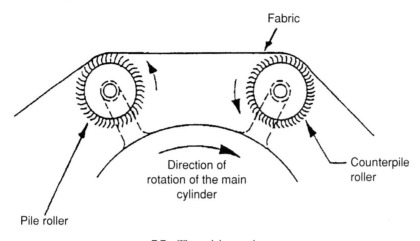

7.7   The raising action.

The most common raising action uses a combination of both of these actions, producing an even raise over the whole of the fabric surface. Control of the raising action has been achieved by measurement of the surface roughness of the raised fabric.[4] It is therefore possible to control the exact height of the nap on the surface of fabrics.

7.3.3   Shearing

This is a process by which the fibres which protrude from the surface of a fabric, are cut to an even height. The principle used in this technique is similar to a lawn mower in that a spiral cutter rotates against a stationary blade and cuts off any material protruding from the fabric surface.[5]

This principle is illustrated in Fig. 7.8, which shows the fabric passing over the cutting bed and the protruding hairs on the surface being caught between the rotating head of the spiral cutter and the ledger blade. By raising and lowering the height of the cutting bed, the depth of cut may be varied. Obviously the cutting action produces a great deal of cut pile and this must be removed by strong suction otherwise a large amount of fly rapidly accumulates. In order to achieve an even cut and a smooth surface, several passes through the machine are required or a single pass through a multiple head machine is required. Average speeds of about 15 m min⁻¹ are commonly encountered.

One important use for this technique is the production of pile fabrics from a looped terry fabric. When this type of fabric is sheared the top of the loops of the terry fabric are cut off and a velvet like appearance is produced. When knitted loop pile fabrics are sheared, knitted velour is produced that has found a great deal of use in upholstery fabric.

7.3.4   Compressive shrinkage

The shrinkage of fabrics on washing is a well-known phenomenon. It is caused in part by the production and processing stresses on the fabric. Production stresses are introduced into the fabric by the yarn tension and also by the tension which is nec-

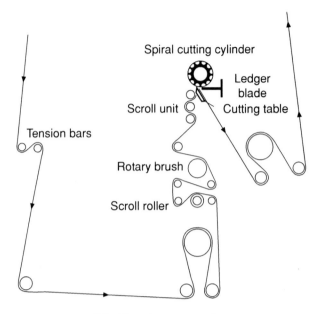

7.8   Shearing or cropping.

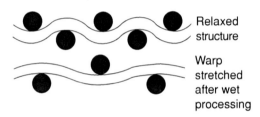

7.9   Fabric relaxation.

essary for the satisfactory production of fabric. Processing stresses are introduced during the bleaching, dyeing and finishing of fabric when the fabric is pulled in the warp direction. This tends to remove the warp crimp from the fabric as illustrated in Fig. 7.9.

In order to replace the warp crimp and thus minimise warp shrinkage, a process known as compressive shrinkage is carried out on the fabric to replace the crimp that has been pulled out in the preparation and finishing processes. This may be illustrated in the following way. A strip of fabric is placed on a convex rubber surface and gripped at each end of the rubber. As the rubber is allowed to straighten, the length of the fabric exceeds that of the rubber as illustrated in Fig. 7.10. However, if the fabric could be stuck to the surface of the rubber then the fabric would be subjected to compression and warp crimp would be introduced.

This principle then is applied to the compressive shrinking machine, where the cloth is fed in a plastic state onto a thick rubber belt at point A as shown in Fig. 7.11. While the belt is in the convex position for A to B the fabric merely lies on the surface, but at point B the belt starts to wrap its way round a large heated cylinder and thus changes from a convex to a concave shape. Thus the surface of the rubber belt contracts and the fabric, which is held onto the surface of the rubber, is

(a)   (b)

7.10   (a) Fabric under tension, (b) tension relaxed.

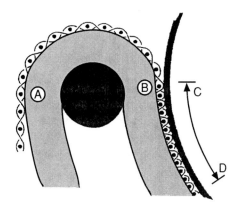

7.11   Shrinkage belt.

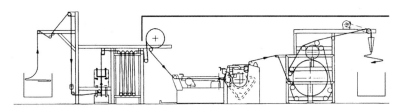

7.12   Complete shrinkage range.

subject to a warp compression over the region C to D. The actual degree of shrinkage which takes place is controlled by the amount of fabric fed onto the rubber belt and the pressure between the heated metal cylinder and the belt, which increases or decreases the concave shape of the rubber belt. The principles of compressive shrinkage have also been reviewed.[6]

A sketch of the complete compressive shrinkage range is shown in Fig. 7.12.

7.4   Heat setting

The main aim of the heat setting process is to ensure that fabrics do not alter their dimensions during use. This is particularly important for uses such as timing and driving belts, where stretching of the belt could cause serious problems. It is important to examine the causes of this loss in stability so that a full understanding can be obtained of the effects that heat and mechanical forces have on the stability of fabrics. All fabrics have constraints placed on them by their construction and method of manufacture, but it is the heat-setting mechanism that occurs within the fibre that will ultimately influence fabric dimensions.

7.4.1   Heat-setting mechanisms

The first attempt to describe the various mechanisms of heat setting synthetic fibres was that of Hearle.[7] He describes the various techniques which have been used to set fabrics into a given configuration and leaving aside the chemical methods of stabilisation, these techniques may be described as influencing the following:

1. chain stiffness
2. strong dipole links
3. hydrogen bonds
4. crystallisation.[8]

all of which are influenced by temperature, moisture and stress.

Hydrogen bonding is the most important of the factors which influence setting, and nylon has a strong hydrogen-bonded structure whereas polyester has not. Thus relaxation of nylon can occur in the presence of water at its boiling point. In fact one of the common tests for the nylon fabrics used in timing belts is a 5 min boil in water.

7.4.2   Fibre structure

All fibre-forming molecules consist of long chains of molecules. In fact, a typical nylon molecule will have a length which is some 5000 times the molecule diameter. X-ray diffraction techniques have confirmed that all synthetic fibres contain crystalline and non-crystalline regions. In nylon and polyester these crystalline regions occupy about 50% of the total space in the fibre.

It can be shown that these ordered or crystalline regions are small compared to the overall length of the polymer chain. To explain this phenomenon the polymer chains cannot be perfectly aligned over the whole length of the molecule, but must pass through alternating regions of order and disorder. The picture that thus emerges is one of short crystalline regions connected by relative regions of disorder. This view of molecular structure is now an accepted model and is known as the fringed micelle model. Figure 7.13 represents this scheme for a disorientated polymer.

The short parallel areas represent regions of order where adjacent chains pack in an ordered fashion. The non-parallel regions are where the chains do not pack together in a regular fashion, creating the non-crystalline regions or amorphous regions in the fibre.

7.4.3   Polymer orientation

During the processing of both polyester and nylon, the fibres are spun through fine holes and have a structure similar to that in Fig. 7.13. However, to develop the

7.13   Undrawn fibre.

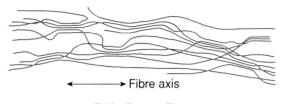

Fibre axis

7.14   Drawn fibre.

strength in the fibre these fibres are then cold drawn to create an orientated structure, which is illustrated in Fig. 7.14. Once the chains have been orientated then the fibres show a much greater resistance to applied loads and a greater stiffness, hence the reason for cold drawing.

7.4.4   Transition temperatures
In the previous section the crystalline and amorphous regions of the polymer were discussed. These do have an effect on two important parameters:

- Glass transition temperature: this represents the temperature at which molecular movement starts in the amorphous regions of the polymer, and was given the name because it is the temperature at which the polymer changes from a glassy solid to a rubbery solid. This is the temperature at which segmental loosening takes place and hence dyeing can only take place above this temperature.
- Melting point: at this point the forces holding the molecules in the crystalline regions of the fibre are overcome by the thermal energy and the polymer melts. In both polyester and nylon these temperatures are separated by about 150 °C.

7.4.5   Heat shrinkage
All textile fibres are subjected to a cold drawing process and hence when they are heated above the point at which molecular motion sets in, they will progressively shrink until they reach the point of thermodynamic equilibrium represented by Fig. 7.13. In other words the cold drawing process is reversed by the application of heat.

7.4.6   Heat setting
If the model of the drawn fibre (Fig. 7.14) is considered then it can be observed that in the drawn fibre, the molecules in the amorphous areas, while still random, are much more parallel than in the non-drawn state. Thus if the fibre is held so that shrinkage is prevented and heated above its glass transition temperature, the

molecules in the amorphous areas start to move, and because of their relative orientation, to crystallise. In practice the fibres are usually heated to about 20–40 °C
below the melting point. It now becomes apparent that the whole process of heat
setting is time dependent, because it takes time for the motion of the molecules to
line themselves up in the first place and then to crystallise. It is possible to heat set
at much lower temperatures than are used in practice. However, if low temperatures are used, the fibres are only stabilised up to the heat set temperature and as
soon as this temperature is exceeded, severe shrinkage occurs. In addition the rate
of crystallisation decreases with decreasing temperature and heat setting would be
a long and tedious process.

Obviously it is the temperature the yarns in the fabric actually achieve that is
important in the setting process. This is dependent not only on the fabric construction but also on the method of heating, because it is essential that the setting temperature is achieved quickly and evenly over the width of the fabric. Commonly,
stenters are used for this purpose, because temperature control to ±1% over a 2 m
width can be obtained. Where contact heating is used, special designs are available
to obtain this degree of control.

7.4.7 Essentials of heat setting

From the previous discussion it can be seen that heat setting is a temperature-
dependent process and for practical purposes the heat setting temperatures vary for
polyester between 190–200 °C and for nylon 6.6 between 210–220 °C. The fibres must
not be allowed to move during the heating process and the heating must be sufficiently long enough to allow crystallisation to take place, after which the fibre must
be cooled down to well below the heat setting temperature before being released.

There is an important difference between the behaviour of the two common
polyamides (nylon 6 and nylon 6.6) and polyester, because of their different behaviour towards water. Polyester is non-absorbent, so the heat setting behaviour is not
affected by water. However, nylon will absorb sufficient water to obtain a temporary set that is based on hydrogen bonding and is destroyed on boiling in water. The
consequence of this is that to obtain a permanent set on nylon, the water has to be
removed from the fibre so that crystallisation can take place. Therefore nylon tends
to be heat set at a higher temperature than polyester.

In summary:

1. Heat the fabric to within about 20–40 °C of the fibre melting point.
2. Hold at this temperature under tension for approximately 20 s.
3. Cool fabric before removing tension.

The time–temperature relationship will vary depending on the polymer, the fabric
weight and construction.[10] The simple way to determine these is to look at the shrinkage of the finished product, if the material has been correctly heat set then it should
show a residual shrinkage of less than 1% on a 5 min immersion in boiling water.

7.5   Chemical processes

It has been suggested that by the end of 2000 some 50% of all textile fibre consumption in industrialised countries will be in technical textiles.[11] A large percentage of this will consist of safety equipment and protective clothing and in fact this

comprises the most significant portion of the technical textiles market. Protective clothing must provide resistance to the elements in the workplace, whilst at the same time providing comfort during wear. The customers for these products demand strict compliance with the regulations designed to protect the wearer. One of the most important properties of this type of clothing is its resistance to small burning sources, thus flame resistance combined with easy cleaning is a most important consideration. The flame retardance must be maintained throughout the lifetime of the garment.

The main regulation governing the use of flame-retardant technical textiles was the Furniture and Furnishing Fire Safety Regulations which were introduced into the UK in 1988.[12]

7.5.1   Durable flame-retardant treatments[13,14]

Fire-retardant technical textiles have been developed from a variety of textile fibres, the choice of which is largely dependent on the cost of the fibre and its end-use. However, the main fibre in this area is cotton and thus treatments of this fibre will be discussed first. Two major flame-retardant treatments are popular. These are Proban (Rhodia, formerly Albright and Wilson) and Pyrovatex (Ciba).[13]

The Proban process uses a phosphorus-containing material, which is based on tetrakis(hydroxymethyl)phosphonium chloride (THPC). This is reacted with urea and the reaction product is padded onto cotton fabric and dried. The fabric is then reacted with ammonia and finally oxidised with hydrogen peroxide. The full reaction scheme is shown in Fig. 7.15.

The Proban process may be summarised as follows:

7.15   Chemistry of the Proban process.

1. Pad the fabric with Proban CC.
2. Dry the fabric to a residual moisture content of 12%.
3. React the fabric with dry ammonia gas.
4. Oxidise the fabric with hydrogen peroxide.
5. Wash off and dry the fabric.
6. Soften the fabric.

The actual chemistry of the process is fairly straightforward and the Proban forms an insoluble polymer in fibre voids and the interstices of the cotton yarn. There is no actual bonding to the surface of the cellulose, but the insoluble Proban polymer is held by mechanical means in the cellulose fibres and yarns. Because of this the Proban treated fabric has a somewhat harsh handle and some softening is usually required before the fabric is fit for sale. The chemistry is shown in Fig. 7.15.

The next method of forming a durable treatment for cellulose is by the use of Pyrovatex. This material is closely related to the crosslinked resins used in textile finishing and is in fact always applied with a crosslinked resin to form a chemical bond to the cellulose. The application scheme is given in Fig. 7.16.

The Pyrovatex process may be summarised as follows:

1. Pad the Pyrovatex mixture.
2. Dry at 120 °C.

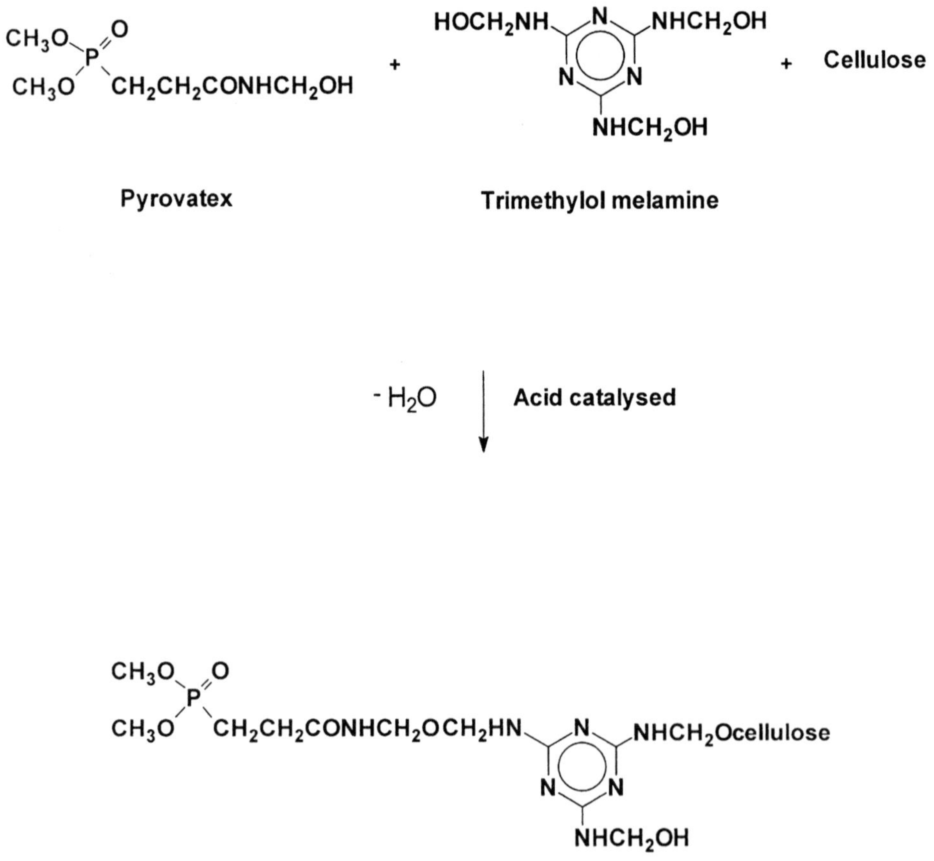

$$CH_3O\text{-}\underset{\underset{CH_3O}{|}}{\overset{O}{\underset{||}{P}}}\text{-}CH_2CH_2CONHCH_2OH$$

Pyrovatex

$+$

Trimethylol melamine structure with ring: HOCH$_2$NH and NHCH$_2$OH substituents, NHCH$_2$OH below.

Trimethylol melamine

$+$ **Cellulose**

$- H_2O$ | **Acid catalysed**

$$CH_3O\text{-}\underset{\underset{CH_3O}{|}}{\overset{O}{\underset{||}{P}}}\text{-}CH_2CH_2CONHCH_2OCH_2HN\text{-}[ring]\text{-}NHCH_2Ocellulose$$

with NHCH$_2$OH below the ring.

7.16   Chemistry of the Pyrovatex process.

3. Cure at 160 °C for 3 min.
4. Wash in dilute sodium carbonate.
5. Wash in water.
6. Dry and stenter to width.

As the reaction is with the cellulose, the flame-retardant substance is chemically bound to the fibre and is therefore durable. However, because the flame-retardant substance has to be applied with a crosslink resin, then the finished fabric has good dimensional stability and also excellent crease-recovery properties making this finish the one preferred for curtains. Unfortunately these desirable properties are not without disadvantages, the main one in the case of Pyrovatex being the loss in tear strength, which occurs with this and all crosslinking systems.

7.5.2   Synthetic fibres with inherent flame-retardant properties

The Furniture and Furnishing (Fire) (Safety) Regulations[12] made it mandatory that all upholstery materials should withstand the cigarette and match test as specified in BS 5852:1979: Pt1. This produced an enormous amount of work in the industry on possible routes which could be used to meet this legislation. These ranged from the use of materials that would not support combustion to chemical treatments and backcoating techniques. It is now clear, however, that backcoating is the main means by which these regulations are being met. Currently, some 5000 tonnes of backcoating formulations are being used in the UK for upholstery covers.

The majority of backcoating formulations are based on the well-known flame-retardant effect of the combination of antimony(III) oxide and a halogen, which is usually bromine, although chlorine is also used to a lesser degree. The synergistic mixture for this is one part of antimony(III) oxide to two parts of bromine containing organic compound. A typical formulation which describes the application level per square metre, is shown in Table 7.1.

To this basic formulation are often added softeners, which modify the fabric handle and antifungal compounds. In addition, foaming agents are used which enable the use of foam application techniques, so that a minimum amount of penetration of the backcoating compound onto the face of the fabric is achieved. The use of foam application also enables higher precision in the weights applied and shorter drying times to be achieved.[15]

Thus Proban, Pyrovatex and backcoating with antimony/bromine compounds represent the major flame-retardant treatments for cellulose.

7.5.3   Water-repellent finishes

The early water-repellent finishes were all based on the application of a mixture of waxes, which were pliable at normal temperatures, applied to tightly woven cotton fabrics. These were, of course, well suited to sail cloth and protective clothing, but problems were encountered when the garments were cleaned. Therefore, the search was on for water-repellent treatments that were simple to apply but would also allow the treated fabrics to be cleaned.

It was noted early on that the heavy metal soaps did have water-repellent properties and therefore the first attempt at the production of a durable treatment was

Table 7.1 Typical backcoating formulation

	Application level ($g\,m^{-2}$)
Antimony(III) oxide	40
Decabromodiphenyl oxide	80
Acrylic binder	120

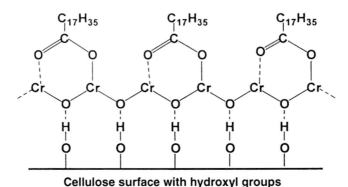

Cellulose surface with hydroxyl groups

7.17   Reaction of the chromium salt of a fatty acid with cellulose.

to use the chromium salt of a fatty acid, which was applied to cotton and then baked. This gave a certain durability to the fabric thus treated and the mechanism is illustrated in Fig. 7.17.

Some of the later treatments involved the use of other fatty acid derivatives and some of these are shown in Fig. 7.18.

The most recent treatments involve the use of the fluorocarbons which are basically esters of polyacrylic acid and a perfluorinated hexanol, as illustrated in Fig. 7.19. A list of finishes is given in Table 7.2.[6]

7.5.4   Antistatic finishes

Static electricity is formed when two dissimilar materials are rubbed together. It cannot be formed if identical materials are rubbed together. Thus when dissimilar materials are rubbed together a separation of charges occurs and one of the materials becomes positively charged and the other negatively charged. The actual sign of the charge depends on the nature of the two materials that are taking part and this may be deduced from the triboelectric series, shown in Table 7.3. The materials at the top of the table will derive a positive charge when rubbed with any of the materials below them.[17]

Cotton is a fibre that has very good antistatic properties on its own and presents few problems. This is because the natural water content of cotton is high (moisture contents of around 8% are commonly quoted), which provides the fibre with sufficient conductivity to dissipate any charge that might accumulate. However, with the advent of synthetic fibres, which had a low water content and were sufficiently non-conductive to hold a static charge on the surface, severe static problems began to arise. Thus some synthetics, particularly polyesters, can sustain such a high charge density on the surface that it can actually ionise the air in the vicinity giving rise to

Methylol derivatives of fatty acid amides

$$O=C\begin{array}{c} \nearrow NH-R \\ \searrow NH-CH_2OH \end{array}$$

where R is a fatty acid, e.g. $C_{17}H_{35}.CO$

Pyridinum salts of fatty acids, for example

$$Cl^-$$

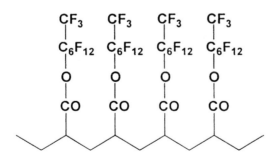

$N^+ - CH_2NHCOC_{17}H_{35}$

will react with cellulose

Cl^-

$N^+ - CH_2NHCOC_{17}H_{35}$ + $HO-$cellulose

$\downarrow$

N +   HCl   + $C_{17}H_{35}CO.NH.CH_2-O-$cellulose

7.18   Reactions of some permanent repellents.

CF_3	CF_3	CF_3	CF_3
C_6F_{12}	C_6F_{12}	C_6F_{12}	C_6F_{12}
O	O	O	O
CO	CO	CO	CO

7.19   Fluorocarbon showing polyacrylate backbone and perfluoro side chain.

a spark, which discharges the static that has been built up. In most cases this results in a mild shock to the person experiencing this static discharge, but where explosive gases might be present it can result in disaster.

Antistatic treatments, therefore, are based on the principle of making the fibre conductive so that high charge densities are dissipated before sparks can form. This

Table 7.2 Common water repellent finishes

Finish	Application method	Trade names
Paraffin wax emulsion	Spray	Mystoline, Ramasit
Paraffin wax plus Al or Cr salt	As above but lower amounts	Mystoline
Metal soap plus fatty acid	Pad, dry, bake	Persistol
Methylol stearamide	Reactive resin, pad, dry, bake	Phobotex
Pyridinium compounds	Pad, dry, bake	Velan (ICI), Zelan (Du Pont)
Reactive silicone resins	Pad, dry, bake	Silicone Finish (ICI)
Fluorocarbon emulsion	Pad, dry, bake	Zepel (Du Pont), Scotchgard (3M)

Table 7.3 The Triboelectric Series

Positive end
Polyurethane
  Nylon
  Wool
  Silk
Rayon and cotton
  Acetate
  Polypropylene
  Polyester
  Acrylic
  PVC (polyvinylchloride)
  PTFE (polytetrafluoroethylene)
Negative end

is done by the application of both anionic and cationic agents to the fibre. Typical structures of these materials are similar to the softening agents used for cotton, which contain a long chain hydrocarbon with an ionic group at the end. One of the most interesting advances in the field of antistatic treatments has been the development of the permanent antistatic finishes, one of which was the Permolose finish developed by ICI. These are actually a series of finishes that consist of block copolymers of ethylene oxide and a polyester. When polyester fibres are treated with Permolose, the polyester block of the copolymer is adsorbed by the polyester fibre but the polyethylene oxide portion is incompatible with the polyester fibre and thus remains on the surface, where it attracts water and forms a conductive surface on the polyester fibre.

7.5.5 Antimicrobial and antifungal finishes
Problems of hygiene are coming more and more to the fore in textile finishing[18] and it is now generally realised that a microbiocidal finish is very valuable in certain textiles for two reasons: as a prophylactic measure to avoid reinfections and as a deodorant.

Perhaps at this stage it might be useful to define some of these terms:

- *Bacteriostatic*: a chemical that inhibits the growth of bacteria. Fabric that has been impregnated with a bacteriostat will stop the growth of germs, which eventually die in time.
- *Fungistatic*: a chemical that inhibits the growth of fungi.

Bactericidal, fungicidal and *microbicidal* all mean that the chemical will kill these three types of microorganism.

Here are just a few of the many microorganisms with the infections they cause:

- *Staphylococcus aureus*: found in mucus membranes, causes boils and abscesses
- *Pseudomonas pyocyanea*: causes spots and boils
- *Trichophylon menagrophytes*: fungus, which causes dermatomycosis of the feet
- *Candida albicans*: yeast-like mould which is the main cause of thrush and foot rot.

7.5.5.1   Areas of use

Microbicidal finishes are mainly used in textiles that are being handled continuously by a large number of people. Locations where these are used include, hotels, hospitals, asylums and student hostels, where mattress ticking, blankets and pillows,

7.20   Microbicidal finishes.

carpets and upholstery all come into contact with a large number of different individuals. The following companies all manufacture microbiocide:

- Bayer: Movin
- Thomson Research Associates: Ultra-Fresh
- British Sanitised: Actifresh
- Sandoz: Antimucin
- Protex: Microcide
- Ciba-Geigy: Fungicide G.

Any one of the following methods can apply all these products:

- exhaust
- pad batch
- continuous
- spray.

The normal add on depends on the efficiency of the particular product, but add-on weights of 1–4% are commonly quoted. Some of these are shown in Fig. 7.20.

References

1. D GILBERT, *Nonwovens Report Int.*, 1998 **327** 21.
2. D BEHR, *Knitting Technol.*, 1996 **18**(1) 311.
3. ANON, *Knitting Technol.*, 1999 (1) 29.
4. M A BUENO *et al.*, *Textile Res. J.*, 1997 **67**(11) 779.
5. ANON, *Knitting Technol.*, 1999 (1) 29.
6. Sanfor Services, *Tinctora*, 1994 **91**(1) 37.
7. J W S HEARLE, in *The Setting of Fibres and Fabrics*, eds J W S Hearle and L W C Miles, Merrow Publishing, Watford, Herts, 1971, 1–25.
8. S RANGANATHAN, *Colourage*, 1985 **32**(25) 29.
9. W J MORRIS, 'The use of heat in the production of synthetic fibre textiles', *Textiles*, 1981 **10**(1) 1–10.
10. J S HEATON, in *The Setting of Fibres and Fabrics*, eds J W S Hearle and L W C Miles, Merrow Publishing, Watford, Herts, 1971, Chapter 8.
11. ANON, *Tessili per impieghi Tecnici*, 1995 **2**(2) 25.
12. UK Furniture and Furnishings (Fire) (Safety) Regulations, The Home Office 1988.
13. M E HALL and A R HORROCKS, *Trends Polym. Sci.*, 1993 **1**(2) 55.
14. M E HALL and A R HORROCKS, *Encyclopedia of Advanced Materials*, ed Bloor *et al.*, Pergamon Press, Oxford, 1994, 231.
15. J KUKULA, '*Progress in Textile Coating and Laminating*', BTTG conference, 2–3 July Manchester, 1990.
16. I HOLME, *Wool Record*, 1996 **155**(3618) 37.
17. V K JOSHI, *Man made Textiles in India*, 1996 **39**(7) 245.
18. J D PAYNE and D W KUDNER, *Textile Chemist and Colorist*, 1996 **28**(5) 28.

8

Coating of technical textiles

Michael E Hall

Department of Textiles, Faculty of Technology, Bolton Institute, Deane Road, Bolton BL3 5AB, UK

8.1   Introduction

This chapter will deal with the chemistry of coatings and their application to various coated technical textiles that are in use. It will leave the reader to make use of the many excellent reviews, which are referenced in this article, for details of individual items.

8.2   Chemistry of coated textiles

Coatings used in the production of technical textiles are largely limited to those products that can be produced in the form of a viscous liquid, which can be spread on the surface of a substrate. This process is followed by a drying or curing process, which hardens the coating so that a non-blocking product is produced. Thus the coatings for these products are limited to linear polymers, which can be coated as a polymer melt or solution and on cooling form a solid film or form a solid film by evaporation of the solvent. There are some types of coatings that can be applied in the liquid form and then chemically crosslinked to form a solid film.

The coatings used in technical textiles are all thermoplastic polymers, which are long chain linear molecules, some of which have the ability to crosslink. The properties of these polymeric materials directly influence the durability and performance of the end product. Therefore, some description of these materials is necessary.

8.2.1   Polyvinyl chloride (PVC)

This polymer is manufactured from the free radical polymerisation of vinyl chloride as shown in Fig. 8.1.

The polymer is a hard rigid solid, which if it is to be used as a coating material for technical textiles needs to be changed to a soft flexible film.[1] This is possible

8.1   Polyvinyl chloride.

because of a remarkable property of PVC, the ability of the powdered polymer to absorb large quantities of non-volatile organic liquids. These liquids are known as plasticisers and a typical plasticiser for PVC is cyclohexylisooctylphthalate. The polymer can absorb its own weight of this plasticiser. However, when the powdered polymer and plasticiser are first mixed, a stable paste is formed which is easily spreadable onto a textile surface. The paste of PVC and plasticiser, known as a plastisol, consists of the partially swollen particles of PVC suspended in plasticiser. When this mixture is heated to 120 °C complete solution of the plasticiser and polymer occurs, which on cooling gives a tough non-blocking film. The flexibility of this film can be varied by the amount of plasticiser added. However, for most uses plasticiser contents of up to 50% are most common. Plasticised PVC forms a clear film, which shows good abrasion resistance and low permeability. The film may be pigmented or filled with flame-retardant chemicals to produce coloured products of low flammability. The coatings are resistant to acids and alkalis but organic solvents can extract the plasticiser, making the coatings more rigid and prone to cracking.

One great advantage of a polymer with an asymmetric chlorine atom is its large dipole and high dielectric strength. This means that the coated product may be joined together by both radiofrequency and dielectric weldings techniques. This factor combined with its low price make it ideal for protective sheetings such as tarpaulins, where the low permeability and good weathering properties make it a very cost effective product.

8.2.2   Polyvinylidene chloride (PVDC)

PVDC is very similar to PVC. As in the case of PVC it is made by the emulsion polymerisation of vinylidene chloride, as illustrated in Fig. 8.2.

The resulting polymer forms a film of low gas permeability to gases, however, the polymer is more expensive than PVC and therefore only tends to be used where flame resistance is required.[2] As may be seen from the formula, PVDC contains twice the amount of chlorine as PVC and this extra chlorine is used in flame-resistant coatings. When a flame heats these materials the polymer produces chlorine radicals which act as free radical traps, thus helping to snuff out the flame.

8.2.3   Polytetrafluoroethylene (PTFE)

PTFE is perhaps the most exotic of the polymers which occur in coated textiles. It is manufactured by the addition polymerisation of tetrafluoroethylene (Fig. 8.3).

Since its discovery by Du Pont in 1941, PTFE has found many uses in coating particularly in the protection of fabrics from the harmful effects of sunlight.[3]

8.2   Polyvinylidene chloride.

8.3   Polytetrafluoroethylene.

One remarkable feature of the polymer is its very low surface energy, which means that the surface cannot be wetted by either water or oil. Textile surfaces treated with this polymer are both water repellent and oil repellent. Hence PTFE is found on diverse substrates which range from conveyer belts used in food manufacture to carpets where stain resistance is required. In addition the polymer shows excellent thermal stability and may be used up to a temperature of 250 °C. It is resistant to most solvents and chemicals, although it may be etched by the use of strong oxidising acids; this latter fact may be used to promote adhesion. In many ways PTFE could be regarded as an ideal polymer, the main drawback to its use being its very high cost compared to the other coating materials.

In order to reduce the cost of fluoropolymers several less expensive compounds have been produced, such as polyvinyl fluoride (PVF) and polyvinylidene fluoride (PVDF), which are analogous to the corresponding PVC and PVDC. However, while these materials are similar to PTFE they are slightly inferior in terms of resistance to weathering.

8.2.4   Rubber

Natural rubber is a linear polymer of polyisoprene, found in the sap of many plants, although the main source is the tree *Hevea brasiliensis*. Rubber occurs as an emulsion, which may be used directly for coating, or the polymer may be coagulated and mixed at moderate temperatures with appropriate fillers.[4] The formula, see Fig. 8.4, shows that the natural polymer contains unsaturated double bonds along the polymer chain.

The double bonds may be crosslinked with sulphur, a process known as vulcanisation, which can give tough abrasion-resistant films or hard ebony-like structures. The flexibility of the rubber may be adjusted by the amount of crosslinking which takes place. However, one great advantage of this process is that the rubber can be mixed at high rates of shear with the appropriate compounding ingredients and spread onto a textile, after which the coated textile is heated to vulcanise the rubber compound. These principles are used in the production of tyres and belting, where the excellent abrasion resistance of natural rubber makes it the material of choice.

Unfortunately, the presence of a double bond makes the polymer susceptible to

8.4   Polyisoprene (rubber).

8.5   Styrene–butadiene rubber (SBR).

oxidation. In addition, the polymer swells in organic solvents although it is unaffected by water and dilute acids and alkalis. Natural rubber is far more sensitive to both oxygen and ozone attack than the other synthetic materials which are described in the next section.

8.2.5   Styrene–Butadiene Rubber (SBR)

SBR is made by the emulsion polymerisation of styrene and butadiene as illustrated in Fig. 8.5.

The formula illustrated implies a regular copolymer but this is not the case and SBR is a random copolymer. The compounding and application techniques are very similar to those for natural rubber although the material is not as resilient as natural rubber and also has a greater heat build-up, which make SBR inferior to natural rubber in tyres. In the case of coated fabrics, the superior weatherability and ozone resistance of SBR, combined with the ease of processing, make this the product of choice. It is estimated that 50% of all rubber used is SBR.[5]

8.2.6   Nitrile rubber

Nitrile rubbers are copolymers of acrylonitrile and butadiene shown in Fig. 8.6.

These materials are used primarily for their excellent oil resistance, which varies with the percentage acrylonitrile present in the copolymer and show good tensile strength and abrasion resistance after immersion in oil or petrol. They are not suitable for car tyres but are extensively used in the construction of flexible fuel tanks and fuel hose.[6]

8.6   Nitrile rubber.

8.7   Butyl rubber.

8.2.7   Butyl rubber

Butyl rubbers are copolymers of isobutylene with a small amount of isoprene to make the copolymer vulcanisable or crosslinked as illustrated in Fig. 8.7.

Because of the low amount of isoprene in the structure, the vulcanised structure contains little unsaturation in the backbone and consequently the rate of oxidation or oxygen absorption is less than that of other elastomers except for the silicones and fluorocarbons. When an elastomer contains double bonds, oxidation leads to crosslinking and embrittlement, whereas in butyl rubbers oxidation occurs at the tertiary carbon atom which leads to chain scission and softening. Further, the close packing of the hydrocarbon chains leads to a structure which is impermeable to gases and its main use is in tyre tubes and inflatable boats.[7]

8.2.8   Polychloroprene (neoprene)

Neoprene rubber was first developed in the United States as a substitute for natural rubber, which it can replace for most applications. It is made during the emulsion polymerisation of 2-chlorobutadiene as illustrated in Fig. 8.8.

Neoprene rubbers can be vulcanised and show tensile properties similar to natural rubber, however, they are perhaps most widely used for their excellent oil resistance.[8] Weathering and ozone resistance is good and the polymer finds its main

$$\underset{\substack{\text{H}\quad\text{Cl}\ \text{H}}}{\overset{\substack{\text{H}\ \text{H}\quad\quad\text{H}}}{\text{C}=\text{C}-\text{C}=\text{C}}} \longrightarrow \left[\underset{\substack{\text{H}\quad\text{Cl}\ \text{H}}}{\overset{\substack{\text{H}\ \text{H}\quad\quad\text{H}}}{-\text{C}-\text{C}=\text{C}-\text{C}-}}\right]_n$$

8.8 Polychloroprene (neoprene).

$$\underset{\text{R}_2}{\overset{\text{R}_1}{\text{HO}-\text{Si}-\text{OH}}} \longrightarrow \left[\underset{\substack{\text{R}_2\quad\text{R}_2}}{\overset{\substack{\text{R}_1\quad\text{R}_1}}{-\text{O}-\text{Si}-\text{O}-\text{Si}-}}\right]_n$$

8.9 Silicone rubbers. R_1 and R_2 are unreactive alkyl or aryl groups.

end-uses in the production of belts and hoses. The neoprene latex can also be used in dipping and coating.

8.2.9 Chlorosulphonated polyethylene (Hypalon)

Treatment of polyethylene with a mixture of chlorine and sulphur dioxide in solution yields a product in which some of the hydrogen atoms in the polyethylene are replaced by chlorine and some by the sulphonyl chloride groups ($-SO_2Cl$). The resulting polymer can be crosslinked via the sulphonyl chloride by the use of metal oxides. Typically the polymer will contain about one sulphur atom for every 90 carbon atoms in the polyethylene chain and about 25% by weight of chlorine.[9]

These products show good resistance to weathering and have excellent ozone resistance, but they do have low elongation. Their main uses in coating are where flame resistance is required; here the synergism between sulphur and chlorine promotes flame retardancy.

8.2.10 Silicone rubbers

Silicone rubbers are polymers which contain the siloxane link Si—O—Si and are formed by the condensation of the appropriate silanol which is formed from the halide or alkoxy intermediate; the final condensation then takes place by the elimination shown in Fig. 8.9.

The groups R_1 and R_2 are normally inert groups such as methyl, but they may include a vinyl group and therefore are capable of crosslinking. It is also normal to fill these polymers with finely divided silica, which acts as a reinforcing filler.[10]

These polymers show outstanding low temperature flexibility and can be used at temperatures as low as $-80\,°C$, while they retain their properties up to $250\,°C$. They also show good resistance to weathering and oxidation. Unfortunately their price is high.

8.2.11 Polyurethanes

Polyurethanes are made by the reaction of a diisocyanate with a diol as shown in Fig. 8.10.[11]

The particular diisocyanate shown is 2,4-toluene diisocyanate and the diol is

8.10   Polyurethane.

pentane diol but any of the analogues may be used. Polyurethanes used for coating textiles are not quite as simple as the one illustrated and the materials are frequently supplied as an isocyanate-tipped prepolymer and a low molecular weight hydroxyl-tipped polyester, polyether or polyamide. The two materials will react at room temperature although this is often accelerated by raising the temperature. The only drawback to this system is that once the components are mixed, crosslinking starts immediately and so the pot life of the system is limited. Stable prepolymers which contain a blocked diisocyanate usually as a bisulphite adduct are now available. These blocked isocyanates will not react at room temperature, but will react at elevated temperatures in the presence of organotin catalysts.

Polyurethane coatings show outstanding resistance to abrasion combined with good resistance to water and solvents, in addition they offer good flexibility. The chemistry of the diol can be varied considerably so as to convey water vapour permeability to the coating. Coatings made from polyurethanes do have a tendency to yellow on exposure to sunlight and they are therefore normally pigmented in use.

8.3   Coating techniques[12]

The original methods of coating were largely based on various impregnating techniques based on an impregnating trough followed by a pair of squeeze rollers to ensure a constant pick-up. The material was then air dried at constant width, usually on a stenter, and rolled. However, when the coating was required on one side of the fabric then total immersion of the fabric in the coating liquor was not possible and other techniques had to be developed.

8.3.1   Lick roll

In this method the fabric was passed over the coating roll which was rotated in a trough of the coating liquor as shown in Fig. 8.11.

There were several variations on this theme, which were developed to ensure a

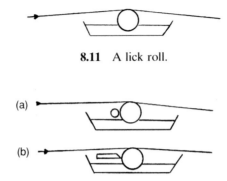

8.11   A lick roll.

8.12   Other lick roll methods. (a) A metering roll and (b) a knife doctor.

more even application of the coating by metering the coating onto the fabric. This was done by two main approaches, the first of which was to use a second roll on the primary coating roll, which picked up a fixed amount. The second was to use a doctor blade on the primary roll, so that again only a fixed amount of liquor was transferred to the fabric. These are also illustrated above in Fig. 8.12.

The main disadvantage of these systems was that the amount of coating on the fabric was dependent on the surface tension and viscosity of the coating fluid and also the surface condition of the fabric. To overcome this problem knife coating was developed, which functions in basically the same way that butter is spread on toast.

8.3.2   Knife coating

In this method the coating fluid is applied directly to the textile fabric and spread in a uniform manner by means of a fixed knife. The thickness of the coating is controlled by the gap between the bottom of the knife and the top of the fabric. The way in which this gap is controlled determines the type of machinery used. The following are the main techniques used:

* knife on air
* knife over table
* knife over roller
* knife over rubber blanket.

In the first of these the spreading blade is placed in direct contact with the fabric under tension and the coating compound is thus forced into the fabric. This is shown in Fig. 8.13.

The main advantage of this technique is that any irregularities in the fabric do not affect the running of the machine. However, this is not the case with the knife over table or knife over roll methods (Fig. 8.14), for although the coating thickness can be accurately controlled, any fabric faults or joints in the fabric are likely to jam under the blade causing fabric breakage.

The problem of metering an accurate amount of coating onto the substrate was finally solved by the use of a flexible rubber blanket, which gives a controlled gap for the coating compound and yet is sufficiently flexible to allow cloth imperfections or sewing to pass underneath the blade without getting trapped and causing breakouts. This is shown in Fig. 8.15.

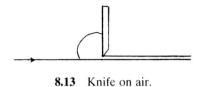

8.13   Knife on air.

(a)

(b)

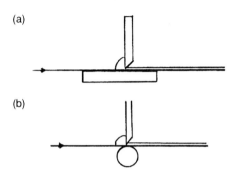

8.14   Coating methods using (a) knife over table and (b) knife over roller.

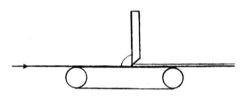

8.15   Knife over blanket.

8.3.2.1   Knife geometry
The geometry of the coating knife and the angle of application also have an important role to play in the effectiveness and penetration of the coating. Obviously if uniform coatings are to be obtained over the width of the fabric then an accurately machined flat blade is mandatory. In addition the profile of the knife can have a marked influence on the coating weights and penetration of the coating. There are three main types of knife profile, illustrated in Fig. 8.16 with many variations in between these three:

- **Pointed blade:** the sharper the blade the more of the coating compound is scraped off and consequently the lower the coating weight.
- **Round blade:** this gives a relatively higher coating weight than a sharp point.
- **Shoe blade:** this gives the highest coating of all the blade profiles; the longer the length of the shoe the higher the coating weight.

In general knife coating fills in any irregularities in the fabric surface giving a smooth finish to the coated surface. The machines which use knife coating are in general simple to operate and can be used for a wide variety of thicknesses from about 1 μm up to 30 μm.

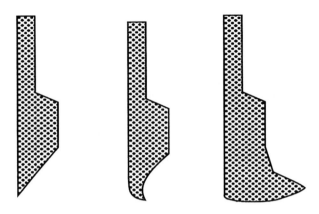

8.16   Three blade profiles.

8.3.2.2   Air knife coating

In discussing knife coating, mention must also be made of the air knife as a method of removal of the excess coating fluid. In this technique a blast of air is used to blow off the excess coating fluid. The viscosity of the fluid is much lower than in the case of conventional knife doctoring and the coating applied follows the profile of the substrate to which it is being applied. The technique is more frequently used in the paper industry, where it is used to coat photographic paper, rather than in the textile industry.

8.3.3   Gravure coating

The use of a gravure roller in coating was developed from the printing industry, where it was used to print designs. The technique involves the use of a solid roller, the surface of which has been engraved with a closely packed series of small hemispherical depressions. These act as metering devices for the coating fluid, which fills the hemispheres with coating fluid from reservoirs of the fluid. The excess fluid is scraped from the roll with a doctor blade, leaving the depressions with an exact amount of fluid in each. This is then transferred to the substrate to be coated. The quantity of fluid transferred depends on the volume of the engraved depressions and the packing on the surface of the roll. However this technique is perhaps the most accurate, in terms of the applied coating weight, of all the techniques discussed. The greatest drawback to this technique is that for a fixed depth of engraving a fixed coating weight is obtained. Thus if a different coating weight is required then a new engraved roll has to be produced. Further, unless the viscosity characteristics of the coating fluid are controlled, the pattern of the printed dot can be seen on the coated substrate. What is required is a printing fluid that will flow and form a flat surface in the drying process. This formation of a flat coating can be greatly improved by the use of offset gravure printing. Here the fluid is printed onto a rubber roller before being transferred onto the substrate.

8.3.4   Rotary screen coating

This technique is similar to the rotary screen printing process that is used to apply coloured patterns to fabric. The applicator is a cylindrical nickel screen, which has a large number of perforated holes. The coating compound is fed into the centre of the screen, from whence it is forced through the holes by either a doctor blade or a circular metal rod. The coating weight can be controlled by the number of holes per unit area and the coating weights are very precise. However, the coatings have a dot configuration and to obtain a continuous coating a wiper blade that spreads the dots into a continuous coat must be used. The two companies associated with this type of coating are Stork Brabant BV, whose system uses a metal doctor blade, and Zimmer, whose system uses the circular metal rod.

8.3.5 Hot-melt coating

In this technique the coating materials must be thermoplastic, so that they melt when heated and in this condition are capable of being spread onto a textile substrate. Thus, in some respects they are similar to paste coating. However, the big difference from paste coating is that the thermoplastic coating has no solvents to evaporate and no water that has to be evaporated, giving the process both economic and ecological advantages. The molten polymer is usually calendered directly onto the textile or in some cases extruded directly from a slotted die. This is followed by contact with a polished chill roller, if a smooth surface is required on the coating or a patterned roll if a patterned effect is required.

One further process, the use of powdered polymers as a coating medium, needs to be mentioned in the area of hot-melt coating. In this technique the powdered polymer is sprinkled onto the substrate, followed by heating with radiant heaters to melt the thermoplastic. The coating is then compacted and rendered continuous by a compaction calender. The main materials used in this are polyethylenes and nylon and these are now being applied in the production of carpets for car interiors, where because of the mouldability of the thermoplastic, a complete car carpet may be pressed out in one operation.

8.3.6 Transfer coating

The final coating technique to be described in this section is transfer coating. In this the coating material is preformed into a continuous sheet which is laminated to the substrate either by the application of heat or by the use of an adhesive known as a tie coat. The great advantage of this technique over all the others is that the coating film may be made completely free of holes and defects before it is applied to the fabric. In general, transfer coating will give the softest coating of all coating techniques in terms of fabric handle and furthermore there is no possibility of the coating bleeding through onto the face of the coated fabric.

8.4   Fusible interlinings

8.4.1   Introduction

Fusible interlinings[13] represent an important field in technical textiles and as such deserve a special mention. A fusible interlining is a fabric that has been coated with an adhesive coating, which under the influence of heat and pressure will melt and form a bond with any other fabric that is pressed against it.

The basic function of a fusible interlining is to control and reinforce the fabric to which it is fused. It does this by giving a degree of stiffness to the fabric by increasing the apparent thickness of the fabric, causing the flexural modulus of the fabric to increase proportionally to the cube power of that thickness. Thus a relatively small increase in thickness will produce a relatively large increase in the stiffness of the laminate.

8.4.2   History

The use of stiffening materials in clothing has been known for many thousands of years. From the tomb drawings in Egypt, for example, we can see how these materials were used to accentuate the rank or social status of the wearer. The Elizabethans used both linen and woven animal hair to reinforce and stiffen the elaborate clothing of the court ladies and gentlemen. This material had to be laboriously sewn into the garments by hand but as long as labour was cheap and servants in plentiful supply, it did not present any problem to the upper classes of the time. The cost of producing these 'hand made' garments was, of course, beyond the reach of any but the moneyed classes and it was not until the development of mechanised methods of garment production and the use of fusible interlinings that it became possible for the man or woman in the street to be able to afford a suit, which carried both elegance and style. Thus it became no longer possible to tell a gentleman by the cut of his jacket.

8.4.3   Development of adhesives

The first synthetic resin to be developed for clothing applications is usually attributed to Alexander Parkes, who at the Great Exhibition of 1862 introduced a nitrocellulose plasticised with camphor, which he called, with all due modesty, Parkesine. This material was a thermoplastic and could thus be moulded by heat and pressure. It was also found that the material was unaffected by water and it was used in the Victorian era for the manufacture of cheap collars and cuffs. Eventually it was given the name 'celluloid' and its use was extended into many areas. It was, for example, the original material on which the early moving pictures were shot. Unfortunately, celluloid had one major disadvantage, it was highly flammable and burned with an almost explosive violence. This prompted the search for a less flammable alternative to celluloid.

The beginning of the 20th century saw the development of another thermoplastic resin from cellulose, cellulose acetate. Benjamin Liebowitz in the USA developed the use of this material in fusible interlinings. He developed a fabric that consisted of both cotton and cellulose acetate woven together. It could be softened in acetone, which produced a very sticky fabric that was used to reinforce the collar of a man's shirt. The Trubenised Process Corporation exploited this invention producing the Trubenised semi-stiff collar, which was washable. Because the cellulose acetate adhesive did not form a continuous glue line in the collar, the collar remained permeable and hence very comfortable to wear. The result was that many millions of men's shirts with Trubenised collars were made in the thirty years from 1930 onwards.

Table 8.1 Typical coating resins

	Dryclean performance	Wash performance
Low density polyethylene	Poor	Very good
High density polyethylene	Very good	Very good
PVA/Novolac	Fair	Fair
PVC/plasticised	Good	Good
Polyamide	Very good	Good
Polyester	Very good	Very good
Reactive	Very good	Very good

8.4.4 Modern adhesive development

In the early 1950s a search began for a fusible adhesive resin which could be coated onto the base fabric and fused by the action of heat and pressure alone, thus avoiding the use of flammable and toxic solvents. The first of these was obtained by the plasticisation of polyvinyl acetate, which was applied to the fabric in the form of a knife coating by an emulsion of plasticised PVA. After the material was dried the coated fabric was wrapped in a release paper for use. The fused products were stiff and were used in the preparation of fabric belts for ladies dresses and suits; this is still their main use.

However, the continuously coated fabrics used as fusibles produced laminates that were rather too stiff for normal clothing use. The use of these materials would have remained in the special belting product, had it not been for the development of the powdered adhesive, for when this was used as a fusible interlining, the resultant stiffness of the laminate could be controlled by the particle size of the adhesive powder and by the amount of powder in the glue line of the laminate. The powder had to be applied to the interlining in an even manner and much thought went into development of machinery to ensure that this was so. The main resins used for these coatings are shown in Table 8.1. However, the most recent development has been the introduction of a reactive adhesive resin. This material melts when heated and then undergoes crosslinking, thus producing a very stable bond.

The initial coatings produced were known as sinter or scatter coatings. New coating techniques were then developed and refined by more precise positioning of the adhesive dot, so that the handle of the fused product could be more accurately controlled. This led to the printing of the adhesive in dot form which can be done either by a paste print through a mesh screen, or by the use of a gravure roller.

All these interlinings are fused by the application of heat and pressure in an electrically heated press, which has developed from the original flat bed system of heated platens to the continuous roller bed presses of the present day.

8.5 Laminating

No discussion on interlinings would be complete without mention of laminating, as this is the ultimate use of these materials. In general, textile laminates are produced by the combination of two or more fabrics using an adhesive. The hot-melt adhesive is environmentally friendly, requires less heat and is now preferred over the more conventional solvent-based adhesives.[14] Flame bonding using a thin layer of polyurethane foam is still being used in some applications, where the bulky appearance of the final product is required by the market. However, strict legislation concerning the gaseous effluent from this process has most manufacturers searching for alternatives that are more environmentally acceptable. It seems likely that hot-melt adhesives will replace most of the other adhesive techniques, either on energy grounds, or environmental grounds. The other driving force behind this change is the continued development of the hot-melt adhesives that are available to the manufacturer, which produce laminates at a higher speed, or more permanently bonded laminates.

An interesting development in the improved efficiency of the process, for example, was the development of Xironet.[15] This is a lightweight net of fusible adhesive, which when sandwiched between two fabrics can effectively laminate the fabrics together by the application of heat and pressure from a heated calender.

To improve the permanence of the adhesive bond, and as mentioned above, a hot-melt adhesive has been developed that will crosslink after the adhesive bond has been formed.[16] These materials will melt at 130 °C and form an adhesive bond and on further heating will crosslink to give an adhesive that is relatively inert.

References

1. J A BRYDSON, *Plastics Materials*, Iliffe Publishers, 1966.
2. T TASIDIM, *Textile Asia*, 1980 **11** 169.
3. ANON, *High Performance Textiles*, 1993 **February** 11.
4. B G CROWTHER, *Natural Rubber Technol.*, 1980 **11**(4) 69.
5. M M HIRSCHLER and R R POLETTI, *J. Coated Fabrics*, 1984 **19** 94.
6. J R WILLIAMS, *Am. Assoc. Industrial Hygiene Quart.*, 1980 **41** 884.
7. N ELLERBROK, US Patent No. 5,375,878.
8. J O STULL, *Industrial Fabrics Products Rev.*, 1996 **72**(12) 42–46.
9. ANON, *Du Pont Magazine*, 1991 **85**(1) 1–3.
10. F BOHIN et al., *J. Coated Fabrics*, 1998 **27** 326.
11. K KRISHNAN, *J. Coated Fabrics*, 1995 **25** 103.
12. F A WOODRUFF, 'Coating and laminating techniques', *Clemson University Presents – Industrial Textiles*, Conference at Clemson University USA, 28th January 1998.
13. ANON, World Clothing Manufacturer, 1996/7 **Dec–Jan 77**(10) 48.
14. I HOLME, *Textile Horizons*, 1997 **June/July** 35–38.
15. G PELTIER, 'Environmentally friendly dry thermoplastic adhesive films and webs for industrial bonding applications', *2nd International Conference on Textile Coating and Laminating: Assessing Environmental Concerns*, Charlotte, USA, Technomic 1992.
16. D FARRELL, 'Reactabond – A reactive response to industry', *8th International Conference on Textile Coating and Laminating*, November 9–10, Frankfurt, Technomic 1998.

9

Coloration of technical textiles

Ian Holme

School of Textile Industries, University of Leeds, Leeds LS2 1JT, UK

9.1   Introduction

Technical textiles are used in a very wide variety of end-uses in which the functional performance requirements are paramount. Thus, technical textiles must possess the requisite physical and mechanical properties to maintain the structural integrity throughout all the manufacturing and fabrication processes and during the service life of the material. For some end-uses, therefore, coloration is not strictly necessary because the aesthetic properties of the technical textile, for example colour and pattern, and also lustre, texture, handle and drape, may not always be required to appeal to the visual and tactile senses of the customer.

Coloration of technical textiles by dyeing or printing is primarily intended for aesthetic reasons but also provides a ready means of identifying different qualities or fineness of the materials. For example, the fineness of a surgical suture and its visibility at the implant site are easily identified by colour. High visibility clothing and camouflage printing clearly provide the extreme ends of the coloration spectrum for technical textiles.

Coloration also introduces other functional properties distinct from the aesthetic appeal of colour. Colorants can hide fibre yellowing and aid fibre protection against weathering, both factors of importance where the physical properties of the technical textile must be maintained over a long service life under adverse end-use conditions.[1] Heat absorption is also increased where black materials are exposed to sunlight, an important factor for baling materials for agriculture.

9.2   Objectives of coloration

The objectives of coloration treatments are first to produce the desired colour in dyeing and colours in the coloured design image in printing on the technical textile.[2] Second, coloration treatments are to ensure that the necessary colour fastness requirements for the end-use are achieved. Third, the whole operation should be carried out at the lowest cost commensurate with obtaining the desired technical performance. After dyeing, the levelness and uniformity of colour across the width and along the length of a technical textile fabric must be within the defined colour tolerances agreed between the dyer and the customer prior to coloration. Preservation of the original appearance and quality of the technical textile prior to coloration is also essential in order to ensure that the technical textile is of marketable quality.

9.3 Coloration of technical textiles

Coloration involves the application of colorants to the technical textile and is a complex field because of the variety of fibres, filaments, yarns, fabrics and other materials requiring coloration and the diverse nature of the end-use and performance requirements. Coloration may be carried out by dyeing the materials to a uniform colour, or by printing to impart a design or motif to the technical textile. Fibres, yarns and fabrics may also be multicoloured by specialised dyeing techniques, for example space dyeing, or by weaving or knitting different coloured yarns.[3]

The colorants used may be either water-soluble (or sparingly water-soluble) dyes, or alternatively water-insoluble pigments.[4] Most dyes are applied to technical textiles in an aqueous medium in dyeing and in printing, although selected disperse dyes may also be dyed in supercritical fluid carbon dioxide (above the critical point under very high pressure) for a few specialised end-uses, for example polyester sewing threads.[5] By contrast, pigments are either physically entrapped within the filaments during synthetic fibre extrusion, for example by mass pigmentation, to give a spun-dyed fibre,[6] or alternatively adhered to technical textiles in pigment printing through the use of an adhesive binder.[7]

Dyeing is normally carried out on textile materials from which surface impurities, for example fibre lubricants, spin finishes, sizes, particulate dirt or natural colouring matters and so on, have been removed by appropriate pretreatments (e.g. by desizing and scouring), and to which a stable whiteness has been imparted by chemical bleaching.[8] However, many synthetic fibres do not normally require chemical bleaching prior to coloration because the fibres may be whitened by incorporation of a fluorescent brightening agent during fibre manufacture.

Printing may be carried out mainly on technical fabrics that may be in their natural state, or chemically bleached, or whitened with a fluorescent brightening agent, or after tinting or dyeing. Conventional dyeing and particularly printing are most conveniently and economically carried out on fabrics, which also allow greater flexibility through the selection of colours late in the technical textile production sequence to meet the varying market requirements.

9.3.1 Dyes

The dyes used commercially to dye technical textiles may be selected from the very wide range of synthetic organic colorants, based upon aromatic compounds derived from petroleum.[4] Dyes are conjugated organic structures containing an alternating system of single and double bonds within the molecule which impart the ability to absorb certain wavelengths of visible light, so that the remaining light scattered by the dyed technical textile is perceived as coloured.[4,9]

The dye structure must contain a chromophore, a chemical group that confers upon a substance the potential to becoming coloured, for example nitro, nitroso, azo and carbonyl groups. To become a useful dye, however, the molecule should contain other chemical groups such as amino, substituted amino, hydroxyl, sulphonic or carboxyl groups which are called auxochromes. These generally modify or intensify the colour, render the dye soluble in water and assist in conferring substantivity of the dye for the fibre.[4,9] High substantivity aids a high degree of dye exhaustion on to the fibres during exhaust dyeing when dyes are progressively and preferentially adsorbed by the fibres from the dyebath to give coloured technical textile materials. High dye-fibre substantivity also generally confers high colour fastness during end-use, for example high colour fastness when exposed to washing and light.

Manufactured dyestuffs are formulated colorants typically containing only 30–40% active colorant. The rest of the dyestuff formulation consists of diluents to dilute the colour strength, electrolytes to improve the dye exhaustion on the fibre, and dispersing agents to facilitate dispersing of the dye in the dyebath.[10] To avoid dusting dyestuffs into the air during weighing and dispensing, granular, grain or pearl forms of dyestuffs are used. Antidusting agents are added to powdered dyestuffs for the same reason. Water-soluble dyes can be produced as true liquid dyestuffs (e.g. basic dyes), but liquid dispersions of disperse and vat dyestuffs are also produced. The latter may contain a viscosity modifier to minimise sedimentation/settling of the dye particles during storage and must be stirred prior to use. Liquid dyestuffs are preferred for continuous dyeing and for printing because of their convenience for use on automated weighing, metering and dispensing systems.

Commercial dyestuffs are subjected to stringent quality control procedures for hue and colour strength, average particle size and particle size distribution, and are dried to a uniform moisture content and packaged in drums or plastic containers suitable for transportation, storage and dispensing.[10,11] Resealable packaging may be used to prevent the ingress of moisture from the atmosphere, and in many colour kitchens the ambient temperature and relative humidity are controlled in order to maintain reproducibility in weighing. Some dyestuffs are mixtures formulated for specific shades such as black, or contain dyes from different application classes suitable for dyeing specific fibre blends, for example 65/35 polyester/cellulosic fibre blends.[12]

9.3.2 Pigments

Pigments are synthetic organic or inorganic compounds that are insoluble in water, although some are soluble in organic solvents.[4,9] The pigment particles are ground down to a fine state of subdivision (0.5–2 µm) and stabilised for use by the addition of dispersing agents and stabilisers. Both powder and liquid (i.e. dispersions of) pigments are used for the coloration of technical textiles, and the pigment finish must be compatible with the application conditions used in mass pigmentation of manufactured fibres or in pigment printing of technical textile fabrics.[13,14]

9.3.3 Fluorescent brightening agents

These are organic compounds that absorb some of the ultraviolet radiation in sunlight or other sources of illumination and re-emit in the longer wavelength

Table 9.1   Percentage distribution of each chemical class between major application ranges

Chemical class	Distribution between application ranges (%)								
	Acid	Basic	Direct	Disperse	Mordant	Pigment	Reactive	Solvent	Vat
Unmetallised azo	20	5	30	12	12	6	10	5	
Metal complex azo	65		10				12	13	
Thiazole		5	95						
Stilbene			98					2	
Anthraquinone	15	2		25	3	4	6	9	36
Indigoid	2					17			81
Quinophthalone	30	20		40				10	
Aminoketone	11			40	8		3	8	30
Phthalocyanine	14	4	8		4	9	43	15	3
Formazan	70						30		
Methine		71		23		1		5	
Nitro, nitroso	31	2		48	2	5		12	
Triarylmethane	35	22	1	1	24	5		12	
Xanthene	33	16			9	2	2	38	
Acridine		92		4				4	
Azine	39	39				3		19	
Oxazine		22	17	2	40	9	10		
Thiazine		55			10			10	25

Source: see ref 4.

blue–violet region of the visible spectrum.[15–17] Such compounds can be applied to technical textiles either by mass pigmentation (in manufactured fibres) or via machinery used for conventional dyeing of all types of materials, and add brightness to the whiteness obtained from chemical bleaching of the textile.

9.3.4   Range of colorants available

The Colour Index International is the primary source of information on colorants and lists the majority of commercial dyestuffs and pigments (both past and present) although the situation continually changes as the manufacture of some colorants ceases, to be replaced by new products.[18] The wide variety of colorants available is based upon many chemical types that make up the major application groups of colorants. The percentage distribution of each chemical class between major application ranges is illustrated in Table 9.1, based upon all the dyes listed where the chemical class is known, but also including products which are no longer used commercially.[4] Azo colorants make up almost two-thirds of the organic colorants listed in the Colour Index International. Anthraquinones (15%), triarylmethanes (3%) and phthalocyanines (2%) are also of major importance. Where the structure of the colorant is known, it is assigned a specific CI (Colour Index) chemical constitution number, although for many new commercial products the structure is undisclosed.

In the Colour Index International dyes are classified by application class, by colour and by number, for example CI Acid Blue 45. The commercial name of the dyestuff will, however, vary according to the dye maker, and dyestuffs, being formulated products, although nominally the same CI number, may vary in active colorant content and in the nature of the additives incorporated within the formulation.[19] Metamerism may also be observed on the dyed textile, that is a difference in colour is observed when two colours are viewed under different illumination conditions, even when using dyestuffs that nominally possess the same colour index number. The Colour Index International is available in book form and on CD–ROM.

Table 9.2 Major classes of dyes and the fibres to which they are applied

Dye class	Major fibre type
Acid (including 1:1 and 1:2 metal complex dyes)	Wool, silk, polyamides (nylon 6, nylon 6.6)
Mordant (chrome)	Wool
Azoic	Cellulosic fibres (cotton)
Direct	Cellulosic fibres (cotton, viscose, polynosic, HWM, modal, cuprammonium, lyocell fibres), linen, ramie, jute and other lignocellulosic fibres
Reactive	Cellulosic fibres (cotton, viscose, polynosic, HWM, modal, cuprammonium, lyocell, linen, ramie, jute), protein (wool, silk)
Sulphur	Cellulosic (cotton)
Vat	Cellulosic fibres (cotton, linen)
Basic	Acrylic, modacrylic, aramid
Disperse	Polyester, cellulose triacetate, secondary cellulose acetate, polyamide, acrylic, modacrylic, polypropylene, aramid

HWM = high wet modulus.

The major classes of dyes and the fibres to which they are applied are illustrated in Table 9.2. The type of dyestuff used for dyeing technical textiles depends upon the fibre(s) present in the material, the required colour and the depth of colour, the ease of dyeing by the intended application method and the colour fastness performance required for the end-use. Some dyestuff classes will dye a number of different fibres, but in practice one fibre will predominate. For example, disperse dyes are mainly used to dye polyester fibres, although they can also be applied to nylon 6, nylon 6.6, acrylic, modacrylic, secondary cellulose acetate, cellulose triacetate and polypropylene fibres, but with limitations upon the depth of colour and the colour fastness that can be attained.[20]

There are, however, many types of manufactured fibres, particularly synthetic fibres, which are currently virtually impossible to dye with conventional dyestuffs using normal dye application methods. This is because filaments which are designed for high strength end-uses are often composed of polymer repeat units that do not contain functional groups that could act as dyesites (e.g. polyethylene, polypropylene, polytetrafluoroethylene), and/or may be highly drawn to produce a highly oriented fibre with a high crystallinity.[21]

Because dyes are considered to diffuse in monomolecular form into the fibre only through non-crystalline regions, or regions of low order (i.e. high disorder), it follows that highly oriented fibres such as meta-aramid, para-aramid, and high strength polyethylene fibres, are extremely difficult to dye. In some cases, specialised dyeing techniques utilising fibre plasticising agents ('carriers') to lower the glass transition temperature (T_g) of the fibre (i.e. the temperature at which increased polymer segmental motion opens up the fibre structure) together with selected dyes may be employed to speed up the diffusion of dyes into the fibres, but the colour depth may be restricted to pale-medium depths.[21–23] In addition, or alternatively, resort may be made to high temperature dyeing in pressurised dyeing machinery at temperatures in the range 130–140 °C for the same purpose.

However, for many high strength fibres, mass pigmentation offers a more satis-factory production route to coloured fibres, provided that the introduction of the pigments does not significantly impair the high strength properties.[6] Some fibres such as carbon fibres and partially oxidised polyacrylonitrile fibres cannot be dyed, simply because they are already black as a result of the fibre manufacturing process. However, contrasting colours may be printed on such fibres using pigment printing methods, or by blending with other dyed fibres.

Microfibres are generally defined as fibres or filaments of linear density less than 1 dtex.[24] Such silk-like fibres pose considerable difficulties for level dyeing of medium to heavy depths of colour. The high surface area per unit volume of microfi-bres increases the light scattering, necessitating the use of greater amounts of dye to achieve the same colour depth as on coarser fibres. The use of bright trilobal filaments, or the incorporation of titanium dioxide delustring agents within the microfibres increases the light scattering, making the difference even more noticeable.[25]

9.4   Dye classes and pigments

9.4.1   Acid dyes

Acid dyes are anionic dyes characterised by possessing substantivity for protein fibres such as wool and silk, and also polyamide (e.g. nylon 6 and nylon 6.6) fibres, or any other fibres that contain basic groups. Acid dyes are normally applied from an acid or neutral dyebath.[24]

As the size of the acid dye molecule generally increases from level dyeing (or equalising) acid dyes to fast acid, half-milling or perspiration-fast dyes to acid milling dyes and thence to supermilling acid dyes, the colour fastness to washing increases because of the increasing strength of the non-polar forces of attraction for the fibre.[26] 1:1 Metal complex and 1:2 metal complex dyes also behave like acid dyes from the viewpoint of application to the fibre, and possess good colour fast-ness to washing and light. Level dyeing acid dyes yield bright colours, while the larger sized milling acid and supermilling acid dyes are progressively duller. 1:1 and 1:2 metal complex dyes also lack brightness, but can provide good colour fastness performance.[27]

Acid dyes contain acidic groups, usually sulphonate groups, either as $-SO_3Na$ or $-SO_3H$ groups, although carboxyl groups ($-COOH$) can sometimes be incorpo-rated. Wool, silk and polyamide fibres contain amino groups ($-NH_2$) which in an acid dyebath are protonated to yield basic dyesites ($-NH_3^+$). The acid dye anion $D.SO_3^-$, where D is the dye molecule, is thus substantive to the fibre and is adsorbed, forming an ionic linkage, a salt link, with the fibre dyesite. Monosulphonated acid dyes can be adsorbed to a greater extent than di-, tri-, and tetrasulphonated acid dyes because the number of basic dyesites in the fibres is limited by the fibre struc-ture. Thus, the colour build-up is generally greatest with monosulphonated dyes on wool. However, the dyestuff solubility in water increases with the degree of dye sulphonation.

Dyed wool fibre quality is improved by pretreatment before dyeing with Valsol LTA-N(TM), an auxiliary which extracts lipids from the cell membrane complex of wool, thereby speeding up the intercellular diffusion of dyes.[28] In this Sirolan (TM)

LTD (low temperature dyeing) process, wool may be dyed for the conventional time at a lower temperature, or for a shorter time at the boil, decreasing fibre degradation in the dyebath.

Levelling agents may be used to promote level dyeing of acid dyes on wool and other fibres.[29] Anionic levelling agents enter the fibre first and interact with basic dyesites, restricting dye uptake. Amphoteric levelling agents, which contain both a positive and a negative charge, block the basic dyesites in the fibre but also complex with the acid dyes in the dyebath, slowing down the rate at which the dyes exhaust on to the fibres. As the dyebath temperature is increased, the anionic (or amphoteric) levelling agent desorbs from the fibres, allowing the dye anions to diffuse and fix within the fibre. In addition, as the dyebath temperature is raised, the dye/amphoteric levelling agent complex breaks down, releasing dye anions for diffusion inside the fibre. The dyes are thus gradually adsorbed by the wool as the dyebath temperature is increased.

In the iso-ionic region (pH 4–5), the cystine (disulphide) crosslinks in the wool are reinforced by salt links formed between charged carboxyl and amino groups in opposing amino acid residues in adjacent protein chain molecules within the wool fibre. This temporarily strengthens the wool and the abrasion resistance of the dyed fabric is improved. Fibre yellowing is decreased and brighter colours may be dyed.[30]

The build-up of acid dyes on silk and polyamide fibres is limited by the fewer number of fibre dyesites compared with wool. Acid dyes are attracted to the amine end groups (AEG) in polyamide fibres such as nylon 6 and nylon 6.6. Nylon 6 has a more open physicochemical structure and a lower glass transition temperature (T_g) compared with nylon 6.6. Thus, acid dyes diffuse more readily into nylon 6 but the colour fastness to washing of a similar dye on nylon 6.6 will generally be superior because of the more compact fibre structure.[31] False twist texturing processes, which use contact heating of the yarns, open up the fibre structure and may modify some of the fibre cross-sections which were in contact with the heater, resulting in a slightly lower colour fastness to washing.

Chemical variations due to changes in the AEG of nylon fibres, can give rise to dyeability variations with acid dyes, a problem often referred to as barré or barriness.[31] Physical variations caused by temperature and/or tension differences in nylon fibres can similarly lead to differences in the uptake of disperse or 1:2 disperse premetallised dyes. Physical variations can be minimised by dyeing at higher temperatures (e.g. up to 120 °C) with nylon 6.6, or prolonging the dyeing time. High temperature dyeing is used with the larger 1:2 metal complex acid dyes in order to achieve better fibre levelling and also fibre penetration, which leads to improved colour fastness to washing. Acid dyes can also be printed on to wool (which has been pretreated by chlorination), degummed silk, polyamide and other fibres using conventional print thickeners in a print-dry-steam fixation-wash off and dry process.[32]

Colour fastness to washing with acid dyes on nylon fibres, especially nylon microfibres, is improved by after-treatments such as syntanning, that is, the adsorption of a sulphonated synthetic tanning agent that provides a physical barrier to desorption by blocking the fibre pore structure in the fibre surface regions and by providing electrostatic (ionic) repulsion to dye desorption.[31] Combining an appropriate syntan with a fluorochemical treatment can provide, in addition, stainblocking properties.[33] However, syntan treatments degrade during high temperature or steaming treatments and tend to yellow the dyed fabric, dulling the colour.

9.4.2   Mordant dyes

Chrome dyes are the only type of mordant dye of any commercial significance.[26] These are used in exhaust dyeing to dye wool or occasionally polyamide fibres to deep dull colours of high colour fastness to wet treatments and light. The fibres are usually dyed by the afterchrome method in which a chrome dye (similar to an acid dye) is dyed on to the fibre and then the dyed fibre is given a treatment at pH 3.5 in a second bath containing sodium or potassium dichromate.[34] The absorption of dichromate ions leads to the formation of a 1:1 and/or 1:2 chromium metal complex dye inside the fibre, which can lead to some problems in batch-to-batch reproducibility of shade, particularly if the pH control in chroming is variable. Low chrome dyeing procedures can promote exhaustion of the chromium on to the fibres, decreasing environmental pollution from the waste water from dyeing.[26,34] A major disadvantage of the afterchrome method is that the final colour is not developed until the chroming stage which can give difficulties in shade matching. Ammonia aftertreatment can improve the colour fastness to washing.

9.4.3   Basic dyes

Basic dyes are cationic dyes characterised by their substantivity for standard acrylic, modacrylic, basic-dyeable polyester and basic-dyeable nylon fibres.[24] Basic dyes can be applied to protein and other fibres, for example secondary cellulose acetate, but the light fastness of basic dyes on hydrophilic fibres is poor.[35] The major outlets for basic dyes are acrylic and modacrylic fibres, on which basic dyes can impart bright colours with considerable brilliance and fluorescence. The ionic attraction between the basic dye and the sulphonic acid dyesites in acrylic fibres is strong, which yields high colour fastness to washing. The close-packed physicochemical nature of acrylic fibres and the strong dye-fibre bonding can result in poor migration and levelling properties during dye application, but impart very high colour fastness to light.[21,35]

Acrylic fibres vary widely in their dyeability because of the different amounts of different co-monomers used with poly(acrylonitrile) that modify the fibre glass transition temperature. This may range from 70–95 °C according to the source of the acrylic fibre manufacturer.[36] Cationic retarders are widely used on acrylic and modacrylic fibres to promote level dyeing and the basic dyes used should preferably be selected with the same compatibility value. The compatibility value may range from 5 (slow diffusing) down to 1 (rapid diffusing). A compatibility value of 3 has been recommended for package dyeing, but for printing, compatibility values are of much less importance. Print fixation is by the print-dry-steam-wash off and dry process, although wet transfer printing techniques are also possible using selected dyes.[35]

The acidic dyesites in acrylic fibres become much more accessible over a narrow band of temperatures in the glass transition temperature range. Therefore, even when using cationic retarders, which initially adsorb on to the fibre blocking the dyesites, the dyebath temperature must be raised slowly through this temperature range to allow the cationic retarder to desorb slowly and allow the basic dye to be adsorbed uniformly.[35,36] Alternatively, constant temperature dyeing methods may be used in exhaust dyeing. Continuous pad-steam-wash off and dry methods can also be used for dyeing acrylic fabrics.

The basic dye uptake is limited by the number of acidic dyesites in the fibre, but approximately 95% of all colours on acrylic fibres are dyed using basic dyes.

However, where very pale colours are to be dyed it is common to use disperse dyes, which have superior migration and levelling properties, in order to attain a level dyeing, because the strong dye-fibre bonding renders this very difficult to achieve using basic dyes.[35] The build-up of disperse dyes on acrylic fibres is limited and the colour fastness to washing and light are generally lower than for basic dyes. The colour fastness of basic dyes on acrylic fibres is superior to similar dyeings on modacrylic fibres. Some modacrylic fibres are prone to delustring during dyeing and may require relustring by boiling for 30 min in a high concentration of electrolyte (e.g. 50–200 g l⁻¹ sodium chloride) or by dry-heat or treatment in saturated steam.[36]

9.4.4   Azoic colouring matters

Azoic colouring matters are formed inside the fibre (usually cellulosic fibres) generally by adsorption of an aromatic hydroxyl-containing compound, such as a naphthol or naphtholate (azoic coupling component) followed by coupling with a stabilised aromatic diazonium compound (the azoic diazo component also termed fast base or salt) to form a coloured insoluble azo compound.[2,37] The application method has to be carried out with care and the final colour is only obtained after soaping off to remove any traces of the azoic dye on the fibre surface that would give inferior colour fastness to washing and rubbing (crocking). Although they are economical for the production of red and black, the colour range is now more limited as some diazo components based upon certain aromatic amines have been withdrawn because of their possible carcinogenic nature. Precise colour matching with such a complex two-bath procedure can also give problems in practice.

9.4.5   Direct dyes

Direct dyes were the first class of synthetic dyes to dye cotton directly without the use of a mordant. Direct dyes are sulphonated bisazo, trisazo or polyazo dyes and are anionic dyes which are substantive to cellulosic fibres when applied from an aqueous bath containing an electrolyte.[2,38] Direct dyes based on stilbene, copper-complex azo, oxazine, thiazole and phthalocyanine structures are also used. Less bright in colour than acid or basic dyes, the brightness diminishes with the molecular complexity of the dye. Phthalocyanine dyes are used for very bright blue and turquoise-blue colours of good colour fastness to light, and copper-complex azo dyes also exhibit good fastness to light, although these dyes yield relatively dull colours.

Electrolyte, in the form of sodium sulphate, is usually added into the dyebath to overcome the negative charge on the cellulosic fibre surface which otherwise would repel the approach of the direct dye anions. The sodium cations from the electrolyte neutralise the negative charge at the fibre surface, allowing the dye anions to be adsorbed and retained within the fibre.[38]

Coulombic attraction, hydrogen bonding and non-polar van der Waals forces may operate depending upon the specific direct dye and the fibre structure. Direct dyes are normally linear and planar molecules allowing multipoint attachment to cellulose chain molecules, but the forces of attraction between dye and fibre are relatively weak when the dyed fibre is immersed in water.[38,39] Thus, the colour fastness to washing is moderate to poor, but the colour fastness to light may vary from excellent to poor depending on the molecular structure.

The colour fastness to washing can be improved by aftertreatment of the dyed fibre, originally by the use of after-coppering (i.e. treatment with copper sulphate to form a metal-complex direct dye) or by diazotisation and development. Both approaches lead to a pronounced shade change and copper in the waste water from dyeing is environmentally undesirable. Modern aftertreatments make use of cationic fixatives that may simply complex with the anionic dye, and/or form a metal complex and/or react with the cellulose fibre hydroxyl groups to form strong covalent bonds. Improvement in colour fastness to washing may be accompanied by some diminution in colour fastness to light.[38,40]

Direct dyes may be used on all cellulosic fibres and selected dyes may also be used to dye wool, silk and nylon fibres in the manner of acid dyes. Addition of electrolyte into the dyebath increases dye aggregation in the dyebath. As the dyeing temperature is increased, the dye aggregation decreases, releasing individual direct dye anions for diffusion and adsorption inside the fibre. Direct dyes have been separated into three classes, Class A – salt-controllable, Class B – temperature-controllable, and Class C – both salt- and temperature-controllable. It is normal to select compatible dyes from within the same class, but selected dyes from Class A and B, and Class B and C may be dyed together in the same dyebath in exhaust dyeing. Direct dyes may also be applied to cellulosic fabrics by continuous pad-steam-wash off and dry methods.[2,38]

The depth of colour obtained using direct dyes is deeper on viscose, lyocell and high wet modulus fibres and on mercerised cotton compared with bleached cotton, and careful dye selection is required to produce level, solid colours when dyeing blends of these fibres.

9.4.6 Reactive dyes

Reactive dyes, sometimes termed fibre-reactive dyes, are a very important class of dyes for dyeing cellulosic fibres[41] and are also used to dye protein fibres such as wool[42] or silk.[43] Although relatively expensive they provide a wide colour gamut of bright colours with very good colour fastness to washing. This is because during dye fixation, usually conducted under alkaline conditions, strong covalent bonds are formed between the dye and the fibre.[2,44] Reactive dyes may react by substitution (e.g. monochlorotriazinyl and dichlorotriazinyl dyes) or by addition (e.g. vinylsulphone dyes).

Typical ranges of reactive dyes and their respective reactive groups are illustrated in Table 9.3. Recent developments have led to the introduction of homobifunctional reactive dyes (e.g. two monochlorotriazine groups) and heterobifunctional reactive dyes (e.g. using monochlorotriazine plus vinyl sulphone) in an attempt to increase the dye fixation on the fibre under alkaline conditions from 50–70% with one reactive group, to 80–95% with two reactive groups. Dye application methods on fabrics include exhaust dyeing, cold pad-batch-wash off or continuous pad-steam-wash off and dry methods. Fixation by dry heat, saturated steam or superheated steam may also be used according to the type of reactive dye employed.[41]

Reactive dyes hydrolyse in contact with water and alkali, and during dyeing some hydrolysed reactive dye is adsorbed by the fibre because it behaves like a substantive direct dye, but with lower colour fastness to washing than the reactive dye that is covalently bonded. Thus, the dyeing stage must always be followed by an extended washing-off treatment to remove hydrolysed reactive dye.[41,45] Provided that the residual hydrolysed reactive dye left within the cellulosic fibre after dyeing or print-

Table 9.3 Important reactive dye systems

System	Typical brand name
Monofunctional	
Dichlorotriazine	Procion MX (BASF)
Aminochlorotriazine	Procion H (BASF)
Aminofluorotriazine	Cibacron F (Ciba)
Trichloropyrimidine	Drimarene X (Clariant)
Chlorodifluoropyrimidine	Drimarene K (Clariant)
Dichloroquinoxaline	Levafix E (DyStar)
Sulphatoethylsulphone	Remazol (DyStar)
Sulphatoethylsulphonamide	Remazol D (DyStar)
Bifunctional	
Bis(aminochlorotriazine)	Procion H-E (BASF)
Bis(aminonicotinotriazine)	Kayacelon React (Nippon Kayaku)
Aminochlorotriazine–sulphatoethylsulphone	Sumifix Supra (Sumitomo)
Aminofluorotriazine–sulphatoethylsulphone	Cibacron C (Ciba)

Source: see ref 41.

ing is at a concentration $<0.1\,g\,m^{-2}$, the best colour fastness to washing will be obtained.[46] If the concentration of hydrolysed reactive dye in the final hot wash bath is $<0.003\,g\,l^{-1}$, staining of white grounds in the fabric will be prevented.[46]

Generally reactive dyes are produced in granular form, but vinyl sulphone dyes are also available as liquids, which are more convenient for use in continuous dyeing and printing. Reactive dyes are sulphonated and highly water soluble and are exhausted on to the fibre using electrolyte (e.g. in the manner of direct dyes) and fixed using an appropriate alkali. The high water solubility creates an environmental problem in waste water treatment plants because generally only 0–30% of the hydrolysed reactive dye is removed by such treatment.[47] Low salt reactive dyes have also been introduced to decrease dyeing costs and to avoid corrosion problems in concrete waste water pipework networks caused by high concentrations of sulphate anions.

9.4.7 Sulphur dyes

Sulphur dyes are chemically complex and are prepared by heating various aromatic diamines, nitrophenols, and so on with sulphur and sodium sulphide. Sulphur dyes are produced in pigment form without substantivity for cellulose.[2] Treatment with a reducing agent (e.g. sodium sulphide or sodium hydrosulphide) in an alkaline dyebath converts the sulphur dye into an alkali-soluble reduced (leuco) form with substantivity for cellulosic fibres. Once absorbed within the fibre, the dye is then oxidised, usually with hydrogen peroxide, back to the insoluble pigment form. Soaping off after dyeing is important to ensure the maximum colour fastness to washing and rubbing is obtained.[2,48]

Sulphur black and navy are the major dyestuffs used and the colour gamut is limited to dull colours of moderate colour fastness to washing and light. The colour fastness to chlorine and to bleach fading is poor when using multiple wash cycles with detergents containing low temperature bleach activators such as TAED (tetraacetylethylenediamine, UK) or SNOBS (sodium nonanoyloxyben-zenesulphonate, USA).[49] Sulphur dyes cost less than many other dyes but the waste water from dyeing may require specialised treatment before release to a conventional waste water treatment plant.

9.4.8   Vat dyes

Vat dyes are water-insoluble, but contain two or more keto groups ($>C=O$) separated by a conjugated system of double bonds that are converted into alkali-soluble enolate leuco compounds ($>C—O^-$) by alkaline reduction, a process called vatting.[2,50] The dye application method involves three stages, namely, alkaline reduction and dissolution of the vat dye (normally using sodium hydroxide and sodium dithionite (hydrosulphite)), absorption of the substantive leuco compound by the fibre, aided by electrolyte (e.g. sodium sulphate and wetting, dispersing and levelling agent), followed by regeneration of the vat dye inside the fibre by oxidation in air or hydrogen peroxide. The dyed fibre is then thoroughly soaped off (washed with soap or special detergent at a high temperature) to remove any surface dye and to complete any dye aggregation inside the fibre in order to obtain the final colour.

Vat dyes are based upon indanthrone, flavanthrone, pyranthrone, dibenzanthrone, acylaminoanthraquinone, carbazole, azoanthraquininone, indigoid and thioindigoid structures.[2,50] Application methods include both batch exhaust dyeing and continuous pad-steam-wash off and dry methods, and apart from sodium dithionite, formaldehydesulphoxylate or hydroxyacetone may be used for the reduction stage.[50]

Vat dyes are relatively expensive but do offer outstanding colour fastness to light and washing on cellulosic fibres. In the unreduced form some vat dyes behave like disperse dyes and can therefore be applied to polyester and polyester/cellulosic blends to achieve pale-medium colour depths. For polyester/cellulosic fabrics a pad-dry-bake-chemical pad-steam-oxidise-wash off and dry production sequence is used. For exhaust dyeing, the novel technique of electrochemical reduction dyeing has been claimed to provide a redox potential of up to $-960\,mV$ suitable for the reduction of vat dyes and also sulphur dyes. Cathodic reduction of vat dyes with the addition of a mediator (a soluble reversible redox system, e.g. an iron [II/III]–amino complex) could potentially decrease the chemical costs for reduction and lower the chemical load in the waste water by 50–75%.[51]

9.4.9   Disperse dyes

Disperse dyes are substantially water-insoluble dyes which have substantivity for one or more hydrophobic fibres (e.g. secondary cellulose acetate) and are usually applied as a fine aqueous dispersion.[2,52,53] The major chemical classes used are aminoazobenzene, anthraquinone, nitrodiphenylamine, styryl (methine), quinophthalone and benzodifuranone-based dyes. Disperse dyes are milled (ground) with a dispersing agent (e.g. polymeric forms of sodium dinaphthylmethane sulphonates) to a fine dispersion (0.5–2 µm) and may be supplied as grains, powders or liquid dispersions.

Disperse dyes are essentially nonionic dyes that are attracted to hydrophobic fibres such as conventional polyester, cellulose triacetate, secondary cellulose acetate and nylon through non-polar forces of attraction.[33,52,53] In the dyebath, some of the disperse dye particles dissolve to provide individual dye molecules that are

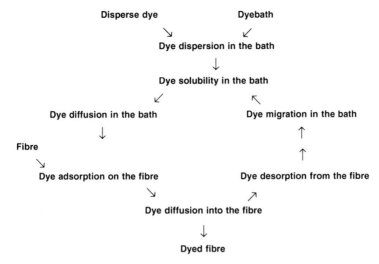

9.1   General dyeing mechanism for disperse dyes (source: see ref 35, p. 126).

small enough to diffuse into the hydrophobic fibres. The aqueous solubility of disperse dyes is low, for example 0.2–100 mg l^{-1} at 80 °C, but increases with increase in
dyebath temperature, in turn increasing the concentration of soluble dye available
for diffusion so that the dyeing rate increases.

Dispersing agents are essential because they assist in the process of decreasing
the dye particle size and enable the dye to be prepared in powder and liquid forms.[52]
In addition, the dispersing agent facilitates the reverse change from powder to
dispersion during dyebath preparation and maintains the dye particles in fine dispersion during dyeing. This prevents agglomeration of the dye particles in powder
form and aggregation in the dyebath. The dye solubility in the dyebath can be
increased by the use of levelling agents and carriers.

The general mechanism of dyeing with disperse dyes under exhaust dyeing conditions is illustrated in Fig. 9.1. The diffusion and adsorption of dye molecules is
accompanied by desorption of some of the dye molecules from the dyed fibre back
into the dyebath, to facilitate dye migration from fibre to fibre to achieve level
dyeing.[35] The disperse dyes are considered to dye fibres via a solid solution mechanism that involves no chemical change.[23] Each disperse dye dissolves in the fibre
more or less independently of any other disperse dye present. Under constant
dyebath conditions the ratio of the amount of dye on the fibre to the amount of dye
in the bath is a constant at equilibrium, which varies according to the particular dye,
but ultimately a separate different saturation level (or solubility limit) is achieved
for each dye above which no more of that dye can be taken up by the fibre. However,
most commercial dyeings are never carried out to equilibrium.

Apart from high temperature (high pressure) exhaust dyeing at 110–140 °C, disperse dyes may also be exhaust dyed using carriers (e.g. fibre plasticising agents that
decrease the fibre T_g) at 100 °C.[22,52,53] This method is not very popular except where
blend dyeing is practised and where high temperature dyeing, if used, would degrade
one of the fibres. Odourless carriers are preferred, but there are problems with
carrier spotting, and carrier retention by the fibres leads to fuming of the carrier
as it is volatilized off during post-heat setting treatments, while some carriers may

potentally pose environmental pollution problems.[22] Some 50% of disperse dye consumption is used for navy and black shades, and some 70% of disperse dyes are applied to polyester and polyester/cellulose blends by exhaust dyeing.[54]

Disperse dyes may also be applied by a pad-dry-thermofix-wash-dry process in a continuous open width fabric treatment. This utilises the sublimation properties of disperse dyes, which vaporise directly from the solid state without prior melting. Polyester can be dyed at 190–220 °C by such a process, originally called the Thermosol process by DuPont.[23] Sublimation of disperse dyes is also utilised in dry-heat transfer printing,[55] but conventional print application by printing and fixing in saturated or superheated steam is also practised, followed by washing off and drying.[32]

To minimise potential problems linked to the presence of disperse dyes on the fibre surface after dyeing (e.g. poor colour fastness to washing, rubbing and thermomigration), reduction clearing after dyeing with sodium hydroxide, sodium dithionite (hydrosulphite) and a surfactant followed by hot washing is normally practised, particularly with medium-heavy depths of colour.[52] However, some classes of disperse dyes (e.g. di-ester and thiophene azo dyes) are clearable by alkali alone.[56]

9.4.10   Pigment formulations

The range of chemical compounds used as pigments varies widely and includes both inorganic pigments such as titanium dioxide, and the oxides of antimony, iron and zinc, as well as carbon.[57] Organic pigments may be based upon a very wide variety of chemical structures, for example azo, anthraquinone, dioxazine, indanthrone, isoindolinone, perylene, quinacridone, copper phthalocyanine, heterocyclic nickel complexes and many others (see, for example, Table 9.4).[58] The general considerations relating to the manufacture of pigments and their application to technical textiles are separately discussed in Section 9.3.2 (manufacture), Sections 9.5.1–9.5.3 (mass pigmentation), Section 9.5.5 (coating) and Section 9.9.1 (printing). The finishing of pigment formulations is technically complex and strict quality control is required to ensure satisfactory application and performance of the pigment in practice.

9.5   Mass coloration of manufactured fibres

9.5.1   Dyes and pigments for mass coloration

In mass coloration, dyes or pigments are incorporated into the polymer melt (in the case of polyamide, polyester and polypropylene) or into the polymer solution (in the case of viscose, secondary cellulose acetate, cellulose triacetate and acrylic or modacrylic fibres) during fibre manufacture.[6,13] The dyes that are being used pass into a dissolved phase on incorporation into the polymer melt or solution, but pigments remain as finely dispersed particles. In the main, mass pigmentation is much more widely used for colouring manufactured fibres (both synthetic and regenerated fibres)[6,13] except on wet-spun acrylic fibres, where gel dyeing is used (see Section 9.5.4).[35,59] However, mass pigmentation can be used for dry-spun acrylic polymers which are dissolved in a solvent (e.g. dimethyl formamide) and spun into hot air.

Technical criteria that are important for mass coloration include:[13]

Table 9.4   High performance pigments for polypropylene fibres

CI generic name	CI number	Chemical class
CI Pigment Yellow 93	20710	Azo condensation
Yellow 95	20034	Azo condensation
Yellow 109	56284	Isoindolinone
Yellow 110	56280	Isoindolinone
Yellow 155	Not known	Azo condensation
Yellow 181	11777	Monoazo
Yellow 182	Not known	Heterocyclic monoazo
Orange 61	11265	Isoindolinone
Red 122	73915	Quinacridone
Red 144	20735	Azo condensation
Red 149	71137	Perylene
Red 166	20730	Azo condensation
Red 177	65300	Anthraquinone
Red 202	Not known	Quinacridone
Red 214	Not known	Azo condensation
Red 242	20067	Azo condensation
Red 257	Not known	Heterocyclic nickel complex
Violet 19	73900	Quinacridone
Violet 23	51319	Dioxazine
Violet 37	51345	Dioxazine
Blue 15:3	74160	Copper phthalocyanine
Blue 60	69800	Indanthrone
Green 7	74260	Copper phthalocyanine

CI denotes Colour Index. Source: see ref 58.

- pigment particle size and particle size distribution, both in the spinning mass and in the filament
- pigment preparations
- solubility of the colorant in the spinning mass
- colour fastness properties
- ability to remain stable under the mass processing conditions of the polymer.

Fine pigment particles (<1 μm) are required because coarse particles would interfere with the filterability of polymer solutions or melts and could impair the tensile strength of fibres, the diameter of which normally lies in the range 16–45 μm. The pigments used must have good resistance to organic solvents and good heat stability for use in fibre manufacture.

Products suitable for the bath dyeing and mass coloration of manufactured fibres are illustrated in Table 9.5.[13]

9.5.2   Mass coloration methods

There are four main methods of incorporating colorants into manufactured fibres, which depend upon the specific fibre production process.[6,13] These are:

- batch process
- injection process
- chip blending
- chip dyeing.

Table 9.5   Products for the bath dyeing and mass coloration of manufactured fibres

Fibre	Bath dyes	Solvent dyes and pigments for mass coloration
Viscose	Direct Vat Reactive	Aqueous pigment pastes
Acetate	Disperse	Acetone-soluble dyes Pigment dispersions in acetate
Acrylic	Disperse Basic	Pigment dispersions in polyacrylonitrile
Polyamide	Disperse Acid Metal complex	Aqueous dyes Polymer-soluble dyes
Polyester	Disperse	Polymer-soluble dyes Pigment dispersions in polyester
Polypropylene	Acid chelatable	Pigment dispersions in various carriers

Source: see ref 13.

Pigment preparations or solvent dyes may be used in the batch method in which the whole of the spinning mass is coloured, the colorant concentration being equal to that in the coloured filament after extrusion. In the injection process, a coloured concentrate is continuously metered into the spinning mass which is then extruded. Chip blending is suitable only for melt coloration, for example for polyamide, polyester and polypropylene. Polymer chips are homogeneously mixed with the colour concentrate prior to extrusion. Chip dyeing is a more specialised technique used with nylon 6. The polymer chips are precoloured with polymer-soluble dyes and then melt spun.[13]

Mass pigmentation is widely used as the major coloration route for polypropylene fibres.[1,6,13] The flexibility of the smaller scale nature of polypropylene fibre manufacture compared with the large continuous polymerization/extrusion processes for polyamide and polyester fibres, the lower polymer melt temperature of polypropylene and the difficulties in dyeing this hydrophobic, low linear density fibre in conventional dyeing equipment, ensure that mass pigmentation is the major coloration route. Mass coloration of many other melt-spun fibres is not economic unless large fibre weights per colour may be spun for a known end-use, because of the high costs of cleaning out the production machinery to avoid subsequent colour contamination problems. However, attempts continue to find a satisfactory conventional dyeing method for modified polypropylene fibres using exhaust dyeing or continuous dyeing techniques.[1,60]

9.5.3   Mass coloration and colour fastness properties

The colour fastness properties of manufactured fibres produced by mass dyeing, and particularly by mass pigmentation, are generally superior to those obtained on technical textiles by conventional dyeing and printing. Mass pigmentation is used where high colour fastness to light and weathering is required, for example in tenting fabrics, awnings, sun blinds, carpets, synthetic sports surfaces and so on, and where the colour can be economically produced.

The organic pigments and a few inorganic pigments used for mass pigmentation of polyester and polypropylene fibres must exhibit:[58]

- sufficient thermal stability
- colour fastness to light, adequate for the intended end-use
- no migration (no blooming or contact bleed)
- compatibility with other additives, e.g. the UV stabilisers widely used in polyolefins, with no photodegradation effects on the polymer
- no adverse effect on the mechanical properties of the fibres.

High performance organic pigments are used for brilliant colours and good overall colour fastness performance on polypropylene fibres (see Table 9.4).

Inorganic pigments such as titanium dioxide (rutile), zinc oxide, and antimony oxide are used for white, and carbon (small particle size channel black) is used for black, while some iron oxide browns are used for cost reasons. The tensile strength of mass-pigmented polypropylene fibres after 300 hours exposure in an Atlas 600 WRC accelerated light fastness testing equipment is markedly higher than the uncoloured fibre, demonstrating the protective effect of the presence of the pigments against the deleterious effects of ultraviolet radiation on the fibres.[58]

Colour retention in awning fabrics in outdoor applications is critical, and two types of climatic conditions are most critical where pigments are incorporated within acrylic and modacrylic technical textiles:[61]

- hot and dry (typically 38–49 °C, relative humidity <20%) with high average sunshine
- warm and humid (32–38 °C, relative humidity >70%) with average sunshine.

In hot and dry conditions, both the polymer and the pigment may degrade, whereas in warm and humid conditions pigment degradation is more likely. However, pigments that are degraded and fade under warm and humid conditions may often be quite stable under hot dry conditions.

9.5.4   Gel dyeing of acrylic fibres

For wet-spun acrylic fibres, the manufacture of producer-dyed fibres involves the passage of acrylic tow in the gel state (i.e. never-dried state) through a bath containing basic dyes.[35,59] This is utilised in the Courtelle Neochrome process (formerly Courtaulds, now Acordis) to produce dyed acrylic fibres from a continuous fibre production line typically at a speed of 50 m min^{-1} with economic batch weights per colour of 250–500 kg.

The liquid basic dyes are metered in at a rate appropriate to the acrylic tow mass and speed, the recipe being based on a computerised colour match prediction system allowing the selection of a very wide range of colours using a choice of the technically best dyes (i.e. easiest to apply and highest colour fastness), or lowest cost dyes, or least metameric dyes.[62] As the freshly coagulated acrylic tow passes through the dyebath, the basic dyes diffuse inside the gel-state acrylic tow in a matter of seconds. The dyed tow is then drawn and steamed, crimped and cut to the appropriate staple length for use in technical textiles, or may be used alternatively in filament form.[59]

9.5.5   Pigments for technical coated fabrics

Many technical textiles are coated in order to provide the high performance specification demanded by the end-use. The many considerations governing the use of coloured organic pigments in coatings, and the range of pigments available, have been reviewed by Lewis.[63]

9.6   Conventional dyeing and printing of technical textiles

9.6.1   Pretreatments prior to conventional coloration

Before fibres, yarns or fabrics are to be dyed via batch exhaust or continuous pad-fixation methods, or printed by print-dry-steam-wash off-dry or pigment print-dry-cure methods, it is important to remove any natural, added, or acquired impurities from the fibres in order that these impurities do not interfere with coloration. Natural fibres such as cotton, wool, silk and flax (linen) fibres contain appreciable quantities of impurities naturally associated with the fibres, which are removed by scouring and by other treatments, for example degumming of silk to remove sericin (gum) and carbonizing of wool to remove vegetable matter (e.g. burrs, seeds etc.).[64] Particularly important for cotton and flax (linen) fibres is the efficient removal of the natural waxes present mainly on the fibre surface, in order to impart hydrophilicity to the fibres, thereby ensuring satisfactory wetting out, with high levels of uniformity and reproducibility in subsequent dyeing or printing treatments.[65]

Manufactured fibres and filaments, that is, artificial fibres from synthetic or regenerated fibres, are produced under carefully controlled conditions so that water-soluble/emulsifiable spin finishes or fibre lubricants are the main impurities, and these are simply removed by scouring.[64]

9.6.2   Singeing

For many woven or knitted technical fabrics, it may be necessary to singe the fabric surface, by passage through a gas flame or an infrared zone at open width, to remove protruding surface fibres. This gives a clear fabric surface and more uniform coloration, because a hairy fabric surface can impart a lighter coloured surface appearance (termed frostiness) after dyeing or printing. Singeing may be integrated with subsequent wet processes such as desizing.[66]

9.6.3   Desizing

The warp yarns of most woven fabrics are generally sized with a film-forming polymer that adheres fibres together with a more cohesive structure.[67] Sizes are used to increase yarn strength, decrease yarn hairiness and impart lubrication to staple fibre warp yarns in order to minimise the number of warp breaks during weaving. In the weaving of synthetic continuous filament yarns, the size provides good inter-filament binding to prevent filament snagging as well as providing yarn lubrication and antistatic performance to decrease the build-up of electrostatic charges on the yarns during high speed weaving. Yarns for knitting are not sized and a desizing treatment is not therefore required.

All sizes, together with any other sizing components, for example waxes, softeners and lubricants that may be hydrophobic, must be removed by appropriate desizing treatments.[68] Synthetic water-soluble sizes, (e.g. acrylates, polyvinyl alcohol) may be simply removed by washing, whereas natural sizes, (e.g. starch, modified starch) may require chemical degradation treatments, like oxidation, hydrolysis and so on. If considered appropriate, desizing may be integrated with scouring, or with scouring and bleaching to provide a shorter integrated treatment, while in large, vertically integrated plants, undegraded sizes that are removed in desizing may be recycled using ultrafiltration and reused.

After every desizing treatment, all traces of size or degraded size are removed by thorough washing followed by efficient mechanical removal of liquid water (e.g. hot mangling or vacuum extraction). This is followed by thermal drying on steam heated cylinders (called cylinder or can drying) or by hot air drying on a stenter (also known as a tenter, or frame (USA)).

9.6.4   Scouring
Scouring is a critical treatment for natural cellulosic fibres, for example cotton and flax (linen) and their blends with other fibres, because all traces of hydrophobic waxes (whether naturally occurring or applied during the manufacturing sequence) must be effectively removed in order to achieve satisfactory wetting in all subsequent dyeing, printing, finishing, coating, lamination and bonding operations.[65] Scouring is normally accomplished by hot alkaline treatment (e.g. sodium hydroxide) followed by a thorough hot wash off. It is also important for the removal of fatty materials from wool and any manufactured fibres using hot detergent solutions.

9.6.5   Bleaching
Fibres must be uniformly white if pale colours or bright colours are required, and hence natural fibres such as cotton, silk, wool and linen must be chemically bleached in order to achieve a satisfactory stable whiteness.[8,64] Hydrogen peroxide under controlled alkaline conditions is normally used in pad-steam-wash off, pad-batch-wash off or immersion bleaching treatments, although other oxidising agents such as peracetic acid, cold alkaline sodium hypochlorite or sodium chlorite under acidic conditions may also be used.

However, chlorine-based bleaching agents give a poor bleaching environment and chlorine is retained by the fibre, necessitating an antichlor aftertreatment with a reducing agent (e.g. sodium sulphite) followed by washing to remove any residual odour in the fabric. Retained chlorine from chlorine-based bleaches or from chlorinated water used for dyeing can give rise to shade changes when dyeing cotton with reactive dyes. Regenerated cellulosic fibres such as viscose, cuprammonium, polynosic, modal, high wet modulus and lyocell fibres are marketed by the fibre producers with a satisfactory whiteness for dyeing and printing.

9.6.6   Fluorescent brightening
If high whites are required, both natural and regenerated cellulose fibres may be chemically bleached and treated with a fluorescent brightening agent.[15,16] Fluorescent brightening agents are based upon diaminostilbene derivatives, triazoles, aminocoumarins and many other organic compounds, and are absorbed by the fibre.

Fluorescent brightening agents are organic compounds that absorb the ultraviolet radiation present in daylight and re-emit light in the blue–violet region of the visible spectrum. During the absorption and re-emission processes, some energy is lost and hence the light emitted is shifted to a longer wavelength. The effect is to add brightness to the whiteness produced by chemical bleaching. Alternatively, blue tints may be used, for example ultramarine.

9.6.7   Mercerization

Mercerization of cotton is a fibre swelling/structural relaxation treatment that may be carried out on yarns, but more usually on fabrics.[69,70] Hank or warp mercerization of yarns often creates dyeability differences because of yarn tension variations that pertain during mercerization. During mercerization in 22–27% caustic soda solution, both mature and immature cotton fibres swell so that the secondary wall thickness is increased. The fibre surface appearance and the internal structure of the fibre are modified. This improves the uniformity of fabric appearance after dyeing and there is an apparent increase in colour depth after mercerization which has been claimed to give cost savings of up to 30% on pale colours (1–2% dye owf, on weight of fabric) and even 50–70% on heavy depths when using some reactive dyes.[71,72] Dead cotton fibres (i.e. those with little or no secondary wall) are, however, not improved after mercerization.

Woven fabric mercerization is normally carried out under tension on chain or chainless fabric mercerizing ranges,[70] whereas tubular fabric mercerizing ranges are widely used for weft knitted cotton fabrics.[73] Mercerization leads to a number of changes in fibre and fabric properties:[69–73]

- a more circular fibre cross-section
- increased lustre
- increased tensile strength, a major factor for technical textile fabrics
- increased apparent colour depth after dyeing
- improved dyeability of immature cotton (greater uniformity of appearance)
- increase in fibre moisture regain
- increase in water sorption
- improved dimensional stability.

After mercerization, the structure of native cotton fibres, cellulose I, is converted into cellulose II which is the stable fibre form after drying.[72] The sorptive capacity of mercerized cotton is greater where the fabric is mercerized without tension (slack mercerizing) to give stretch properties to the fabric. An increase in drying temperature can also decrease the sorptive capacity, especially at temperatures above 80 °C.[74]

9.6.8   Anhydrous liquid ammonia treatments

This form of cotton fabric pretreatment is much less common than mercerizing and is most widely used in Japan.[72,75] Impregnation in anhydrous liquid ammonia at −38 °C in an enclosed machine followed by a swelling/relaxation stage, and removal of the ammonia by thermal drying and steaming, converts the cellulose I crystalline form back to either cellulose I or into cellulose III, depending upon the structural collapse of the fibre while the final traces of ammonia are removed in the steamer.

This ammonia-dry-steam process can be used to give better improvements in cotton fabric properties than mercerization, although the increase in colour depth after dyeing is usually somewhat lower than that achieved after mercerization. The high capital cost of the machinery for anhydrous liquid ammonia treatment and ammonia recovery, and environmental considerations, have limited the wider exploitation of this technique.

9.6.9 Heat setting

Synthetic thermoplastic fibres, yarns and fabrics may be heat set, steam set or hydro set in order to obtain satisfactory dimensional stability during subsequent hot wet treatments.[76] Fabrics may be preset prior to coloration or postset after coloration.[31] Hydrosetting in hot water is rarely carried out, but false twist textured yarn can be steam set in an autoclave using a double vacuum-steam cycle to attain satisfactory removal of air and hence uniformity of temperature in the treatment. Steam setting avoids the slight fibre yellowing that can occur in fabric form through fibre surface oxidation during hot air setting on a stenter, and the handle is softer.

Heat setting on modern stenters is often carried out by first drying and then heat setting in one passage through the stenter. The temperature and time of heat setting must be carefully monitored and controlled to ensure that consistent fabric properties are achieved. During heat setting, the segmental motion of the chain molecules of the amorphous regions of the fibre are generally increased leading to structural relaxation within the fibre structure. During cooling, the temperature is decreased below the fibre glass transition temperature (T_g) and the new fibre structure is stabilised. Because the polymer chain molecules have vibrated and moved into new equilibrium positions at a high temperature in heat setting, subsequent heat treatments at lower temperatures do not cause the heat-set fibres to relax and shrink, so that the fabric dimensional stability is high.

Presetting of fabric prior to dyeing alters the polymer chain molecular arrangement within the fibres, and hence can alter the rate of dye uptake during dyeing.[23,31,53] Process variations (e.g. temperature, time or tension differences) during heat setting may thus give rise to dyeability variations that become apparent after dyeing. Fabric postsetting after coloration can lead to the diffusion of dyes such as disperse dyes to the fibre surface and to sublimation, thermomigration and blooming problems, all of which can alter the colour and markedly decrease the colour fastness to washing and rubbing of technical textiles containing polyester fibres.[56]

9.6.10 Quality control in pretreatment

In all the pretreatments given to fabrics, it is necessary to control the process carefully in order to minimise fibre degradation and yellowing.[8] Mechanical damage (i.e. holes, poor dimensional stability and fibre damage through overdrying) must also be avoided. Fibre yellowing makes it very difficult to dye pale bright colours while any fibre degradation may be subsequently increased during dyeing and printing treatments and result in inferior physical properties (e.g. low tensile strength, tear strength and abrasion resistance) or to poor colour fastness performance. Inadequate pretreatment can lead to poor wettability, uneven coloration and inferior adhesion of coatings in technical textiles.

9.7   Total colour management systems

9.7.1   Specification of colours and colour communication

The time-honoured method of specifying colours involves sending physical standards (e.g. dyed or coloured patterns) to the dyehouse followed by dyeing samples to match the colour in the laboratory.[77] The samples are submitted to the colour specifier for acceptance or rejection. If they are not accepted by the colour specifier, this process must be repeated in order to obtain satisfactory colour matching to the standard. The approved laboratory sample will then be used as the basis for the initial dye recipe for bulk dyeings. A sample from bulk dyeing may then be submitted to the specifier for final approval prior to delivery. Similar principles apply to the production of laboratory strike-offs of each colour in a print for approval by the colour specifier prior to machine printing.

All this is a time-consuming procedure which has now been shortened considerably in order to avoid colour changes through storage and handling of physical samples and to achieve a quick response and provide just-in-time delivery to the customer. The use of a colour specifier program can dramatically decrease the cycle time for approval. This involves the use of instrumental colour measurement for formulation and quality control, accreditation procedures to enable self approval of colours and colour specification using reflectance data.[77–79] In the most advanced colour management systems it is now possible to visualise a colour on a colour monitor so that colour communication between the colour specifier and the dyer is vastly simplified. A number of standardised colour specification systems, for example Pantone, Munsell Chip systems, can also be used.[79]

9.7.2   Colour measurement

Objective colour matching using a spectrophotometer for colour measurement is superior to visual colour matching assessment because instrumental colour measurement allows the colour to be specified in terms of reflectance data and the precision of colour matching to the standard reflectance data can then be calculated numerically in colour difference units (ΔE).[80]

Colour tolerancing systems can be used to provide better agreement between visual assessment and the instrumentally measured colour difference, and colour acceptability limits can be agreed in advance in order to facilitate decisions on colour.[80–82] Colour acceptability limits are numerical values at which the perceived colour differences become unacceptable to the specifier, that is, single number shade passing systems. Instrumental shade sorting systems are widely used to separate fabric lengths on fabric rolls into similar colours.[83]

The spectrophotometers currently in use may give different measured reflectance values according to the instrument design. The geometry, wavelength scale, bandwidth or light source may differ between different makes of spectrophotometer.[77,80,81] The two most common geometries used for direction of illumination/

direction of view are: diffuse /8° and 45°/0°. Readings given by these two geometries are not compatible unless the ideal perfect diffuser is measured. Illumination is provided either by a tungsten filament halogen cycle incandescent lamp or by a pulsed xenon discharge lamp.

Tungsten filament lamps do not have significant ultraviolet emission compared with pulsed xenon discharge lamps. The measurement of samples containing fluorescent brightening agents or some fluorescent materials is thus affected by inclusion or exclusion of the ultraviolet component of the illumination.[77]

Spectrophotometers use either single-beam, double-beam or dual-beam optics. The latter are now more widely used to compensate for the variability of the spectrum of pulsed xenon discharge lamps.[77,84] The collection optics may gather light across the sample image, focussing it on to the spectrophotometer aperture, or alternatively it is imaged at infinity, gathering light along a narrow range of angles. This latter method is claimed to provide a greater depth of field and insensitivity to sample surface imperfections. Considerable progress has been made by the manufacturers of spectrophotometers to ensure that the instruments give reliable, repeatable measurements, and the inter-instrument agreement (i.e. the colour difference values for a set of colour standards measured with two or more instruments of the same model) is now very low (typically 0.04 ΔE). Inter-instrument agreement and precision of colour measurement are now very important for the colour measurement of some technical fabrics where the colour matching has to be very close to the standard, as in automotive fabrics where reproducibility to within 0.3 ΔE is typically demanded.

The spectrophotometers in use may utilise 16 data points at 20 nm intervals over the 400–700 nm range, 31 data points from 400 to 700 at 10 nm intervals, or 40 data points from 360 to 750 nm at 10 nm intervals.[77] In addition, the bandwidth may be 20 nm or 10 nm or less. As a result, it is difficult to convert measurements from one format to the other. Although conversion is possible, the colour difference measured at two different bandwidths on the same sample can be as high as 2.0 CIELAB units. However, the largest aperture should be used for colour measurement because this should provide more repeatable measurements.

Colour measurement of dyed textiles should be carried out on conditioned samples because the colour may change with temperature and moisture content and the colour change may exceed the pass/fail tolerance.[77] The sample presentation should also specify the number of fabric layers and the backing material to be used because these factors also affect the perceived colour. Colour measurement of loose fibres generally gives rise to greater variations in measurement than on woven fabrics so that it is advisable to use multiple measurements to improve the repeatability. There are considerable differences in colour between cut and uncut pile yarn, where measurements are made on the fibre tips of a cut hank of yarn or on the side of the yarn.[80]

Portable colour measurement systems utilise hand-held spectrophotometers with data storage and extensive memory capacity for monitoring the colour batch-to-batch and within batch (e.g. side-centre-side, and end-to-end colour variations within a fabric roll).[85] These are very convenient for quality control purposes both within the laboraratory and in the production plant, but sophisticated on-line colour measurement systems are also employed on continuous fabric dyeing ranges.[86] A typical total colour control system is illustrated in Fig. 9.2.[80]

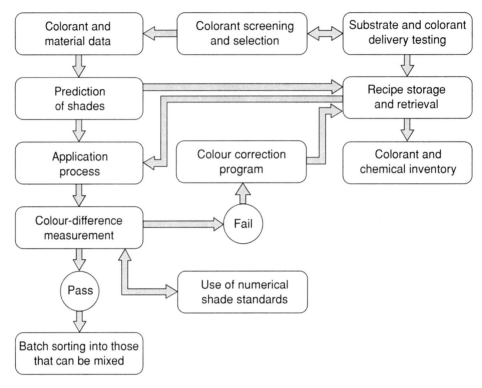

9.2   Total colour control (source: see ref 80, p. 45).

9.7.3 Laboratory matching

In the modern dyehouse, dyestuffs and chemicals are purchased to an agreed quality standard.[19,80] Dyestuffs for use in laboratory matching in many dyehouses are dissolved or dispersed in stock solutions above an automated laboratory dispensing unit. The colour specifier may supply a physical sample, the reflective curve of which is measured on the spectrophotometer in the laboratory or reflectance data are supplied directly to the laboratory. Computer match prediction is then carried out using the appropriate database of the dyes used in the dyehouse on the substrate to be dyed.[80] If the substrate to be dyed in bulk has a different dyeability to that originally used to construct the database, adjustment of the recipe must be made. The colorist in the laboratory must then select from the computer match prediction the most appropriate recipe, taking into account such factors as recipe cost, the anticipated technical performance in bulk dyeing, colour fastness, metamerism, and the closeness of the predicted colour match to the specified colour.

The appropriate dye recipe is then dispensed manually or on a laboratory dye dispensing unit and a sample of the material is then dyed in an automated laboratory dyeing machine under conditions that will simulate those to be used in bulk dyeing, for example the same pH, chemicals and auxiliaries, liquor ratio and temperature–time relationship. The laboratory matching is given the same rinsing or after-treatment that will be used in bulk dyeing, and is dried, conditioned, and the colour measured on the spectrophotometer. If the colour is a commercial match to the specified colour, then bulk dyeing is initiated. If the colour is outside the commercial colour tolerance owing to differences in substrate dyeability and so on, a corrected recipe is predicted and the process repeated to obtain a commercial colour match followed by bulk dyeing.[80,81,87]

Where there is close control over the colour strength of the dyestuffs and consistent substrate dyeability, it is often possible to operate so-called 'blind dyeing' in which the computed dye recipe in the laboratory is used immediately for bulk dyeing.[87] This shortens the time required, decreases dyehouse costs and offers quick response and rapid delivery to customers. Where repeat dyeings of the same colour are required, it is usually possible to input the reflectance data gained from bulk dyeings in order to refine the database and thereby achieve a greater level of right-first-time dyeings. Right-first-time, right-on-time, right-every-time dyeing is the goal of the dyer, because this is the lowest cost dyeing system that provides quick response for customers.[87]

If a bulk dyeing proves to be off-shade and the original recipe requires a correction (i.e. an addition) that results in extending the dyeing time, the dyer suffers a considerable financial penalty. If the colour is too dark and the bulk dyeing must be stripped and redyed, this imposes further cost penalties and often impairs the quality, physical properties and surface appearance of the dyed material. In general, the shorter the processing time under hot wet conditions in the dyebath, the lower will be the fibre degradation and hence the quality of the dyed material will be superior, all important considerations for many technical textiles.

9.8   Dyeing machinery

Apart from mass coloration that has been discussed in Section 9.5, conventional dyeing machinery is used for the dyeing of technical textile materials in the form of all types of fibres, tows, yarns (e.g. warp, hank, package), fabric, garments, carpets and so on.[13,88–91] Fabric may be dyed in rope form (i.e. as a strand) or at open width (i.e. flat), or circular weft knitted fabrics may be dyed in tubular form. Modern trends are to dye in yarn or fabric form because this allows the decision on colour to be made later in the manufacturing chain and the cost of dyeing is lower. Many technical fabrics are dyed at open width to avoid inserting creases into the fabric. Such creases can be difficult to remove in synthetic fibre fabrics because of hydrosetting occurring during dyeing.

9.8.1   Batch dyeing machinery for exhaust dyeing

In conventional exhaust dyeing, dye molecules are transported to the fibre surface where the substantivity of the dye for the fibre ensures that adsorption takes place to produce a higher surface concentration of dye.[92] This promotes dye diffusion into the fibre, the rate of dye diffusion being dependent upon the dye concentration. Dye molecules diffuse within the disordered regions of the fibres, to provide adequate fibre penetration.[21] Level dyeing is attained through the migration of dye, by desorption from inside the fibre back into the dye liquor and thence adsorption and diffusion in another part of the fibre, this levelling process being aided by the relative motion of the dye liquor and the fibre. After an appropriate time, a level, well-penetrated dyeing is obtained. This process is facilitated if compatible dyes, that is, dyes that diffuse at the same rate, are initially selected for use together with a

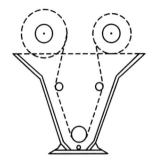

9.3   Jig dyeing machine (source: see ref 89, p. 29).

selected levelling agent. Dye fixation is ensured either through (i) electrostatic attraction between the dye and dyesites inside the fibre, (ii) by covalent bond formation between the dye and the fibre, (iii) by increasing the size of the dye or by dye aggregation leading to mechanical entrapment, or (iv) by conversion of the water-soluble dye into a water-insoluble form.

The three types of dyeing machinery for batch or exhaust dyeing are based on machines in which:[88,89,91]

1   the material moves, but the liquor is stationery, e.g. jig and winch (beck) machines for fabric dyeing
2   the liquor moves, but the material is stationery, e.g. hank or package dyeing of yarns and beam dyeing of fabrics
3   both the liquor and the material move, e.g. jet, softflow and overflow jet dyeing machines for fabrics.

In general, a lower liquor ratio decreases the water and energy consumption and decreases the volume of waste water, decreasing effluent treatment costs. It can also facilitate a more rapid dyeing cycle and increase the dye exhaustion on the fibre, because the total dye liquor is circulated more rapidly.

In the jig (or jigger), fabric is dyed at open width by traversing fabric from one roll through a dyebath and on to a second roll (see Fig. 9.3).[89] When the second roll is full, the rolls are stopped and the fabric motion is reversed, and the procedure repeated as required. The dye liquor is added in portions to ensure satisfactory levelness from end-to-end of the fabric. A hood prevents the release of steam and helps to maintain the fabric temperature on the rolls where the majority of the dyeing takes place. Most jigs operate at 98–100 °C (i.e. under atmospheric conditions) but pressurised jigs operating at up to 140 °C are used for some technical fabrics, for example polyester sailcloths. In the latter machine, the rolls and the dyebath are enclosed in an autoclave (i.e. pressure vessel) which can be closed with a ring seal. Jig dyeing operates at a liquor ratio of 3–5:1 and rinsing/washing off is more rapid on modern twin bath jig designs in which a combination of water spray and vacuum is used for removal of loose dyestuff.[93]

Jig dyeing is particularly suitable for technical fabrics that may be subjected to creasing and hence are preferably dyed at open width. In addition it is preferred for many technical fabrics that have a dense structure and where it is difficult to pump

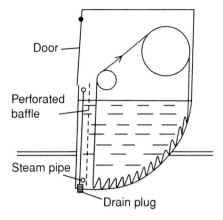

9.4 Winch dyeing machine (source: see ref 89, p. 38).

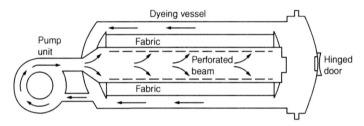

9.5 Sectional diagram of a high temperature beam dyeing machine (source: see ref 27, p. 117).

dye liquor through multiple fabric layers, as in beam dyeing (e.g. sailcloths, filter fabrics etc.).

The winch or beck dyeing machine may be used for dyeing in rope form, normally at temperatures of up to 98–100 °C (see Fig. 9.4).[89] High temperature winches have also been used in the past, but have now been replaced by high temperature (i.e. pressurised) jet dyeing machines. The latter are less likely to give rise to fabric creasing and have improved liquor interchange which aids the more rapid attainment of level dyeing by promoting dye migration.

In beam dyeing, the fabric is wound at open width on to a perforated stainless steel beam, and kept in place by the use of end plates (see Fig. 9.5).[89] Dye liquor is pumped through the multiple layers of fabric, usually from out to in, but in-to-out flow is also used. The fabric winding tension must be low to avoid stretching the fabric, and to avoid high pressure being applied to the innermost fabric layers, otherwise the fault known as watermarking or moiré may be observed. This is usually seen as a shimmering surface pattern on dark colours, the pattern changing with the orientation of the fabric relative to the observer. Too low a winding tension can, however, lead to telescoping and movement of the fabric on the beam. It is normal to wind a fent (or short sacrificial fabric length) on to the beam first to prevent perforation marks on the fabric layers next to the beam. These show as a pattern of dark spots corresponding to the perforated holes in the beam and are often caused

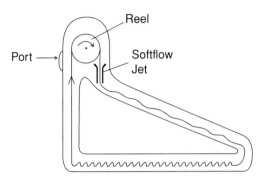

9.6   Softflow jet dyeing machine.

by poor dispersing or aggregation of disperse dyes. The liquor ratio in beam dyeing is around 10:1 although this can be decreased by the use of spacers, and most beam dyeing machines are pressurised machines capable of dyeing at temperatures of 130–140 °C. Beam dyeing is particularly effective for dyeing thin permeable fabrics composed of synthetic filaments, e.g. nylon 6.6 and polyester fabrics.

Pressurised machines are generally used for dyeing technical textiles composed of synthetic fibre materials. Softflow jets provide a more gentle mechanical action on the fabric rope, the main fabric transportation being carried out over a winch reel followed by a softflow jet system (see Fig. 9.6).[89,94] In some jet dyeing machinery, the liquor flow is split between two jets which exert a lower pressure on the fabric. This is claimed to decrease linting (loss of fibre) in staple fibre fabrics, and surface distortion. Garments are generally dyed in atmospheric or pressurised rotary drum machines or in atmospheric overhead paddle dyeing machines.[89,95]

Aerodynamic fabric transport systems utilise a blower to aid the fabric circulation through the jet dyeing machine. These enable liquor ratios as low as 1.3:1 to be used on hydrophobic filament fabrics such as polyester, although higher liquor ratios in the range 5–10:1 are used in conventional jets on hydrophilic cellulosic fabrics.[94]

9.8.2   Pad-batch dyeing

Both woven and knitted fabrics can be dyed using this method with the padding (impregnation) stage being conducted under ambient or hot conditions.[41,96] With reactive dyes, fabric prepared for dyeing is usually padded under ambient conditions and then wound up on a roll rotating on an A-frame. The dye concentration and the pick up of dye liquor by the fabric must be controlled to ensure uniform fabric coloration. The length of padded fabric on the roll may be covered with polyethylene film to prevent evaporation and is then rotated slowly (to avoid migration of the dye liquor to the outer layers of the fabric roll) for an appropriate time to achieve dye fixation. The time required is dependent upon the depth of colour, the reactivity of the reactive dye, the temperature and the alkalinity, but is generally in the range 4–24 hours (often conveniently overnight). This is then followed by a thorough hot wash off to remove the unfixed dyestuff and the alkali, and the fabric is then dried.

9.8.3   Continuous dyeing machinery

Continuous dyeing can be carried out by dye impregnation, fixation, washing off and drying on ranges designed for handling woven fabrics at open width, and there are also carpet and warp dyeing ranges.[90] Continuous dyeing was originally designed for efficient and economical dyeing of long runs of fabric, and many modifications have been made to facilitate more rapid turn around and less down-time in between a greater number of shorter production runs per colour. Automatic metering of liquid dyes to control the dye concentration in the dyebath and intelligent padding systems are used to control the pick up of dye liquor by the fabric. Thermofixation (dry heat), superheated steam or saturated steam fixation units in which the temperature is monitored and controlled may be used to obtain dye fixation. Reproducible colour is obtained after washing off loose dyestuffs and drying. Colour change during a production run is accomplished on-line within minutes using automated wash down systems. Continuous colour monitoring on the output dyed fabric on some ranges can be utilised to adjust the colour being dyed on the input fabric. Either the whole width of the fabric may be continuously monitored, or a traversing spectrophotometer measuring side-centre-side variations may be employed.

For the continuous dyeing of woven polyester/cotton fabric, the polyester fibres are first dyed with disperse dyes using a pad-dry-thermofix treatment.[90] The cotton is then dyed using direct, reactive or vat dyes using a pad-steam process. The dyed fabric is cleared of loose surface dyestuff by treatment on an open width washing range and then dried over steam-heated cylinders and batched. A scray is used at the fabric entry to the range to accumulate fabric so that the range continues to run at $50–150\,\mathrm{m\,min^{-1}}$ while the next roll of fabric is sewn on to the fabric being dyed.

9.9   Printing

9.9.1   Printing styles

Printing leads to the production of a design or motif on a substrate by application of a colorant or other reagent, usually in a paste or ink, in a predetermined pattern. In printing, each colour is normally applied separately and a wide variety of printing techniques is used. In direct printing, dyes are applied in a print thickener containing auxiliaries and are diffused into the fibres and fixed in a steamer or high temperature steamer.[57] Surface dye and thickener are then thoroughly washed off in a continuous open width washer, and the materials dried.

In pigment printing, water-insoluble pigments are applied with a heat-curable binder system, followed by drying and curing, and the physical properties of the pigment print depend greatly upon the adhesive properties of the print binder system. Pigment printing is a simple, popular method particularly for printing blended fibre fabrics.[7,57]

Two other major styles of printing are employed, namely, discharge and resist printing.[97] In discharge printing, a dyed fabric is printed with a discharging agent (a reducing agent) which decolorises the dye leaving a white motif against a dyed ground. Alternatively, a dye resistant to discharging can simultaneously be applied to give a second colour in the discharged area of the motif. In resist printing, the resisting agent (which may act mechanically or chemically) is printed on to the undyed fabric. This prevents the fixation of the ground colour which is then developed by dyeing, padding or overprinting. A white resist, or a coloured resist, can be achieved if a selected dye or pigment is added to the resist paste and is then subsequently fixed.

9.9.2   The print image

The design image may be painted by hand on to transparent film for photographic development of diapositive images, or sophisticated CAD (computer-aided design) systems are used in conjunction with a design input scanner. The design image is then manipulated on screen, the colours are electronically separated and the digitised design images for each colour transferred to an engraving system for transfer on to a rotary or flat screen, or for direct use in digital ink jet printing.[98]

9.9.3   Fabric printing machinery

Rotary screen printing is the dominant printing method now employed for 60–70% of printed fabric production.[99] The design motif for each colour is developed as open mesh on the rotary screen by the use of film, laser or black wax engraving systems. Both lacquer screen and galvano screens are used. Each rotary screen secured in end rings is fixed in bearings and rotated in precise register and in a predetermined sequence. A separate colour is supplied to each rotary screen and pushed by a squeegee (e.g. a blade or a roller) through the open mesh of the screen on to the fabric which is temporarily gummed on to a print table with a moving blanket (apron).[100] Machines capable of continuously printing up to 24–36 colours are available, although most designs involve less than eight colours, and rotary screen printing on textile materials up to 5 m in width can be carried out, at speeds up to 80 m min^{-1}, but typically at 40 m min^{-1}.

Automatic or semiautomatic flat screen (flat bed) printing is used for many designs where very precise images are required, using polyester, polypropylene or polyamide monofilament fabric as the screen material.[100] In automated flat screen printing, the fabric is gummed to the print table and progressed forward intermittently, one design repeat at a time, after the lowering, printing and lifting of the screens, to print all the colours sequentially, using a squeegee system to force the print paste into the fabric. The printing speed is lower than that normally obtained in rotary-screen printing. Modern rotary and flat screen printing machines may be supplied with enough colour for each screen, with all the operations of weighing, metering, dispersing and mixing of dyes and auxiliaries with stock thickener (suitably diluted), or of pigments with binders and softeners controlled by a robotised colour kitchen based upon a computed print recipe. The colours are supplied in drums and supply pipes are used to furnish the colour for each screen in rotary screen printing.

At the end of the print run, excess print paste is removed and stored, and this may then be reused in a subsequent print run, or disposed of to landfill. Print monitoring systems are available for monitoring and comparing the fabric print with the electronically stored digitised design image. In the most sophisticated machines, it is possible to use such monitoring systems for correction of faults online, and many machines are designed for quick changeover of colours, for example by screen and blanket washing and drying facilities on-line, and also for rapid changeover of screens at a design change.[101] Automated setting up, monitoring, control and correction systems and washing and drying on-line have all dramatically improved the productivity and repeatability in rotary and flat screen printing.

The latest advance in textile printing is the introduction of digital ink jet printing machines, capable of printing fabrics up to 2 m in width using acid, reactive or disperse dye inksets.[102–104] The fabric is normally pretreated and placed in the machine in roll form, printed and then the dyes are fixed, usually by steaming in a separate machine, washed off and dried. Both piezoelectric and bubble jet printing systems may be used, with any unused colour being diverted back to the ink reservoir and recycled. Generally four, seven–eight, and up to 12 colours may be printed; the greater the number of colours the greater the colour gamut that can be printed.

In general, digital ink jet printing systems are designed principally for use with natural fibres, (e.g. cellulosic, wool and silk fabrics) and also polyester fabrics.[102–104] On some machines, the inkjets are periodically cleaned with solvent, automatically, to avoid jet blockage, particularly with disperse dye systems. Such printing machines may be run overnight without operator supervision, the design images being preloaded, and design changeover being essentially instantaneous. Other systems are already being used for printing flags and banners and clearly have great potential for printing short production runs of advanced technical textile fabrics.[105] Ultimately, some form of reprographic printing method may be developed and research and development along these lines is continuing.[106]

Transfer printing may be carried out by sublimation transfer, melt transfer or film release methods.[55] In sublimation (or dry heat) transfer printing, volatile dyes (typically disperse dyes) are preprinted on to a paper substrate and are heated in contact with the textile material, typically polyester fabric. The dyes sublime and are transferred from the vapour phase into the fabric in this dry-heat transfer printing method. This may be assisted by the application of a vacuum. Melt transfer is principally used on garments whereby a waxy ink is printed on paper and a hot iron applied to its reverse face to melt the wax on to the fabric surface. In the film release method, the print design is held in an ink layer which is transferred completely to the textile from a release paper using heat and pressure. The design is held on to the textile by the strong forces of adhesion between the film and the textile.

9.10 Colour fastness of technical textiles

The performance of a dyed or printed technical textile when exposed to various agencies during end-use is normally assessed by appropriate colour fastness testing.[107,108] High standards of quality and performance in such tests are often related to the higher cost of the dyes and pigments used which possess superior colour fastness properties. There are national standards, for example British Standards (BS), European Standards (EN) and International Standards (ISO), but within most countries now there is a move towards harmonisation of test methods and performance standards, so that EN ISO standards will be used in the future. In North America, there are ASTM (American Society for Testing and Materials) and AATCC (American Association of Textile Chemists and Colorists) test methods. In addition, there are also test methods that have been devised by industry for use for specific applications, for example automotive textiles.

Many colour fastness tests have been devised to simulate the likely end-use conditions, and the Society of Dyers and Colourists in the UK and the AATCC in the USA have been at the forefront of colour fastness testing developments. The colour fastness test usually defines a standard test method to be followed, and gives the method of assessment to be used, but the performance level which is satisfactory for the end-use has to be agreed between the dyer or printer and the colour specifier. Companies often set their own in-house performance criteria for their ranges of dyed or printed technical textiles, based upon their general working knowledge and experience in the field, through technical liaison with dye makers, through field testing under actual end-use conditions and by assessment of materials which are the subject of complaints.

It is not possible to discuss the many colour fastness tests that are now in use worldwide, but complete details are available in many other publications, which give details of a wide range of standard tests.[107,108] The major types of colour fastness test relate to the colour fastness to wet treatments, (e.g. to washing), and to light and weathering, to rubbing (crocking), atmospheric contaminants, and organic solvents, (e.g. dry cleaning).

The colour fastness performance in standardised wash tests is rated by visual assessment of the change in colour of the coloured material and the degree of staining on to adjacent materials in the wash liquor (e.g. a multifibre fabric test strip,[109] or specific adjacent fabrics), using ISO Grey Scales under standardised lighting conditions against a neutral grey background in a viewing cabinet. The Grey Scale ratings range from 5 (no change, i.e. excellent performance) through half point ratings, for example 4–5, down to 1 (large change, i.e. poor performance). The colour change may also be measured objectively using a spectrophotometer and the colour difference converted into a Grey Scale rating.[110]

Colour fastness to light is normally assessed using a high intensity filtered xenon arc lamp to simulate natural daylight, sample strips being mounted, part covered, on cards which can be individually rotated in a bank of holders that may be rotated around the accelerated fading lamp, depending on the type of machine, in an atmosphere of controlled temperature and relative humidity. A set of blue wool standard fabric strips that fade at known rates must also be used in each test. The set-up conditions for specific test methods used for many types of technical textiles are pre-programmed on modern light fastness testing machines.[111]

The degree of fade in British and European Standards is based upon visual assessment of the degree of colour fade on the test sample compared with the equivalent degree of fade on the blue wool samples. The light fastness rating changes from 8 (highest) down to 1 (lowest). Because of difficulties with the supply of blue wool standards, the Society of Dyers and Colourists is introducing a set of blue pigment-printed light fastness standards that will replace the blue wool standards, but in addition may also extend the light fastness rating range beyond 8 and below 1.[112] Weathering may be conducted in field trials, with samples exposed to sunlight either covered or uncovered by glass, or alternatively by exposure for a standard number of hours in an accelerated fading machine.

Colour fastness to rubbing (crocking) is assessed by rubbing a standard white fabric against the dyed sample under a constant pressure for an agreed number of strokes. The test may be conducted under wet or dry conditions, and the machine may be operated by hand, or in the latest machines performed automatically.[113] Assessment of the degree of staining on the white fabric is assessed using the ISO Grey Scale for staining. For pigment printed materials, the rub fastness is dependent upon the properties of the adhesive binder used.

References

1. J SHORE, 'Coloration of polypropylene', Rev. Prog. Coloration, 1975 **6** 7–12.
2. C B STEVENS, in *The Dyeing of Synthetic-polymer and Acetate Fibres*, ed. D M Nunn, The Dyers Company Publications Trust, Bradford, 1979, 1–75.
3. K-H FLUSS, 'Space dyeing – survey of methods', Bayer Farben Revue, E1976 **26** 19–33.
4. J SHORE, *Colorants and Auxiliaries, Organic Chemistry and Application Properties, Volume 1 – Colorants*, The Society of Dyers and Colourists, Bradford, 1990.
5. W SAUS, 'SFD-Dry dyeing of polyester in CO_2', Textile Technol. Int., 1995 145–6, 148, 150.
6. P ACKROYD, 'The mass coloration of man-made fibres', Rev. Prog. Coloration, 1974 **5** 86–96.
7. V GIESEN and R EISENLOHR, 'Pigment printing', Rev. Prog. Coloration, 1994 24 26–30.
8. W S HICKMAN, in *Cellulosics Dyeing*, ed. J Shore, The Society of Dyers and Colourists, Bradford, 1995, 81–151.
9. H ZOLLINGER, *Colour Chemistry, Synthesis, Properties and Applications of Organic Dyes and Pigments*, VCH, New York, 1987.
10. I HOLME, 'The provision, storage and handling of dyes and chemicals in dyeing and finishing plants', *J. Soc. Dyers Colourists*, 1978 **94**(9) 375–394.
11. G BOOTH, *The Manufacture of Organic Colorants and Intermediates*, The Society of Dyers and Colourists, Bradford, 1998.
12. J SHORE, *Blends Dyeing*, The Society of Dyers and Colourists, Bradford, 1998.
13. G CLARKE, *A Practical Introduction to Fibre and Tow Coloration*, The Society of Dyers and Colourists, Bradford, 1983.
14. K MCLAREN, *The Colour Science of Dyes and Pigments*, 2nd edn, Adam Hilger, Bristol, 1986.
15. A K SARKAR, *Fluorescent Whitening Agents*, Merrow, Watford, 1971.
16. R WILLIAMSON, *Fluorescent Brightening Agents*, Elsevier, Amsterdam, 1980.
17. R ANLIKER and G MÜLLER, 'Fluorescent whitening agents', *in Environmental Quality and Safety*, eds. F Coulston and F Korte, Suppl Vol IV, Thieme, Stuttgart, 1975.
18. Colour Index International, 3rd edn, 4th revision, Books and CD-ROM Issue 3, Colour Index-Pigments and Solvent Dyes, Book and CD-ROM, 1997.
19. I HOLME, *The Provision, Storage and Handling of Dyes and Chemicals for Textile Dyeing, Printing and Finishing*, UNIDO Textile Monograph UF/GLO/78/115, Vienna, 1980.
20. D M NUNN (ed), *The Dyeing of Synthetic-polymer and Acetate Fibres*, The Dyers Company Publications Trust, Bradford, 1979.
21. I HOLME, 'Fibre physics and chemistry in relation to coloration', Rev. Prog. Coloration, 1970 **1** 31–43.
22. A MURRAY and K MORTIMER, 'Carrier dyeing', Rev. Prog. Coloration, 1971 **2** 67–72.
23. S M BURKINSHAW, *Chemical Principles of Synthetic Fibre Dyeing*, Blackie, London, 1995.
24. J E MCINTYRE and P N DANIELS (eds.), *Textile Terms and Definitions*, 10th edn, The Textile Institute, Manchester, 1995.
25. J HILDEN, 'The effect of fibre properties on the dyeing of microfibres', Int. Textile Bull., Dyeing/Printing/Finishing, 1991 number (3), 19, 22, 24, 26.
26. P A DUFFIELD, in 'Wool Dyeing', ed. D M Lewis, The Society of Dyers and Colourists, Bradford, 1992, 176–195.
27. W INGAMELLS, *Colour for Textiles – A User's Handbook*, The Society of Dyers and Colourists, Bradford, 1993.
28. ANON, 'A revival of interest in low-temperature dyeing', Wool Record, 1996 **155**(3618) 35.
29. A C WELHAM, in 'Wool Dyeing', ed. D M Lewis, The Society of Dyers and Colourists, Bradford, 1992, 88–110.
30. D M LEWIS, 'Damage in wool dyeing', Rev. Prog. Coloration, 1989 **19** 49–56.
31. P GINNS and K SILKSTONE, in *The Dyeing of Synthetic-polymer and Acetate Fibres*, ed. D M Nunn, The Dyers Company Publications Trust, Bradford, 1979, 241–356.
32. L W C MILES (ed.), *Textile Printing*, 2nd edn, The Society of Dyers and Colourists, Bradford, 1994.
33. T F COOKE and H-D WEIGMANN, 'Stain blockers for nylon fibres', Rev. Prog. Coloration, 1990 **20** 10–18.
34. A C WELHAM, 'Advances in the afterchrome dyeing of wool', *J. Soc. Dyers Colourists*, 1986 **102**(4) 126–131.
35. I HOLME, 'Dye–fibre interrelations in acrylic fibres', Chimia, 1980 **34** 110–130.

36. W BECKMANN, in *The Dyeing of Synthetic-polymer and Acetate Fibres*, ed. D M Nunn, The Dyers Company Publications Trust, Bradford, 1979, 359–392.
37. J SHORE, in *Cellulosics Dyeing*, ed. J Shore, The Society of Dyers and Colourists, Bradford, 1995, 321–351.
38. J SHORE, in *Cellulosics Dyeing*, ed. J Shore, The Society of Dyers and Colourists, Bradford, 1995, 152–188.
39. T VICKERSTAFF, *The Physical Chemistry of Dyeing*, 2nd edn, Oliver and Boyd, London, 1954.
40. J A HOOK and A C WELHAM, 'The use of reactant-fixable dyes in the dyeing of cellulosic blends', *J. Soc. Dyers Colourists*, 1988 **104**(9) 329–337.
41. J SHORE, in *Cellulosics Dyeing*, ed. J Shore, The Society of Dyers and Colourists, Bradford, 1995, 189–245.
42. D M LEWIS, in *Wool Dyeing*, ed. D M Lewis, The Society of Dyers and Colourists, Bradford, 1992, 222–256.
43. M L GULRAJANI, 'Dyeing of silk with reactive dyes', Rev. Prog. Coloration, 1993 **23** 51–56.
44. P RYS and H ZOLLINGER, in *The Theory of Coloration of Textiles*, 2nd edn, ed. A Johnson, The Society of Dyers and Colourists, Bradford, 1989, 428–476.
45. M J BRADBURY, P S COLLISHAW and S MOORHOUSE, 'Exploiting technology to gain competitive advantage', Int. Dyer, 1996 **181**(4) 13, 14, 17, 20, 22–23.
46. R SCHNEIDER, 'Minimization of water consumption in washing-off processes', 18th IFATCC (International Federation of Textile Chemists and Colorists) Congress, Copenhagen, Denmark, IFATCC, 1999, 10–15.
47. B D WATERS, in *Colour in Dyehouse Effluent*, ed. P Cooper, The Society of Dyers and Colourists, Bradford, 1995, 22–30
48. C SENIOR, in *Cellulosics Dyeing*, ed. J Shore, The Society of Dyers and Colourists, Bradford, 1995, 280–320.
49. D A S PHILLIPS, M DUNCAN, A GRAYDON, G BEVAN, J LLOYD, C HARBON and J HOFFMEISTER, 'Testing colour fading of cotton fabrics by activated oxygen bleach-containing detergents: an inter-laboratory trial', *J. Soc. Dyers Colourists*, 1997 **113**(10) 281–286.
50. F R LATHAM, in *Cellulosics Dyeing*, ed. J Shore, The Society of Dyers and Colourists, Bradford, 1995, 246–279.
51. T BECHTOLD, 'Electrochemistry in vat dyeing and sulphur dyeing – concepts and results', 18th IFATCC (International Federation of Textile Chemists and Colorists) Congress, Copenhagen, Denmark, IFATCC, 1999, 42–46.
52. D BLACKBURN, in *The Dyeing of Synthetic-Polymer and Acetate Fibres*, ed. D M Nunn, The Dyers Company Publications Trust, Bradford, 1979, 77–130.
53. R BROADHURST, in *The Dyeing of Synthetic-Polymer and Acetate Fibres*, ed. D M Nunn, The Dyers Company Publications Trust, Bradford, 1979, 131–240.
54. A T LEAVER, 'Novel approaches in disperse dye design to meet changing customer needs', American Association of Textile Chemists and Colorists 1999 International Conference, Charlotte, USA, AATCC, 1999, 367–374.
55. I D RATTEE, in *Textile Printing*, 2nd edn, ed. L W C Miles, The Society of Dyers and Colourists, Bradford, 1994, 58–98.
56. P W LEADBETTER and A T LEAVER, 'Recent advances in disperse dye development. Introducing a new generation of high fastness disperse dyes', 15th IFATCC (International Federation of Textile Chemists and Colorists) Congress, Lucerne, Switzerland, IFATCC, 1990.
57. H GUTJAHR and R R KOCH, in *Textile Printing*, 2nd edn, ed. L W C Miles, The Society of Dyers and Colourists, Bradford, 1994, 139–195.
58. B KAUL, C RIPKE and M SANDRI, 'Technical aspects of the mass-dyeing of polyolefin fibres with organic pigments', Chem. Fibres Int., 1996 **46**(4) 126–129.
59. H EMSERMANN and R FOPPE, in *Acrylic Fiber Technology and Applications*, ed. J C Masson, Marcel Dekker, New York, 1995, 285–312.
60. L RUYS and F VANDEKERCKHOVE, 'Breakthrough in dyeable polypropylene', Int. Dyer, 1998 **183**(8) 32–36.
61. A LULAY, in *Acrylic Fiber Technology and Applications*, ed. J C Masson, Marcel Dekker, New York, 1995, 313–339.
62. I HOLME, 'Dispensing system enables continuous quick response', Int. Dyer, 1991 **176**(1) 10–11.
63. P A LEWIS, 'Coloured organic pigments for coating fabrics', J. Coated Fabrics, 1994 **23**(3) 166–201.
64. E R TROTMAN, *Textile Scouring and Bleaching*, Griffin, London, 1968.
65. I HOLME, I A PANTI, B D PATEL and H XIN, 'Chemical pretreatment of cotton fabrics for higher quality and performance', West-European Textiles Tomorrow, International Symposium, University of Ghent, Belgium, 1990, 35–59.
66. H DRIVER, 'Fabric singeing – the vital first step', Textile Technol. Int., 1993, 178–180.
67. I HOLME, 'Sizing for high speed weaving', Textile Horizons, 1985 **5**(6) 42, 44.
68. I HOLME, 'Chemical pretreatment – current technology and innovations', Textile Horizons Int., 1993 **13**(4) 27–29.

69. J T MARSH, *Mercerising*, Chapman and Hall, London, 1951.
70. R FREYTAG and J-J DONZÉ, in *Handbook of Fiber Science and Technology: Volume 1. Chemical Processing of Fibers and Fabrics, Fundamentals and Preparation Part A*, eds. M Lewin and S B Sello, Marcel Dekker, New York, 1983, 93–165.
71. P F GREENWOOD, 'Piece mercerizing: a modern process for knitted cottons', Textile Inst. Ind., 1976 **14**(12) 373–375.
72. I HOLME, in *The Dyeing of Cellulosic Fibres*, ed. C Preston, The Dyers Company Publications Trust, Bradford, 1986, 106–141.
73. G EUSCHER, 'Medium knit mercerizing', Textile Asia, 1982 **13**(9) 57–60, 62.
74. I GAILEY, 'Causes of unlevel dyeing of cotton cellulose. The influence of mercerizing and bleaching processes on the fine structure of cellulose', *J. Soc. Dyers Colourists*, 1951 **67** 357–361.
75. C V STEVENS and L G ROLDAN (-GONZALEZ), in *Handbook of Fiber Science and Technology: Volume 1. Chemical Processing of Fibers and Fabrics, Fundamentals and Preparation Part A*, eds. M Lewin and S B Sello, Marcel Dekker, New York, 1983, 167–203.
76. J W S HEARLE and L W C MILES, *The Setting of Fibres and Fabrics*, Merrow, Watford, 1971.
77. K C LAU, 'Dynamic response to colour specifications', *J. Soc. Dyers Colourists*, 1995 **111**(5) 142–145.
78. C SARGEANT, 'Colour range management', *J. Soc. Dyers Colourists*, 1995 **111**(9) 272–274.
79. C SARGEANT, 'Colour visualisation and communication – a personal view', Rev. Prog. Coloration, 1999 **29** 65–70.
80. J PARK, *Instrumental Colour Formulation – A Practical Guide*, The Society of Dyers and Colourists, Bradford, 1993.
81. R MCDONALD (ed.), *Colour Physics for Industry*, 2nd edn, The Society of Dyers and Colourists, Bradford, 1997.
82. X-Rite, *A Guide to Understanding Color Tolerancing*, Grandville, Michigan, USA, X-Rite Inc, 1994.
83. Y S W LI, C W M YUEN, K W YEUNG and K M SIN, 'Instrumental shade sorting in the last three decades', *J. Soc. Dyers Colourists*, 1998 **114** 203–209.
84. D BATTLE, in *Colour Physics for Industry*, 2nd edn, ed. R McDonald, The Society of Dyers and Colourists, Bradford, 1997, 57–80.
85. D S REININGER, 'Textile applications for hand-held colour measuring instruments', Textile Chem. Colorist, 1997 **29**(2) 13–17.
86. K VAN WURSCH, 'On-line colorimetry in continuous dyeing', *J. Soc. Dyers Colourists*, 1995 **111**(5) 139–141.
87. B GLOVER, P S COLLISHAW and R F HYDE, 'Creating wealth from textile coloration', 15th IFATCC (International Federation of Textile Chemists and Colorists) Congress, Lucerne, Switzerland, IFATCC, 1990.
88. J PARK, *A Practical Introduction to Yarn Dyeing*, The Society of Dyers and Colourists, Bradford, 1981.
89. D H WYLES, in *Engineering in Textile Coloration*, ed. C Duckworth, The Dyers Company Publications Trust, Bradford, 1983, 1–137.
90. J PARK and S S SMITH, *A Practical Introduction to the Continuous Dyeing of Woven Fabrics*, Roaches (Engineering), Upperhulme, Leek, UK, 1990.
91. G W MADARAS, G J PARISH and J SHORE, *Batchwise Dyeing of Woven Cellulosic Fabrics*, The Society of Dyers and Colourists, Bradford, 1993.
92. R MCGREGOR and R H PETERS, 'Effect of rate of flow on dyeing, I – diffusional boundary layer in dyeing', *J. Soc. Dyers Colourists*, 1965 **81**(9) 393–400.
93. E HENNINGSEN, 'Serious alternative to continuous dyeing', Textile Month, 1998 February 19–20.
94. M WHITE, 'Developments in jet dyeing', Rev. Prog. Coloration, 1998 **28** 80–94.
95. J A BONE, P S COLLISHAW and T D KELLY, 'Garment dyeing', Rev. Prog. Coloration, 1998 **18** 37–46.
96. M R FOX and H H SUMNER, in *The Dyeing of Cellulosic Fibres*, ed. C Preston, The Dyers Company Publications Trust, Bradford, 1986, 142–195.
97. C BERRY and J G FERGUSON, in *Textile Printing*, 2nd edn, ed. L W C Miles, The Society of Dyers and Colourists, Bradford, 1994, 196–239.
98. I HOLME, 'Quick response printing', African Textiles, 1997/98 Dec/Jan 20, 30.
99. H A ELLIS, 'Printing techniques – the choice', Textile Horizons, 1985 **5**(4) 37–38, 40.
100. C J HAWKYARD, in *Textile Printing*, 2nd edn, ed. L W C Miles, The Society of Dyers and Colourists, Bradford, 1994, 18–57.
101. I HOLME, 'Right first time', African Textiles, 1998 August/September 39–40.
102. W C TINCHER, Q HU and X LI, 'Ink jet systems for printing fabrics', Textile Chem. Colorist, 1998 **30**(5) 24–27.
103. B SIEGEL, S ERVINE and K SIEMENSMEYER, 'Ink jet: the future of textile printing', 18th IFATCC (International Federation of Textile Chemists and Colorists) Congress, Copenhagen, Denmark, 1999, IFATCC, 144–148.
104. T L DAWSON, 'Jet Printing', Rev. Prog. Coloration, 1992 **22** 22–31.
105. ANON, 'Digital printing points the way', Textile Horizons, 1997 **16**(8) 25–26.

106. W W CARR, F L COOK, W R LANIGAN, M E SIKORSKI and W C TINCHER, 'Printing textile fabrics with xerography', Textile. Chem. Colorist, 1991 **23**(5) 33–41.
107. Methods of Test for Colour Fastness of Textiles and Leather (BS 1006:1990. ISO 105).
108. AATCC Technical Manual 1999, Vol 74, Research Triangle Park, North Carolina, USA, American Association of Textile Chemists and Colorists, 1998.
109. W BEAL, 'New multifibre from the Society', J. Soc. Dyers Colourists, 1987 **103**(5/6) 177.
110. K J SMITH, in Colour Physics for Industry, 2nd edn, ed. R McDonald, The Society of Dyers and Colourists, Bradford, 1997, 121–208.
111. I HOLME, 'Colour fastness testing', Textile Horizons, 1998 **18**(6) 12–13.
112. I HOLME, 'Dyed materials: getting the colour right', ATA J., 1998 **9**(1) 70–73.
113. P J SMITH, 'Colour fastness testing methods and equipment', Rev. Prog. Coloration, 1994 **24** 31–40.

10

Heat and flame protection

Pushpa Bajaj

Department of Textile Technology, Indian Institute of Technology, Hauz Khas, New Delhi – 110016, India

10.1 Introduction

With industrialisation, the safety of human beings has become an important issue. A growing segment of the industrial textiles industry has therefore been involved in a number of new developments in fibres, fabrics, protective clothing.[1–10] Major challenges to coatings and fabrication technology for production in the flame-retardant textile industry have been to produce environmentally friendly, non-toxic flame-retardant systems that complement the comfort properties of textiles.[11–14] The 1990s, therefore, saw some major innovations in the development of heat-resistant fibres and flame-protective clothing for firefighters, foundry workers, military, aviation and space personnel, and for other industrial workers who are exposed to hazardous conditions.

For heat and flame protection, requirements range from clothing for situations in which the wearer may be subjected to occasional exposure to a moderate level of radiant heat as part of his/her normal working day, to clothing for prolonged protection, where the wearer is subject to severe radiant and convective heat, to direct flame, for example the firefighter's suit. In the process of accomplishing flame protection, however, the garment may be so thermally insulative and water vapour impermeable that the wearer may begin to suffer discomfort and heat stress. Body temperature may rise and the wearer may become wet with sweat. Attempts have therefore been made to develop thermal and flame protective garments which can be worn without any discomfort.

In this chapter, various factors affecting the flammability, development of new inherently heat-resistant fibres, and the flame-retardant finishes for both natural and man-made fibre fabrics along with the relevant test methods are described.

10.2   What constitutes flammability?

Ease of ignition, rate of burning and heat release rate are the important properties of textile materials which determine the extent of fire hazard. The other factors that influence the thermal protection level include melting and shrinkage characteristics of synthetic fibre fabrics, and emission of smoke and toxic gases during burning. So, while selecting and designing flame protective clothing, the following points should be kept in mind:

*   the thermal or burning behaviour of textile fibres
*   the influence of fabric structure and garment shape on the burning behaviour
*   selection of non-toxic, smoke-free flame-retardant additives or finishes
*   design of the protective garment, depending on its usage, with comfort properties
*   the intensity of the ignition source
*   the oxygen supply.

10.3   Thermal behaviour of fibres

The effect of heat on a textile material can produce physical as well as chemical change.[8,15,16] In thermoplastic fibres, the physical changes occur at the second order transition (T_g), and melting temperature (T_m), while the chemical changes take place at pyrolysis temperatures (T_p) at which thermal degradation occurs. Textile combustion is a complex process that involves heating, decomposition leading to gasification (fuel generation), ignition and flame propagation.

A self-sustaining flame requires a fuel source and a means of gasifying the fuel, after which it must be mixed with oxygen and heat. When a fibre is subjected to heat, it pyrolyses at T_p (Fig. 10.1) and volatile liquids and gases, which are combustible, act as the fuels for further combustion. After pyrolysis, if the temperature is equal to or greater than combustion temperature T_c, flammable volatile liquids burn in the presence of oxygen to give products such as carbon dioxide and water. When a textile is ignited, heat from an external source raises its temperature until it degrades. The rate of this initial rise in temperature depends on the specific heat

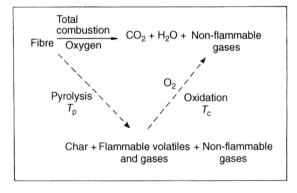

10.1   Combustion of fibres.

of the fibre, its thermal conductivity[17] and also the latent heat of fusion (for melting fibres) and the heat of pyrolysis.

Frank-Kamenetzky[18] demonstrated the influence of the nature of a reactive (flammable) material and its environment. The heat generation and heat loss was plotted as a function of temperature (Fig. 10.2). The plot shows that the loss of heat is approximately proportional to the difference in temperature between the combustion zone and the environment, and can be represented by an approximately straight line. The equilibrium between the heat generation and the heat loss is realized at the points of intersection of I and II (Fig. 10.2(a)). Point A represents the ambient temperature, point C represents the stationary temperature and both are stable while B is unstable. To the left of B, the heat loss exceeds the heat generation, while to the right of B this is just the reverse. Therefore, the temperature corresponding to B is the ignition temperature. During a fire accident, the material must be heated to such an extent that it reaches the ignition temperature. The temperature at B is also considered to be the self-extinguishing temperature; at lower temperatures heat loss exceeds heat generation. Figure 10.2(b) shows three materials with different degrees of flammability but in the same environment. The first (Ia) is highly flammable, the second (Ib) moderately flammable while the third (Ic) is flame resistant under these conditions. Figure 10.2(c), on the other hand, represents a material in different environments. An increased heat loss may be caused by a higher rate of air flow, less insulation and so on. From the above, it is thus evident that the flammable material may be barely flammable or even non-flammable under different environments.

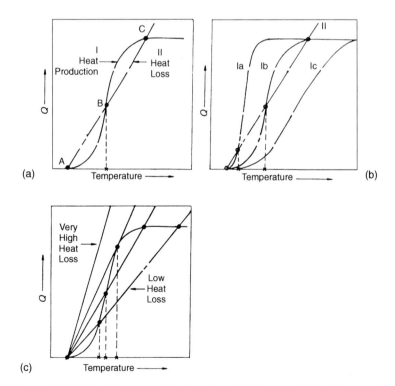

10.2   Schematic Stability diagram of Combustion. (a) Flame stability diagram, (b) three degrees of flammability, (c) four different surroundings.

In protective clothing, it is desirable to have low propensity for ignition from a flaming source or, if the item ignites, a slow fire spread with low heat output would be ideal. In general, thermoplastic-fibre fabrics[19,20] such as nylon, polyester fibre, and polypropylene fibres fulfil these requirements because they shrink away from flame and, if they burn, they do so with a small slowly spreading flame and ablate. For protective clothing, however, there are additional requirements, such as protection against heat by providing insulation, as well as high dimensional stability of the fabrics, so that, upon exposure to the heat fluxes that are expected during the course of the wearer's work, they will neither shrink nor melt, and if they then decompose, form char. The above mentioned requirements cannot be met by thermoplastic fibres and so recourse must be made to one of the so-called high-performance fibres such as aramid fibre (e.g. Nomex, DuPont), flame-retardant cotton or wool, partially oxidised acrylic fibres, and so on. It may also be noted that the aramid fibres, in spite of their high oxygen index and high thermal stability, have not been found suitable for preventing skin burns in molten-metal splashes because of their high thermal conductivity.

From the foregoing discussion, it may be noted that the mode of decomposition and the nature of the decomposition products (solid, liquid, and gaseous products) depend on the chemical nature of the fibre, and also on the type of finishes or coatings applied to the fabrics. If such decomposition products are of a flammable nature, the presence of atmospheric oxygen gives rise to ignition, with or without flames. When the heat evolved is higher than that required for thermal decomposition, it can spread the ignition to cause the total destruction of the material (Fig. 10.3).

In addition to the fibre characteristics and fabric finish, several garment characteristics also influence thermal protection. For a given fabric thickness, the lower the density, the greater the thermal resistance. This applies to fibres such as cotton, wool, and so on, which produce an insulating char on heating. Hence, thicker fabrics made from cotton, wool and other non-melting fibres give good thermal protection, whereas the thicker thermoplastic-fibre fabrics produce more severe burns.

Flame retardance of materials is normally expressed as oxygen index:[16]

$$(OI) = (OI)_m + f(FR) \tag{1}$$

where $(OI)_m$ is the oxygen index of the virgin material, and $f(FR)$ is the function of the flame retardant. Thermal properties of textile fibres including OI values are given in Table 10.1.

Miller et al.[21] have also studied the extinguishability of fabrics by determining the burning rate as a function of the environmental oxygen concentration. They quoted the intrinsic oxygen index, $(OI)_o$, values at extrapolated zero burning rates for cotton, wool, modacrylic, and aramid fibre for top ignition of vertically oriented samples, and for bottom ignition of both vertically oriented and 45°-inclined samples. Intrinsic oxygen index values were, however, lower than normal limiting oxygen index (LOI) values, for example, $(OI)_o$ for top-ignited cotton is 13 whereas the normal LOI is 18.

In another study, Van Krevelen[16] has established a good correlation between the chemical composition and the LOI value of the polymers. According to him, the composition parameter (CP) describing the combined effect of hydrogen and halogen content should be less than one for flame retardant materials. Figure 10.4 shows the relationship between CP and LOI for some textile materials

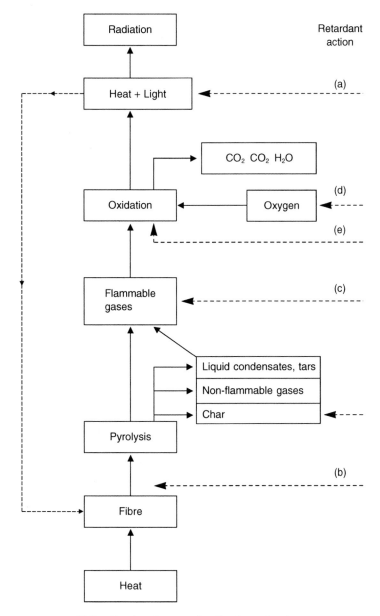

10.3 Combustion as a feedback mechanism.

$$CP = (H/C) - 0.65(F/C)^{1/3} - 1.1(Cl/C)^{1/3} - x(Br/C)^{1/3} \qquad (2)$$

where (H/C), (F/C) and (Cl/C) are the atomic ratios of the respective elements in the polymer composition, and coefficient x is probably 1.6, but is still uncertain owing to the lack of sufficient data. Although there is some data scatter, as shown in Fig. 10.4, there is a correlation between the LOI and CP.

- if $CP \geq 1$ then $LOI \approx 17.5$
- if $CP \leq 1$ then $LOI \approx 42.5$ to 60.

Table 10.1   Thermal and flame-retardant properties of some fibres[8]

Fibre	T_g (°C) Glass transition	T_m (°C) Melt	T_p (°C) Pyrolysis	T_c (°C) Combustion	LOI (%)
Wool	–	–	245	600	25
Cotton	–	–	350	350	18.4
Viscose	–	–	350	420	18.9
Triacetate	172	290	305	540	18.4
Nylon 6	50	215	431	450	20–21.5
Nylon 6,6	50	265	403	530	20–21
Polyester	80–90	255	420–477	480	20–21.5
Acrylic	100	>320	290	>250	18.2
Polypropylene	–20	165	469	550	18.6
Modacrylic	<80	>240	273	690	29–30
PVC	<80	>180	>180	450	37–39
PVDC	–17	180–210	>220	532	60
PTFE	126	>327	400	560	95
Oxidised acrylic	–	–	>640	–	55
Nomex	275	375	310	500	28.5–30
Kevlar	340	560	590	>550	29
PBI	>400	–	>500	>500	40–42

PBI = polybenzimidazole. For other abbreviations, see Fig. 10.4.
T_g is the second order transition, T_m is the melting transition, T_p is the pyrolysis temperature, T_c is the combustion temperature.

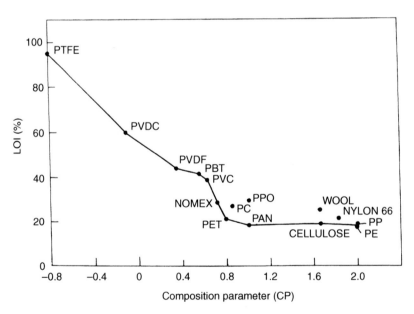

10.4   Correlation between oxygen index and elemental composition. PTFE = poly(tetrafluoroethylene), PVDC = polyvinylidene chloride, PBT = poly(butylene terephthalate), PVDF = polyvinylidene fluoride, PVC = poly(vinyl chloride), PET = poly(ethylene terephthalate), PC = polycarbonate, PPO = poly(phenylene oxide), PAN = poly(acrylonitrile), PP = polypropylene, PE = polyethylene.

(H/C) could be considered as a measure of the 'aromaticity' in polymers not containing halogen. This indicates that flame-retardant (FR) polymers and fibres should be intrinsically aromatic in nature with a CP value of <1.

Dependence of LOI on the environmental temperature, sample thermal history, moisture content, fabric sample dimensions, fabric area density and so on has been well documented. Horrocks *et al.*[22] have demonstrated the use of extinction oxygen index (EOI) rather than the LOI as a measure of textile flammability. By considering the potential fabric extinguishability in terms of an EOI parameter, the authors showed that the influence of ignition may be removed by simple extrapolation to zero ignition time to generate $(EOI)_o$.

10.4   Selection of fibres suitable for thermal and flame protection

The fibres could be classified into two categories:

*   Inherently flame-retardant fibres, such as aramid, modacrylic, polybenzimidazole (PBI), Panox (oxidised acrylic) or semicarbon, phenolic, asbestos, ceramic etc.
*   chemically modified fibres and fabrics, for example, flame retardant cotton, wool, viscose and synthetic fibres.

10.4.1   Inherently flame-retardant fibres

For some 2000 years, there was only one type of naturally occurring mineral fibre, asbestos which could not be completely destroyed by fire. Asbestos has many desirable properties and is cheap as well. However, the fibres are so fine that they can be breathed into the lungs and can promote fatal cancerous growth.

Glass fibres are also heat-resistant materials. In earlier times such fibres were used for printed circuit boards. Now developments in the texturing of glass fibres have provided a material that could substitute for the asbestos fibres to some extent. Unlike asbestos fibres, glass fibres with high diameter are non-respirable. They have an upper temperature resistance of about 450 °C. They spin well, knit or braid easily and can be coated with rubber, polyacrylate or silicones. Glass fibres have also good electrical and insulation properties. However, they cause skin irritation, which limits their application in protective clothing.

A number of ceramic fibres, SiC, silicon or boron nitride, polycarbosilicones, alumina and so on, have been developed. These ceramic fibres can withstand temperatures between 1000–1400 °C. However, the biggest problem with ceramic in staple form is that it is very abrasive and may wear out the processing machinery at a very high rate. Ceramic fibres generally contain a mixture of components, for example, Nicalon (Nippon Carbon Co.) is silicon carbide fibre containing up to 30% of silica, and carbon, while Nextel (DuPont) is a 70/28/2 mixture of $Al_2O_3/SiO_2/B_2O_3$. Tyranno (Ube Industries Ltd., Japan) is a ceramic fibre composed of silicon–titanium, carbon and oxygen (Si-Ti-C-O), which is reported to have a non-crystalline microstructure. Altex (Sumitomo, Japan) produces continuous α-alumina fibres with a diameter of 8–10 μm. The tenacity of these commercial ceramic fibres is in the range of 0.32–3.2 N tex^{-1} and breaking extension is only 0.4–5.4%. An important aspect of high temperature fibres that should be considered is their ability to insulate from thermal flux or conductivity. Silica-based fibres have high rates of thermal conductivity, a property that may be valuable in heat dissipation in some uses but in situations like hot metal splashes, where the heat is transmitted to the person by conduction, they will cause more burn injuries instead of protecting the skin. Thus, the selection of the fibre for making thermally protective clothing should be decided on the basis of the environment to which a worker is exposed, namely, whether the heat will be transmitted to the person by conduction, convection or radiation. Despite their high temperature resistance, ceramic fibres have poor aesthetic characteristics, high densities and are difficult to process.

10.4.1.1   Aramids[23-29]

Aromatic polyamides such as poly(metaphenylene isophthalamide) char above 400 °C and may survive short exposures at temperatures up to 700 °C. Nomex (DuPont), Conex (Teijin), Fenilon (Russian) and Apyeil (Unitika) meta-aramid fibres have been developed for protective clothing for fighter pilots, tank crews, astronauts and those working in certain industries. Para-aramid fibres like Kevlar (DuPont), Twaron (Akzo Nobel) and Technora (Teijin) are also being used for ballistic and flame protection. Nomex nonwovens are used for hot gas filtration and thermal insulation.

Aramids are resistant to high temperatures, for example at 250 °C for 1000 hours the breaking strength of Nomex is about 65% of that before exposure. They begin to char at about 400 °C with little or no melting. Generally, meta-aramids are used in heat protective clothing, however, in intense heat, Nomex III (a blend of Nomex and Kevlar 29 (95:5 by wt) is preferred, in order to provide a greater mechanical stability to the char. Teijin[23] has introduced a new fabric, X-fire, a combination of Teijin Conex (meta-aramid) and Technora (para-aramid) fibres. This fabric is capable of resisting temperatures up to 1200 °C for 40–60 s.

Nomex can also be blended with FR fibres, for example FR wool and FR viscose. Karvin (DuPont) is a blend of 30% Nomex, 65% FR viscose and 5% Kevlar. Kevlar blends were formerly used by Firotex Co. UK (now defunct) with partially carbonised viscose in fabric form. This blend was developed as a fire blocking fabric for aircraft seats but found little favour because of the poor abrasion resistance of the carbonised viscose component (see Section 10.4.1.6 on semicarbon fibres).

Other examples of such blends include Fortafil and Fortamid needle felt NC580, which comprise aramid and FR viscose. This material is useful for gloves and mittens in which temperatures may reach up to 350 °C. The outer working surface of the aramid fibre is needled through a reinforcing polyester fibre scrim over an inner layer of FR viscose.

Another aromatic copolyamide fibre developed by Lenzing AG is P84. This fibre does not melt but becomes carbonized at temperatures in excess of 500 °C and has an LOI value of 36–38%. The basic fibre is golden yellow in colour but Lenzing AG offers it as spun material dyed in limited colours. P84 fibres have irregular cross-section, which provides a higher cover factor at lower weights of fabrics made from it. Its extensibility is >30% with good knot and loop strength. The applications of high performance P84 include protective clothing, as a sealing or packing material, for hot gas filtration and in aviation and space including covers for aircraft seats.[28] Mitsui Toatsu Chemical Co. has also claimed the development of a more heat-resistant aramid fibre. This fibre has been made from an aromatic isocyanate and an aromatic carboxylic acid.

Table 10.2 Properties of selected high temperature fibres[10]

Fibre properties	Aramid[a]	Carbon[b]	Glass (type)	PBI	Novoloid phenolic	PPS	Polyacrylate	PTFE	Polyimide	Polyamide-imide
Tensile and physical properties										
Tensile strength (Gpa)	0.6M / 3.4P	4	3.5E / 4.6S	0.37	0.26	0.42	0.22	0.18	0.35	0.32
Modulus (Gpa)	17.0M / 99P	230	72E / 87S	5.7	3.0	7.3	4.36	4.5	6.2	N/A
Elongation at break (%)	22M / 3.0P	1.8	4.8E / 5.4S	30	30	40	20–30	19–140	19–21	15–20
Moisture regain (%)	6.5M / 4.0P	9	<0.1	15	6–7.3	0.6	12	0	3.0	3.4
Density	1.38M / 1.45P	1.40	2.55E / 2.48S	1.43	1.27	1.37	1.50	2.10	1.41	1.34
Abrasion resistance	Good M / Poor P	Poor	Poor	Good	Poor	Good	Fair	Good	Good	Good
Resilience	Excellent	Poor	Poor	Excel	Fair	Good	Good	Poor	Fair	N/A
Chemical resistance										
Acids concentrated	Fair P	Poor	Excel	Excel	Poor	Good	Excel	Excel	Excel	Fair
Alkalis concentrated	Good P	Poor	Fair	Good	Excel	Excel	Excel	Excel	Poor	Good
Ultraviolet	Poor	Good	Excel	Good	Excel	Excel	Excel	Excel	Good	Good
Thermal properties										
LOI	30M / 29P	55	>100	38	33	34	43	>95	40	32
Thermal conductivity (BTU-in hr^{-2}°F^{-1})	0.26M / 0.30P	<0.03	7.20	0.26	0.28	0.30	0.31	0.20	N/A	0.08
Usable temperature (°C)	315–370	V.High	V.High	>595	400	260	[c]	430	<485	>420
Short term continuous	230	500	315	315	205	205	160	288	260	250
Smoke emission density	1.0	N/A	Low	Trace	<0.30	NA	Trace	Low	<1.0	<2.0

PPS = Poly(phenylene sulphide).
[a] M = poly meta-aramid (Nomex); P = poly para-aramid (Kevlar129); modulus (GPa) = (gpd × density) /11.33. [b] Celion 3000 (HS). [c] Auto ignition at 450°C and 100% oxygen.

10.4.1.2 Poly(amide-imide) fibres

Rhone-Poulenc produces polyamide–imide fibre called Kermel. This is available in two forms: 234 AGF and 235 AGF. Type 234 is a staple fibre for use in both cotton and worsted spinning systems, and is produced in five spun-dyed colours. Type 235 is intended for nonwovens applications. In France, Kermel is used by firefighters and military personnel where the risk of fire is higher than usual. Its LOI is 31–32%, and it resists up to 250 °C exposure for a long duration. At 250 °C after 500 hours exposure, the loss of mechanical properties is only 33%. Kermel fibre does not melt but carbonises. During its carbonization it generates very little opacity. Blends of 25–50% Kermel with FR viscose offer resistance to ultraviolet (UV) radiation and price advantage also compared with 100% Kermal fabrics. Blending with 30–60% wool also produces more comfortable woven fabrics with enhanced drape. In the metal industry, the 50:50 blend gives very good results, but a 65:35 Kermel/viscose blend is preferred for such applications. Kermel-based fabrics are now used both on-shore and off-shore by leading petrochemical groups. The army, navy and air-forces are also using Kermel in woven and knitted forms.

10.4.1.3 Polybenzimidazole (PBI) fibres

Celanese developed PBI,[28,30,31] a non-combustible organic fibre. Its LOI is 41% and it emits little smoke on exposure to flame. PBI can withstand temperatures as high as 600 °C for short-term (3–5 s) exposures and longer term exposure at temperatures up to 300–350 °C. It provides the same protection as asbestos while weighing half as much. It also absorbs more moisture than cotton. The current area of interest in PBI is in the replacement of asbestos-reinforced rubbers used in rocket motors and boosters to control ignition. Its other applications include fire blocking fabrics in aircraft seats, firefighter suits and racing-car driver suits. Studies of subjective wearer evaluations have shown that PBI fibre exhibits comfort ratings equivalent to those of 100% cotton.

Ballyclare Special Products, UK[31] has recently developed a fire-resistant garment assembly for firefighter's safety. The outer fabric of the garment is made from Pbi Gold[(R)], a fire-resistant fabric from Hoechst Celanese. This fabric, which was originally developed for the US Apollo space programme, combines the comfort, thermal and chemical resistance of polybenzimdazole (PBI) with the strength of aramid fibre. Pbi Gold is stable even under simulated flash conditions at 950 °C. The fabric is also resistant to puncturing, tearing and ripping.

10.4.1.4 Poly(phenylene sulphide) PPS fibres

Ryton (Sulfar) fibres (Tm 285 °C) produced by Amoco Fabrics and Fibres Co. are nonflammable. They do not support combustion under normal atmospheric conditions, and the LOI is 34–35%. Chemical resistance and the ability to retain physical properties under extremely adverse conditions make the fibre valuable for protective clothing.

10.4.1.5 Polyacrylate (Inidex)

Polyacrylate[10,30] is a crosslinked copolymer of acrylic acid and acrylamide. Its LOI is 43%, and when subjected to flame, it neither burns nor melts. It emits virtually no smoke or toxic gases. Because of its low strength and brittleness, it can be used in nonwovens although the durability of fabrics made from this fibre may not be adequate for some apparel uses. As Inidex also offers protection from attack by chemicals, including strong acids and alkalis, it may be found useful in filtration of liquids and hot gases.

10.4.1.6 Semicarbon/Panox fibres

These fibres[32] are produced by thermal treatment (thermo-oxidative stabilization) of either viscose or acrylic fibres. Asgard and Firotex are produced from viscose while Panox, Pyromex, Fortasil, Sigratex and so on are made from acrylic precursors. The acrylic fibres can be oxidised in the fibre, filament or fabric form at 220–270 °C in air, but the viscose fibres are generally partially carbonized in the fabric form in a nitrogen atmosphere.

These semicarbon fibres have excellent heat resistance, do not burn in air, do not melt and have outstanding resistance to molten metal splashes. After exposure to flame, there is no afterglow and fabrics remain flexible. In view of their outstanding properties, the Panotex fabrics (Universal Carbon Fibres) made from Panox (RK Textiles), for example, are ideal for use in protective clothing where protection against the naked flame is required. Currently, this range of fabrics is probably the most common and versatile of oxidised acrylic-based materials. Panotex fabrics can withstand flame temperatures in excess of 1000 °C, display very little shrinkage and yet, breathe like wool. However, the fabrics have relatively poor abrasion resistance. Therefore, a blend of oxidised acrylic fibre and an aramid used in a honeycomb woven fabric has been considered an ideal material as a fire blocker for aircraft seats and military tank crews. The honeycomb weave, which is intermittently tight and slack in construction, provides a spongy stretch fabric, which is easy to cut and fit around difficult shapes.

Panox/wool blends are suitable for flying suits. However, Panotex fabrics have high thermal conductivity and are non-reflecting. It is, therefore, necessary to have a suitable underwear to protect the skin while using Panotex in the outer layer of protective clothing. For this purpose a 60% Panox 40% modacrylic fibre double jersey fabric and a 60% wool 40% Panox core fibre have been recommended. In some cases, the heat conduction of Panotex fabrics can be of advantage in the construction of covers for aircraft seats; a fabric with a Zirpro-treated wool face and Panox back will probably spread the heat from a localized igniting source thereby delaying the ignition of underlying foam.

To prevent transfer of radiant heat, Panotex fabrics may be aluminized. An aluminized Panotex fabric is thus suitable for fire-proximity work but not for fire entry. It has been demonstrated that with a heat flux of $3\,W\,cm^{-2}$, an aluminium coating will ignite, but a stainless steel coating can withstand such a situation for a prolonged period. Multiple layers of Panotex fabric tend to protect a polyvinyl chloride (PVC)-simulated skin against irradiance as high as $170\,W\,cm^{-2}$ applied for 2 s.

Another advantage of Panotex outer fabric is the shedding of burning petrol, and it can even withstand several applications of napalm.

10.4.1.7 Phenolic or novoloid fibres

Kynol is a well-established novoloid heat-resistant fibre which is produced by spinning and postcuring of phenol formaldehyde resin precondensate. The fibre is soft and golden coloured with a moisture regain of 6%. When strongly heated, Kynol fabric is slowly carbonised with little or no evolution of toxic gases or smoke. However, its poor strength and abrasion properties preclude its application for making apparel. To upgrade its mechanical properties, Kynol fibres can be blended with Nomex or FR viscose to produce flame-protective clothing.

Another phenolic fibre, Philene has also been developed, for example Philene 206 (0.9 den) and Philene 244 (2.1 den). The moisture regain of the fibre is 7.3% and is said to be non-flammable and self-extinguishing, with an LOI of 39%. It does not show any change in tensile properties after being heated for 24 hours at 140 °C (or for 6 hours at 200 °C). A charred Philene fabric is claimed to form a thermal insulating barrier that retains its initial form.

BASF[33] has also developed Basofil melamine staple fibres of 2.2 dtex with a tenacity of 2–4 cN dtex^{-1} and an elongation of 15–20%. It has LOI values of 31–33 and moisture regain is about 4%. Basofil can be used in continuous service at 200 °C. Above 370 °C, thermal degradation results in char formation rather than a molten drip. For protective wear, DREF-2 yarns comprising a 34 tex glass-fibre core is spun to Nm 12/2 (83 tex/2), sheathed with a blend of Basofil and para-aramid (80/20) to produce 400 g m^{-2} or 580 g m^{-2} fabrics required for foundries where these are constant hazards from molten metal splashes and sparks. Such materials may also be used to make proximity suits to protect against intense radiation or to make entry suits.

Kotresh et al.[34] have developed a flame-retardant fabric from DREF-2 friction spun core yarn of kevlar and FR viscose for anti-G suits (AGS) outer garment applications. This fabric has also been recommended for firefighters.

10.4.1.8   Modacrylic

Flame-retardant modacrylic under different brand names, such as Velicren FR (Montefibre, Italy) and SEF (Solutia Inc.) is a copolymer of acrylonitrile, vinyl chloride or vinylidene chloride in the ratio of 60:40 (w/w) along with a sulphonated vinyl monomer. It has an LOI in the range of 26–31%.

Kaneka Corporation has also developed Kanecaron, an FR modacrylic with an LOI value in the range of 30–35%. Fabrics from Kanecaron (e.g. Protex M) blended with cotton meet the requirements of BS 6249 Index B, while maintaining the softness and comfort of cotton.

10.4.1.9   Chlorofibre

Rhone-Poulenc's chlorofibre, Rhovylon FR and Clevyl is used in furnishing fabrics, nightwear and institutional blankets. It has quite a high LOI value of around 45%. Extensive testing has shown that chlorofibre meets International Standards for furnishing fabrics and it is in use in French high speed trains (TGVs) and a passenger liner (QE2).

10.4.2   Flame retardation of conventional textile fibres

10.4.2.1   FR viscose

Inherently flame-retardant viscose fibres are produced by incorporating FR additives/fillers in the spinning dope before extrusion. For example, Sandoflam 5060 (Sandoz), polysilicic acid or polysilicic acid and aluminium (Sateri).

Sandoflam 5060 contains both phosphorus and sulphur as shown in Structure I, bis (2-thio-5,5-dimethyl-1,3,2-dioxaphosphorinyl) oxide, below:

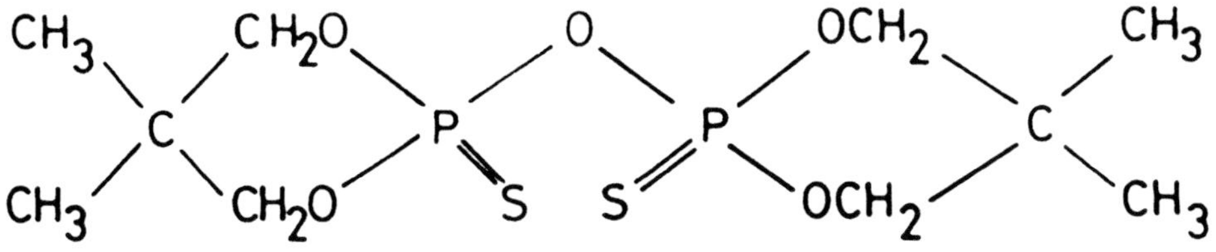

Structure I

Aqueous dispersions of pyro(thio)phosphates(II) containing polyoxyalkylenes have also been used as dope additives (15.8 parts) in 200 parts of 9% cellulose xanthate. The rayon produced from this mixture had an LOI value of 27.5 compared with 18 without the phosphate (II) dispersion (see Structure II).

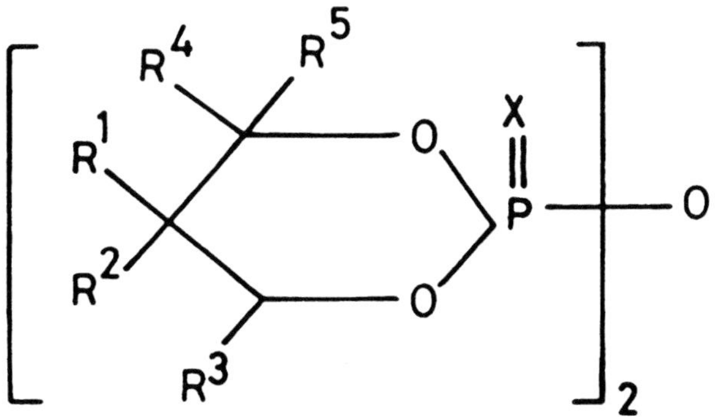

Structure II

R^1, R^2 = H, alkyl, CH_2Cl, CH_2Br, alkoxymethyl or R^1 = phenyl, R^3, R^5 = H, alkyl, R^4 = H, Me, X = O or S

Hoechst AG has previously offered an FR viscose staple fibre under the brand name Danufil CS. Here the FR additive was Sandoflam 5060 used in the viscose dope before wet spinning. The products offered were 1.7 and 3.3 dtex for short-fibre spinning and 4.0 dtex for long-fibre spinning. Its LOI value was 27%.

In 100% form, the fibre can be used for mattress covers and fire extinguishing blankets. A blend with aramid is suitable for protective clothing, such as army and fire service uniforms. Further, blending with wool shows a synergistic effect, and this blend is used for upholstery fabrics.

Lenzing AG currently produces Lenzing Viscose FR which contains Sandoflam 5060. It has been demonstrated that a wool/viscose FR blend performs better than either fibre alone and this is being used as seat covers in the Airbus 310.

DuPont and Lenzing AG have jointly developed a special blend consisting of 65% Viscose FR, 30% Nomex and 5% Kevlar for industrial wear under the brand name of Karvin[R].

Sateri (formerly Kemira) Fibres, Finland has developed an environmentally friendly hybrid viscose fibre containing cellulose and polysilicic acid under the brand name Visil[R], and viscose containing polysilicic acid and aluminium under the name

VisilAP$^{(R)}$. This hybrid fibre is a cellulosic fibre containing molecular chains of poly-silicic acid produced by wet spinning of water glass and alkaline cellulose xanthate during the coagulation process, the cellulose component is regenerated simultane-ously with the polymerization of polysilicic acid; $nSi (OH)_4 \rightarrow$ polysilicic acid.

The incorporation of polysilicic acid enhances the fire resistance of the hybrid by the following mechanism:[35–37]

- An inherently incombustible char is formed on the fibre surface.
- The temperature at which water is released from the fibre is lowered.
- The hydrated nature of the inorganic component suppresses the flame and exerts a self-extinguishing effect.

In another patent,[38] it is stated that polysilicic acid that contains aluminium silicate sites, a blend of viscose and water glass, after spinning, tow stretching and subse-quent washing with $40\,\mathrm{g}\,\mathrm{l}^{-1}$ Na aluminate solution and carding, showed an LOI value of 32. Properties of Visil fibre are given in Table 10.3.

Anand and Garvey[39] have also demonstrated the use of Visil/modacrylic and Visil/wool blends for use in protective clothing. In Visil/modacrylic blends, it has been shown that the level of flame retardancy is dependent not only on the fibre content, but the physical structure of the yarn as well. The blended ring spun yarn fabrics exhibited synergism relative to rotor spun yarn fabrics at each blend ratio (Visil/modacrylic: 16.7/83.3 to 83.3/16.7) in both char and LOI values. In contrast, Visil/wool blended fabrics failed in the BS5438 strip test despite the higher LOI values (25.7–36%) of the blends. This anomalous behaviour has been attributed to the lofty wool structures which are known to burn extensively because of high oxygen accessibility to the fibre surface in the blended fabrics. Further, it may also be noted that in the LOI test, burning of the fabric takes place downward but in the vertical strip test upward burning conditions apply, which would favour the enhanced access of oxygen from the convective effects of the flame.

In another paper, Horrocks et al.[40] have shown that flame-retardant textiles can be developed in composite structures having fibre-intumescent interactive systems. The interactive pyrolysis of both components creates a fibrous char-reinforced intu-mescent charred derivative of the original structure, which offers a toughened and oxygen impermeable barrier to heat, flame and subsequent oxidation. The compos-ite structure comprised a nonwoven core ($200\,\mathrm{g}\,\mathrm{m}^{-2}$) of Visil in which an ammonium polyphosphate-based intumescent was dispersed and resin bonded up to 50% w/w. This core was sandwiched between a flame-retardant cotton face and backing woven fabric (Fig. 10.5). Exposure of this composite structure at around 500 °C in a furnace

Table 10.3 Physical and mechanical properties of Visil[36]

Linear density (dtex)	1.7, 3.5, 5.0, 8.0
Length (mm)	40 and 80 currently available
SiO$_2$ loading (%)	30–33, depending on grade
Cross-section	Kidney bean, irregular
Tenacity (cN/dtex)	1.5–2.0
Elongation-at-break (%)	22–27
Moisture regain (%)	9–11
Water imbibition (%)	50–60
Limiting oxygen index (%)	26–33, depending on grade or textile structure

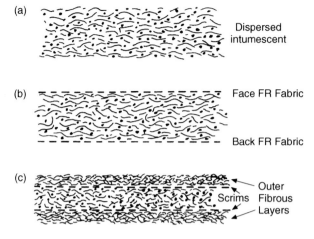

(a) Dispersed intumescent

(b) Face FR Fabric

Back FR Fabric

(c) Outer Fibrous Layers
Scrims

10.5 Textile/fibre applications of intumescent systems. (a) Random fibre web with dispersed intumescent, (b) same as (a) plus front and back FR fabrics, (c) similar to (b) but with fabrics located as reinforcing scrims.

showed that both the intumescent and organic cellulose component of the Visil charred along with the FR facing and backing fabrics. Simultaneously expanding the intumescent increased the thickness of the core fabric by a factor of 2. On further heating above 500 °C, the carbonized expanded composite slowly oxidised and the polysilicic acid content of the Visil was converted fully to silica.

These charred residues could survive 10 min exposure at 1100 °C while leaving the coherent inorganic silica residue at a similar thickness to that of the original fabric. The authors[41] also cite their patent in which they have disclosed the unique nature of such a composite structure for thermal protection, constructed from five layers as follows:

1   200 g m^{-2} Visil web,
2   120 g m^{-2} FR cotton woven fabric,
3   200 g m^{-2} Visil web containing the intumescent,
4   120 g m^{-2} FR cotton woven fabric,
5   200 g m^{-2} Visil web.

The intumescent content was varied from 100–250 g m^{-2} and the bonding resin (10% w/w on the basis of intumescent concentration) was added to it to produce the 'sandwich' structure.

10.4.2.2   Flame-retardant polyester
There are three methods of rendering synthetic fibres flame retardant:

• use of FR comonomers during copolymerization,
• introduction of an FR additive during extrusion,
• application of flame retardant finishes or coatings.

The first two methods would give inherently flame-retardant polyester fibres.

Trevira CS$^{(R)}$ and Trevira FR$^{(R)}$ produced by Hoechst are flame-retardant polyesters. Both are manufactured by copolymerizing a bifunctional organophosphorus compound based on phosphinic acid derivative (Table 10.4):

Table 10.4   Flame-retardant additives for polyester fibres[4]

Additives/Co-monomor	LOI	Reference
Phosphorus-based		
Phosphinic acid derivative (Trevira CS)	29–30	81
HO-P(O) XYCOOH		
X = H or alkyl		
Y = alkyl		
Bisphenol S oligomer (Toyobo GH)	28	81
Cyclic phosphonate (dimeric)	27–28	
(Antiblaze 1045)		
37.5: bisphenol S +	29	
55.35: Neopentyl glycol		JP 79,80,355(1979)
chlorophosphate →		*Chem. Abstr.* **91**,
5 parts FR ester + PET (100)		194591r (1979)
100 parts: PET	Good FR with no	50
7 parts: cresyl diphenyl phosphate	dripping fibre web	
5 parts: triallyl cyanurate	irradiated with	
	electron beam	
100 parts: PET	Resistant to	51
7 parts: diphenyl cresyl phosphonate	heating and melting	
[2,5-bis(2-hydroxyethoxy) phenyl]-	30	*Chem. Abstr.* **122**,
diphenyl phosphine oxide		83672h (1995)
mpt 195 °C		
Tetrakis (hydroxymethyl) phosphonium	28	*Chem. Abstr.* **122**,
chloride		12081c (1995)
Halogen-based		
p-bromophenoxycyclophosphazene	28	*Chem. Abstr.* **112**,
		8647t (1990)
Decabromodiphenyl oxide	29	*Chem. Abstr.* **99**,
		6796e (1983)
Neopentyl glycol chlorophosphate	29	*Chem. Abstr.* **91**,
		194591r (1979)

The LOI of Trevira CS fabric, having 0.6% phosphorus, is 28% and the burning fabric does not give rise to burning molten droplets. Subject to fabric construction, dyeing and finishing processes, interior textiles made from flame retardant Trevira (containing 0.6% w/w phosphorus) can be anticipated to pass all the stringent international test standards[42], such as BS 5867 part 2 type C for textiles, curtains and drapes.

Hoechst claims that the production of flame retardant Trevira is environmentally a clean process compared to other artificial and natural fibre fabrics. Tests on the toxicity of burning fabrics at increasing temperatures demonstrate the superiority of Trevira CS polyester fabric which is even less toxic as compared to treated FR cotton (Fig. 10.6). Trevira CS and FR are chlorine-free, unlike modacrylics (which contain up to 50% PVC) and PVC chlorofibres. They also liberate extremely low levels of toxic gases and smoke when exposed to an ignition source. There is zero hydrogen cyanide released, for example, in marked contrast to both smouldering and flaming FR wool, or modacrylics. Hence fumes released from Trevira polyester when it is subjected to 700 °C temperature, indicate a mortality rate of only 8% com-

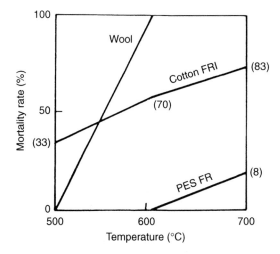

10.6   Toxicity of fabrics at different burning temperatures. PES = polyester.

pared with cotton FR which results in a mortality rate of 83% in experimental animal tests. Thus, Trevira CS/FR has been approved for the Oeko-Tex standard 100 certificate as containing no harmful substances.

Toyobo GH also licensed a flame retardant polyester to Montefibre, which is believed to be a sulphone–phosphonate copolymer. Ma *et al.*[43] have reported the synthesis and properties of intumescent, phosphorus-based flame-retardant polyesters. Spirocyclic pentaerythritol di(phosphate acid monochloride) was used as a comonomer (Structure III):

$$Cl - \overset{\overset{O}{\parallel}}{P} \begin{matrix} OCH_2 \\ \\ OCH_2 \end{matrix} C \begin{matrix} CH_2O \\ \\ CH_2O \end{matrix} \overset{\overset{O}{\parallel}}{P} - Cl$$

Structure III

The LOI value was found to be 27–30 and the analysis of the SEM photographs of the copolymer chars indicated the presence of solid phase intumescence.

Unitika[44,45] has also developed new flame-retardant melt spun polyester using [2,5-bis(2-hydroxyethoxy)phenyl]diphenyl phosphine oxide and bis(β-hydroxy ethyl)terephthalate copolymer. The filament is of the sheath–core type with regular polyester as the sheath and copolymer polyester as the core. This sheath–core structure provides stability to heat in the texturing process and facilitates texturing. The LOI value of this FR polyester is 30%. It resists burning as it has a self-extinguishing property. Moreover, it generates no hazardous gases while burning.

Horrocks *et al.*[46] have developed an analytical model for understanding the environmental consequences of using flame-retardant textiles. An environmental rank value is given at each stage in the manufacturing process and product life of each flame-retardant fibre textile. The results show that each of the eleven generic fibres analysed showed an environmental index value within the range of 32–51%, where 100% denotes the worst environmental position possible.

Zubkova[47] has recommended the use of poly(vinyltriethoxy)silane microencapsulated T-2 fire retardant in polyethylene terephthalate (PET) melt prior to extrusion, to reduce the combustibility of polyester and its blends.

In the patent literature, Japan Exlan Co.,[48] has also disclosed the production of flame-retardant polyester conjugate fibres. These are polyesters containing 0.5–5.0 mol% (based on total acid components) inorganic boron compounds as the core and polyester containing phosphorus, Mg or Si compounds as the sheath. Thus, ethylene glycol (I) and terephthalic acid (II) were esterified and polymerized in the presence of 1.5 mol% (on the basis of total acid component) B_2O_3 to give polyester (III), and I was polycondensed with II in the presence of 1.5 mol% $Mg(OAc)_2$ to give III. Then III containing B_2O_3 as core and III containing $Mg(OAc)_2$ as sheath were cospun at 295 °C and drawn to give conjugated fibres with excellent fire resistance even after washing.

Japan Exlan[49] has also disclosed the use of OH-terminated poly(dimethyl siloxane) or silane coupling agents as the sheath component for producing FR polyester bicomponent fibres.

Unitika[50,51] recommended the use of cresyl diphenyl phosphate or phosphonate (7 parts) and triallyl cyanurate (5 parts) in 100 parts PET prior to melt spinning. The fibres were then woven and irradiated with electron beam to 20 Mrad dose which showed good flame retardance and no dripping. Some of the FR additives/comonomers disclosed in the patent literature for polyester are listed in Table 10.4.

10.4.2.3   Flame-retardant nylon

Nylons have a self-extinguishing property due to extensive shrinking and dripping during combustion. Problems arise in blends with natural fibres like cellulosics which will char and form a supporting structure (the so-called scaffolding effect) which will then hold the molten polymer.

Introduction of flame or combustion retarders into polyamide melts before spinning appears to be an economical and feasible process if they are stable. Butylkina et al.[52] compared the performance of non-melting type compounds (e.g. lead methylphosphonate and a complex of alkylphosphonic acid and antimony) with highly viscous FR compounds like phosphorylated pentaerythritol (Fostetrol[(R)]) and phosphorus-based Borofos[(R)] as dope additives during melt spinning of nylon 6. The oxygen index value of about 50% was found in FR nylon 6 having (2 w/w% of antimony) as an antimony complex of alkyl phosphonic acid but the carbonized residue/char was found to be highest when Borofos was used as melt additive.

In another study, Tyuganova et al.[53] have used a ternary system of flame retardant, namely, boric acid, brominated pentaerythritol and antimony oxide mixed into the polymer melt prior to extrusion. It has been demonstrated that boron compounds display condensed phase active mechanisms and increase the yield of water and carbonized residue, while halogen-containing compounds are effective inhibitors of free radical reactions in the gas phase. A nylon sample having 2.28 w/w% Boron, 3.09 w/w% Br and 2.26 w/w% Sb showed the LOI value of 29.2.

Allied Signal[54] has disclosed in a patent the use of zinc (0.01–2.9 w/w%), molybdenum (0.002–0.58 w/w%) as calcium zinc molybdate and 0.05–1.3 w/w% chlorine, as chlorinated ethylene in nylon 6 melt for producing flame-retardant fibre for carpets. The dyed carpets were backed with regular SBR latex and Actionbak secondary backing. Flammability was tested by the ASTM E-648 Flooring Radiant Panel Test. The sample with compound B (as Kem Gard™ 425) from Sherwin-Williams as molybdenum-based flame retardant (0.3 w/w%) and compound A (chlorinated polyethylene with 25% chlorine) at 0.7 w/w% had an average of greater than 1.13 W cm^{-2} compared to 0.43 W cm^{-2} for the control carpet. Thus, FR carpet developed by Allied Signal was a class I carpet, while the control was a class II carpet as per the ASTM-E 648 test method.

Day et al.[55] have made a systematic study of the role of individual constructional components in the flammability characteristics of carpets. It was shown that carpets that are glued to cement asbestos board are less easy to ignite, burn at a slower rate, and give off less smoke and heat compared to carpets not glued down.

Stoddard et al.[56] have also studied the effect of construction parameters, face yarn additives and backing materials on the flame retardancy of nylon 6.6 carpets. A number of halogen-containing FR additives were introduced into nylon 6.6 melt before extrusion. The melt-stable additives were Dechlorane+25^R, hexabromobiphenyl and decabromobiphenyl oxide. Alumina trihydrate in the carpet latex and application of polyvinyl chloride to the face side of the primary backing were also used. It has been observed that FR carpets containing lower levels of Dechlorane/antimony oxide (2/1 w/w%) burnt for longer periods but extinguished between 2–9 min, while carpets containing 3 w/w% decabromobiphenyl oxide under similar test conditions extinguished in 5.5–10.5 min.

Organophosphorus compounds were also used as melt additives in nylon 6.6. These phosphorus derivatives in most cases reduced polymer viscosity to too low a degree for fibre formation and they also had very poor wash resistance.

Reduced flammability of nylon by use of potassium iodide[57] has also been disclosed. KI (<5 w/w%) as an additive in nylon textile fibres improved the heat, light and dye stability of fibres. The iodides used are Na, Li and other inorganic iodides and a combination of antimony oxide and inorganic iodides and inorganic phosphates.

Phosphorus-containing polyamide fibres with increased stain and fire resistance have also been disclosed by Monsanto Co.[58] Thus, bis(2-carboxyethyl)phosphinic acid neutralised with NaOH, was polymerized with nylon 6.6 salt and hexamethylenediamine to give phosphorus-containing polyamide fibres with enhanced flame resistance.

Levchik et al.[59] have suggested the role of inorganic fillers (talc, $CaCO_3$, $ZnCO_3$, MnO_2) in improving the flame retardance of a nylon 6/ammonium polyphosphate (APP) blend. It has been observed that MnO_2 oxidises nylon 6 thereby enhancing the char yield from the polymer. Furthermore, these fillers react, which increases the solid residue, improves thermal shielding of the char and gives inorganic glasses, which hinder propagation of the flame by combustible gases.

10.4.2.4   Flame-retardant acrylic fibres
Like other synthetic fibres, acrylic fibres shrink when heated, which can decrease the possibility of accidental ignition. However, once ignited, they burn vigorously accompanied by black smoke. Thus, many efforts have been devoted to improve the flame resistance of acrylic fibres.[60–67] Among these studies, halogen-based and particularly bromine derivatives or halogen- or phosphorus-containing comonomers, are the most effective flame retardants used in acrylic fibres.

Self-extinguishing modacrylic fibres have been produced from vinylidene chloride and acrylonitrile copolymers or terpolymers.[68–69]

A number of spinning dope additives are also known to render acrylic fibres flame retardant (Table 10.5), for example esters of antimony, tin and their oxides, SiO_2, halogenated paraffins, halogenated aromatic compounds, phosphorus compounds,[70–74] and so on. Bromine compounds have been found to be most effective flame retardants for acrylic fibres. The flame-retardant mechanism of these halogen compounds is believed to be associated with the interaction of halogen with reactive moieties of the flame itself.

While it is often considered that hydrogen halide reduces the concentration of the free radicals OH˙ and H˙, which help in propagation of the flame, thus following a gas phase reaction mechanism, this seems not to be the case following the pyrolysis of PAN (polyacrylonitrile) and its copolymers under burning conditions. Hall et al.[66] have demonstrated the role of various inorganic and organic phosphorus and nitrogen, or sulphur or halogen-containing derivatives with or without antimony oxide (Table 10.6). From thermogravimetric analysis (TGA), differential scanning calorimetry (DSC) and residual char-forming techniques, they have shown that flame retardancy relates directly to char-forming tendency for all the flame retardants and their ability to reduce the flammable volatiles formed during the first stage of acrylonitrile copolymer pyrolysis. Ammonium dihydrogen phosphate has been found to be the most char-forming flame retardant.

Various stages of acrylic polymer pyrolysis have been systematically studied by the same authors.[66] It is interesting to note that apparent activation energies of the acrylonitrile polymer cyclization reaction have been reduced and the lowest values are for $NH_4H_2PO_4$ containing copolymers. This suggests that the more effective, char-promoting flame retardants may initially modify the cyclization reaction to favour production of char-forming precursors. From the thermal data, a positive linear relationship between LOI and the char yield obtained at 500 °C at respective LOI values (Table 10.7) has been observed (Fig. 10.7) using the following linearly regressed equation under isothermal conditions

$$LOI = 17.0 + 0.40 \text{ (char w/w\%)} \tag{3}$$

Table 10.5 Possible dope additive for making flame-retardant (FR) acrylic fibres[8]

Polymer	Additive	Property	Reference
P(AN-vinyl chloride)	5% Dibutyl tin ethyl maleate	FR fibres	8
Acrylonitrile (AN) copolymer with 3.5% Cl (chlorinated aliphatic acid ester)	0.6% Sb_2O_3 and 0.07% Bu_2SnO	FR fibres with good dyeability, strength and transparency	8
PAN	Finely divided inorganic tin compound	FR fibres with improved gloss, transparency, whiteness, and dyeability	8
P(AN-vinylidene chloride)	$SnCl_4$ and NaOH	Fibres with good transparency and lustre	8

Table 10.5 *Continued*

Polymer	Additive	Property	Reference
P(AN-vinylidene chloride-vinyl chloride)	Compounds containing Sb, Al, Sn, or Zn	FR fibres	8
P(AN-vinylidene chloride-vinyl chloride)	Dispersion of Sb_2O_3 particles in the dope	FR fibres	8
Polymer with AN $\geq$ 50%	15–50% of a metal oxide filler with mean particle size 1–10 nm	FR fibres	8
PAN	Cyanoethyl (*N,N*-dimethyl amido)-phosphate, tris(β-chloroethyl) phosphate	FR fibres	8
P(AN-methyl acrylate-sodium methallyl sulphonate)	0.5–5.0 wt% of cyclic phosphonitrile compound containing phenoxy groups	LOI = 24.1	8
PAN	12.5% Tetrabromobisphenol A	LOI > 30	8
PAN	Polychlorotrifluoroethylene	Fibres with low flammability	8
P(AN-vinyl acetate)	Bu_3PO_4 Tris(dibromopropyl) phosphate or $Al(OH)_3$ gel	LOI ~ 29 synergistic flame retardant	64
P(AN-vinylidene chloride)	Ca phosphate and Sb_2O_3		
P(AN-vinyl acetate)	15% Hexabromocyclododecane (Great Lakes CD-75P) 15% poly(dibromostyrene) (Great Lakes PDBS)	LOI 41–45.5 LOI 43–48	67
P(AN-vinyl chloride and/or vinylidene chloride/sodium methallyl sulphonate) 70 parts + 15 parts cellulose diacetate	37.5 parts colloidal soln. of Sb_2O_5 in dimethyl acetamide + 380 parts dimethyl acetamide	LOI = 36 Ts = 2.8 g/d Elongation = 36%	71
P(AN-Na styrene sulphonate – vinyl chloride – vinylidene chloride) 30% solution in Me_2CO	30% SnO_2 (av. diameter 0.05 μm) + 4% HCl + 0.15 part bisphenol-A-diglycidyl ether	LOI 35.5	72, 73

Table 10.6 Selected flame retardants[66]

Flame retardants	Flame-retardant elements
Ammonium polyphosphate (APP) (Amgard MC) (Albright & Wilson Ltd.)	P and N
Ammonium dihydrogen phosphate (ADP)	P and N
Diammonium hydrogen phosphate (DAP)	P and N
Proban cc (NH$_3$ condensate of THPC urea) (Albright & Wilson Ltd.)	Organic P and N
Sodium dihydrogen phosphate	P
Hexabromocyclododecane (hexa) (Schill & Seilacher Ltd.)	Aliphatic Br
Decabromodiphenyloxide (deca) (Schill & Seilacher Ltd.)	Aromatic Br
Antimony trioxide (Aldrich Chemical Co.)	Sb
Ammonium thiocyanate	S and N
Urea	N
Melamine	N
Sandoflam-5060 (Sandoz)	S and P
Ammonium chloride	Cl and N
Thiourea	S and N
Red Phosphorus (Amgard CRP) (Albright & Wilson Ltd.)	P
Diammonium sulphate	S and N
Zinc phosphate (Aldrich Chemical Co.)	P and Zn
Zinc borate (Aldrich Chemical Co.)	B and Zn
Flacavon-TOC (Schill & Seilacher Ltd.)	organic Br + Sb
Amgard CUS (Albright & Wilson Ltd.)	P and N
Phosphonitrilic chloride trimer (Aldrich Chemical Co.)	P, Cl and N

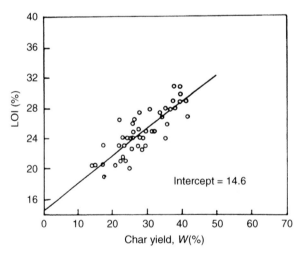

10.7 LOI–char relationship of acrylic polymers with various flame retardants under LOI burning conditions.

Table 10.7 LOI values and char production of acrylic polymers with various flame retardants[66]

Polymer (100 parts)	Flame retardant (parts)	LOI %	Char yield at LOI	Char yield at 31% O$_2$	Char yield[a] at 500°C
PAN	–	19.0	17.3	11.8	19.7
PAN	Ammonium dihydrogen phosphate (15)	27.0	34.4	32.0	–
P(AN-VA), 10% VA (vinyl acetate)	–	20.4	14.9	13.2	22.0
	Ammonium dihydrogen phosphate (15)	31.0	39.8	39.8	52.0
	Ammonium polyphosphate (15)	31.0	38.1	38.1	62.1
	Diammonium hydrogen phosphate (15)	30.0	–	–	47.5
	Sandoflam-5060 (15)	27.0	–	–	42.5
P(AN-MA) 10% MA (methyl acrylate)	–	20.4	20.8	15.9	27.7
	Ammonium polyphosphate (15)	29.0	39.1	37.7	65.2
	Ammonium dihydrogen phosphate (15)	28.0	36.7	35.0	58.9
	Diammonium hydrogen phosphate (15)	27.0	41.7	40.0	60.4
	Antiblaze CUS (15)	27.0	34.3	33.0	–
	Red phosphorus (15)	26.5	22.0	20.6	40.1
	Sandoflam-5060 (15)	26.0	35.8	33.3	40.3
	Flacavon-TOC (15)	26.0	29.6	27.3	42.5
	Proban CC (15)	24.0	28.9	27.8	43.6
	Urea (15)	23.0	17.0	16.1	36.6
	Sodium dihydrogen phosphate (15)	23.0	21.9	20.4	29.2
	Thiourea (15)	23.0	26.0	25.0	42.5
	Zinc phosphate (15)	22.7	23.0	22.0	–
	Zinc borate (15)	22.6	24.5	23.5	–
	Hexa[b] (10) + antimony oxide (III) (5)	27.5	27.8	–	41.5
	Hexa (15)	26.5	26.4	–	33.0
	Deca[c] (15)	25.0	26.2	–	33.5
	Deca (10) + antimony oxide (III) (5)	25.2	27.6	–	40.0

[a] Char yield obtained from isothermal pyrolysis at 500°C and calculated at 200s. [b] Hexabromocyclododecane. [c] Decabromodiphenyl oxide.

On this basis, it may be concluded that char-promoting flame retardants, such as ammonium phosphates, are more effective than vapour phase inhibiting bromine-containing flame retardants.

Fire-resistant acrylic fibres[74–76] have also been manufactured by coating partially oxidised acrylic fibres (heated in air at 250°C under tension for 15min) with phosphoric acid (2.0 vol% in methanol) or 2.5% guanidine phosphate. After coating with the latter the LOI increased to 42. In another patent, Japan Exlan disclosed the use of hydrizine hydrate ($30\,g\,l^{-1}$) for crosslinking with $ZnCl_2$ after hydrolysing the fibres with 30% NaOH. The fibres, P(AN-MA) (poly(acrylonitrile–methylacry late)) thus cross-linked with 5% $ZnCl_2$ solution for 30 min at 20 °C gave fibres with an LOI value of 34.

10.5   Fire-retardant finishes

A number of fire-retardant finishes and their modes of application have been well documented by Drake and Reeves,[77] Holme,[78] Wakelyn et al.,[79] Barker and Drews[80] and Horrocks.[81]

10.5.1   Cellulosic fibre fabrics

An early review of cellulose phosphorylation by Reid and Mazzeno[82] showed the typical structure of cellulose phosphate ester, Structure IV:

$$Cell-CH_2O-\overset{\overset{\displaystyle O^-NH_4^+}{|}}{\underset{\underset{\displaystyle O}{\parallel}}{P}}-O^--NH_4^+$$

However, exchange with hardness ions such as Ca^{2+} and Mg^{2+}, available in hard water, gave a salt, Structure V:

$$Cell-CH_2O-\overset{\overset{\displaystyle O^-}{|}}{\underset{\underset{\displaystyle O}{\parallel}}{P}}-O^-\quad Ca^{2+}$$

This salt is stable and prevents the release of phosphoric acid on heating, the effectiveness of the flame-retardant finish thereby being reduced.

Phosphorylation of cellulose by means of diammonium phosphate (DAP)/urea gives rise to a finish that resists exchange of Ca^{2+} ions, although it is saponified by alkaline washes. The use of titanium, zirconium, or tin salts as a post-treatment of phosphorylated cotton also minimizes the exchange with hardness ions.[83]

The application of various finishes has been summarized well by Horrocks[81] (Table 10.8). The commercially most successful finishes are the N-methylol dialkylphosphonopropionamides, from which Pyrovatex CP (Ciba-Geigy) is derived.

Another product, Pyrovatex7572$^{(R)}$ (N-dimethylol dimethylphosphonopropionamide) has also been developed to obtain improved reactivity with the fibre. With environmental and safety issues in mind, an improved Pyrovatex CP reduces formaldehyde emissions during processing by over 50% and there is a remarkable reduction of tar build-up in the stenter/curing chamber during curing. This finishing agent carries the PA111 certificate and also meets the international Oekotex Standard 100.

Table 10.8 Selected retardant-finish formulations for cotton[8,81]

Finishes and coatings	Application
Finishes	
$(CH_2OH)_4P^+X^-$ tetrakis(hydroxymethyl)-phosphonium (THP) salts, where $X = Cl, OH, (1/2)(SO_4)^{2-}$ often as a condensate with urea, e.g. THPC-urea-NH_3, Proban CC (Albright and Wilson)	(i) Apply with trimethylolmelamine, heat cure at 160 °C (ii) Apply and cure with ammonia gas and oxidise with H_2O_2
$(CH_3O)_2POCH_2CH_2CONHCH_2OH$ *N*-methylol dimethyl phosphonopropionamide, Pyrovatex CP (CIBA)	Apply with melamine resins, cure at 160 °C
Pyrovatex 7572	Pad-dry-cure, chemical cross-linking
Ammonium polyphosphate	Non- or semidurable depending on '*n*'
Diammonium phosphate	Non-durable
Ammonium sulphamate + urea or urea-based cross-linking agent	Curing at 180–200 °C for 1–3 min. Treated fabrics pass the vertical flame test (VST) even after 50 hard alkaline launderings
Coatings	
Sb_2O_3/chlorinated paraffin wax	Apply from solvent
Sb_2O_3 or Sb_2O_5 + decabromodiphenyl oxide or hexabromocyclododecane + acrylic resin e.g. Myflam (B F Goodrich, formerly Mydrin)	Semi to fully durable

Other durable treatments include the use of ProbanCC[(R)] (Albright & Wilson). It involves padding of tetrakis (hydroxymethyl) phosphonium chloride (THPC)/ urea solution onto the fabric, curing with ammonia in a specially designed reactor to generate a highly crosslinked phosphorus–nitrogen three-dimensional polymer network. Cotton fabric finished with ProbanCC is subsequently treated with hydrogen peroxide which converts the P^{3+} to the P^{5+} state and enhances the durability of the finish. Pyrovatex CP is generally used for curtains while ProbanCC, which retains greater strength, is used for hospital bed sheets and so on. However, Le Blanc[84] has observed the loss of phosphorus content in Pyrovatex CP new treated cotton during storage at room temperature and after steam sterilization. It is assumed that the loss of phosphorus, consequently leading to failure in flame retardant properties, could be due to the following:

- Hydrolysis of the methyl ester groups of *N*-methylol dimethylphosphonopropionamide (MDPPA) molecules, which are not attached to cellulosic hydroxyl groups, producing – $H_2C\,P(O)(OH)_2$ groups which are acidic in nature.
- Acidity of these groups further accelerates the hydrolysis of more methyl ester groups and also cleaves the ether linkages between the MDPPA and cellulose.

The fabrics treated by modified process retained their FR properties, even up to 100 launderings. Morrison from Albright & Wilson Ltd. has given an excellent report concerning the application of FR treatments after the introduction of flammability regulations in the UK for furnishing fabrics.

Both Proban and Pyrovatex can satisfy the requirements of the UK Furniture and Furnishings (Fire Safety) Regulations, 1990 when applied at lower application levels. Pad bath additives normally consist of Pyrovatex CP new, orthophosphoric acid catalyst, and a low formaldehyde melamine resin and a softening agent. However, for some furnishing fabrics, a combination of melamine resin and a tetramethylol acetylene diurea-based resin is added to the pad bath.

New systems which provide extended durability to soaking include Amgard LR1[R] and Amgard LR2[R] for application to cellulosic or cellulosic-rich blends. Pad bath solutions consist of Amgard LR1 (ammonium polyphosphate-based solution), a fluorocarbon, water and soil repellent, softening agent and volatilisable wetting agent and water.

Many coating systems have also been recommended for the production of FR upholstery fabrics. New generation coating systems include Amgard LR4[R] and Amgard LR3[R] phosphorus-based flame retardants that are halogen and heavy metal free. The Amgard LR4[R] system is a mixture, a low solubility ammonium polyphosphate and Amgard LR2 (ammonium polyphosphate-based solution), applied in combination with an antifoaming agent and an acrylic-based latex.

Horrocks and co-workers[81,85] have demonstrated the role of char-forming and intumescent systems in the flame retardation of textiles. It is interesting to note that the majority of intumescent systems are based on ammonium polyphosphate (an acid source), melamine and its derivatives (as blowing agents) and pentaerythritol derivatives as char-forming agents.

Camino and Costa[86] studied the mechanism of intumescent char formation of ammonium polyphosphate (APP), pentaerythritol and pentaerythritol diphosphate combination. According to them,

1   APP starts decomposing at 210 °C onwards and phosphorylation of polyol occurs without elimination of gaseous products.
2   Subsequent phosphorylation produces cyclic phosphate esters.
3   Between 280–330 °C, the polyol phosphates decompose to char with considerable formation of phosphoric acid, which subsequently polymerises to polyphosphoric acid.

The degree of char expansion depends on the rates of gas/volatiles evolution, the viscosity of the liquefied pyrolysis products and the transformation of the latter to solid char which acts as a barrier.

Char studies of flame-retardant cellulose have indicated that most phosphorus remains in the char. However, Drews and Barker[87] showed that the phosphorus retention in chars depends on the reactivity of phosphorus moieties with cellulose.

Faroq et al.[88] have also established that the simple competitive pyrolysis mechanism in cellulose proposed by Kilzer and Broido[89] is influenced by the nature of the flame retardants. Figure 10.8 shows this modified mechanism and includes the 'activated cellulose' intermediate state as postulated by Bradbury et al.[90] Lewin[91] has reported the use of sulphamates as flame retardants. Cotton fabrics were treated with ammonium sulphamate in conjunction with urea or a urea based crosslinking agent, as a coadditive by pad dry cure method. Treated fabrics, both cotton and woollens passed the vertical strip test (VST) even after 50 hard water alkaline launderings.

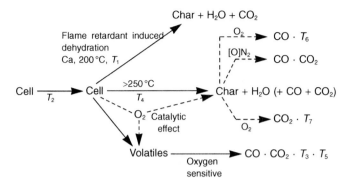

10.8   Modified Kilzer and Broido mechanism for cellulosic pyrolysis; for values of T_1–T_7 see Kandola *et al.*[85]

El-Alfy *et al.*[92] have reported the application of semidurable flame retardants for cotton. Urea $(400\,g\,l^{-1})$-phosphoric acid $(100\,g\,l^{-1})$ solution was applied to cotton in the presence of various catalysts by the pad dry cure method. The catalytic efficiency was found to be in the following order: ammonium molybdate > zinc acetate > copper sulphate.

In another communication, Kurose and Shirai[93] have studied the flame retardation of cotton fabric treated with Ni^{2+}-phos-PVA (30 mol% phosphorus). In particular, cotton fabric treated with Ni^{2+}-phos-PVA (polyvinyl alcohol) complex (molar Ni^{2+}/phos-PVA ratio $= 1.25 \times 10^{-1}$) showed the LOI value of 57.5.

Use of Caliban FR/P-44, one of the first antimony trioxide/DBDPO (decabromodiphenyl oxide) retardants used in a resin category, has also been made in producing durable FR cotton and polyester fibre blends. They are used as flame-retardant protective garments for aluminium foundry workers.

10.5.2   Flame-retardant finishes for polyester

There has been some developments in flame retardant finishes[94–98] for polyester fabric and its blends. Flame-retardant finishes for synthetic fibres should either promote char formation by reducing the thermoplasticity or enhance melt dripping so that the drops can be extinguished away from the igniting flame. For protective clothing, char forming finishes would be desirable.[80]

Day and co-workers[94,95] have studied in depth the flammable behaviour of polyester fibre by using a series of phosphorus- and bromine-containing flame retardants as both additives and finishing agents. Alternatives to Tris-BP (tris(2,3-dibromopropyl) phosphate), a known carcinogen, were applied from tetrahydrofuran solution. The chemicals used were tris(2,3-dibromo-2-methyl propyl)phosphate, tris(2,3-dibromo-3,3-dimethylpropyl)phosphate, and so on. The pyrolysis and gaseous combustion of PET incorporating poly(4-bromostyrene), poly(vinyl bromide) and poly(vinylidene bromide) applied via topical treatment or radiation grafting showed significant release of HBr, which is capable of inhibiting gas-phase combustion reactions. Condensed phase interactions that are capable of altering gaseous pyrolysates were also noted.

Thermal stability data of the above-mentioned flame-retardant systems suggest that, although the aliphatic bromides are excellent sources of HBr, they cannot be

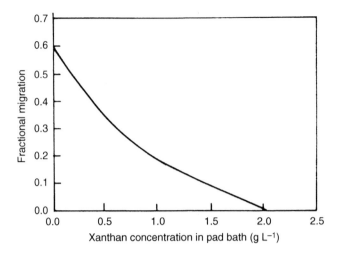

10.9 Fractional migration of liquid flame retardant on 100% polyester fabric as a function of xanthan concentration in a pad bath.

considered ideal flame retardants because of the toxicity of halogen halides. Furthermore, they are not char forming.

Mischutin[96] has recommended the application of low melting solid or liquid brominated compounds in combination with colloidal antimony oxide and a suitable binder by a coating technique – Caliban FR/P-44 is an example here. Simultaneous dyeing and flame retardation of 100% polyester fabric on a Beck dyeing machine or by the thermosol process have also been suggested. Flame-retardant treatments applied by the thermosol process are found to be extremely durable and they can withstand multiple launderings and dry cleanings. In a classical paper,[97] the problem of migration of liquid flame retardant in continuous thermal fixation finishing of polyester fabric by the pad-dry-cure process has been addressed. The experimental results reveal that migration control occurs by a gel formation mechanism and not by the particle flocculation mechanism that exists in the thermal fixation of disperse dyes. Xanthan gums have been suggested to control the migration of liquid flame retardants onto polyester fabrics (Fig. 10.9).

Toray Industries[98] and Nikka Chemical Industry[99] have used tetraphenoxydiaminocyclotriphosphazene dispersions in the dyeing bath,[98] and hexa bromocyclododecane along with Disperse Red 127 for simultaneous dyeing and flame retardance.[99] Akovali and Takrouri[100] have recommended the generation of crosslinking reactions by cold plasma treatment of polyester fabric in the presence of several flame-retardant volatile monomers (Fig. 10.10). Unitika[101] has also demonstrated the application of low temperature oxygen plasma onto polyester fabric padded with 10% polysiloxane, PSR-10(R) solution. After heat treatment, the fabric exhibited good resistance to melting in contact with a burning cigarette for 5 min.

10.5.3 Flame-retardant finishes for polyester/cellulosic blends

Flame-retardation of polyester/cellulosic blends is still a complex problem owing to the differential thermal behaviour of polyester and cellulosic components. Most of the approaches documented in the literature[102–104] have limitations because of toxi-

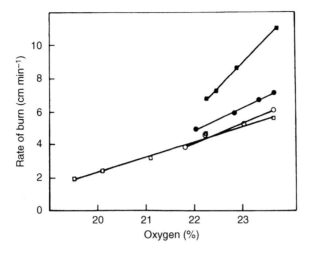

10.10 Rate of burning as a function of environmental oxygen concentration for untreated polyester fabric and plasma treated fabric at 20 W, 60 min. ■, untreated polyester fabric; ●, plasma treatment under vacuum; ○, tetrachloroethylene plasma-treated fabric; □, HMDS plasma-treated fabric (see Akovali and Takrouri[100]).

city, or difficulty in maintaining desirable aesthetic and performance properties. Miller et al.[102] observed that for a combination of untreated polyester and cotton, the system ignites sooner, burns faster, and evolves more volatile hydrocarbons than would be anticipated from the individual behaviour of polyester or cotton independently. On ignition cellulose chars and provides a scaffold for the molten polyester, preventing its escape from the flaming zone. It has been suggested that the interaction could be of a chemical nature (based on vapour phase interaction of pyrolytic products) or of physical origin, which alters the heat transfer characteristics.

Flame retardants that are active in both the condensed phase and vapour phase have been found to be very efficient on polyester/cotton blends. The performance of various phosphorus, nitrogen and antimony/halogen compounds on the flame retardance of polyester/cellulose blends has been reported by Bajaj et al.,[104] Holme and Patel,[105] Horrocks et al.,[4] Shukla and Singh.[106]

10.5.4   Flame-retardant finish for wool

Wool is not as flammable as cotton, and wool fabric was the traditional material for thermal protection except for the more arduous conditions where asbestos was required. However, for thermal protective clothing a Zirpro(IWS) finish,[8] based on hexafluorotitanates and hexafluorozirconates, has been developed, which is extremely stable in acid solutions and exhausts onto wool well below the boil. The Zirpro finish produces an intumescent char, which is beneficial for protective clothing, where thermal insulation is a required property of a burning textile. Work attributed to Benisek[107] has been reported to the effect that the addition of tetrabromophthalic acid (TBPA) to the basic Zirpro treatment produces a finish suitable for end-uses in which low afterflame times are required. However, TBPA increases the smoke density. A multipurpose finish incorporating both Zirpro (as a titanium complex) and fluorocarbon in a single bath application that makes wool flame-retardant, as well as oil, water and acid repellent, has also been developed.

In one study, shirts made from 100% cotton, Firestop cotton, Flampro wool, and Nomex III were evaluated for their protective and wear-life performance. The greatest protection was provided by fire-resistant cotton and wool fabrics. Nomex aramid-fibre fabrics gave less protection and untreated cotton gave the least.

10.5.5 Glass-fibre fabrics

Haigh[108] has dealt with the special finishes for glass fabrics and heat reflective coatings. In one finishing treatment, colloidal graphite was used, together with silicone oil, to provide protection at higher temperatures. Clothes treated in this way can be used at 400 °C or higher if exposure times are in minutes rather than days or in the absence of oxygen.

Another feature of glass fibre is that it melts at around 1000 °C, so that in the untreated form, it is unsuitable for applications at higher temperatures. However, it can be treated to improve its resistance to such temperatures, by incorporating finely dispersed vermiculite and another involving aluminium salts. At high temperatures, the aluminium will react with the glass fibre to raise its melting point above 1500 °C.

Most hot environments are created by a mixture of convective and radiant heat. Glass fabrics provide good protection against the former, because they generally have low coefficients of thermal conductivity (around $0.6 \, W \, m^{-1} \, K^{-1}$). Their performance against radiant heat can be greatly improved by the application of an aluminium reflective layer to one surface. It can be applied directly to the fabric, either as a very thin foil or supported on a thin polyester film.

The market for FR coatings is growing because back coating is proving to be one of the most versatile techniques for producing FR furnishing fabrics.[109]

In order to understand the environmental impact of FR systems, BTTG's Fire Technology Services division is actively engaged in modifying the existing FR systems in order to minimise the toxic emissions during end-use, and also to develop new FR systems employing less hazardous materials.

10.6 Flame-retardant test methods

It is important to test fabrics to ascertain if they are likely to be suitable for the application for which they are required. ASTM and BSI methods are available for such tests, in some cases, ISO standards also exist.

The first group of tests answer the question 'does the fabric burn when a flame is applied to it?' Both ASTM and BSI have a vertical specimen test for fabrics for thermal protection. ASTM D3659 resembles BS 3119, which has been virtually replaced by BS 5438. If the fabric ignites, these tests can determine the spread of flame and the burning rate. Fabrics for thermal protection should not ignite, so the other aspects are not needed. Tests to determine the critical oxygen content of textiles or LOI are ASTM D2863, which is identical to Method 141 of BS 2782.

The second group of tests answer the question 'how well will fabric protect a person wearing it?' ASTM D4108 and BS 3791 deal with this aspect. ISO6942 test has been drafted for gas-heated radiant panels. The ASTM test uses convective heat while the ISO test uses purely radiant heat.[108]

10.6.1   Standards

The formulations of fire safety standards within Europe[109,110] and the new construction products directive operative within EU member countries have been discussed, and this was followed by a review of the CEN/TC 127 fire safety in buildings in the Technical Committee. Some of the important test methods/standards are discussed below:

10.6.1.1   Vertical strip test

The vertical strip test has achieved the widest acceptance for determining flammability. British Standard[111] BS 5438:1989 describes the test method for measuring the ease of ignition and rate of flame spread of vertically oriented test samples. A specified small butane flame is applied to the bottom edge of a vertical test specimen (200 mm long × 80 mm wide) for prescribed times and the minimum ignition time is determined. In another set of experiments, a small igniting flame is applied for 10 s to the bottom edge of a vertical test specimen and flame spread times are measured by recording the severance of marker threads in seconds (Test 3b).

10.6.1.2   Fire tests for upholstered composites and mattresses

BS 5852 (part 1) describes the test method[112] for assessing the ignitability of material assemblies of upholstered composites for seating when subjected to either a smouldering cigarette or a lighted match. The test materials shall be representative of cover, filling and any other components used in the final assembly. The cover size needed for each test is 800 ± 10 mm × 650 ± 10 mm. The upholstery filling required for each test is two pieces, one $450 \times 300 \times 75$ mm thick and the other piece $450 \times 150 \times 75$ mm thick. Some cushioning assemblies may consist of several layers of felt, wadding or different foams. In these cases the test pieces shall reproduce the upper 75 mm of the cushioning assembly.

A smouldering cigarette is placed along the junction between the vertical and horizontal test pieces, allowing at least 50 mm from the nearest side edge to the cigarette. The progress of combustion is observed using the clock and any evidence of progressive smouldering or flaming in the interior and/or cover is recorded.

BS 5852: Part 2:1982 covers ignition sources between 2 and 7, although in practice 5 and 7 are specified (two butane flames and four burning wooden cribs), which together with the butane flame source of BS 5852: Part 1, form a sequence of increasing thermal output from that approximating to a match burning to four sheets of full size newspaper burning.[112]

The objective of this test is to subject an assembly of upholstered composites, arranged to represent the junction between a seat and back, as is typical in chairs, to six flaming sources selected to cover the intensities of actual sources that might be encountered in various end environments.

Progressive smouldering failure can be detected for any composite that produces externally noticeable amounts of smoke, heat or glowing for 30 min after the removal of the burner for sources 2 or 3. However, for sources 4, 5, 6 and 7, any composite that produces externally detectable amounts of smoke, heat or glowing for 60 min after ignition of the crib.

British Standard BS 6807:1996 describes the method for assessing the ignitability of mattresses, upholstered divans and bed bases when subjected to flaming types of primary and secondary ignition sources of different severities.[113]

Holme cites the review of Bryson which considers these standards from the fabric producer's viewpoint. According to Bryson the basic standards, BS 5852 Part 1 (for cigarette and match ignition) and BS 5852 Part 2 (covering ignition sources anywhere between 2 and 7, but in practice 5 and 7), state that the cigarette test could be carried out over the infill used in the furniture, but the match test has to be carried out over $22 \, kg \, m^{-3}$ polyurethane foam, which could not legally be used in domestic furniture. However, BS 5852 Part 2 might be carried out over any foam. Water soak tests and the use or non-use of interliners create further complexities.

Peter Cook International[115] has also claimed that the development of the Partex System for upholstery fabrics, interlinings, drapes, workwear fabrics and mattress tickings meets the fire retardancy regulations. This system allows for fabric treatment to meet BS 5852 Part 1 (after the soaking test) and can take natural fabrics through to Crib 5.

Cigarette ignition of upholstered chairs has been analysed by Braun et al.[116] at the National Bureau of Standards, Washington DC. The objective of this work was to observe the smouldering behaviour of upholstered chairs varying in cover fabrics and filling materials, after cigarette ignition. The chairs which smouldered only contained polyester batting under the cover fabric in the area where the cigarette was placed, those which first smouldered and then burst into flames contained polyurethane foam or cotton batting.

10.6.1.3 Ignitability of fabrics used in tented structures

British Standard BS 7157:1989 describes the method for testing the ignitability of fabrics used in the construction of tented structures such as marquees, large tents, awnings or flexible membrane.[117]

The range of ignition sources used in this test are pinewood cribs: ignition source 4 to 7 as described in BS 5852:Part 2. The test frame consists of a metallic rod arranged to represent the main and side walls and roof of the tent structure. The frame shall be constructed slightly smaller than the test specimen (as a mini tent) but by not more than 10 mm less than any dimension. The fabric specimens and the cribs shall be conditioned before performing the test for 72 hours in indoor ambient conditions. The ignitability test should be performed in a draught-free environment with a temperature of $20 \pm 5 \, °C$ and a relative humidity of $55 \pm 20\%$.

The crib is ignited as described in BS 5852:Part 2 and progress of combustion is recorded along with the formation of flaming droplets or glowing, if any.

10.6.1.4 Ignitability of bedcovers and pillows by smouldering and flaming

In this test (BS 7175:89),[118] the test specimen is placed on a mineral wool fibre pad (MWFP) and subjected to smouldering and flaming ignition sources placed on top of and/or below the test specimen. A cigarette and the test specimen are placed on the MWFP so that the cigarette lies adjacent to the pillow. Light the cigarette and draw air through it until the tip glows brightly. Light another cigarette and place it on the upper surface of the pillow. Start the clock and observe the specimen for progressive smouldering or concealed smouldering. In the case of a quilt, observation continues for 60 min after the placement of a third cigarette.

If ignition of the test specimen is observed, extinguish the text specimen and record that ignition occurred for the ignition source used. For ignition sources 1 to 3, a butane flame test is used.

The requirements for contract upholstered furniture in the UK have generally been that the cover fabric, in conjunction with the actual filling to be used, be able to pass an ignitability test to BS 5852 source 5 (flaming wood crib) level. With a wide range of different fabrics, and to a lesser extent fillings, available on the contract market, manufacturers have been put in the difficult position of having to test very large numbers of fabric-filling combinations.

However, BS 7176:1995, the 'predictive testing' approach used in this method, recommends the use of a particular grade of non-FR polyurethane foam.[119] The rationale is that if a fabric will pass for ignition source 5 test over this foam, it should pass over any CMHR (combination modified high resilience) foam.

BS 7176 also lists four hazard classes categorised as low, medium, high and very high[120] (Table 10.9). The examples cited in Table 10.9 for each hazard category[120,121] cannot be exhaustive and some of the examples appear in more than one hazard category. This reflects the range of the hazards possible under different circumstances.

10.6.1.5   Evaluation of textile floor coverings
Assessment of the textile floor coverings tested in accordance with BS 4790 is valuable in determining the ease with which the textile floor covering will ignite when a burning cigarette, a hot coal or a similar source of ignition is dropped on it.[122,123] It is applicable to all types of textile floor coverings used in the horizontal position. BS 4790 is not intended to give an overall indication of the potential fire hazard under actual conditions of use. The radiant panel test described in ASTM-E648:1978, however, can be used to assess the performance of textile floor coverings in a fire situation.

Table 10.9 BS7176:1991 Hazard Classifications and Ignitability Performance Requirements[120]

	Low hazard	Medium hazard	High hazard	Very high hazard
Test methods	Section 4 of BS5852:1990	Section 4 of BS5852:1990	Section 4 of BS5852:1990	Section 4 or Section 5 of BS5852:1990 (or other as specified)
Requirements	Resistant to ignition sources 0, 1	Resistant to ignition sources 0, 1, 5	Resistant to ignition sources 0, 1, 7	At the discretion of the specifier but at least high hazard requirements
Typical examples	Offices, schools, colleges, universities, museums, exhibitions, day centres	Hotel bedrooms, public buildings, restaurants, services' messes, places of public entertainment, public halls, public houses, bars, casinos, hospitals, hostels	Sleeping accommodation in certain hospital wards and in certain hostels, offshore installations	Prison cells

Three methods (loose-laid, fully adhered or loose-laid with underlay) for mounting the samples can be used for testing floor coverings. The selection of which method will depend on how the floor coverings are laid in actual use. According to its observed behaviour, a specimen can be considered satisfactory if the radius of the affected area is not greater than 75 mm.

The hot metal nut method[122] has also been used in BS 4790:1987 to study the effect of a small source of ignition on textile floor coverings. In this test, the nut is heated to a temperature of $900 \pm 20\,^{\circ}C$ in a muffle furnace. It is then placed centrally on the specimen in a chamber within 3 s of its removal from the furnace. The sliding panel is opened and the nut removed from the specimen after it has been in contact with it for 30 ± 2 s. The sliding panel is closed after all the effects of ignition have ceased.

In another British Standard, BS 6307:1982, the methenamine tablet test has been used to study the influence of a small source of ignition on textile floor coverings.[124] However, this method is used only to assess the properties of materials in response to heat and flame under controlled laboratory conditions. This method is generally used for acceptance testing in the trade.

10.6.1.6   Evaluation of protective clothing
The European Committee for Standardization has recommended the EN366:1993 standard for evaluation of materials and material assemblies exposed to a source of radiant heat, while EN367:1993 is for determination of the heat transmission on exposure to flame. EN532:1993 standard has been used to test the specimens for limited flame spread. The requirements and test methods for protective clothing for firefighting[125] have been covered in EN469:1995. This standard specifies the requirements for clothing to be worn during structural firefighting operations. Heat transfer (flame) by multilayer clothing assemblies for firefighters tested as per EN367 standard shall give a mean heat transmission index, $HTI_{24} \geq 13$, while heat transfer (radiation) is measured in accordance with Method B of EN366 at a heat flux density of $40\,kW\,m^{-2}$, and after the pretreatment should give a mean $t_2 \geq 22$ s and a mean transmission factor $\leq 60\%$.

The complete garment assembly may be additionally tested for typical scenarios encountered by a firefighter under the conditions given in Table 10.10.

However, testing under emergency conditions must be carried out on an instrumented mannequin.

Krasny *et al.*[126] have also discussed the heat flux conditions measured in seven room fires for protecting firefighters. The fires ranged from just short of flashover through rapid build-up to considerable postflashover burning. The standard, NFPA 1971, Protective Clothing for Structural Fire Fighting, requires that the turnout coat or pants assembly must protect the wearer against second degree burns when a heat flux of $84\,kW\,m^{-2}$ is applied to its outside surface for a minimum of 17.5 s. The

Table 10.10   Firefighting scenarios

Conditions	Exposure time	Temperature (°C)	Heat flux density (kW m^{-2})
A. Normal	8 h	40	1
B. Hazardous	5 min	250	1.75
C. Emergency	10 s	800	40

results imply that firefighters have only 10 s or less to escape under most flashover conditions.

Welder's protective garments requirements and evaluation[127] has been discussed in EN470-1(1995). Welder's protective garments may be designed to provide protection for specific areas of the body, for example, sleeves, aprons, and gaiters. Electrical conduction from the outside to the inside of the garment should also be avoided. As per this standard, the garment when tested for flame spread in accordance with EN532 after washing/dry cleaning shall meet the following requirements:

- No specimen should give flaming to the top or either side edge.
- No specimen should produce a hole or give flaming or molten debris.
- The mean value of after flame time and afterglow time shall be ≤ 2 s.

European Standard EN531:1995 is applicable to protective clothing for industrial workers exposed to heat[128] (excluding firefighter's and welder's clothing). The heat may be in the form of convective heat, radiant heat, large molten metal splashes or a combination of these heat hazards. This European Standard specifies the performance requirements and methods of test and gives design recommendations for the clothing wherever necessary. When tested in accordance with EN366 method B at a heat flux density of $20 \, kW \, m^{-2}$, all clothing assemblies offering protection against radiant heat shall meet at least level C_1, that is, mean time to level t_2 between 8 s (min) to 30 s (max). For convective heat protection, the clothing assemblies if tested as per EN367 shall meet at least B1, that is, the range of HT1 values should be between 3 min (minimum) and 6 min (maximum).

The performance requirements of limited flame spread materials and material assemblies used in protective clothing[129] are given in EN533:1997. All material assemblies claiming compliance with this European Standard should have a limited flame spread index of 2 or 3 when tested in accordance with EN532 with the flame applied to the outer face and to the inner face.

Whiteley[130] has made some important observations on the measurement of ignitability and on the calculation of 'critical heat flux'. According to him there is no universally accepted model for the correlation of ignition time with incident radiant heat flux.

Whiteley et al.[131] and others[132,133] have demonstrated the use of a cone calorimeter test in studying the heat release from flame-retardant polymeric materials. This test gives a better correlation between room scale testing and large scale fire testing and provides a pyrolysis profile under ambient oxygen conditions.

In an excellent review, Weil et al.[134] have commented on the oxygen index test for evaluating the flammability of polymeric materials and discussed what the oxygen index correlates to. According to them, this test can be improved by using it with bottom ignition rather than with standard top ignition.

10.6.1.7 British Textile Technology Group manikin test systems, RALPH
The influence of external heat flux is observed in various manikin tests. These methods are typified by the manikin developed by the British Textile Technology Group (BTTG) known as RALPH (Research Aim Longer Protection Against Heat). Improved standards for fire protective clothing are the objective of BTTG (Fire Technology Services). The reported differences between RALPH II and DuPont's Thermo-man are part of the debate for a possible European standard based on manikin testing for heat and flame protective clothing.[135,136]

The basis of the RALPH manikin was a 'shop' display manikin composed of glass fibre/polyester resin. This composition is insufficiently inert to direct flame contact and was therefore coated with a ceramic material of approximately 5 mm thickness that also served to create greater thermal inertia. This coating was chosen, after comparison with other materials, because it had the best combination of flame retardancy, resistance to cracking when heated/cooled, ease of application and adhesion to the glass fibre/polyester resin. The coated manikin was then painted with an intumescent material to impart further protection.

The heat sensing equipment consists of 32 linearized thermocouple amplifiers connected to a computer. Each sensor was a Type T copper–constantin thermocouple soldered to a copper disc 8 mm in diameter and of 0.2 mm thickness. These sensors were then implanted into the manikin with all wiring exiting via the head to the amplifier/computer.

RALPH II carries a total of 57 heat sensors, instead of the original 32. Improved computer manipulation and display of data with the existing RALPH facility permits any half of the manikin to be exposed to flames at approximately 60 kW m^{-2}. The facility has been extended to increase this to at least 84 kW m^{-2}, by adding a second array of burners so that RALPH can be enveloped in flames, and to provide an alternative purely radiant heat source.

To calibrate the sensors, two approaches were adopted. For 'time to pain' the pain response of volunteers was measured for heat fluxes of 1–31 kW m^{-2}, this data then being compared with the temperature recorded by a single sensor when covered with various types and thicknesses of thermoplastic tape. By this method, the tape producing the best 'match' response with volunteers' 'time to pain' was achieved. For burn injury prediction, reference was made to the work carried out by Stoll and Chianta for the US Air Force.[137]

The University of Alberta has also developed mannequin (nicknamed 'Harry Burns') with its accompanying flash fire system.[138] The male mannequin, size 40R, is made of 3 mm thick fibre glass to provide sufficient strength for handling and to have a thick layer for easy mounting of heat flux sensors. The thermal protective quality of a garment is judged on estimates of the extent of skin damage resulting from a controlled flash fire.

It has been claimed that with the mannequin and flash fire system, real life exposure conditions can be more closely simulated, and the whole garment assembly can be tested together, taking into account many variables which otherwise cannot be included in small scale testing.

10.7 Summary

Most flame-retardant textiles are designed to reduce the ease of ignition and also reduce the flame propagation rates. Conventional textiles can be rendered flame retardant by chemical after-treatments as co-monomers in their structures or use of FR additives during extrusion. High performance fibres with inherently high levels of flame and heat resistance require the synthesis of all aromatic structures, but these are expensive and used only when performance requirements justify cost. In addition, while heat and flame-resistant textiles have been reviewed and compared, the mechanisms of char formation and the role of intumescents and plasma treatment have also been highlighted. The increasing need to use environmentally friendly FR finishes has been emphasized. Improved standards for fire and heat protective clothing including more realistic tests, such as instrumented manikins, have also been discussed.

References

1. E D WEIL, in *Flame Retardancy of Polymeric Materials*, eds. W C Kuryla and A J Papa, Marcel Dekker, New York, 1975, Vol. 3.
2. Shirley Publication S45, *Protective Clothing*, Shirley Institute, Manchester, UK, 1982.
3. D JACKSON, 'An overview of thermal protection', *Protective Clothing* Conference, Clemson University, SC, June 9–11, 1998.
4. A R HORROCKS, M TUNC and D PRICE, 'The burning behaviour of textiles and its assessment by oxygen index methods', *Textile Prog.*, 1989 **18**(1/2/3).
5. D STAN, 'FR fibres – The European scene', *Textile Horizons*, June 1986 **6** 33–35.
6. R JEFFRIES, 'Clothing for work and protection', *Textile Asia*, 1988 **19**(11) 72–82.
7. W C SMITH, 'Protective clothing in the US', *Textile Asia*, 1989 **20**(9) 189–194.
8. P BAJAJ and A K SENGUPTA, 'Protective clothing', *Textile Prog.*, 1992 **22**(2/3/4) 1–110.
9. D B AJGAONKAR, 'Flame/Fire retardant/Thermostable clothing', *Man-made Textiles India*, Oct. 1994 465–469.
10. W C SMITH, 'High temperature fibres, fabrics, markets – An overview', *Man-made Textiles India*, Feb. 1994 47–55.
11. BTTG, 'Burning issues – Environmentally friendlier flame retardant systems', *Int. Dyer*, Oct. 1991 17–20.
12. T CARROLL, 'Revitalising the market for safety apparel with high performance disposables', *Tech. Textiles Int*, Nov. 1997 17–21.
13. D L ROBERTS, M E HALL and A R HORROCKS, 'Environmental aspects of flame retardant textiles – an overview', *Rev. Prog. Col.*, 1992 **22** 48–57.
14. ANON, 'Environmental and safety focus', *Textile Month*, April 1997 37.
15. M LEWIN, in *Handbook of Fibre Science and Technology*, Vol. II: *Chemical processing of fibres and fabrics: functional finishes*, Part B, eds. M Lewin and S B Sello, Dekker, New York, 1984.
16. D W VAN KREVELEN, 'Flame resistance of chemical fibres', *J. Appl. Polym. Sci., Appl. Polym. Symp.*, 1977 **31** 269–292.
17. J H ROSS, 'Thermal conductivity of fabrics as related to skin burn damage', *J. Appl. Polym. Sci., Appl. Polym. Symp.*, 1977 **31** 293–312.
18. D A FRANK – KAMENETZKY, *Diffusion and Heat transfer in Chemical Kinetics*, Plenum Press, New York, 1969.
19. M M GAUTHIER, R D DEANIN and C J POPE, 'Man made fibres: Flame retardance and flame retardants', *Polym. Plast. Technol. Eng.*, 1981 **16** 1–39.
20. J KUMAR, A A VAIDYA and K V DATYE, 'Flame retardant Textiles of synthetic fibres', *Man-made Textiles India*, Part I, Dec. 1980 617–22, and Part II, Jan. 1981 23–37.
21. B MILLER, B C GOSWAMI and R TURNER, 'The concept and measurement of extinguishability as a flammability criterion', *Textile Res. J.*, 1973 **43** 61–67.
22. A R HORROCKS, D PRICE and M TUNC, 'Studies on the Temperature dependence of Extinction Oxygen Index values for cellulosic fabrics', *J. Appl. Polym. Sci.*, 1987 **34** 1901.
23. Teijin Ltd., 'Super FR cloth', *Textile Horizons*, June 1989 **9** 31.
24. D T WARD, 'High Tech. Fibres featured as Frankfurt show', *Int. Fibre J.*, Aug. 1991 **6** 89–93.
25. H MINICHSHOFER, 'New Developments in flame retardant protective clothing', *Textile Horizons*, March 1990 **10** 7.
26. L-K PETER, 'Aramid growth spectacular', *Textile Month*, Dec. 1994 29–32.
27. F LENZING, 'Lenzing fibre combines comfort and protection', *High Performance Textiles*, Nov. 1995 3.
28. S DAVIES, 'FR Fibres-The European scene', *Textile Horizons*, June 1986 **6** 33–35.
29. R A SPEER, 'Fire fighting blanket', *High Performance Textiles*, Sept. 1997 10.
30. L JARECKI, 'New Fibres in Western Europe', *Amer. Textiles Int.*, July 1988 50–54.
31. R RUSSELL, 'Ballyclare special products Ltd.', *Tech. Textiles Int.*, Nov. 1997 **6** 8.
32. N SAVILLE and M SQUIRES, 'Multiplex panotex textiles', International Conference *Industrial and Technical Textiles*, University of Huddersfield, UK, 6–7 July 1993.
33. P LENNOX – KERR, 'Friction spinning creates hybrid yarns for improved thermal protection', *Tech. Textiles Int.*, Oct. 1977 **6** 18–20.
34. T M KOTRESH, A S K PRASAD, K THAMMAIAH, V N JHA and L MATHEW, 'Development and evaluation of flame retardant outer garment for anti G suit', *Man-made Textiles India*, May 1997 206–210.

35. w RAINER (Sandoz-patent-G.m.b.H.), 'Aqueous dispersions of cyclic pyro(thio)phosphate esters as fire proofing agents', Ger. offen. DE 4,128,638, 5 Mar. 1992 (*Chem. Abstr.*, 1992 **117** 9754r).

36. s HEIDARI, 'Visil: a new hybrid technical fibre', *Chemifasern/Textile Industrie*, Dec. 1991 **41/93** T224/E186.

37. s HEIDARI, A PAREN and P NOUSIANINEN, 'The mechanism of fire resistance in viscose/silicic acid hybrid fibres', *J. Soc. Dyers Colorist*, 1993 **109**, 201–203.

38. A PAREN and P VAPAAOKSA (Kemira Oy), 'Cellulosic product containing silicon dioxide as a method for its preparation', *PCT Int Appl WO 93, 13, 249*, June 1993 (*Chem. Abstr.*, 1993 **121** 37540u).

39. s C ANAND and s J GARVEY, 'Flame retardancy', *Textile Horizons*, 1994 **14** 33–35.

40. A R HORROCKS, s C ANAND and D SANDERSON, 'Fibre-intumescent interactive systems for barrier textiles', 6th International Conference *Flame Retardants'94*, UK Interscience Communications, 1994 117–128.

41. A R HORROCKS, s C ANAND and D SANDERSON, 'Responsive barrier fabrics up to 1200 °C', *Conference Techtextil Symposium*, Textil, Frankfurt, 1993 56–58.

42. P LEWIS, 'Polyester safety fibres for a fast growing market', *Textile Month*, April 1997 39–40.

43. Z MA, W ZHAO, Y LIU and J SHI, 'Synthesis and properties of intumescent phosphorus-containing flame-retardant polyesters', *J. Appl. Polym. Sci.*, 1997 **63** 1511–1515.

44. ANON, 'Unitika develops new flame retardant polyester', *Japan Textile News*, April 1994 24.

45. N NOGUCHI, A MATSUNAGA and Y YONEZAWA (Unitika), Fire resistant nonwoven fabric composites with good softness and their manufacture, Japanese Patent 09 78, 433, 25 Mar. 1997 (*Chem. Abstr.*, 1997 **127**(2) 19499g).

46. A R HORROCKS, M E HALL and D ROBERTS, 'Environmental consequences of using flame-retardant textiles – a simple life cycle analytical model', *Fire Mater.*, 1997 **21**(5) 229–234.

47. N S ZUBKOVA, 'A highly effective domestic fire retardant for fire-proofing fibrous textile materials', *Fibre Chem.*, 1997 **29**(2) 126–129.

48. s NAGAI, Y KAGAWA, T MATSUMOTO and B IMURA (Japan Exlan), Flame retardant polyester conjugated fibres, Japanese Patent 62, 41, 317, 23 Feb. 1987 (*Chem. Abstr.*, 1987 **107** 116890x).

49. K KAMEYAMA, T MATSUMOTO, Y KAGAWA (Japan Exlan), Flame retardant polyester conjugate fibres, Japanese Patent 62, 45, 723, 27 Feb. 1987 (*Chem. Abstr.*, 1987 **107** 116891y).

50. T KOTANI, M FUJII, T MORYAMA (Unitika), Nondripping flame retardant polyester fibres, Japanese Patent 05, 09, 808, 19 Jan. 1993 (*Chem. Abstr.*, 1993 **119** 10394w).

51. T KOTANI and M FUJII (Unitika), Manufacture of polyester fibres resistant to melting and fire, Japanese Patent 05, 78, 978, 30 Mar. 1993 (*Chem. Abstr.*, 1994 **120** 10284f).

52. N G BUTYLKINA, A YA IVANOVA and M A TYUGANOVA, 'Polycaproamide fibres with reduced combustibility', *Khim. Volokna*, May–June 1988 (No.3) 14–15.

53. M A TYUGANOVA, N G BUTYLKINA, E G YAVORSKAYA and V P TARAKANOV, 'Preparation of fire-resistant polycaproamide fibres', *Khim. Volokna*, Jan–Feb. 1988 (No.1) 51–52.

54. R L WELLS and C J COLE (Allied Signal Corp N J), Method of producing flame retardant polyamide fibre, US Patent, Pat No 4,719,066, Jan. 1988.

55. M DAY, T SUPRUNCHUK and D M WILES, 'A systematic study of the effects of individual constructional components on the flammability characteristics of a carpet', *Textile Res. J.*, 1979 **40** 88–93.

56. J W STODDARD, O A PICKETT, C J CICERO and J H SAUNDERS, 'Flame-retarded nylon carpets', *Textile Res. J.*, 1975 **36** 474–483.

57. H E STEPNICZKA, 'Flame retarded nylon textiles', *Ind. Eng. Chem. Prod. Res. Develop.*, 1973 **12**(1) 29–40.

58. J W STODDARD (Monsanto Co. USA), Phosphorus-containing polyamide fibers with increased strain and fire resistance, US Patent, Pat No 5,545,833, 13 Aug. 1996 (*Chem. Abstr.*, 1996 **125**(18) 224462q).

59. G F LEVCHIK, S V LEVCHIK and A I LESNIKOVICH, 'Mechanisms of action in flame retardant reinforced nylon 6', *Polym. Degrad. Stability*, 1996 **54**(2–3) 361–363.

60. R C NAMETZ, 'Flame-retarding synthetic textile fibres', *Ind. Eng. Chem.*, 1970 **62**(3) 41–53.

61. Y P KHANNA and E M PEARCE, in *Flame retardant polymeric materials*, eds. M Lewin, S M Atlas and E M Pearce, Plenum Press, New York, 1978, Vol. 2.

62. P BAJAJ and K SURYA, 'Modification of acrylic fibres: an overview', *J. Macromol. Sci.-Rev. Macromol. Chem. Phys.*, 1987 **C27**(2) 181–217.

63. N D SHARMA and R MEHTA, 'Flame-retardant acrylic fibre through copolymerization', *Indian Textile J.*, Aug. 1990 66–75.

64. J S TSAI, 'The effect of flame-retardants on the properties of acrylic and modacrylic fibres', *J. Mater. Sci.*, 1993 **28** 1161–1167.

65. M E HALL, A R HORROCKS and J ZHANG, 'The flammability of polyacrylonitrile and its copolymers', *Polym. Degrad. Stability*, 1994 **44** 379–386.

66. M E HALL, J ZHANG and A R HORROCKS, 'The flammability of polyacrylonitrile and its copolymers III: Effect of Flame retardants', *Fire Mater.*, 1994 **18** 231–241.

67. s CHOU and C-J WU, 'Effect of brominated flame retardants on the properties of acrylonitrile/vinyl acetate copolymer fibres', *Textile Res. J.*, 1995 **65**(9) 533–539.

68. J S TSAI and C H WU, 'Selection of spinneret for modacrylic fiber, Part III: Effect of chlorine content', *J. Mater. Sci. Lett.*, 1993 **12** 548–550.

69. J S TSAI, 'Thermal characterization of acrylonitrile – vinylidene chloride copolymers of modacrylic fibres', *J. Mater. Sci. Lett.*, 1991 **10** 881–883.

70. P BAJAJ, D K PALIWAL and A K GUPTA, 'Modification of acrylic fibres for specific end uses', *Indian J. Fibre Textile Res.*, 1996 **21** 143–154.

71. Y NISHIHARA, H HOSOKAWA, Y FUJII, T KOBAYASHI and S OOISHI (Mitsubishi Rayon Co.), Fire resistant acrylic composite fibres, Japanese Patent 06,212,514, 2 Aug. 1994 (*Chem. Abstr.*, 1994 **121**(22) 257823d).

72. T OGAWA, S HASEBE, N NISHI and K TOMIOKA (Kanegafuchi Chemical Industry), Method of making fire resistant acrylic fibres containing tin oxide and epoxy compound, Japanese Patent 62,263,313, 16 Nov. 1987 (*Chem. Abstr.*, 1987 **108** 152063j).

73. T OGAWA, S HASEBE, N NISHI and K TOMIOKA, Method of making fire resistant acrylic fibres containing tin oxide, Japanese Patent 62,263,312, 16 Nov. 1987 (*Chem. Abstr.*, 1987 **108** 152064k).

74. M ARAI and K YOSHIDA (Kanebo), Fire resistant acrylonitrile-based fibres and their manufacture, Japanese Patent 04,316,616, 9 Nov. 1992 (*Chem. Abstr.*, 1992 **118** 214846j).

75. D CHO, 'Protective behaviour of thermal oxidation in oxidised PAN fibres coated with phosphoric acid', *J. Mater. Sci.*, 1996 **31** 1151–1154.

76. J TAKAGI and T SUMIYA (Japan Exlan), Manufacture of fire resistant acrylic fibres, Japanese Patent 02,84,528, 26 Mar. 1990 (*Chem. Abstr.*, 1990 **113** 61214w).

77. G L DRAKE and W A REEVES, in *High Polymers*, Vol. V, Part V, eds. N M Bikales and L Segal, Interscience, New York, 1971.

78. I HOLME, 'Challenge and change in functional finishes for cotton', *Textile Horizons*, 1997 **16**(7) 17–24.

79. P J WAKELYN, W REARICK and J TURNER, 'Cotton and flammability – overview of new developments', *Am. Dyestuff Reptr.*, 1998 **87**(2) 13–21.

80. R H BARKER and M J DREWS, in *Cellulose Chemistry and its Applications*, eds. T P Nevell and S H Zeronian, Ellis Horwood, Sussex UK, 1985.

81. A R HORROCKS, 'Developments in flame retardants for heat and fire resistant textiles – the role of char formation and intumescence', *Polym. Degrad. Stability*, 1996 **54** 143–154.

82. J D REID and L W MAZZENO, 'Preparation and properties of cellulose phosphates', *Ind. Eng Chem.*, 1949 **41** 2828–2830.

83. P BAJAJ, S CHAKRAPANI, N K JHA and A JAIN, 'Effect of hardness ions on the flammability of phosphorylated and after treated polyester/viscose blend fabrics', *Textile Res. J.*, 1984 **54** 854–862.

84. R B LEBLANC, 'The Durability of flame retardant treated fabrics', *Textile Chem. Colorist*, 1997 **29**(2) 19–20.

85. B K KANDOLA, A R HORROCKS, D PRICE and G V COLEMAN, 'Flame-retardant treatments of cellulose and their influence on the mechanism of cellulose pyrolysis', *J. Macromol. Sci.-Rev. Macromol. Chem. Phys.*, 1996 **C36**(4) 721–794.

86. G CAMINO and L COSTA, *Rev. Inorg. Chem.*, 1986 **8**(1/2) 69.

87. M J DREWS and R H BARKER, *J. Fire Flammability*, 1974 **5** 116.

88. A A FAROQ, D PRICE, G J MILNES and A R HORROCKS, *Polym. Degrad. Stability*, 1991 **33** 155.

89. F J KILZER and A BROIDO, *Pyrodynamics*, 1965 **2** 155.

90. A G W BRADBURY, Y SAKAI and F SHAFIZADEH, *J. Appl. Polym. Sci.*, 1979 **23** 3271–3280.

91. M LEWIN, 'Flame retarding of polymers with sulfamates, Part I: Sulphation of cotton and wool', *J. Fire Sci.*, 1997 **15**(4) 263–276.

92. E A EL-ALFY, S H SAMAHA, F M TERA and E S SALEM, 'Finishing of cotton fabric with flame retardants', *Colourage*, 1997 **49**(7) 19–23.

93. A KUROSE and H SHIRAI, 'Flammability of cotton fabric treated with Ni (II) partially phosphorylated polyvinyl alcohol complexes', *Textile Res. J.*, 1996 **66**(3) 184.

94. M DAY, T SUPRUNCHUK, J G OMICHINSKI and S O NELSON, 'Flame retardation studies of polyethylene terephthalate fabrics treated with tris-dibromo alkyl phosphates', *J. Appl. Polym. Sci.*, 1988 **35** 529–535.

95. M DAY, T SUPRUNCHUK, J D COONEY and D M WILES, 'Flame retardation of polyethylene terephthalate containing poly(4-bromo styrene) poly(vinyl bromide) and poly(vinylidene bromide)', *J. Appl. Polym. Sci.*, 1987 **33** 2041–2052.

96. V MISCHUTIN, 'Application of a clear flame retardant finish to fabrics', *J. Coated Fabrics*, Jan. 1993 **22** 234–252.

97. Southeastern section, ITPC Committee, 'The influence of polymeric padbath additives on flame retardant fixation on polyester fabric', *Textile Chem Colorist*, 1995 **27**(12) 21–24.

98. I MASAMI, A JIRO (Toray Industries, Japan), Polyester fibres treated by phosphazenes as fireproofing agents and their manufacture, Japanese Patent 08,291,467, 5 Nov. 1996 (*Chem. Abstr.*, 1997 **126**(7) 90685z).

99. B HIROTOMO and U SHIGEJI (Nikka Chemical Industry Japan), Fireproofing of polyester based fibre fabrics by using hexabromocyclododecane and detergents used in the process, Japanese Patent 09,195,163, 29 July 1997 (*Chem. Abstr.*, 1997 **127**(15) 206911x).

100. G AKOVALI and F TAKROURI, 'Studies on modification of some flammability characteristics by Plasma II: Polyester Fabric', *J. Appl. Polym. Sci.*, 1991 **42** 2717–2725.
101. S TOKAO (Unitika, Japan), Finishing synthetic fibre fabrics for improved resistance to melting by fire and good hygroscopicity, Japanese Patent 09,143,884, 3 June 1997 (*Chem. Abstr.*, 1997 **127**(7) 96545w).
102. B MILLER, J R MARTIN, C H JR. MEISER and GATGUILLO, 'The flammability of polyester cotton mixtures', *Textile Res. J.*, July 1976 **46** 530–538.
103. J E HENDRIX, G L JR. DRAKE and W A REEVES, 'Effect of temperature on oxygen index values', *Textile Res. J.*, April 1971 **41** 360.
104. P BAJAJ, S CHAKRAPANI and N K JHA, 'Flame retardant finishes for polyester/cellulose blends: an appraisal', *J. Macromol. Sci.-Rev. Macromol. Chem. Phys.*, 1985 **C25**(2) 277–314.
105. I HOLME and S R PATEL, 'The effect of *N*-methylolated resin finishes on the flammability of 67-33 P/C fabric', *J. Textile Inst.*, 1983 **74** 182–190.
106. L SHUKLA and R K SINGH, 'An approach on flame retardancy development of polyester-cotton blend fabric using antimony and bromine compounds', *Textile Dyer Printer*, Aug. 1993 24–26.
107. A D R STAFF, 'Review of flame retardant products used in Textile wet processing', *Amer. Dyestuff Reptr.*, 1997 **86**(1) 15–33.
108. H HAIGH, 'Fabrics for thermal protection', *Textile Horizons*, Oct. 1987 **7** 69–72.
109. ANON, 'Recent developments in flame retardancy', *Textile Dyer Printer*, Dec. 1991 29–32.
110. I HOLME, 'Developments in fire resistant textiles', *Textile Month*, July 1989 54–55.
111. BS 5438:1989, 'Flammability of textile fabrics when subjected to a small igniting flame applied to the face or bottom edge of vertically oriented specimens'.
112. BS 5852:1979: Fire tests for furniture Part 1, 'Ignitability by smokers' materials of upholstered composites for seating'.
     BS 5852:1982:Part 2, 'Ignitability of upholstered composites for seating by flaming sources'.
113. BS 6807:1996, 'Assessment of the ignitability of mattresses, upholstered divans and upholstered bed bases with flaming types of primary and secondary sources of ignition'.
114. I HOLME, 'Fighting flammability now', *Textile Horizons*, June 1989 52–53.
115. ANON, Peter Cook International, 'Focus on flame retardancy', *Textile Month*, September 1989 72.
116. E BRAUN, J F KRASNY, R D PEACOCK, M PAABO, G F SMITH and A STOLTE, 'Cigarette ignition of upholstered chairs', *Consumer Product Flammability*, Dec. 1982 **9** 167–183.
117. BS 7157:1989, 'Ignitability of fabrics used in the construction of large tented structures'.
118. BS 7175:1989, 'The ignitability of bedcovers and pillows by smouldering and flaming ignition sources'.
119. BS 7176:1995, 'Resistance to ignition of upholstered furniture for non-domestic seating by testing composites'.
120. P M EATON, 'Flammability and Upholstery Fabrics', *Textiles*, 1992 **1** 20–24.
121. BS 7177:1996, Specification for Resistance to ignition of mattresses, divans and bed bases'.
122. BS 4790:1987, 'Determination of the effects of a small source of ignition on textile floor coverings (hot metal nut method)'.
123. BS 5287:1988, 'Assessment and labelling of textile floor coverings tested to BS 4790'.
124. BS 6307:1982, 'Determination of the effects of a small source of ignition on textile floor coverings (methenamine tablet test)'.
125. EN 469:1995, 'Protective clothing for firefighters – requirements and test methods for protective clothing for firefighting'.
126. J KRASNY, J A ROCKETT and D HUANG, 'Protecting Fire-fighters exposed in room fires: comparison of results of bench scale test for thermal protection and conditions during room flashover', *Fibre Technol.*, Feb. 1988 5–19.
127. EN 470–1:1995, 'Protective clothing for use in welding and allied processes–Part 1: General requirements'.
128. EN 531:1995, 'Protective clothing for industrial workers exposed to heat (excluding firefighters' and welders' clothing)'.
129. EN 533:1997, 'Protective clothing – Protection against heat and flame – Limited flame spread materials and material assemblies'.
130. R H WHITELEY, 'Some comments on the measurement of ignitability and on the calculation of critical heat flux', *Fire Safety J.*, 1993 **21**(2) 177–183.
131. R H WHITELEY, M D SAWYER and M J MCLOUGHLIN, 'Cone calorimeter studies of the flame retardant effects of decabromodiphenyl ether and antimony trioxide in cross-linked polyethylene', 6th Conference *Flame Retard '94*, London, UK, Interscience Communications, 1994.
132. J P REDFERN and M HILL, 'The use of the cone calorimeter in studying the heat release from flame retarded plastics', 2nd International Symposium/exhib *Flame Retard*, Beijing, Geol. Publishing House, 1993.
133. L J GOFF, 'Testing of polymeric and flame retardant polymeric materials using the cone calorimeter', Fall Conference *Fire Retard Chem Assoc*, Lancaster, PA., 1992.

134. E D WEIL, N G PATEL, M M SAID, M M HIRSCHLER and S SHAKIR, 'What does oxygen index correlate to?', International Conference *Fire Safely*, Polytech University, Brooklyn, NY, 1992.
135. N SORENSEN, 'Evaluation of heat and flame protective clothing – the role of Manikin tests', International Congress *Industrial and Technical Textiles*, University of Huddersfield, UK 6–7 July 1993.
136. ANON, 'Treated cotton vs. inherently FR fibres', *Textile Month*, April 1993 50.
137. A M STOLL and M A CHIANTA, *Aerospace Medicine*, 1969, **40** 1232–1238.
138. E M CROWN and J D DALE, 'Built for the hot seat', *Canadian Textile J.*, Mar. 1993 16–19.

11

Textile-reinforced composite materials

Stephen L Ogin

School of Mechanical and Materials Engineering, University of Surrey, Guildford, GU2 7XH, UK

11.1 Composite materials

Textile-reinforced composite materials (TRCM) are part of the general class of engineering materials called composite materials. It is usual to divide all engineering materials into four classes: metals, polymers, ceramics and composites. A rigorous definition of composite materials is difficult to achieve because the first three classes of homogeneous materials are sometimes heterogeneous at submicron dimensions (e.g. precipitates in metals). A useful working definition is to say that composite materials are characterised by being multiphase materials within which the phase distribution and geometry has been deliberately tailored to optimise one or more properties.[1] This is clearly an appropriate definition for textile-reinforced composites for which there is one phase, called the matrix, reinforced by a fibrous reinforcement in the form of a textile.

In principle, there are as many combinations of fibre and matrix available for textile-reinforced composites as there are available for the general class of composite materials. In addition to a wide choice of materials, there is the added factor of the manufacturing route to consider, because a valued feature of composite materials is the ability to manufacture the article at the same time as the material itself is being processed. This feature contrasts with the other classes of engineering materials, where it is usual for the material to be produced first (e.g. steel sheet) followed by the forming of the desired shape.

The full range of possibilities for composite materials is very large. In terms of reinforcements we must include S-glass, R-glass, a wide range of carbon fibres, boron fibres, ceramic fibres (e.g. alumina, silicon carbide) and aramid fibres, and recognise that the reinforcement can come in the form of long (or continuous) fibres, short fibres, disks or plates, spheres or ellipsoids. Matrices include a wide ranges of polymers (epoxides, polyesters, nylons, etc), metals (aluminium alloys, magnesium alloys, titanium, etc) and ceramics (SiC, glass ceramics, etc). Processing methods include hand lay-up, autoclave, resin transfer moulding (RTM), injection moulding for polymer matrices, squeeze casting and powder metallurgy routes for metals, chemical vapour infiltration and prepregging routes for ceramics. A reader interested in a general introduction to composite materials should consult one of a number of wide ranging texts (e.g. Matthews and Rawlings,[2] Hull and Clyne,[3]). A good introduction to the fabrication of polymer matrix composites is provided by Bader *et al.*[4]

The market for composite materials can be loosely divided into two categories: 'reinforced plastics' based on short fibre E-glass reinforced unsaturated polyester resins (which account for over 95% of the volume) and 'advanced composites' which make use of the advanced fibres (carbon, boron, aramid, SiC, etc), or advanced matrices (e.g. high temperature polymer matrices, metallic or ceramic matrices), or advanced design or processing techniques.[1] Even within these loosely defined categories, it is clear that textile composites are 'advanced composites' by virtue of the manufacturing techniques required to produce the textile reinforcement. This chapter will be mostly concerned with textile-reinforced polymeric matrices. The reader should be aware that ceramic fibres in a textile format which reinforce ceramic matrices are also under investigation (e.g. Kuo and Chou,[5] Pryce and Smith[6]).

11.2   Textile reinforcement

11.2.1   Introduction

Textile-reinforced composites have been in service in engineering applications for many years in low profile, relatively low cost applications (e.g. woven glass-reinforced polymer hulls for minesweepers). While there has been a continual interest in textile reinforcement since around 1970, and increasingly in the 1980s, the recent desire to expand the envelope of composite usage has had a dramatic effect on global research into, and usage of, textile reinforcement. In addition to the possibility of a range of new applications for which textile reinforcement could replace current metal technology, textile reinforcement is also in competition with relatively mature composite technologies which use the more traditional methods of prepregging and autoclave manufacture. This is because TRCMs show potential for reduced manufacturing costs and enhanced processability, with more than adequate, or in some cases improved, mechanical properties. Those economic entities within which composite materials have been well developed, notably the European community (with about 30% of global composite usage), the USA (with about 30%) and Japan (with about 10%) have seen a growing interest in textile reinforcement in the 1990s, with China, Taiwan, Russia, South Korea, India, Israel and Australia being additional major contributors. In the last years of the 20th century, conferences devoted to composite materials had burgeoning sessions on textile reinforcement.

Of the available textile reinforcements (woven, braided, knitted, stitched), woven fabric reinforcement for polymer matrices can now be considered to be a mature application, but many textiles are still the subject of demonstrator projects. For example, a knitted glass fabric drawn over a mould and injected with a resin (using the RTM technique) has been used to manufacture a door component for a helicopter with the intention of replacing the current manufacturing route based on autoclave processing of carbon fibre/epoxy resin prepreg material.[7] Several textile techniques are likely to be combined for some applications. For example, a combination of braiding and knitting can be used to produce an I-shaped structure.[8]

For structural applications, the properties which are usually considered first are stiffness, strength and resistance to damage/crack growth. The range of textiles under development for composite reinforcement is indicated in the schematic diagram shown in Fig. 11.1 from Ramakrishna.[9] The intention of the following sections is to give an introduction to textile-reinforced composite materials employing woven, braided, knitted or stitched textile reinforcement. For more information, the reader is referred to the relevant cited papers in the first instance. However, before discussing textile-reinforced composites, it is necessary to provide an indication of the degree of complexity of the mechanical properties of the more traditional continuous fibre reinforcement of laminated composites. This discussion will also be useful when textile reinforcement is discussed subsequently.

11.2.2   Basic mechanics of composite reinforcement

11.2.2.1   Composites fabricated from continuous unidirectional fibres
It is important to recognise that the macroscopic elastic stress–strain relationships that are valid for isotropic materials are not valid for composite materials, except in rare cases when isotropy has been deliberately engineered (e.g. quasi-isotropic laminates loaded in-plane) or is a natural consequence of the material microstructure (e.g. transverse isotropy in the plane perpendicular to the fibre direction in a lamina). In composite materials texts, the basic mechanics always begin with continuous unidirectional fibres reinforcing a matrix, with the explicit (or implicit) assumption of a strong bond between matrix and fibre to enable good load transference from the matrix into the fibres (the detailed chemistry and properties of the 'interphase' region between fibre and bulk matrix is the subject of much research). This is both a logical and a practical starting point because much traditional composite fabrication uses sheets of reinforcing fibres preimpregnated with a resin which is partially cured to facilitate handling. These 'prepreg' sheets, which are usually about 0.125 mm thick, are stacked in appropriate orientations (depending on the expected loading) and cured, usually in an oven under load or applied pressure (autoclave processed), to produce the required component or part (Fig. 11.2).

The Young's modulus of a composite lamina parallel to the fibres, E_1, is to a good approximation (which ignores the difference in Poisson's ratio between matrix and fibre) given by the 'rule of mixtures' expression (sometimes called the Voigt expression), which is:

$$E_1 = E_f V_f + (1 - V_f) E_m \tag{11.1}$$

where, V_f is the fibre volume fraction in a void-free composite, and E_f and E_m are the fibre and matrix moduli, respectively. Perpendicular to the fibres, the modulus is given by:

$$E_2 = \frac{1}{\dfrac{V_f}{E_f} + \dfrac{1 - V_f}{E_m}} \tag{11.2}$$

which, for a given fibre volume fraction, is much lower than the rule of mixtures expression. This is because the longitudinal modulus is fibre dominated and the transverse modulus is matrix dominated.

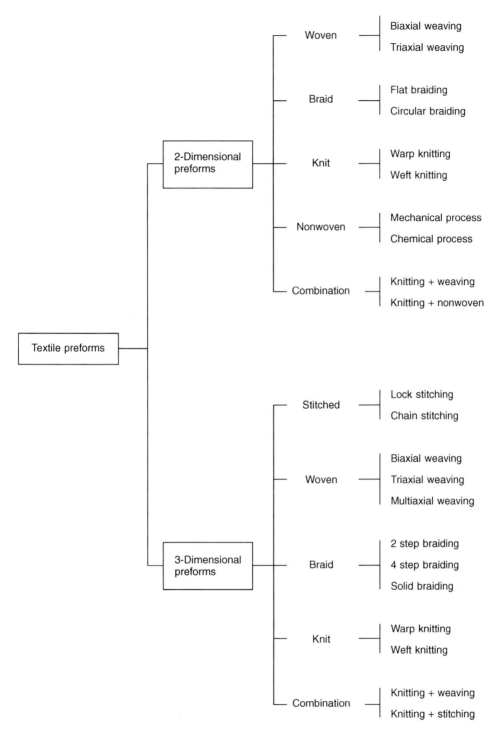

11.1 Textile techniques under development for composite materials.
Reprinted from S Ramakrishna, *Composites Sci. Technol.*, 1997, **57**, 1–22,
with permission from Elsevier Science.[9]

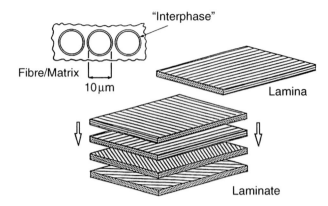

11.2 Schematic of the interphase around a fibre, a lamina (or prepreg sheet, typical thickness 0.125 mm) and laminae stacked at different orientations to form a lamina. Reproduced courtesy of Bader.[1]

The longitudinal strength of a composite lamina is also described by rule of mixtures expressions, though the precise form depends on which of the strains to failure, matrix or fibres, is the larger. For example, if the strain to failure of the matrix is larger, and the fibre volume fraction is typical of the range of engineering composite materials (i.e. over 10% and up to about 70%), the composite strength, σ_c, is given by:

$$\sigma_c = \sigma_{fu} V_f \tag{11.3}$$

where σ_{fu} is the fibre strength.

Laminated composites will usually combine laminae with fibres at different orientations. To predict the laminate properties, the stress–strain relations are required for loading a lamina at an angle θ to the fibre direction, and for loading both in-plane and in bending. Composite mechanics for laminated composites is well developed and many textbooks deal with the subject (e.g. Jones,[10] Matthews and Rawlings,[2] Agarwal and Broutman[11]). For example, the modulus, E_x, of a ply loaded at an angle θ to the fibre direction is given by:

$$\frac{1}{E_x} = \frac{1}{E_1}\cos^4\theta + \left(\frac{1}{G_{12}} - \frac{2v_{12}}{E_1}\right)\sin^2\theta\cos^2\theta + \frac{1}{E^2}\sin^4\theta \tag{11.4}$$

where E_1 and E_2 have been defined above, v_{12} is the principal Poisson's ratio of the lamina (typically 0.3) and G_{12} is the in-plane shear modulus of the lamina. Unlike isotropic materials, which require two elastic constants to define their elastic stress–strain relationships, the anisotropy of a composite lamina (which is an orthotropic material, i.e. it has three mutually perpendicular planes of material symmetry) needs four elastic constants to be known in order to predict its in-plane behaviour. The stress–strain relationships for a laminate can be predicted using laminated plate theory (LPT), which sums the contributions from each layer in an appropriate way for both in-plane and out-of-plane loading. Laminated plate theory gives good agreement with measured laminate elastic properties for all types of composite material fabricated from continuous unidirectional prepreg layers (UD). Predicting laminate strengths, on the other hand, is much less reliable, except in some simple cases, and is still the subject of ongoing research. Because composite structures are usually designed to strains below the onset of the first type of visible damage in the structure (i.e. to design strains of about 0.3–0.4%), the lack of ability to predict the ultimate strength accurately is rarely a disadvantage.

Ply orientations in a laminate are taken with reference to a particular loading direction, usually taken to be the direction of the maximum applied load, which, more often than not, coincides with the fibre direction to sustain the maximum load, and this is defined as the 0° direction. In design it is usual to choose balanced symmetric laminates. A balanced laminate is one in which there are equal numbers of $+\theta$ and $-\theta$ plies; a symmetric laminate is one in which the plies are symmetric in terms of geometry and properties with respect to the laminate mid-plane. Hence a laminate with a stacking sequence 0/90/+45/–45/–45/+45/90/0, which is written $(0/90/\pm45)_s$ is both balanced and symmetric. Balanced symmetric laminates have a simple response. In contrast, an unbalanced asymmetric laminate will, in general, shear, bend and twist under a simple axial loading.

11.2.2.2   Overview of composite moduli for textile reinforcements
One of the simplest laminate configurations for continuous unidirectional fibre reinforced composites is the cross-ply laminate, for example $(0/90)_s$, which is 0/90/90/0. For such a laminate, the Young's moduli parallel to the 0° and 90° directions, E_x and E_y, are equal and, to a good approximation, are just the average of E_1 and E_2.

Yang and Chou[12] have shown schematically the change in these moduli, E_x and E_y, for a carbon fibre-reinforced epoxy laminate with a range of fibre architectures, but the same fibre volume fraction of 60% (see Fig. 11.3). This diagram provides a

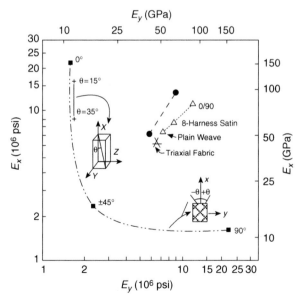

11.3   Predicted E_x and E_y moduli for a range of reinforcement architectures; $\pm\theta$ angle ply (for $\theta = 0$ to ±45 to 90), cross-ply (0/90), eight-harness satin and plain woven, triaxial woven fabric, braided ($\theta = 35°$ to 15°) and multiaxial warp knit ($\bullet$-$\bullet$), for the same fibre volume fraction of 60%. Reprinted, with minor changes, from Yang and Chou, *Proceedings of ICCM6/ECCM2*, ed. F L Matthews *et al.*, 1987, 5.579–5.588, with permission from Elsevier Science.[12]

good starting point for the discussion of textile-reinforced composites. The cross-ply composite has E_x and E_y moduli of about 75 GPa. In the biaxial weaves of the eight-harness satin and the plain weave, the moduli both fall to about 58 GPa and 50 GPa, respectively. These reductions reflect the crimps in the interlaced woven structure, with more crimps per unit length in the plain weave producing a smaller modulus. The triaxial fabric, with three sets of yarns interlaced at $60°$ angles, behaves similarly to a $(0/\pm60)_s$ angle-ply laminate. Such a configuration is quasi-isotropic for in-plane loading, that is, it has the same Young's modulus for any direction in the plane of the laminate. The triaxial fabric shows a further reduction in E_x and E_y to about 42 GPa, but this fabric benefits from a higher in-plane shear modulus (which is not shown in the diagram) than the biaxial fabrics. The anticipated range of properties for a multiaxial warp-knit fabric (or multilayer multidirectional warp-knit fabric) reinforced composite is also shown, lying somewhere between the triaxial fabric and above the cross-ply laminate (at least for the modulus E_x), depending on the precise geometry. Here warp, weft and bias yarns (usually ±45) are held together by 'through-the-thickness' chain or tricot stitching. Finally, a three-dimensional braided composite is shown, with braiding angles in the range $15°$ to $35°$. This type of fibre architecture gives very anisotropic elastic properties as shown by the very high E_x moduli (which are fibre dominated) and the low E_y moduli (which are matrix dominated). In the following sections, the properties of these textile reinforcements (woven, braided, knitted, stitched) will be discussed in more detail.

11.3   Woven fabric-reinforced composites

11.3.1   Introduction
Woven fabrics, characterised by the interlacing of two or more yarn systems, are currently the most widely used textile reinforcement with glass, carbon and aramid reinforced woven composites being used in a wide variety of applications, including aerospace (Fig. 11.4). Woven reinforcement exhibits good stability in the warp and

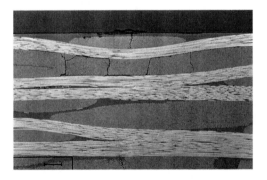

11.4   Optical micrograph of an eight-harness woven CFRP laminate showing damage in the form of matrix cracks and associated delaminations. The laminate is viewed at a polished edge. The scale bar is 200 μm. Reprinted from F. Gao *et al.*, *Composites Sci. Technol.*, 1999, **59**, 123–136, 'Damage accumulation in woven fabric CFRP (carbon fibre-reinforced plastic) laminates under tensile loading: Part 1 – Observations of damage,' with permission from Elsevier Science.[15]

weft directions and offers the highest cover or yarn packing density in relation to fabric thickness.[13] The possibility of extending the useful range of woven fabrics was brought about by the development of carbon and aramid fibre fabrics with their increased stiffness relative to glass. Prepreg manufacturers were able, by the early 1980s, to supply woven fabrics in the prepreg form familiar to users of nonwoven material.[14]

There are a number of properties that make woven fabrics attractive compared to their nonwoven counterparts. They have very good drapability, allowing complex shapes to be formed with no gaps. Manufacturing costs are reduced since a single biaxial fabric replaces two nonwoven plies and the ease of handling lends itself more readily to automation. Woven fabric composites show an increased resistance to impact damage compared to nonwoven composites, with significant improvements in compressive strengths after impact. These advantages are gained, however, at the expense of lower stiffness and strength than equivalent nonwoven composites.

11.3.2   Mechanical behaviour

11.3.2.1   Mechanical properties
Bishop and Curtis[16] were amongst the first to demonstrate the potential advantages of woven fabrics for aerospace applications. Comparing a five-harness woven fabric (3k tows, which means 3000 carbon fibres per tow) with an equivalent nonwoven carbon/epoxy laminate, they showed that the modulus of the biaxial (0/90) woven laminate was slightly reduced compared to the nonwoven cross-ply laminate (50 GPa compared to 60 GPa, respectively). The compressive strength after a 7 J impact event was increased by over 30%. Similar results have been found by others. For example, Raju et al.[17] found a decreasing modulus for carbon/epoxy laminates moving from eight-harness (73 GPa) to five-harness (69 GPa) to plain weave (63 GPa). These results are in line with the moduli changes indicated in Fig. 11.3. The tensile strengths of woven composites are also slightly lower than the nonwoven equivalents. Bishop and Curtis[16] for example, found a 23% reduction in the tensile strength compared to UD equivalent laminates. Triaxial woven fabric composites, naturally, have further reduced longitudinal properties, as mentioned earlier. Fujita et al.[18] quote a Young's modulus and tensile strength of 30 GPa and 500 MPa, respectively, for a triaxial woven carbon/epoxy.

Glass-reinforced woven fabrics give rise naturally to composites with lower mechanical properties because of the much lower value of the glass fibre modulus compared to carbon. Amijima et al.[19] report Young's modulus and tensile strength values for a plain weave glass/polyester ($V_f = 33\%$) of 17 GPa and 233 MPa, respectively, while Boniface et al.[20] find comparable values for an eight-harness glass/epoxy composite, that is, 19 GPa and 319 MPa, respectively ($V_f = 37\%$).

Clearly, the mechanical properties of woven fabric-reinforced composites are dominated by the type of fibre used, the weaving parameters and the stacking and orientation of the various layers. However, there are additional subtleties which also affect composite performance. For example, some authors have noted the possibility of slightly altered mechanical properties depending on whether the yarns are twisted prior to weaving,[21] and work in this area has shown that damage accumulation under static and cyclic loading is different in laminates fabricated from twisted or untwisted yarn.[22]

11.3.2.2   Damage accumulation

Damage under tensile loading in woven composites is characterised by the development of matrix cracking in the off-axis tows at strains well above about 0.3–0.4%. Most investigations of damage have considered biaxial fabrics loaded in the warp direction. Cracks initiate in the weft bundles and an increasing density of cracks develops with increasing load (or strain). The detailed crack morphology depends on whether the tows are twisted or untwisted. Twisted tows lead to fragmented matrix cracks; untwisted tows lead to matrix cracks, which strongly resemble the 90 ply cracks that develop in cross-ply laminates.[22,23] The accumulation of cracks is accompanied by a gradual decrease in the Young's modulus of the composite. In woven carbon systems, the matrix cracking can lead to considerable delamination in the region of the crimps in adjacent tows which further reduces the mechanical properties.[15] Damage modelling has been attempted using finite element methods (e.g. Kriz,[24] Kuo and Chou[5]) or closed-form models (e.g. Gao et al.[25]).

11.3.3   Analyses of woven composites

The majority of closed-form analyses of woven fabric composites have a substantial reliance on laminated plate theory. Numerical methods rely on the finite element method (FEM).

In a series of papers in the early 1980s by Chou, Ishikawa and co-workers (see Chou[26] for a comprehensive review) three models were presented to evaluate the thermomechanical properties of woven fabric composites. The mosaic model treats the woven composite as an assemblage of assymetric cross-ply laminates, ignoring the fibre continuity and undulation. The fibre undulation model takes these complexities into account by considering a slice of the crimped region and averaging the properties with the aid of LPT. This model is particularly appropriate for plain and twill weave composites. For five-harness and eight-harness satins, the fibre undulation model is broadened in the bridging model. These essentially one-dimensional models have been extended to two dimensions by Naik and co-workers (e.g. Naik and Shembekar[21]).

The finite element method is a powerful tool that makes use of a computer's ability to solve complex matrix calculations very quickly. When applied to analysing textile composites, the procedure consists of dividing the composite into a number of unit cells interconnected at nodal points. If the force-displacement characteristics of an individual unit cell are known, it is then possible to use FEM to evaluate the stress fields and macroscopic responses to deformation of the entire structure. The difficulty for FEM methods is that they are expensive and ideally need to be reapplied for even small changes in reinforcement architecture. For woven reinforcements in particular, where adjacent layers have a great degree of lateral freedom to move during fabrication, the results need to be treated with caution. Examples of this approach to investigation of the distribution of stresses and strain energy densities in woven fabric composites can be found in papers by Glaessgen and Griffin[27] and Woo and Whitcomb.[28]

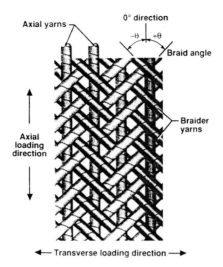

11.5 Braided two-dimensional reinforcement; the pattern is a 2 × 2 braid. Reprinted from Naik *et al.*, *J. Composite Mater.*, 1994 **28**, 656–681, with permission from Technomic Publishing Co., Inc, copyright, 1994.[29]

11.4   Braided reinforcement

11.4.1   Introduction

Braided textiles for composites consist of intertwined two (or more) sets of yarns, one set of yarns being the axial yarns. In two-dimensional braiding, the braided yarns are introduced at ±θ directions and the intertwining is often in 1 × 1 or 2 × 2 patterns (see Fig. 11.5).[29,30] However, for significant improvements in through-the-thickness strength, three-dimensional braided reinforcement is an important category (e.g. Du *et al.*[31]). The braided architecture enables the composite to endure twisting, shearing and impact better than woven fabrics. Combined with low cost fabrication routes, such as resin transfer moulding, braided reinforcements are expected to become competitor materials for many aerospace applications (where they may replace carbon prepreg systems) or automobile applications (e.g. in energy absorbing structures), although realisation in practice is currently limited.

A variety of shapes can be fabricated for composite applications from hollow tubular (with in-laid, non-intertwined yarns) to solid sections, including I-beams. The stability or conformability of the braided structure depends on the detailed fibre architecture. With in-laid yarns, for example, stability in the 0° direction in tension is improved, though the axial compressive properties may be poor.[13] In general terms, the mechanical properties of composites fabricated using braided reinforcement depend on the braid parameters (braid architecture, yarn size and spacing, fibre volume fraction) and the mechanical properties of fibre and matrix.

11.4.2   Mechanical behaviour

In this section, two-dimensional braided reinforcement will be considered primarily, since it lends itself to direct comparison with laminated composites with a 0/±θ construction and such comparisons have been made by a number of authors. For example, Naik and co-workers[29] manufactured braided carbon fibre-reinforced epoxy resin composites with a number of fibre architectures while maintaining a constant fibre volume fraction ($V_f = 56\%$) overall. By keeping the axial yarn content constant, but varying the yarn size or braid angle, the effect of each variable on composite properties could be investigated. An insensitivity to yarn size was found (in the range of 6–75 k tow size), but the braid angle had a significant effect, as anticipated. A modest increase in longitudinal modulus (from 60–63 GPa) occurred in moving from a braid architecture of 0/±70 to 0/±45, with a much larger fall in transverse modulus (from 46–19 GPa).

The strengths of braided reinforced composites are lower than their prepregged counterparts. Norman et al.[32] compared the strengths of 0/±45 braided composites with an equivalent prepreg (UD) system, finding that the prepreg system had a tensile strength that was some 30% higher than the braided two-dimensional composite (849 MPa compared to 649 MPa). Similar results found by Herszberg et al. (1997) have been attributed to fibre damage during braiding. Norman et al.[32] also found the braided reinforcement to be notch insensitive for notch sizes up to 12 mm, whereas equivalent UD laminates showed a significant notch sensitivity in this range. Compression after impact tests also favour braided composites when normalised by the undamaged compression strengths, in comparison with UD systems. Indeed, the ability to tailor the braided reinforcement to have a high energy absorbing capability may make them of use in energy-absorbent structures for crash situations.[33] A review by Bibo and Hogg[34] discusses energy-absorbing mechanisms and postimpact compression behaviour of a wide range of reinforcement architectures, including braided reinforcement.

11.4.3   Analyses of braided reinforcement

The potential complexity of the braided structure, particularly the three-dimensional architectures, is such that the characterisation of structures is often taken to be a major first step in modelling the behaviour of the reinforced material. The desired outcome of this work is to present a three-dimensional visualisation of the structure (e.g. Pandey and Hahn[35]) or to develop models to describe the structural geometry (e.g. Du et al.[31]). Analytical models for predicting properties are frequently developments of the fibre-crimp model developed by Chou[26] and colleagues for woven reinforcements, extended in an appropriate way by treating a representative 'unit cell' of the braided reinforcement as an assemblage of inclined unidirectional laminae (e.g. Byun and Chou[36]). Micromechanics analyses incorporated into personal computer-based programs have also been developed (e.g. the Textile composite analysis for design, TEXCAD; see e.g. Naik[37]).

11.5   Knitted reinforcement

11.5.1   Introduction

The major advantages of knitted fabric-reinforced composites are the possibility of producing net shape/near net shape preforms, on the one hand, and the exceptional drapability/formability of the fabrics, which allows for forming over a shaped tool of complex shape, on the other. Both of these features follow from the interlooped

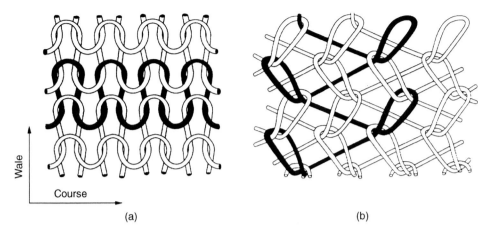

11.6 Schematic diagrams of (a) weft-knitted and (b) warp-knitted reinforcement. Reprinted from S Ramakrishna, *Composites Sci. Technol.*, 1997, **57** 1–22, with permission from Elsevier Science.[9]

nature of the reinforcing fibres/yarns which permits the fabric to have the stretchability to adapt to complex shapes without crimp (Fig. 11.6). However, the advantages which the knitted fibre architecture brings also lead to the disadvantages, which are the reduced in-plane stiffness and strength of the composites caused by the relatively poor use of the mechanical properties of the fibre (glass, carbon or aramid). Weft and warp knits can, however, be designed with enhanced properties in certain directions by the use of laid-in yarns.[13]

Both warp-knitted and weft-knitted reinforcements are under investigation. In general terms, the weft-knitted structures are preferred in developmental work owing to their superior formability (based on their less stable structure) and warp-knitted structures are preferred for large scale production (owing to the increased production rate allowed by the knitting of many yarns at one time).[7]

11.5.2   Mechanical behaviour

11.5.2.1   Mechanical properties
The tensile and compressive properties of the knitted fabrics are poor in comparison with the other types of fabric already discussed, but they are more likely to be chosen for their processability and energy-absorbing characteristics than their basic in-plane properties.

The detailed fibre architecture of knitted fabric reinforcement leads to in-plane properties which can either be surprisingly isotropic or very anisotropic. For example, Bannister and Herszberg[38] tested composites manufactured using both a full-milano and half-milano knitted glass-reinforced epoxy resin. The full-milano structure was significantly more random in its architecture than the half-milano, with the consequence that the tensile strengths in both the wale and the course directions were approximately the same. Typically, the stress–strain curve is approximately linear to a strain of about 0.6%,[39] followed by a sharp knee and pseudoplastic behaviour to failure. The tensile strengths were proportional to the fibre volume fraction (in a manner which is understandable based on a rule-of-mixtures prediction of composite strength; and see Section 5.3 below), with a typical value being about 145 MPa for a fibre volume fraction of 45%. However, the strains to failure were not only very large (in the range from about 2.8% for seven cloth layers to about 6.6% for 12 cloth layers) but also increased with number of layers/fibre volume fraction. The reasons for this variation are presumably related to the detailed manner in which the damage accumulates to produce failure in the composites. In contrast to the relatively isotropic full-milano reinforcement, the half-milano knitted architecture, which has a higher degree of fibre orientation, showed tensile strengths which varied by 50% in the two directions and difference in strains to failure which were even larger (about a factor of two).

Knitted carbon reinforcement has been investigated by Ramakrishna and Hull.[40] In general, the weft-knitted composites showed moduli which increased roughly linearly with fibre volume fraction, being typically 15 GPa when tested in the wale direction and 10 GPa when tested in the course direction, for a fibre volume fraction of about 20%. Tensile strengths also increase in a similar fashion for the wale direction (a typical value is 60 MPa for a 20% volume fraction), whereas the course direction strengths are reasonably constant with fibre volume fraction at around 34 MPa. These differences are related to the higher proportion of fibre bundles oriented in the wale direction.

In compression, the mechanical properties are even less favourable. For both the half-milano and full-milano glass-reinforced composites[39] the compression strengths showed features which are a consequence of the strong dominance of the matrix in compression arising from the highly curved fibre architecture. These features are manifest as compression strengths that were approximately the same in both wale and course directions and as a compression strength that only increased by about 15% as the fibre volume fraction increased from 29–50% (interestingly, the compression strengths were found to be consistently higher than the tensile strengths, by up to a factor of two). In the light of these results, it is not surprising that deforming the knitted fabric by strains of up to 45% prior to infiltration of the resin and consolidation of the composite has virtually no effect on the composite compressive strength.[41]

Similar findings have been reported by others. Wang *et al.*[42] tested a 1×1 rib-knit structure of weft-knitted glass-reinforced epoxy resin, finding compressive strengths which were almost twice as high as the tensile strengths. The relatively isotropic nature of this fibre architecture led to Young's modulus values and Poisson's ratio values which were also approximately the same for testing in both the wale and course direction.

11.5.2.2 Damage accumulation

There are a large number of potential sites for crack initiation in knitted composites. For example, observations on weft-knitted composites tested in the wale direction suggest that cracks initiate from debonds which form around the needle and sinker loops in the knitted architecture. Similarly, crack development in fabrics tested in tension in the course direction is believed to occur from the sides (or legs) of loops.[39,40] It appears likely that crack linking will occur more readily for cracks initiated along the legs of the loops (i.e. when the composite is loaded in the course direction) than when initiation occurs at the needle and sinker loops.

The damage tolerance of knitted fabrics compares favourably with other reinforcement architectures. For example, it has been found that a higher percentage of impact energy in the range 0–10 J is absorbed by a weft-knitted glass reinforced composite (V_f = 50%) than was absorbed by an equivalent woven fabric. Observations indicated, in addition, that the damaged area was approximately six times larger for the knitted fabric than for the woven fabric, presumably reflecting the increased availability of crack initiation sites in the knitted architecture. Compression after impact (CAI) strengths were decreased by only 12% for the knitted fabric in this impact energy range, whereas the woven fabric CAI values fell by up to 40%.[38]

11.5.3 Analyses of knitted composites

Models for the elastic moduli and tensile strengths of knitted fabric reinforced composites have been developed (e.g. Ramakrishna,[9] Gommers et al.[43,44]). Ramakrishna, for example, divides a weft-knitted fabric architecture into a series of circular arcs with each yarn having a circular cross-section. It is then possible to derive an expression for the Young's modulus of the composite by integrating the expression for the variation in Young's modulus with angle (equation 11.4) along the required directions. Indeed, all the elastic moduli can be calculated in a similar fashion, although the predictions were about 20% higher than the experimental results. The predictions of tensile strength depend on the expression for the strength of an aligned fibre composite modified by terms which attempt to account for the average orientation of the yarns with respect to the loading direction and the statistical variation of the bundle strengths. The tensile strengths are predicted to scale in proportion to the fibre volume fractions in both the wale and course directions, which is exactly the result found by Leong et al.[39] Gommers et al.[43,44] use orientation tensors to represent fibre orientation variations in the fabric.

11.6 Stitched fabrics

11.6.1 Introduction

Stitching composites is seen as a direct approach to improving the through-the-thickness strength of the materials. This in turn will improve their damage tolerance, and particularly the CAI behaviour, where failure is usually triggered by microbuckling in the vicinity of a delamination. In its simplest form, stitching of composites adds one further production step with the use of a sewing machine to introduce lock stitches through the full thickness of the laminate. The stitching can be performed on unimpregnated fibres or fibres in the prepreg form, although the latter is usually to be avoided owing to excessive fibre damage. Stitching in this way can be carried out with carbon, glass or aramid fibre yarns. In its more sophisticated form, chain or tricot stitches are used to produce a fabric which consists of warp (0°), weft (90°) and (optionally) bias (±θ) yarns held together by the warp-knitted stitches, which usually consist of a light polyester yarn (Fig. 11.7). The resulting fabric is called a non-crimp fabric (NCF) or a multiaxial warp-knit fabric (MWK) (see e.g. Hogg et al.,[45] Du and Ko[46]). Whatever the terminology, the warp-knitted fabrics are highly drapable, highly aligned materials in which the tow crimp associated with woven fabrics has been removed almost completely (though some slight misalignment is inevitable). The fabric can be shaped easily and it remains stable when removed

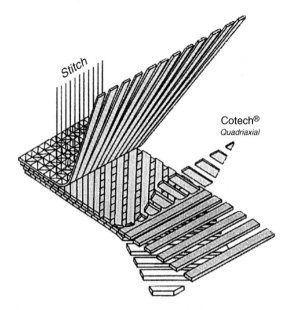

Stitch

Cotech®
Quadriaxial

11.7   Schematic of a quadriaxial non-crimp fabric (courtesy of BTI Europe Ltd).

from a tool owing to the ability of the stitching to allow sufficient relative movement of the tows.[47] With the potential for combining the fabric with low-cost fabrication routes (e.g. RTM), these fabrics are expected both to broaden the envelope
of composite usage and to replace the more expensive prepregging route for many
applications. The ability to interdisperse thermoplastic fibres amongst the reinforcing fibres also provides a potentially very attractive manufacturing route.[47] Hence,
this brief introduction will concentrate on the warp-knitted materials. A comprehensive review of the effect of all types of stitching on delamination resistance has
been published by Dransfield *et al.*[48]

11.6.2   Mechanical behaviour

11.6.2.1   Mechanical properties
The basic mechanical properties of NCFs are somewhat superior to the equivalent
volume fraction of woven roving-reinforced material. For example, Hogg *et al.*[45]
find the Young's modulus and tensile strength of a biaxial NCF glass-reinforced
polyester, volume fraction 33%, to be 21 GPa and 264 MPa, respectively, which are
values some 13 and 20% higher than those found for an equivalent volume fraction
of plain woven-reinforced composite (see Section 11.3.2.1; Amijima *et al.*,[19]).
Quadriaxial reinforcement of the same fibre volume fraction gave similar results
(24 GPa and 286 MPa, respectively). The improvement in properties compared
to woven-reinforced composites is emphasised by the work of Godbehere *et al.*[49]
in tests on a carbon fibre-reinforced NCF epoxy resin and equivalent unidirectional (UD) laminates. All the composites had 0/±45 orientations. Although the
NCF laminates had poorer properties than the UD laminates, the reduction
was small (e.g. less than 7%) in the 0° direction. For example, the UD equivalent
laminate gave values of Young's modulus and tensile strength of 58 GPa and

756 MPa, respectively, compared to NCF values of 56 GPa and 748 MPa (for fibre volume fractions of 56%).

The increases in through-the-thickness reinforcement achieved by NCFs have been demonstrated by a number of authors. For example, Backhouse et al.[50] compared the ease of delaminating polyester stitched 0/±45 carbon fibre NCF with equivalent carbon fibre/epoxy UD laminates. There were large increases, some 140%, in the measured parameters used to quantify resistance to delamination (the mode I and mode II toughness values) for the NCF fabrics compared to the UD material.

11.6.2.2 Damage accumulation

Owing to the fact that the fibres in each layer in an NCF-reinforced composite are parallel, it is to be expected that the damage accumulation behaviour is very similar to equivalent UD laminates. Indeed, Hogg et al.[45] found the matrix cracking in biaxial glass NCF to be very similar to matrix cracking in the 90° ply of cross-ply UD laminates. There are, however, microstructural features introduced because of the knitting yarn which do not have parallels in UD laminates. Local variations in fibre volume fraction, resin-rich pockets and fibre misalignment provide significant differences. In biaxial reinforced NCFs, for example, transverse cracks can initiate preferentially where the interloops of the knitted yarn intersect the transverse ply.[51]

11.6.3 Analyses of non-crimp fabrics

For in-plane properties of NCF composites, it is likely that there is sufficient similarity to UD materials to enable similar analyses to be used (although Hogg et al.[45] suggest that the properties of NCF composites may exceed the in-plane properties of UD equivalents). However, detailed models of the three-dimensional structure of NCF-based composites for manufacturing purposes (i.e. for determining process windows for maximum fibre volume fractions, for example) and for the prediction of mechanical properties, are being developed (e.g. Du and Ko[46]).

11.7   Conclusion

The 1990s saw a growing mood of cautious optimism within the composites community worldwide that textile-based composites will give rise to new composite material applications in a wide range of areas. Consequently, a wide range of textile-reinforced composites are under development/investigation or in production. Textile reinforcement is thus likely to provide major new areas of opportunity for composite materials in the future.

References

1. M G BADER, Short course notes for 'An introduction to composite materials,' University of Surrey, 1997.
2. F L MATTHEWS and R RAWLINGS, Composite Materials: Engineering and Science, Chapman and Hall, London, 1994.
3. D HULL and T W CLYNE, An Introduction to Composite Materials, Cambridge University Press, Cambridge, 1996.
4. M G BADER, W SMITH, A B ISHAM, J A ROLSTON and A B METZNER, Delaware Composites Design Ency-

clopedia – Volume 3, Processing and Fabrication Technology, Technomic Publishing, Lancaster, Pennsylvania, USA, 1990.

5. w-s kuo and t-w chou, 'Elastic response and effect of transverse cracking in woven fabric brittle matrix composites', J. Amer. Ceramics Soc. 1995 **78**(3) 783–792.

6. a w pryce and p a smith, 'Behaviour of unidirectional and crossply ceramic matrix composites under quasi-static tensile loading', J. Mater. Sci., 1992 **27** 2695–2704.

7. k h leong, s ramakrishna and h hamada, 'The potential of knitting for engineering composites', in Proceedings of 5th Japan SAMPE Symposium, Tokyo, Japan, 1997.

8. a nakai, m masui and h hamada, 'Fabrication of large-scale braided composite with I-shaped structure', in Proceedings of the 11th International Conference on Composite Materials (ICCM-11), Gold Coast, Queensland, Australia, published by Australian Composites Structures Society and Woodhead Publishing, 1997, 3830–3837.

9. s ramakrishna, 'Characterization and modeling of the tensile properties of plain weft-knit fabric-reinforced composites', Composites Sci. Technol., 1997 **57** 1–22.

10. r m jones, Mechanics of Composite Materials, Scripta (McGraw-Hill), Washington DC, 1975.

11. b d agarwal and l j broutman, Analysis and Performance of Fiber Composites, John Wiley and Sons, New York, 1980.

12. j-m yang and t-w chou, 'Performance maps of textiles structural composites', in Proceedings of Sixth International Conference on Composite Materials and Second European Conference on Composite Materials (ICCM6/ECCM2) eds F L Matthews, N C R Buskell, J M Hodgkinson and J Morton, Elsevier, London, 1987, 5.579–5.588.

13. f scardino, 'An introduction to textile structures and their behaviour', in Textile Structural Composites, Chapter 1, Composite Materials Series Vol 3, eds T W Chou and F K Ko, Elsevier, Oxford 1989.

14. j a baillie, 'Woven fabric aerospace structures', in Handbook of Fibre Composites, eds C T Herakovich and Y M Tarnopol'skii, Elsevier Science, Oxford 1989, Vol 2, 353–391.

15. f gao, l boniface, s l ogin, p a smith and r p greaves, 'Damage accumulation in woven fabric CFRP laminates under tensile loading. Part 1: Observations of damage; Part 2: Modelling the effect of damage on macromechanical properties', Composites Sci. Technol., 1999 **59** 123–136.

16. s m bishop and p t curtis, 'An assessment of the potential of woven carbon fibre reinforced plastics for high performance applications', Composites, 1984 **15** 259–265.

17. i s raju, r l foye and v s avva, 'A review of analytical methods for fabric and textile composites', in Proceedings of the Indo-US Workshop on Composites for Aerospace Applications: Part 1, Bangalore, India, 1990, 129–159.

18. a fujita, h hamada and z maekawa, 'Tensile properties of carbon fibre triaxial woven fabric composites', J. Composite Mat., 1993 **27** 1428–1442.

19. s amijima, t fujii and m hamaguchi, 'Static and fatigue tests of a woven glass fabric composite under biaxial tension-tension loading', Composites, 1991 **22** 281–289.

20. l boniface, s l ogin and p a smith, 'Damage development in woven glass/epoxy laminates under tensile load', in Proceedings 2nd International Conference on Deformation and Fracture of Composites, Manchester, UK, Plastics and Rubber Institute, London 1993.

21. n k naik and p s shembekar, 'Elastic behaviour of woven fabric composites: I – lamina analysis', J. Composite Mater., 1992 **26** 2196–2225.

22. w marsden, l boniface, s l ogin and p a smith, 'Quantifying damage in woven glass fibre/epoxy laminates,' in Proceedings FRC '94, Sixth International Conference on Fibre Reinforced Composites, Newcastle upon Tyne, Institute of Materials, 1994, paper 31, pp. 31/1–31/9.

23. w marsden, 'Damage accumulation in a woven fabric composite', PhD Thesis, University of Surrey, 1996.

24. r d kriz, 'Influence of damage on mechanical properties of woven fabric composites', J. Composites Technol. Res., 1985 **7** 55–58.

25. f gao, l boniface, s l ogin, p a smith and r p greaves, 'Damage accumulation in woven baric CFRP laminates under tensile loading. Part 2: Modelling the effect of damage on macro-mechanical properties', Composites Sci. Technol., 1999 **59** 137–145.

26. t w chou, Microstructural Design of Fiber Composites, Cambridge Solid State Science Series, Cambridge University Press, 1992.

27. e h glaessgen and o h griffin jr, Finite element based micro-mechanics modeling of textile composites, NASA Conference Publication 3311, Part 2: Mechanics of Textile Composites Conference, Langley Research Centre, eds C C Poe and C E Harris, 1994, 555–587.

28. k woo and j whitcomb, 'Global/local finite element analysis for textile composites', J. Composite Mater., 1994 **28** 1305–1321.

29. r a naik, p g ifju and j e masters, 'Effect of fiber architecture parameters on deformation fields and elastic moduli of 2-D braided composites', J. Composite Mater., 1994 **28** 656–681.

30. p tan, l tong and g p steven, 'Modelling for predicting the mechanical properties of textile composites – A review', Composites, 1997 **28A** 903–922.

31. G-W DU, T-W CHOU and POPPER, 'Analysis of three-dimensional textile preforms for multidirectional reinforcement of composites', *J. Mater. Sci.*, 1991 **26** 3438–3448.
32. T L NORMAN, C ANGLIN and D GASKIN, 'Strength and damage mechanisms of notched two-dimensional triaxial braided textile composites and tape equivalents under tension', *J. Composites Technol. Res.*, 1996 **18** 38–46.
33. I HERSZBERG, M K BANNISTER, K H LEONG and P J FALZON, 'Research in textile composites at the Cooperative Research Centre for Advanced Composite Structures Ltd', *J. Textile Inst.*, 1997 **88** 52–67.
34. G A BIBO and P J HOGG, 'Role of reinforcement architecture on impact damage mechanisms and post-impact compression behaviour – a review', *J. Mater. Sci.*, 1996 **31** 1115–1137.
35. R PANDEY and H T HAHN, 'Visualization of representative volume elements for three-dimensional four-step braided composites,' *Composites Sci. Technol.*, 1996 **56** 161–170.
36. J-H BYUN and T-W CHOU, 'Modelling and characterization of textile structural composites: a review', *J. Strain Anal.*, 1989 **24** 65–74.
37. R A NAIK, 'Failure analysis of woven and braided fabric reinforced composites', *J. Composite Mater.*, 1995 **29** 2334–2363.
38. M BANNISTER and I HERSZBERG, 'The manufacture and analysis of composite structures from knitted preforms', in *Proceedings 4th International Conference on Automated Composites*, Nottingham, UK, Institute of Materials, 1995.
39. K H LEONG, P J FALZON, M K BANNISTER and I HERSZBERG, 'An investigation of the mechanical performance of weft knitted milano rib glass/epoxy composites', *Composites Sci. Technol.*, 1998 **58** 239–251.
40. S RAMAKRISHNA and D HULL, 'Tensile behaviour of knitted carbon-fibre fabric/epoxy laminates – Part I: Experimental', *Composites Sci. Technol.*, 1994 **50** 237–247.
41. M NGUYEN, K H LEONG and I HERSZBERG, 'The effects of deforming knitted glass preforms on the composite compression properties', in *Proceedings 5th Japan SAMPE Symposium*, Tokyo, Japan, 1997.
42. Y WANG, Y GOWAYED, X KONG, J LI and D ZHAO, 'Properties and analysis of composites reinforced with E-glass weft-knitted fabrics', *J. Composites Technol. Res.*, 1995 **17** 283–288.
43. B GOMMERS, I VERPOEST and P VAN HOUTTE, 'Analysis of knitted fabric reinforced composites: Part 1. Fibre distribution', *Composites*, 1998 **29A** 1579–1588.
44. B GOMMERS, I VERPOEST and P VAN HOUTTE, 'Analysis of knitted fabric reinforced composites: Part II. Stiffness and strength', *Composites*, 1998 **29A** 1589–1601.
45. P J HOGG, A AHMADNIA and F J GUILD, 'The mechanical properties of non-crimped fabric-based composites', *Composites*, 1993 **24** 423–432.
46. G-W DU and F KO, 'Analysis of multiaxial warp-knit preforms for composite reinforcement,' *Composites Sci. Technol.*, 1996 **56** 253–260.
47. P J HOGG and D H WOOLSTENCROFT, 'Non-crimp thermoplastic composite fabrics: aerospace solutions to automotive problems', in *Proceeding of 7th Annual ASM/ESD Advanced Composites Conference, Advanced Composite Materials: New Developments and Applications* Detroit, Michigan, 1991, 339–349.
48. K DRANSFIELD, C BAILLIE and Y-W MAI, 'Improving the delamination resistance of CFRP by stitching – a review', *Composites Sci. Technol.*, 1994 **50** 305–317.
49. A P GODBEHERE, A R MILLS and P IRVING, Non crimped fabrics versus prepreg CFRP composites – a comparison of mechanical performance, in Proceedings Sixth International Conference on Fibre Reinforced Composites, *FRC '94*, University of Newcastle upon Tyne, Institute of Materials Conference, 1994, pp 6/1–6/9.
50. R BACKHOUSE, C BLAKEMAN and P E IRVING, 'Mechanisms of toughness enhancement in carbon-fibre non-crimp fabrics', in *Proceedings 3rd International Conference on Deformation and Fracture of Composites*, held at University of Surrey, Guildford, UK, published by Institute of Materials, 1995, 307–316.
51. S SANDFORD, L BONIFACE, S L OGIN, S ANAND, D BRAY and C MESSENGER, 'Damage accumulation in non-crimp fabric based composites under tensile loading', in *Proceedings Eighth European Conference on Composite Materials (ECCM-8)*, ed I Crivelli-Visconti, Naples, Italy, Woodhead Publishing, 1997, Vol 4, 595–602.

12

Waterproof breathable fabrics

David A Holmes

Faculty of Technology, Department of Textiles, Bolton Institute, Deane Road, Bolton BL3 5AB, UK

12.1   What are waterproof breathable fabrics?

Waterproof breathable fabrics are designed for use in garments that provide protection from the weather, that is from wind, rain and loss of body heat. Clothing that provides protection from the weather has been used for thousands of years. The first material used for this purpose was probably leather but textile fabrics have also been used for a very long time. Waterproof fabric completely prevents the penetration and absorption of liquid water, in contrast to water-repellent (or, shower-resistant) fabric, which only delays the penetration of water. Traditionally, fabric was made waterproof by coating it with a continuous layer of impervious flexible material. The first coating materials used were animal fat, wax and hardened vegetable oils. Nowadays synthetic polymers such as polyvinylchloride (PVC) and polyurethane are used. Coated fabrics are considered to be more uncomfortable to wear than water-repellent fabric, as they are relatively stiff and do not allow the escape of perspiration vapour. Consequently they are now used for 'emergency' rainwear. Water-repellent fabric is more comfortable to wear but its water-resistant properties are short lived.

The term 'breathable' implies that the fabric is actively ventilated. This is not the case. Breathable fabrics passively allow water vapour to diffuse through them yet still prevent the penetration of liquid water.[1] Production of water vapour by the skin is essential for maintenance of body temperature. The normal body core temperature is 37 °C, and skin temperature is between 33 and 35 °C, depending on conditions. If the core temperature goes beyond critical limits of about 24 °C and 45 °C then death results. The narrower limits of 34 °C and 42 °C can cause adverse effects such as disorientation and convulsions. If the sufferer is engaged in a hazardous pastime or occupation then this could have disastrous consequences.

During physical activity the body provides cooling partly by producing insensible perspiration. If the water vapour cannot escape to the surrounding atmosphere the relative humidity of the microclimate inside the clothing increases causing a corresponding increased thermal conductivity of the insulating air, and the clothing

Table 12.1 Heat energy produced by various activities and corresponding perspiration rates[3]

Activity	Work rate (Watts)	Perspiration rate (g day^{-1})
Sleeping	60	2280
Sitting	100	3800
Gentle walking	200	7600
Active walking	300	11500
With light pack	400	15200
With heavy pack	500	19000
Mountain walking with heavy pack	600–800	22800–30400
Maximum work rate	1000–1200	38000–45600

becomes uncomfortable. In extreme cases hypothermia can result if the body loses heat more rapidly than it is able to produce it, for example when physical activity has stopped, causing a decrease in core temperature. If perspiration cannot evaporate and liquid sweat (sensible perspiration) is produced, the body is prevented from cooling at the same rate as heat is produced, for example during physical activity, and hyperthermia can result as the body core temperature increases. The heat energy produced during various activities and the perspiration required to provide adequate body temperature control have been published.[2,3] Table 12.1 shows this information for activities ranging from sleeping to maximum work rate.

If the body is to remain at the physiologically required temperature, clothing has to permit the passage of water vapour from perspiration at the rates under the activity conditions shown in Table 12.1. The ability of fabric to allow water vapour to penetrate is commonly known as breathability. This property should more scientifically be referred to as water vapour permeability. Although perspiration rates and water vapour permeability are usually quoted in units of grams per day and grams per square metre per day, respectively, the maximum work rate can only be endured for a very short time.

During rest, most surplus body heat is lost by conduction and radiation, whereas during physical activity, the dominant means of losing excess body heat is by evaporation of perspiration. It has been found that the length of time the body can endure arduous work decreases linearly with the decrease in fabric water vapour permeability. It has also been shown that the maximum performance of a subject wearing clothing with a vapour barrier is some 60% less than that of a subject wearing the same clothing but without a vapour barrier. Even with two sets of clothing that exhibit a small variation in water vapour permeability, the differences in the wearer's performance are significant.[4] One of the commonest causes of occupational deaths amongst firefighters is heart failure due to heat stress caused by loss of body fluid required to produce perspiration. According to the 1982 US fire death statistics, only 2.6% were due to burns alone whereas 46.1% were the result of heart attacks.[5] Firefighters can lose up to 4 litres (4000 g) of fluid per hour when in proximity to a fire.[6]

In 1991 Lomax reported that modern breathable waterproof fabrics were being claimed to be capable of transmitting more than 5000 g m^{-2} day^{-1} of water vapour.[2] By 1998 it was common to see claims of 10000 g m^{-2} day^{-1}.

Thus, waterproof breathable fabrics prevent the penetration of liquid water from outside to inside the clothing yet permit the penetration of water vapour from inside

Table 12.2 Applications of waterproof breathable fabrics[2,7]

Leisure	Work
Heavy duty, foul weather clothing: Anoraks, cagoules, packs, over-trousers, hats, gloves, gaiters	Foul weather clothing: Survival suits, special military protective clothing, clean-room garments, surgical garments, hospital drapes, mattress and seat covers, specialised tarpaulins, packaging, wound dressings, filtration
Fashionable weather protection: Rainwear, skiwear, golf suits, walking boot linings, panels and inserts, sport footwear linings, panels and inserts	Domestic and transport: Non-allergic bedding, car covers, fire smoke curtains in ships, cargo wraps in aircraft
Tents	
Sleeping bag covers	

the clothing to the outside atmosphere. Examples of over 25 applications of waterproof breathable fabrics have been published.[2,7] Table 12.2 lists some examples of the applications of waterproof breathable fabrics with some additions by the author.

12.2   Types of waterproof breathable fabric

There are several methods which can be used to obtain fabrics which are both breathable and waterproof. These can be divided into three groups:

- densely woven fabrics
- membranes
- coatings.

12.2.1   Densely woven fabrics

Probably the first effective waterproof breathable fabric was developed in the 1940s for military purposes and is known as Ventile (Fig. 12.1). The allied forces were losing aircrew that were shot down or had to ditch in the North Atlantic Ocean. This is a particularly hazardous environment, particularly in winter. A fabric was needed that would allow the personnel to be comfortable whilst carrying out their normal flying duties and prevent penetration of water if they were immersed in the sea. Ventile fabric was carefully engineered to make it effective.[8] The finest types of long staple cottons are selected so that there are very small spaces between the fibres. The cotton is processed into combed yarn, which is then plied. This improves regularity and ensures that the fibres are as parallel as possible to the yarn axis, and that there are no large pores where water can penetrate. The yarn is woven using an Oxford weave, which is a plain weave with two threads acting together in the warp. This gives minimum crimp in the weft, again ensuring that the fibres are as parallel as possible to the surface of the fabric.

When the fabric surface is wetted by water, the cotton fibres swell transversely reducing the size of the pores in the fabric and requiring very high pressure to cause

12.1   Scanning electron micrograph of Ventile fabric.

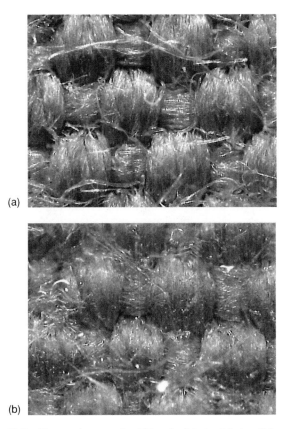

(a)

(b)

12.2   Photomicrograph of Ventile fabric, (a) dry, (b) wet.

penetration (Fig. 12.2). The fabric is thus rendered waterproof without the need for any water-repellent finishing treatment. It was first made for military applications but the manufacturers are now producing a range of variants to widen the market appeal.[9] The military variants use thread densities as high as 98 per cm. Fabric for

12.3   Scanning electron micrograph of mirofilament fabric.

other applications uses much lower thread densities, necessitating a water-repellent finish to achieve the waterproof properties.

Densely woven fabric can also be made from synthetic microfilament yarns. The individual filaments are less than 10μm in diameter, so that fibres with very small pores can be engineered. Microfilaments are usually made from polyamide or polyester. The latter is particularly useful as it has inherent water-repellent properties. The water penetration resistance of the fabric is improved by application of silicone or fluorocarbon finish.

Although fabrics made from microfilaments have a soft handle many of them are windproof, but not truly waterproof as the synthetic filaments do not swell when wet (Fig. 12.3).

The use of very fine fibres and filaments and dense construction (sett) results in fabrics with very small pore size compared with conventional fabrics. Typical pore size for a waterproof fabric is about 10μm compared with 60μm for conventional fabric. Ventile fabric has a pore size of about 10μm when dry and 3–4μm when wet.[2] Fabric made from microfilaments is claimed to have up to 7000 filaments per centimetre. The author has estimated that the military variant of Ventile fabric has about 6000 fibres per centimetre.

12.2.2   Membranes

Membranes are extremely thin films made from polymeric material and engineered in such a way that they have a very high resistance to liquid water penetration, yet allow the passage of water vapour. A typical membrane is only about 10μm thick and, therefore, is laminated to a conventional textile fabric to provide the necessary mechanical strength. They are of two types, microporous and hydrophilic.

12.2.2.1   Microporous membranes

The first and probably the best known microporous membrane, developed and introduced in 1976 by W Gore, is known as Gore-Tex (http://www.gorefabric.com). This is a thin film of expanded polytetrafluoroethylene (PTFE) polymer claimed to contain 1.4 billion tiny holes per square centimetre. These holes are much smaller

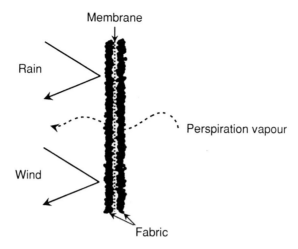

12.4 Schematic diagram of a typical membrane system.

than the smallest raindrops (2–3 μm compared with 100 μm),[10] yet very much larger than a water vapour molecule (40×10^{-6} μm). Other manufacturers make similar membranes based on microporous polyvinylidene fluoride (PVDF) cast directly on to the fabric.[12] The hydrophobic nature of the polymer and small pore size requires very high pressure to cause water penetration. Contamination of the membrane by various materials including body oils, particulate dirt, pesticide residues, insect repellents, sun tan lotion, salt and residual detergent and surfactants used in cleaning have been suspected of reducing the waterproofing and permeability to water vapour of the membrane. For this reason microporous membranes usually have a layer of a hydrophilic polyurethane to reduce the effects of contamination.[11]

Figure 12.4 is a schematic diagram of a fabric incorporating a microporous membrane. Figure 12.5(a) is the polyurethane surface of the bicomponent microporous membrane and (b) shows the polyurethane layer partly removed to reveal the microporous fibrilar nature of the PTFE underneath.

12.2.2.2 Hydrophilic membranes
Hydrophilic membranes are very thin films of chemically modified polyester or polyurethane containing no holes which, therefore, are sometimes referred to as non-poromeric. Water vapour from perspiration is able to diffuse through the membrane in relatively large quantities. The polyester or polyurethane polymer is modified by incorporating up to 40% by weight of poly(ethylene oxide).[2] The poly(ethylene oxide) constitutes the hydrophilic part of the membrane by forming part of the amorphous regions of the polyurethane polymer system. It has a low energy affinity for water molecules which is essential for rapid diffusion of water vapour.[1] These amorphous regions are described as acting like intermolecular 'pores' allowing water vapour molecules to pass through but preventing the penetration of liquid water owing to the solid nature of the membrane.

Figure 12.6 is a diagrammatic representation of the hydrophilic polymer vapour transport mechanism. Figure 12.7 is a scanning electron micrograph of a hydrophilic membrane.

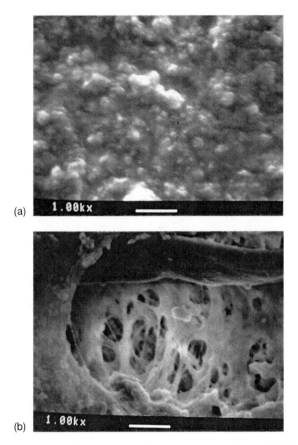

12.5   Scanning electron micrograph of microporous membrane. (a) Hydrophilic surface layer, (b) hydrophilic layer partly removed showing PTFE layer.

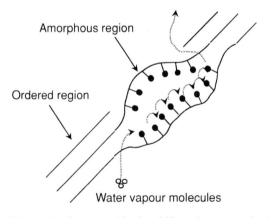

12.6   Schematic diagram of hydrophilic polymer mechanism.

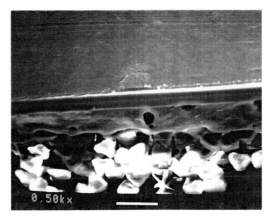

12.7   Scanning electron micrograph of hydrophilic membrane.

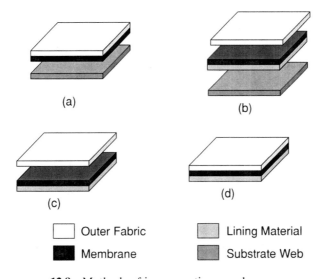

12.8   Methods of incorporating membranes.

12.2.2.3   Methods of incorporation

Membranes have to be incorporated into textile products in such a way as to maximise the high-tech function without adversely affecting the classical textile properties of handle, drape and visual impression.[10] There are four main methods of incorporating membranes into textile articles. The method employed depends on cost, required function and processing conditions:[10]

1   Laminate of membrane and outer fabric (Fig. 12.8a) – The membrane is laminated to the underside of the outer fabric to produce a two-layer system. This method has the disadvantage of producing a rustling, paper-like handle with reduced aesthetic appeal but has the advantage of having very effective protective properties of wind resistance and waterproofing. This method is mainly used for making protective clothing.

2   Liner or insert processing (Fig. 12.8b) – The membrane is laminated to a light-

weight knitted material or web. The pieces are cut to shape from this material, sewn together and the seams rendered waterproof with special sealing tape. This structure is then loosely inserted between the outer fabric and the liner. The three materials (outer, laminate and lining) are joined together by concealed stitch seams. If high thermal insulation is required then the lightweight support for the membrane is replaced by a cotton, wool or wadding fabric. This method has the advantage of giving soft handle and good drape. The outer fabric can also be modified to suit fashion demands.

3   Laminate of membrane and lining fabric (Fig. 12.8c) – The laminate is attached to the right side of the lining material. The functional layer is incorporated into the garment as a separate layer independent of the outer fabric. This method has the advantage that the fashion aspects can be maximised.

4   Laminate of outer fabric, membrane and lining (Fig. 12.8d) – This produces a three-layer system, which gives a less attractive handle and drape than the other methods and, therefore, is not commonly used.

12.2.3   Coatings

These consist of a layer of polymeric material applied to one surface of the fabric. Polyurethane is used as the coating material. Like membranes, the coatings are of two types; microporous and hydrophilic. These coatings are much thicker than membranes.

12.2.3.1   Microporous coatings

Microporous coatings have a similar structure to the microporous membranes. The coating contains very fine interconnected channels, much smaller than the finest raindrop but much larger than a water vapour molecule (Figs. 12.9 and 12.10).

Methods of production of microporous coatings have been described in differing detail in a number of publications.[2,10] Example recipes and processing conditions for producing microporous coatings have also been published.[10]

- *Wet coagulation*: Polyurethane polymer is dissolved in the organic solvent dimethyl formamide to produce a solution insoluble in water. This is then coated on to the fabric. The coated fabric is passed through a conditioning chamber containing water vapour. As the organic solvent is miscible with water, it is diluted and solid polyurethane precipitates. The fabric is then washed to remove the solvent, which leaves behind pores in the coating. Finally the coated fabric is mangled and dried. This method is not very popular as it requires high capital cost for machines and solvent recovery is expensive.

- *Thermocoagulation*: Polyurethane is dissolved in an organic solvent and the resulting solution mixed with water to produce an emulsion. The emulsion 'paste' is coated on to one side of the fabric. The coated fabric then goes through a two-stage drying process. The first stage employs a low temperature to remove the organic solvent, precipitating the polyurethane. The coating is now a mixture of solid polyurethane and water. The second stage employs a higher temperature to evaporate the water leaving behind pores in the coating.

- *Foam coating*: A mixture of polyurethane and polyurethane/polyacrylic acid esters are dispersed in water and then foamed. The foam is stabilised with the aid of additives. The foam is then coated on to one side of the fabric. The coated

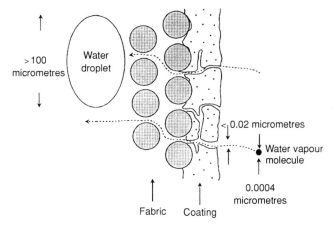

> 100 micrometres

Water droplet

< 0.02 micrometres

Water vapour molecule

0.0004 micrometres

Fabric   Coating

12.9   Schematic diagram of a microporous coating.

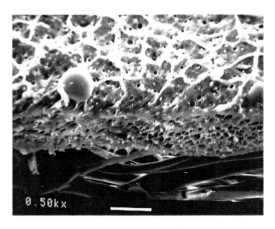

0.50kx

12.10   Scanning electron micrograph of a microporous coating.

fabric is dried to form a microporous coating. It is important that the foam is open cell to allow penetration of water vapour but with small enough cells to prevent liquid water penetration. The fabric is finally calendered under low pressure to compress the coating. As the foam cells are relatively large, a fluorocarbon polymer water-repellent finish is applied to improve the water-resistant properties. This type of coating production is environmentally friendly as no organic solvents are used.

12.2.3.2 *Hydrophilic coatings*

Hydrophilic coatings[11] (Fig. 12.11) use the same basic water vapour permeability mechanism as the hydrophilic membranes. The difference between microporous materials and hydrophilic materials is that with the former, water vapour passes through the permanent air-permeable structure whereas the latter transmit vapour by a molecular mechanism involving adsorption–diffusion and desorption.[11] These coatings are all based on polyurethane, which has been chemically modified by incorporating polyvinyl alcohols and polyethylene oxides. These have a chemical

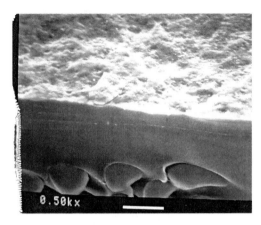

12.11   Scanning electron micrograph of a hydrophilic coating.

affinity for water vapour allowing the diffusion of water vapour through the amorphous regions of the polymer (see Fig. 12.5). The balance between hydrophilic and hydrophobic components of the polymer system has to be optimised to give acceptable vapour permeability, flexibility, durability and insolubility in water and dry cleaning solvents. Swelling of the membrane is encouraged to assist water vapour diffusion yet it also has to be restricted to prevent dissolution or breakdown in water or in the other solvents with which the polymer is likely to come into contact.[11] Poly(ether–urethane) coatings and membranes have excellent integrity. This can be conferred in two ways:

1   by a high degree of hydrogen bonding, principally between polar groups in the hydrophobic segments of adjacent polymer chains
2   by forming covalent crosslinks between adjacent polymer chains. The effective length and density of the crosslinks are variables affecting polymer swelling and thus vapour permeability.[11] Hydrophilic polyurethanes are discussed and formulations for the Witcoflex range of hydrophilic coatings are given by Lomax.[11]

12.2.3.3   Methods of applying coatings
The conventional method of applying coatings to fabric is to use direct application using the knife over roller technique.[10] The fabric is passed over a roller and liquid coating is poured over it. Excess liquid is held back by a 'doctor blade' set close to the surface of the fabric. The thickness of the coating is determined by the size of the gap between the blade and the surface of the fabric. The coated fabric is passed through a dryer to solidify the coating. Sometimes the coating is built up in several layers by a number of applications. In order to achieve thinner coatings and, therefore, more flexible fabric and to apply coating to warp knitted, nonwoven, open weave and elastic fabric, transfer coating is used. The liquid coating is first applied to a silicone release paper using the knife over roller technique. This is then passed through an oven to solidify the coating. A second coating is then applied and the textile fabric immediately applied to this. The second coating, therefore, acts as an adhesive. This assembly is passed through an oven to solidify the adhesive layer. The coated fabric is stripped from the release paper, which can be reused.

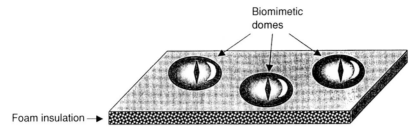

Biomimetic
domes

Foam insulation →

12.12   Stomatex biomimetic material.

12.2.4   Biomimetics

Biomimetics is the mimicking of biological mechanisms, with modification, to produce useful artificial items. The British Defence Clothing and Textile Agency instituted a three-year research programme to study natural systems which could be used to equip service personnel to survive a number of threats, one of which is weather and climate.[13] These workers consider that there is potential for improving the vapour permeability of fabric coatings by incorporating an analogue of the leaf stomata which opens when the plant needs to increase moisture vapour transpiration and closes when it needs to reduce it. They modelled an opening pore comprising flaps of two laminated materials with different moisture uptakes.

Biomimetics has now become a commercial reality. Akzo Nobel is marketing what they claim to be 'the most comfortable clothing and footwear systems in the world today', under the trade name of Stomatex.[14] This is closed foam insulating material made from neoprene incorporating a series of convex domes vented by a tiny aperture at the apex. These domes mimic the transpiration process that takes place within a leaf, providing a controlled release of water vapour to provide comfortable wear characteristics. Stomatex is claimed to respond to the level of activity by pumping faster as more heat is produced, returning to a more passive state when the wearer is at rest. Stomatex is used in conjunction with Sympatex, Akzo Nobel's waterproof breathable membrane, to produce a breathable waterproof insulating barrier for use in clothing and footwear (Fig. 12.12).

12.2.5   Market for waterproof breathable fabrics

In the UK alone there are about 800 000 climbers and walkers, many of whom require performance garments to protect them from the elements.[15] When Toray introduced Dermizax hydrophilic membrane laminated fabric in 1995 the initial sales volume was estimated as 300 000 m² rising to 1.2 million m² in three years.[16] Akzo state that 3 million jackets and coats were made from fabric containing Sympatex hydrophilic membrane in 1992.[17] D & K Consulting carried out a survey of the Western European market between November 1996 and July 1997.[18] The survey predicted that the market would grow to about 45 million linear metres per year by the year 2000. Since the beginning of the 1980s when the market started to expand rapidly almost 40 million m of waterproof breathable fabric have been used in Europe with a value of more than £270 million. Initially laminated fabrics dominated the market but coated fabrics had increased to 55% of the market by 1996. The UK market, accounting for more than 30% of the total European market, has been shown to be different from that of the rest of Europe. About 50% of the UK market is workwear/protective clothing and about 70% of fabrics are coated.

12.3   Assessment techniques

Assessment of the effectiveness of waterproof breathable fabrics requires measurement of three properties:

- resistance to penetration and absorption of liquid water
- wind resistance
- water vapour permeability.

12.3.1   Measurement of resistance to penetration and absorption of liquid water

These measurements are conducted by two kinds of test, simulated rain tests and penetration pressure tests. The simulated rain tests include the Bundesmann rain tester, the WIRA shower test, the Credit Rain simulator and the AATC rain test. They simulate showers of rain, usually of relatively short duration, under controlled laboratory conditions.

12.3.1.1   Bundesmann rain tester

This apparatus and procedure was developed in 1935.[19] Water is fed from the mains through a filter and deioniser to an upper reservoir. The reservoir has a large number of jets of defined size in the base. The pressure of water in the reservoir causes water to flow out through the jets, which produce drops of simulated rain. Specimens of fabric are placed over four inclined cups and sealed at the edge. The cups contain rotating wipers, which rub the underside of the specimens, simulating the action of rubbing, which may occur in use due to the movement of the wearer. Any water penetrating the fabric collects in the cups which have taps so that the penetrated water can be run out and collected and its volume measured. The percentage of water retained by the fabric is also determined on a mass basis. The Bundesmann apparatus (Fig. 12.13) has been criticised on the grounds that it does not produce a realistic simulated shower.[20] Table 12.3 compares the drop sizes and velocities of the Bundesmann shower with different types of rain. It can be seen that the Bundesmann produces drops which have twice the diameter and, therefore, eight times the mass of drops in a cloudburst. The kinetic energy of the drops is 5.8 times the value for a cloudburst and 21 000 times the value of light rain. Owing to this criticism this apparatus fell into disuse but there has been sufficient renewed interest in it for it to be adopted for British, European and International standards.[21,22]

12.3.1.2   WIRA (Wool Industries Research Association) shower tester

A standard volume of water is placed in a funnel, which acts as a reservoir. The water flows slowly out of the funnel into a transparent reservoir with a perforated base made from PTFE. A filter paper is placed on top of the perforated base to slow down the flow of water through the perforations. The water then produces separate drops, which fall onto the fabric specimens placed a standard distance below the base of the reservoir. The fabric specimens are placed under tension over ribbed

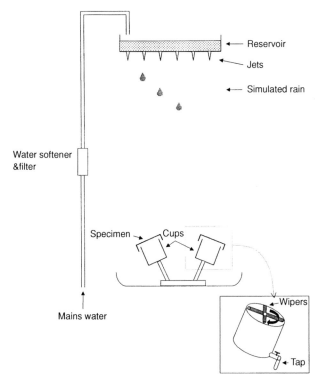

12.13   Diagram of Bundesmann apparatus.

Table 12.3   Comparison of Bundesmann simulated shower and actual rain[20]

Type of drop	Diameter (mm)	Terminal velocity (cm s^{-1})	Kinetic energy (J $\times$ 10^{-6})
Cloud burst	0.30	700	346
Excessive rain	0.21	600	87
Heavy rain	0.15	500	22
Moderate rain	0.10	400	4.2
Light rain	0.045	20	0.095
Drizzle	0.02	75	0.0012
Bundesmann	0.64	540	2000

glass plates forming the top surface of an inclined box. Any water which penetrates the fabric runs down the ribbed plates into the box and then into a 10 cm^3 measuring cylinder. If the measuring cylinder becomes full it overflows into a beaker so that the total volume of penetrated water can be measured. The apparatus has been adopted as a British Standard[24] in which three results may be determined:

1   percentage absorption on a mass basis
2   the total volume of water that penetrates the fabric
3   the time taken for the first 10 cm^3 to penetrate.

Figure 12.14 is a diagram of the WIRA apparatus.

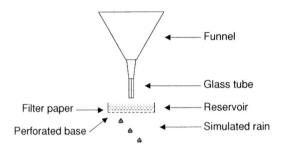

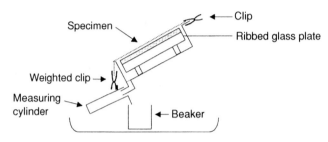

12.14   Diagram of WIRA simulated shower tester.

12.3.1.3   The Credit Rain Simulation tester

When waterproof fabric is made into garments, the seams can become the weak link unless properly constructed. The Credit Rain Simulator[24] was designed to test the effectiveness of seams. It consists of a small water reservoir, the base of which contains jets, which allow the water to flow out slowly forming drops of simulated rain. The drops hit a wire gauze 'drop splitter' which breaks the drops into random sized droplets. The seamed fabric specimen is placed over a semicylindrical printed circuit board (PCB) a standard distance below the base of the reservoir. The water is made electrically conductive by dissolving a small amount of mineral salt in it. Any water penetrating the seam completes an electrical circuit and a read-out indicates the elapsed time to penetration and locates the position of the penetration (Fig. 12.15).

12.3.1.4   AATCC (American Association of Textile Chemists and Colorists) rain test[25]

Many ASTM (American Society for Testing and Materials) fabric specifications[26] stipulate the use of this procedure for testing water penetration resistance. A column of water maintained at a constant height in a vertical glass tube is used to supply pressurised water to a horizontal spray nozzle containing a specified number of holes of specified size. The fabric specimen is placed vertically a specified distance in front of this nozzle backed by a slightly smaller piece of standard blotting paper. This specimen assembly is exposed to the standard spray for 5 min. The gain in mass of the blotting paper is an indication of the penetration of water through the fabric. The severity of the simulated rain can be altered by changing the height of the column of water to give pressures ranging from 60–240 cm water gauge. A complete overall picture of the performance of the fabric is obtained by determining:

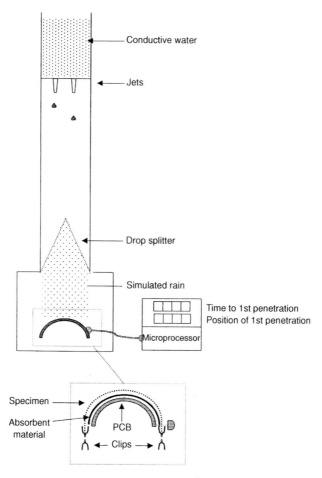

12.15   Diagram of Credit Rain Simulator.

- the maximum pressure at which no penetration occurs
- the change in penetration with increasing pressure and
- the minimum pressure required to cause the penetration of more than 5 g of water, described as 'breakdown'.

Figure 12.16 is a diagram of the AATCC rain test apparatus.

12.3.1.5   Penetration pressure tests

High performance waterproof fabrics are designed so that a high pressure of water is required to cause penetration and their effectiveness is usually measured on this basis. Tests can be carried out in two ways:[27]

- By subjecting fabric to water under pressure for a long period of time and noting if any penetration occurs
- By subjecting the fabric to increasing pressure and measuring the pressure required to cause penetration.

The British Standard[28] has adopted the second alternative. Fabric is placed over a recessed base filled with water so that the face exposed on the outside of the

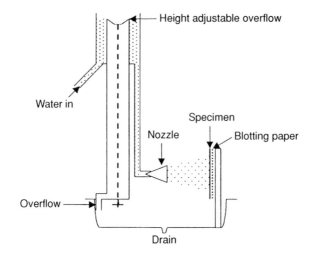

12.16   Diagram of AATCC rain test apparatus.

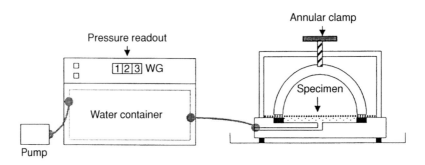

12.17   Diagram of hydrostatic head apparatus.

garment is in contact with the water. The fabric is clamped using an annular clamp. The recessed base is connected to a container of water. The pressure of the water is increased at a standard rate and the surface of the fabric is observed for signs of water penetration. Penetration pressure is determined as the pressure when the third penetration occurs, measured in centimetres water gauge.

Figure 12.17 shows a diagram of apparatus for determining penetration pressure.

12.3.2   Measurement of wind resistance

Before the development of modern waterproof breathable fabrics it was considered that vapour permeability was proportional to air permeability and could be used as a measure of breathability. This is not the case even for conventional woven fabrics. Also, theoretically, the air permeability of hydrophilic membranes and coatings should be zero. Wind resistance is usually assessed by measuring air permeability. This is the rate of air flow per unit area of fabric at a standard pressure difference across the faces of the fabric. Suitable apparatus consists of a pump to provide moving air, a manometer to establish standard pressure and flowmeters to measure

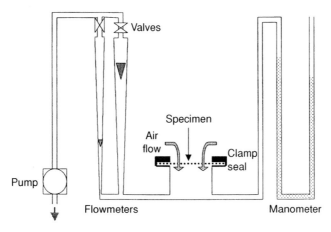

12.18 Typical air-permeability apparatus.

the rate of air flow. Several flowmeters with different ranges are usually incorporated to enable the instrument to deal with a wide range of fabrics (Fig. 12.18). The ASTM[29] procedure determines the volume rate of air flow per unit area of fabric in cubic centimetres per square centimetre per second. The British, European and International standard procedure[30] determines the velocity of air of standard area, pressure drop and time in millimetres per second. The velocity is calculated from measurements of volume rate of air flow. The standard pressure specified in the ASTM standard procedure is 125 Pa (12.7 mm water gauge) whereas that specified in the British Standard procedure is 100 Pa for apparel fabrics and 200 Pa for industrial fabrics. Results obtained using the two procedures are, therefore, not comparable. It must be pointed out that these are very low pressure differences and may not realistically simulate the pressure differences produced by actual wind.

12.3.3 Measurement of water vapour permeability

Techniques for measuring water vapour permeability can be divided into two types:

1 those which simulate sweating bodies and skin and which are mainly used for research work
2 those which are used for routine quality control, fabric development and marketing purposes.

Hong, Hollies and Spivak[31] consider that there are two approaches to measuring moisture vapour transfer:

1 Dynamic methods: these are concerned with moisture transfer prior to the time to reach equilibrium.
2 Equilibrium methods: these deal primarily with the moisture transfer after equilibrium has been established.

Hong *et al.* consider that short term dynamic moisture holds most promise for explaining wetness and moisture-related subject sensations in relation to the level of human comfort in clothing.

12.3.3.1 Research techniques

These are too large in number and too complex to review in detail here. They range from a thermoregulation model of the human skin for testing entire garments in which the microclimate in the vicinity of the skin can be reproduced,[32] through sweating hot plates with eight lines feeding water for simulated perspiration at different rates,[33] and the use of simulated skin made from wetted chamois leather,[31] to the adoption of highly sophisticated instrumental analysis techniques such as differential scanning calorimetry (DSC).[34]

Most research methods have been used to study ordinary non-waterproof woven and knitted apparel fabrics rather than waterproof breathable fabrics, although there is no reason why they cannot be used to study the latter. Similarly, standard routine methods have been used for research purposes.

12.3.3.2 Standard routine tests

The British and US standards all employ the same basic principle, which is an adaptation of the Canadian Turl Dish method.[35] A shallow impermeable dish is used to contain distilled water. The vapour evaporating from the surface of this water represents perspiration. The fabric specimen is placed over the mouth of the dish and sealed round the rim with a suitable impermeable, non-volatile material such as wax. An air gap is left between the surface of the water and the lower surface of the fabric to prevent them from coming into contact. It has been found that the size of this air gap has an influence on the results obtained below about 10–15 mm.[36] It is assumed that the relative humidity in the air space is 100%. The assembly is placed in a standard atmosphere for a length of time. A low-velocity current of air is passed over the upper surface of the fabric to remove the microclimate that develops as a consequence of water vapour passing through the fabric from inside the dish and that would otherwise suppress the passage of further vapour. The temperature of the water contained in the dish is the same as that of the atmosphere outside the dish. The loss in mass of the assembly after a certain time has elapsed is equal to the mass of water vapour that has passed through the fabric. Dolhan[36] compared several dish methods and concluded that none of them are appropriate for testing fabrics with hydrophilic coatings.

Figure 12.19 is a schematic diagram of the dish assembly.

12.3.3.2.1 ASTM methods

ASTM E96-80[37] is not specific to textile fabric. It is a standard for testing a range of materials where transmission of water vapour may be important. ASTM E96-80 includes two methods, one of which has two variants. The size and shape of the dish is not specified other than that it must have a minimum mouth area of 3000 mm^2.

In one method the dish is partly filled with desiccant, such as calcium chloride or silica gel so that the relative humidity inside the dish is very low. The specimen is placed over the mouth and sealed round the rim. The assembly is weighed at regular intervals until its mass has increased by a standard percentage. In this method, therefore, the water vapour is transmitted through the fabric from the outside atmosphere to inside the dish.

In the other method the dish is partly filled with distilled water, the fabric placed over the mouth and sealed round the rim. The assembly is weighed at regular intervals. Provision is made for the assembly to be inverted so that the water is in contact

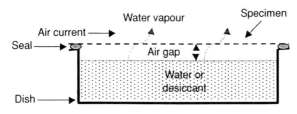

12.19 British Standard and ASTM test assembly.

with the fabric during the test. This is intended for testing materials normally in contact with water during use. A graph is plotted of weight against elapsed time. The linear section of the graph represents nominally steady state conditions. The slope of this linear portion is used to calculate two results.

1 Water vapour transmission (WVT): this is the rate of water vapour transmission in grams per hour per square metre. It is an inherent property of the material tested (sic).
2 Permeance: this is the WVT rate per unit vapour pressure difference between the faces of the fabric in kilograms per pascal per second per square metre. It is a performance evaluation of the material and not an inherent property (sic).

As the standard allows for a wide range of test conditions such as dish shape and size, depth of water, depth of air space between water surface and fabric lower surface, atmospheric conditions, and so on, two laboratories will not necessarily obtain the same results for a given fabric sample. The only parameter specified is an atmospheric relative humidity of 50%. A temperature of 32.2 °C is recommended but not specified.

Many manufacturers quote the water vapour transmission rates of their products for marketing purposes but do not quote the test conditions employed. Therefore, it is difficult for customers to compare products from different manufacturers.

12.3.3.2.2 British Standard method
The British Standard[38] includes only one method and there is no choice of conditions as there is in the ASTM method. The impermeable dish of standard dimensions is partly filled with distilled water leaving a 10 mm gap between the surface of the water and the underside of the fabric. Sealant is placed on the rim of the dish and the fabric specimen is placed over it supported by wire to prevent it from sagging and touching the water. Several dishes are placed on a turntable rotating slowly to provide the air current which removes the microclimate above the fabric (Fig. 12.20). One hour is allowed for equilibrium conditions to develop and each assembly is weighed. The procedure is then continued for a period of at least 5 hours and the assemblies are reweighed. The procedure is carried out in a standard atmosphere of 20 +/– 2 °C and 65 +/– 2% relative humidity. From the mass loss of the specimen the WVP in grams per square metre per day is calculated. The WVP index is also determined. This is the WVP of the test fabric relative to the WVP of a standard reference fabric woven from high tenacity monofilament polyester and having a standard percentage open area.

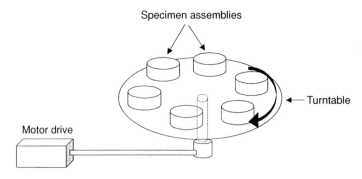

12.20   BS 7209 turntable system.

12.3.3.2.3   European and international standards
EN 31092[39] and ISO 11092[40] are identical. The apparatus and procedure are more complex than those used in the British and ASTM standards. The apparatus is described as a sweating guarded hotplate. In other words the apparatus simulates sweating skin. It consists of a two-layer metal block maintained at a temperature of 35 °C. The top layer of the heated block is porous and the top surface of the bottom layer contains channels allowing water to be fed into the porous layer. The upper face of the porous layer is covered in a cellophane membrane that is impervious to liquid water but permeable to water vapour. Water is fed to the porous plate and kept at a constant level by a controlled dosing device such as a motor driven burette. This maintains a constant rate of evaporation. The sides and bottom of the plate are surrounded by a heated guard allowing heat loss only from the top surface of the plate. The apparatus is used in an enclosure with the atmosphere controlled at 20 °C and 40% relative humidity. A current of turbulent air is passed over the top surface of the apparatus. The electrical power required to maintain the plate at 35 °C is measured (Fig. 12.21).

   Two measurements of power consumption are made after steady state conditions have been obtained, one with and one without a fabric specimen covering the top surface of the plate. The difference between these two measurements is the heat loss from the fabric, which is also equal to the heat required to evaporate the water producing the vapour that passes through the fabric. As the latent heat of vaporisation of water is known, this equates to the mass of water vapour transmitted by the fabric. The two measurements are used to calculate three values related to water vapour transmission properties:

1   Water vapour resistance: this is the water vapour pressure difference across the two faces of the fabric divided by the heat flux per unit area, measured in square metres pascal per watt. As standard plate temperature and atmosphere temperature and relative humidity are used, the vapour pressure difference has a standard value.
2   WVP: this is the mass of water vapour passing through the fabric per unit area unit time, unit vapour pressure difference, measured in grams per square metre per hour per pascal
3   WVP index: this is the ratio of dry thermal resistance to water vapour resistance. A value of zero implies that the fabric is impermeable to water vapour and a

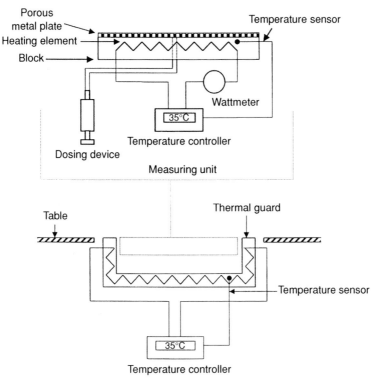

12.21   EN and ISO water vapour resistance apparatus.

value of one indicates that it has the same thermal and water vapour resistance of a layer of air of the same thickness. This WVP index is not the same as the index determined by the British Standard method.[38]

The apparatus can also be used to determine water vapour permeability characteristics under transient conditions and under different wear and environmental conditions but the standard only relates to steady state conditions.

Alternative definitions and explanations of water vapour transmission rate, water vapour resistance and resistance to evaporative heat flow have been given by Lomax.[11] The measurement of water vapour resistance in thickness units (mm) is explained. This is the thickness of a still air layer having the same resistance as the fabric. The use of thickness units facilitates the calculations of resistance values for clothing assemblies comprising textile and air layers.

12.4   Performance of waterproof breathable fabrics

Several research workers have compared the performance characteristics of different types of waterproof breathable fabrics. As Salz[41] points out,

When it comes to waterproofing and comfort, an objective comparison is almost impossible. Many fabrics can withstand a hydrostatic head of 100 cm (9800 Pa) which makes them seem more than adequate for all but the most severe conditions. But are they? The vapour transmission rates are often difficult to

Table 12.4 Typical water vapour resistance of fabrics[11]

Fabric	WVR (mm still air)
Outer (shell) material	
Neoprene, rubber or PVC coated	1000–1200
Conventional PVC coated	300–400
Waxed cotton	1000+
Wool overcoating	6–13
Leather	7–8
Woven microfibre	3–5
Closely woven cotton	2–4
Ventile L28	3.5
Other Ventiles	1–3
Two-layer PTFE laminates	2–3
Three-layer laminates (PTFE, polyester)	3–6
Microporous polyurethane (various types)	3–14
Open pores	3–5
Closed pores	6–14
Hydrophilic coated, e.g.	
Witcoflex Staycool	
on nylon, polyester	9–16
on cotton/polyester	5–10
Superdry	6–14
on cotton, polyester/cotton	4–7
on liner fabric	3–4
Waterproof, breathable liners (coated and laminated)	2–4
Inner clothing	
Vests (cotton, wool)	1.5–3
Shirtings (cotton, wool)	0.8–3
Pullover (lightweight wool)	3–5

compare because of the variety of the test methods, but also because the conditions during the tests do not even remotely reflect the conditions of rainy weather for which the fabrics are designed.

Lomax[11] quotes the example of a two-layer PTFE laminate having a water vapour permeability of only $500 \, \text{g m}^{-2} \text{day}^{-1}$ by one recognised method, whereas exactly the same test specimen can have a permeability of $20\,000 \, \text{g m}^{-2} \text{day}^{-1}$ by another method. This large difference is due to the difference in test conditions.

In standard water vapour permeability tests the inside of the cup is at the same temperature as the outside laboratory atmosphere (e.g. 20 °C), whereas in use, conditions inside the clothing will be at a temperature approaching 37 °C. When it is raining the outside relative humidity is likely to be near saturation rather than 50 or 65% as used in standards. Some water vapour resistance data on different types of outerwear fabrics obtained using a dish method are presented in Table 12.4 and compared with typical values for normal clothing layers.[11] The conditions under which the tests were carried out are not stated but the data are still useful for comparison purposes. In general the following conclusions can be drawn:

• Breathable materials are very much better than fabrics coated with conventional waterproof materials.

Table 12.5 Water vapour transport (WVT) through rainwear fabrics under different conditions[41]

Material	WVT (g h^{-1} m^{-2})	
	Dry	Rain
Microporous PU coated fabric A	142	34
Microporous PU coated fabric B	206	72
Two-layer PTFE laminate	205	269
Three-layer PTFE laminate	174	141
Hydrophilic PU laminate	119	23
Microporous AC coated fabric	143	17
Microfibre fabric	190	50
PU coated fabric	18	4

- Breathable fabrics have a higher resistance to vapour transport than ordinary woven and knitted apparel fabrics but in some cases this difference is not very large.
- In a limited number of cases waterproof breathable fabrics have a lower vapour resistance than some ordinary apparel fabrics.
- The vapour resistance of breathable membranes and coatings is influenced by the fabric substrate to which they are applied.

Attempts have also been made to study performance under simulated 'actual' weather conditions rather than standard conditions.

Saltz[41] developed a laboratory method by using a heated cup method in combination with an artificial rain installation. The simulated rain tester combined mechanical action and wicking effects. The water in the cup was maintained at 36 °C and the water for the shower at 20 °C. The shower duration was one hour, much longer than standard simulated rain tests. A series of fabrics including two- and three-layer PTFE laminates, normal and microporous coated, hydrophilic polyurethane laminated, Ventile and microfibre, were tested under dry and rain conditions. The water vapour transport results are presented in Table 12.5.[41]

The authors make no attempt to carry out a general comparison of the various fabrics but it can be seen that the microporous coatings and laminates have higher vapour transport properties than the hydrophilic laminate. All are significantly higher than the normal coated fabric. The microfibre fabric has one of the highest values. It is concluded that rain has a major influence on vapour transport. In most cases wetting by rain reduces the vapour transmission rate. The two-layer PTFE laminate was an exception to this general rule, showing an increase in vapour transport when exposed to rain. This increase is claimed to be caused by the condensation of escaping water vapour on the reverse side of the fabric caused by the cooling effect of the rain. Thus, the air layer, which forms a vapour resistance in the dry test, is bridged in rain conditions. It is assumed that vapour transport in rain is influenced by the following factors:

- Microporous fabrics can become virtually impermeable in rain owing to blocking of the micropores by water.
- If the pores in the fabric are very small and highly hydrophobic, then blocking will not occur.

- Saturation of hydrophilic membranes with rain water can prevent the absorption of water vapour from the heated cup.

Although the best fabrics studies allowed water vapour transport of up to $200 \, g \, m^{-2} \, h^{-1}$ in dry conditions, rain reduced the value of most fabrics to a maximum of $50 \, g \, m^{-2} \, h^{-1}$. This may prove to be insufficient, particularly for persons working in strenuous conditions. The cooling effect of rain can cause condensation of the escaping vapour causing the wearer to become wet owing to backward wicking. Mechanical action, such as movement of limbs, was found to reduce the rain resistance of some microporous fabrics owing to the pumping action and wicking. Some fabrics requiring very high water penetration pressure when tested using a standard hydrostatic pressure test had a poor resistance rating compared with the other fabrics when exposed to simulated rain under mechanical action.

Holmes, Grundy and Rowe[42] studied the effect of atmospheric conditions on the water vapour permeability characteristics of various types of waterproof breathable fabrics. A modification of the BS 7209 was used. The water in the cups was maintained at a temperature of 37 °C by partial immersion in a water bath whose temperature was controlled by heating and cooling circuits. The whole assembly was placed in an environmental chamber to control the atmosphere outside the fabric at a range of temperatures and relative humidities. In general it was concluded that the fabrics could be ranked in decreasing order of vapour permeability as follows:

- Tightly woven:
  tightly woven synthetic microfilament
  tightly woven cotton
- Membranes:
  microporous membrane
  hydrophilic membrane
- Coatings:
  hydrophilic coating.

This rank agrees with Ruckman's rank[43] and the order in which Salz's results can be ranked.[41] It was found that atmospheric conditions have a considerable effect on vapour permeability, which can be much less than that measured under standard conditions. In fact, at high temperatures and relative humidities the vapour can pass from the outside atmosphere to the inside of the fabric. Water vapour permeability is a function of the vapour pressure difference between the two faces of the fabric. Vapour pressure is, in turn, a function of both temperature and relative humidity. The relationship between vapour permeability and vapour pressure difference is complex. The best fit is given by a third degree polynomial (i.e. $WVP = ax + bx^2 + cx^3 + C$, where a, b, and c are constants and x is the vapour pressure difference). The graphs of vapour permeability against vapour pressure differences are presented in Fig. 12.22.

The regression equations presented in Table 12.6 were used to predict the permeabilities of two fabrics at opposite ends of the performance spectrum under 'realistic' conditions. Such conditions are represented by the atmosphere inside the clothing at 37 °C and saturated with vapour, and the outside atmosphere relatively cold and also saturated, as may be found during rain, and at virtually zero as may be found at subzero temperatures. The total breathability of a suit made from

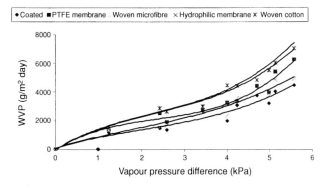

12.22   Water vapour permeability plotted against vapour pressure difference.

Table 12.6   Regression equations and correlation coefficients[42]

Fabric type	Regression equation	Correlation coefficient
Densely woven cotton	$Y = 1788x - 4401x^2 + 62.0x^3 - 25$	0.9876
Woven polyamide Hydrophilic PU coating	$Y = 929x - 194x^2 + 30.4x^3 + 58$	0.9325
Two-layer: woven polyamide Hydrophilic polyester membrane	$Y = 928x - 148x^2 + 25.5x^3 + 78$	0.9434
Three-layer: knitted polyamide Microporous PTFE membrane	$Y = 1856x - 580x^2 + 81.9x^3 - 38$	0.9703
Woven polyester microfilament	$Y = 1835x - 473x^2 + 68.8x^3 - 11$	0.9877

Y = water vapour permeability $(g\,m^{-2}\,day^{-1})$; x = vapour pressure difference (kPa)

breathable fabric was estimated. The results are presented in Table 12.7. Comparing this data with published perspiration rates, it was concluded that the best fabric could not cope with activity beyond active walking with a light pack and the worst could not cope with activity beyond gentle walking.

Ruckman has published a series of investigations into water vapour transfer in waterproof breathable fabrics under steady state, rainy, windy, and rainy and windy conditions.[43–45] Twenty-nine samples of microfibre, cotton Ventile, PTFE laminated, hydrophilic laminated, polyurethane-coated and poromeric polyurethane-coated fabrics of differing weights per unit area and thickness were studied. A dish method was used in which the water in the dish was controlled at a temperature of 33 °C to simulate body temperature and with ambient temperatures ranging from −20 to +20 °C. It was found that water vapour permeability was influenced by ambient temperature, the permeability reducing as the ambient temperature decreased. There is very little difference between the various fabrics at low ambient temperatures although there is, in fact, a factor of about four between the lowest and highest. It is shown that the effect is caused by differences in vapour pressure difference between the faces of the fabric. Graphs of water vapour transmission (water vapour permeability per unit vapour pressure difference) show very complex behaviour possibly due to condensation on the inner surface of the fabric at ambient temperatures below 0 °C.

Table 12.7 Water vapour permeability under 'realistic' conditions[42]

	Temperature (°C)	Relative humidity (%)	Vapour pressure (kPa)	
Inside	37	100	6.28	
Outside	0	0	0.56	
	6	100		0.93
Vapour pressure difference between inside and outside atmospheres (kPa)			5.84	5.33

	Water vapour permeability (WVP) $(g\,m^{-2}\,day^{-1})$	
Vapour pressure difference (kPa)	5.33	5.84
Woven cotton fabric	6300	7080
Hydrophilic coated fabric	4200	4530
	Calculated Total WVP $(g\,day^{-1})$	
Woven cotton suit	15000	17000
Hydrophilic coated suit	10000	10900

The effect of windy conditions was studied by using a fan to produce a low turbulence current of air parallel to the face of the fabric.[44] Experiments were carried out using wind velocities between zero and $10.0\,m\,s^{-1}$ and air temperatures between zero and 20 °C. It was found that wind increases the vapour permeability of the fabric by causing an increase in the vapour pressure difference across the fabric. The effect is greatest with the microfibre and cotton Ventile fabrics and increases as air temperature decreases. Condensation of water vapour at low temperatures reduces the rate of vapour transfer. It is suggested that careful consideration should be given to the suitability for intended end-use when selecting waterproof breathable fabrics for manufacture into sportswear and foul weather garments, especially garments to be used under windy conditions. In particular it should be taken into account that different types of product behave differently under windy conditions in terms of formation of condensation.

The effect of rainy and windy conditions was studied using the dish method, modified so that the specimen was held at an angle that allowed rain to run off but did not affect the hot plate controlling the water temperature.[45] Simulated rain at various temperatures was provided by a set of shower heads complying with the requirements of the AATCC standard rain test.[25] Experiments were carried out in rain, wind driven rain and prolonged rain. All fabrics continued to breathe under rainy conditions except microfibre fabrics which ceased to breathe in less than 24 hours of rain. Although no rain penetrated the fabrics, the cooling effect of rain caused condensation on the inner surface of the fabrics. This effect was the least with PTFE laminated fabrics.

The following conclusions were drawn:

1   Water vapour transfer in waterproof breathable fabrics decreases as rain temperature increases.
2   Waterproof breathable fabrics continue to breathe under rainy conditions. However, the breathability of most of them ultimately ceases after long exposure to prolonged severe rainy conditions.

The time of cessation of breathability can be ranked in the following increasing order:

- microfibre
- cotton Ventile
- poromeric polyurethane laminate
- PTFE laminate
- polyurethane coated
- hydrophilic laminate.

3 More condensation occurs on all fabrics under rainy conditions than under dry conditions except for PTFE laminated fabric.

4 The water vapour transfer rate is reduced under wind driven rainy conditions compared to that under rainy conditions for all fabrics owing to the effect of both rain and condensation.

Ruckman draws general conclusions from her extensive studies discussed above.[45] These are that water vapour transfer of breathable fabrics depends very much on atmospheric conditions. In general, wind increases and rain decreases the water vapour transfer rate of fabric, giving in descending order of water vapour transfer performance: windy, dry, wind driven rainy, rainy. The findings suggest that careful consideration should be given when choosing the appropriate waterproof breathable fabric for manufacture of sportswear and foul weather garments. The end-use envisaged for the garment and the environment it will be used in should always be taken into account.

The effect of subzero temperatures on the water vapour transfer properties of breathable fabrics has been studied by Oszevski[46] but was limited to the hydrophilic component of Gore-Tex II. The experiment was designed to simulate diffusion of water vapour through a clothing shell from a coating of ice on its inner surfaces as can occur when the fabric is used in very cold weather conditions. A procedure similar to the ASTM desiccant method[37] was used with one face of the fabric specimen in contact with or close to a block of ice. The experiments were carried out in an environmental chamber at various temperatures as low as $-24\,°C$. It was found that the water vapour diffusion resistance increased exponentially as temperature decreased because the vapour pressure over ice has a very low value, thus creating a very low vapour pressure difference across the fabric. It is predicted that the diffusive flux of water vapour through the fabric at $-10\,°C$ will only be about 4% of its value at room temperature. It is speculated that the same is probably true of any hydrophilic rainwear fabric although high-tech waterproof materials should dry more quickly than non-breathable materials.

The effect of other garment layers and condensation in clothes systems incorporating waterproof breathable fabric has been studied by Gretton and co-workers.[47,48] Samples of every type of breathable fabric were used with a standard clothing system comprising a polyester lining fabric, a polyester fleece and a cotton jersey T-shirt fabric. A heated evaporative dish procedure based on BS 7209[38] was employed with ambient temperatures ranging from 5–20 °C and a constant relative humidity. The results obtained were confirmed by the use of field trials. It was found that condensation occurred at ambient temperatures below 10 °C. There are several mechanisms of condensation formation and dispersal within clothing systems. The mechanism involved depends on the type of breathable fabric, and fabric layers underneath also interact with the condensation and alter the transport properties of the clothing system.

A large amount of condensation accumulates in microporous coatings significantly impeding the transport of water vapour. Condensation cannot be removed easily from such coatings because the liquid water cannot travel through the pore network of the polymer. Hydrophilic coatings accumulate less condensation than the microporous coatings because liquid water is absorbed into and transported through the hydrophilic polymer without the need for re-evaporation. The high water content of the polymer caused by condensation plasticises and swells the polymer improving the water vapour transport at relatively high temperatures. However, low temperatures offset the benefits of plasticisation.

Condensation can occur within the pore structure of microporous PTFE membranes but the hydrophilic layer in the bicomponent versions reduces this to a small value and also prevents condensation forming on the surface. However, once condensation has occurred it is slow to re-evaporate. Laminated fabrics were found to perform better than coated fabrics under cold ambient conditions, transmitting more water vapour and preventing condensation from forming for a longer period of time.

The presence of a lining fabric was found to influence the manner in which the waterproof breathable fabric dealt with condensation. Lining fabrics actively drawing condensation away from microporous polymers improve the vapour transport properties, whereas linings not drawing condensation away provide an improvement in vapour transport of hydrophilic polymers.

The incorporation of other clothing layers reduced the vapour transport properties of the breathable layer. This is claimed to be because the additional insulation maintains the inner layer of the breathable at a relatively low temperature, reducing the vapour driving force across it and the mobility of its polymer molecules. During physical activity that produces sensible perspiration, removal of an insulating layer, such as a fleece garment, should improve the comfort of the clothing system by increasing the temperature of the breathable layer, improving its vapour transport and reducing condensation. The use of a laminated fleece (presumably meaning the wind-resistant type) promoted the formation of condensation on the fleece and the waterproof breathable layer owing to restriction of ventilation within the clothing system. This severely reduced the thermal insulation of the fleece. It was found possible for a fleece fabric to deal effectively with condensation if the pile is worn facing inwards towards the body so that condensation was transported away from the body into the knitted backing of the fleece fabric. Like other research workers,[45] Gretton[48] concludes that assessment of waterproof breathable fabrics in isolation using standard test methods does not give a true indication of their performance in actual wear situations. It is important to assess the transport properties of performance clothing systems as a whole entity under the range of ambient conditions the garments are likely to experience in use.

Gretton et al.[49] also carried out a study of simulated clothing systems using a wider range of under layer fabrics than the 'standard' described above. The full range of under layer fabrics was: woven lining, double-sided fleece, lightweight double-sided fleece, single-sided fleece, laminated fleece, honeycomb knitted under layer and single jersey T-shirt fabric. The fibre content was 100% polyester except for the T-shirt, which was cotton. The 'standard' clothing layer reduced the moisture vapour transmission rate of the breathable outer layer samples by 14%. The effect

Table 12.8 Water vapour permeabilities of membranes under different conditions[51]

Membrane type	Water vapour permeability ($g\,m^{-2}\,day^{-1}$)	
	dry	wet
Microporous PTFE	4000	3200
Hydrophilic	4400	4300

was the same on microporous and hydrophilic polymers. It was found possible to predict the water permeability index of a multiple layer clothing assembly from the indices of the individual layers using Equation (1):

$$\text{Multiple layer \% MVP} = \frac{\text{T-shirt \% MVP}}{100} \times \frac{\text{fleece \% MVP}}{100}$$
$$\times \frac{\text{lining \% MVP}}{100} \times \text{outer fabric \% MVP} \qquad (1)$$

where MVP is the moisture vapour permeability.

Marxmeier[50] carried out a limited study on the effects of wetting on breathable materials. Only two membranes were used, a microporous PTFE type and a hydrophilic type. Tests were carried out on the membranes alone so as to exclude the influence of textile fabric and lamination method. A dish method was employed with the water at a temperature of 37 °C and the ambient atmosphere at 25 °C and 50% relative humidity. Two experiments were carried out, one with the membrane dry and the other with the membrane covered in a continuous film of water. This was achieved by covering the membranes with porous water absorptive polyester fabric with its ends dipped into reservoirs of water at 25 °C. The results obtained are presented in Table 12.8.

Under dry conditions there is little difference between the permeabilities of the membranes. However, under wet conditions the permeabilities of the membranes were reduced, that of the hydrophilic type by only a small amount, but that of the microporous type was reduced dramatically. The difference in behaviour is explained with reference to the mechanism of vapour transport. The water swells the hydrophilic membrane, increasing its diffusion coefficient, whereas the PTFE microporous membrane is hydrophobic. The swelling compensates for the reduction in partial vapour pressure difference across the fabric caused by the presence of water. It is concluded that hydrophilic membranes have an advantage over microporous membranes under aggravated conditions. It is also stated that hydrophilic membranes have advantages in terms of after care (dry cleaning) and durability (abrasion resistance).

Water penetration can occur at points on garments subjected to pressure loading (for example, at knees or elbows, by rucksack straps, carrying handles and so on). In adverse circumstances pressures of 10 bar can occur. This is particularly important when the membrane material is to be used in gloves and footwear. In footwear especially, high loading is a necessary consequence of a good fit. Creases at the toe and heel are a problem. Uedelhoven and Braun[51] built a modified hydrostatic head apparatus enabling them to test breathable materials under pressure loading. They tested 24 materials including PTFE and polyester membranes alone and in two- and three-layer constructions, as well as polyurethane coatings. It was found that the microporous PTFE membranes had the highest pressure resistance of over 20 bar and the polyurethane materials had the lowest at only 0.5 bar. It was also found that repeated pressure loading reduced the water penetration pressure of all the materials significantly. The general conclusion drawn was that hydrostatic penetration resistance depends, not only on the membrane material, but also especially on the construction of the finished garment. Apart from better protection from damage, lamination onto woven or knitted fabric enables hydrostatic pressure resistance to be significantly improved. Inappropriate laminate construction can have an adverse influence on hydrostatic pressure resistance, especially with repeated cycles of pressure loading.

Most studies on the performance of waterproof breathable fabrics concentrate on the vapour permeability characteristics. Weder[52] carried out an extensive study of the performance of breathable rainwear materials with respect to protection, physiology, and durability. Twenty-three samples of waterproof breathable fabrics from the following categories were tested:

- PU coatings
- PTFE membranes
- PES membranes
- cotton Ventile

A non-breathable PVC-coated fabric and an ordinary polyester apparel fabric were included for comparison. The samples were subjected to a large number of simulated performance tests including:

- vapour permeability
- rain resistance
- water pressure resistance
- wind resistance
- mechanical strength
- abrasion resistance
- effect of mechanical pressure
- effect of wind on water penetration resistance
- effect of laundering
- effect of diagonal pulling
- effect of weathering
- effect of flexing
- effect of low temperatures
- occurrence of condensation
- effect of thorn pricking
- drying behaviour
- complete jacket tests.

Not every possible combination of sample and test was discussed in detail. The various categories of sample were compared by constructing three-dimensional plots of weather protection, physiology and durability to give what is described as 'cobweb diagrams' (radar diagrams). These are reproduced in Fig. 12.23.

It is concluded that:

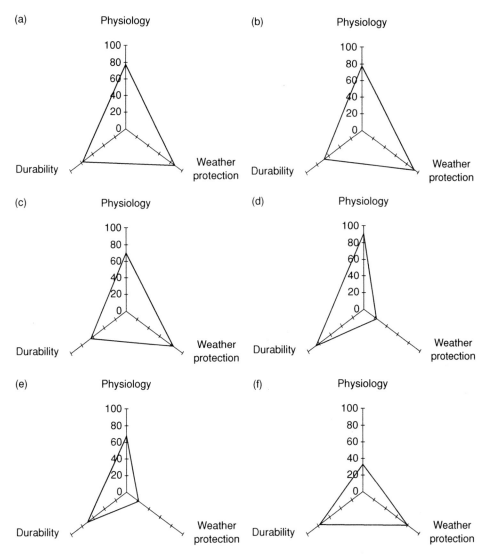

12.23   Attributes of different types of breathable and conventional fabrics.[52]
(a) PTFE laminate, (b) PES laminate, (c) PU coating, (d) PES/cotton blend, (e) Ventile,
(f) PVC coating.

1   There was no significant overall difference between the various categories of
     breathable fabric.
2   The main parameter contributing to durability is the outer material to which the
     waterproof polymer is attached.
3   It makes no sense to design leisurewear that is used only occasionally to meet
     the same rigorous standards as professional clothing.
4   Polyurethane coatings have better resistance to mechanical damage than PTFE
     and PES laminates.
5   Almost all the samples had sufficient to good vapour transport properties
     but ventilation openings in garments are important because the vapour perme-

ability capabilities of the fabrics are not sufficient for moderate physical activity.

6 Correct manufacture of the final product is as important as the properties of the fabric for water penetration resistance.

7 The question 'which product performs the best' cannot be answered. The conditions of application and corresponding requirements imposed on a product are quite different.

References

1. W O LOTENS, 'What Breathability Do You Need', Symposium, '*Breathe*', Stratford-upon-Avon, June, 1991.
2. G R LOMAX, *Textiles*, 1991, No. 4, 12.
3. J H KEITHLEY, *J. Coated Fabrics*, 1985 **15**(October) 89.
4. L BENISEK *et al.*, *Performance of Protective Clothing*, eds. R L Barker and G C Coletta, ASTM Special Technical Publication 900, Philadelphia, 1986, p 405.
5. B V HOLCOMBE and B N HOSHCKE, *Performance of Protective Clothing*, eds. R L Barker and G C Coletta, ASTM Special Technical Publication 900, Philadelphia, 1986, p 327.
6. N SAVILLE and N SQUIRES, 'Multiplex Panotex Textiles', *International Conference on Industrial and Technical Textiles*, Huddersfield, 6–7th July, 1993.
7. S J KRISHNAN, *J. Coated Fabrics*, 1991 **21**(July) 20.
8. Ventile Technical Literature, Harris Watson Investments Ltd, Talbot Mill, Froom Street, Chorley, Lancashire, England.
9. ANON, *World Sports Activewear*, 1996 (Winter) 8.
10. W MAYER, U MOHR and M SCHUIERER, *Int. Textile Bull.*, 1989, N°2, 18.
11. J LOMAX, *Coated Fabrics*, 1990 **20**(October) 88.
12. ANON, *Design News*, 1988, **44**, N°13 (July) 48.
13. DAWSON *et al.*, Biomimetics in Textiles, *Textiles Engineered for Performance*, UMIST, Manchester, April, 1998.
14. ANON, *Textile Horizons*, 1998 (May) 22.
15. ANON, *Textile Month*, 1982 (July) 14.
16. ANON, *JTN*, 1995, No 484 (March) 42.
17. C CLARK, *Apparel Int.*, 1993, **24**, N°1 (July/August) 19.
18. ANON, *Textile Month*, 1998, January, 23.
19. Bundesmann, *Textilberichte*, 1935 **16** 128.
20. S BAXTER and A B D CASSIE, *JTI*, 1945 **36** T67.
21. BS EN 29865: 1993, *Textiles, Determination of Water Repellency of Fabrics by the Bundesmann Rainshower Test*, British Standards Institution, 1993.
22. ISO 9865: 1991, *Textiles, Determination of Water Repellency of Fabrics by the Bundesmann Rainshower Test*, International Standards Organisation, 1991.
23. BS 5066: 1974, *Method of Test for the Resistance of Fabric to an Artificial Shower*, British Standards Institution.
24. Technical Literature, John Godrich, Ludford Mill, Ludlow, Shropshire, England.
25. AATCC Test Method 35-1980, *Water Resistance: Rain Test*, AATCC Technical Manual, Volume 57, 1981/82.
26. *ASTM Standard Performance Specifications for Textile Fabrics*, 2nd Edition, American Society for Testing and Materials, 1988.
27. M BONAR , *Textile Quality*, Texilia, 1994, p. 353.
28. BS EN 20811: 1992, *Resistance of Fabric to Penetration by Water (Hydrostatic Head Test)*, British Standards Institution, 1992.
29. ASTM D737, *Standard Test Method for Air Permeability of Textile Fabrics*, American Society for Testing and Materials.
30. BS EN ISO 9237: 1995 *Textiles – Determination of the Permeability of Fabrics to Air*, British Standards Institution, 1995.
31. K HONG, N R S HOLLIES and S M SPIVAK, *Textile Res. J.*, 1988 **58** 697.
32. K H UMBACH, *Melliand Textilberichte*, 1987 **11** 857.
33. B FARNWORTH and P A DOLHAN, *Textile Res. J.*, 1985 **55** 627.
34. M DAY and Z STURGEON, *Textile Res. J.*, 1986 **57** 157.
35. CAN2-4.2-M77: 1977, *Method of Test for Resistance of Materials to Water Vapour Diffusion (Control Dish Method)*, 1977.

36. P A DOLHAN, *J. Coated Fabrics*, 1987 **17**(October) 96.
37. ASTM Method E96-80, *Standard Test Method for Water Vapor Transmission of Materials*, American Society for Testing and Materials.
38. BS 7209: 1990, *British Standard Specification for Water Vapour Permeable Apparel Fabrics. Appendix B. Determination of Water Vapour Permeability Index*, British Standards Institution, 1990.
39. EN 31092: 1993E, *Textiles – Determination of Physiological Properties – Measurement of Thermal and Water-vapour Resistance under Steady-state Conditions (Sweating Guarded-hotplate Test)*, 1993.
40. ISO 11092: 1993, *Textiles – Determination of Physiological Properties – Measurement of Thermal and Water-vapour Resistance under Steady-state Conditions (Sweating Guarded-hotplate Test)*, International Standards Organisation, 1993.
41. P SALZ, 'Testing the quality of breathable textiles', *Performance of Protective Clothing: Second Symposium*, ASTM Special Technical Publication 989, eds. F Z Mansdorf, R Sagar and A P Nielson, American Society for Testing and Materials, Philadelphia, 1988, p 295
42. D A HOLMES, C GRUNDY and H D ROWE, *J. Clothing Technol. Management*, 1995 **12**(3) 142.
43. J E RUCKMAN, *Int. J. Clothing Sci. Technol.*, 1997 **9**(1) 10.
44. J E RUCKMAN, *Int. J. Clothing Sci. Technol.*, 1997 **9**(1) 23.
45. J E RUCKMAN, *Int. J. Clothing Sci. Technol.*, 1997 **9**(2) 141.
46. R J OSZEVSKI, *Textile Res. J.*, 1996 **66**(1) 24.
47. J GRETTON et al, *Textile Res. J.*, 1998, In print.
48. J GRETTON, 'Condensation in Clothing Systems', *Survival 98 Conference*, University of Leeds, England, June, 1998
49. J C GRETTON et al., *J. Coated Fabrics*, 1997 **26**(January) 212.
50. H MARXMEIER, *Chemifarsen/ Textilindustrie*, 1986 **36/88**(July/August) 575.
51. W UEDELHOVEN and W BRAUN, *Melliand Textilberichte*, 1991 **72**(3) E71.
52. M WEDER, *J. Coated Fabrics*, 1997 **27**(October) 146.

13

Textiles in filtration

Edwin Hardman

Madison Filters (formerly Scapa Filtration), Haslingden,
Rossendale, Lancashire, UK

13.1 Introduction

The separation of solids from liquids or gases by textile filter media is an essential part of countless industrial processes, contributing to purity of product, savings in energy, improvements in process efficiency, recovery of precious materials and general improvements in pollution control. In fulfilling their tasks, the media may be expected to operate for quite lengthy periods, frequently in the most arduous of physical and chemical conditions. As performance is crucial to the success of an operation, fabric failure during use could result in heavy penalties, for example, owing to loss of product, maintenance and lost production costs and possibly environmental pollution costs.

The final products of processes which involve filtration by textile filter media may ultimately find their way into our everyday lives, some examples being edible products such as sugar, flour, oils, fats, margarine, beer and spirits, and other products such as dyestuffs and pigments (as used in clothing, furnishings and paints), viscose rayon fibres and films, nickel, zinc, copper, aluminium, coal, cement, ceramics, soaps, detergents, fertilisers and many more. In addition to assisting in the refinement of products for our general everyday use, textile filter media are also engaged in the purification of both industrial and domestic effluents, thereby contributing to a cleaner environment.

The purpose of this chapter is to provide the reader with a general introduction to the more common types of solid–gas (dust collection) and solid–liquid filtration mechanisms, the raw materials, polymers, fibres and different types of fabric construction which are employed in media manufacture and some typical fabric finishing processes.

13.2   Dust collection

13.2.1   Introduction

Gas-borne dust particles arise wherever solid materials are handled. Examples include conveyors, smelting processes, hopper filling, pulverising processes, combustion processes, milling operations, bag filling and so on. The dusts may create environmental pollution problems or other control difficulties caused by their toxicity, flammability and possibly risk of explosion. The particles in question may simply require removal and be of no intrinsic value or alternatively may constitute part of a saleable product, for example sugar or cement. Typically in the range 0.1–25 µm they may be collected by one of several techniques, viz. settling chambers, cyclones, granulate filters, electrostatic precipitators and fabric collectors. Of these, arguably the most efficient and versatile is the fabric collector, especially when processing very fine particles, which are slow to settle and, by virtue of their greater light scatter, more visible to the naked eye.

13.2.2   Dust collection theory and principles

Much has been written on the various mechanisms by which particles are arrested by unused filter media.[1] These are normally explained in terms of the effect of a spherical particle on a single fibre and may be summarised as gravitational, impaction, interception, diffusion (Brownian motion) and electrostatic. These mechanisms are shown diagrammatically in Fig. 13.1.

The theories behind these mechanisms notwithstanding, it has been argued[2] that although they may be valid for certain air filtration applications where total particle capture is vital, for the purposes of industrial dust collection, they are of limited value. A sieving mechanism is probably more appropriate wherein the size of the apertures in the medium assumes a more dominant role, at least until the fibres have accumulated a layer of dust which then takes over the sieving action.

13.2.3   Practical implications

In operation, fabric dust collectors work by drawing dust laden gas through a permeable fabric, usually constructed in the form of tubular sleeves, longitudinal envelopes or pleated elements. As the gas passes through the fabric, the particles in

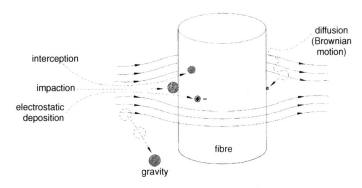

13.1   Particle collection mechanisms.

the gas stream are retained, leading to the formation of a layer of dust on the surface. This is normally referred to as a 'dust cake'. After a period of time, the accumulated dust leads to a reduction in the permeability of the material, and creates an increased pressure drop on the outlet side of the fabric. Consequently the fabric must be cleaned at appropriate intervals to return the pressure drop to a more acceptable level. Dust is then again collected and the filter continues through cycles of dust accumulation and cleaning. This mechanism is shown graphically in Fig. 13.2.

From the graph it will be observed that, after cleaning, the pressure drop does not return to the original level. This is because the fabric still retains a fraction of dust that actually assists in filtration by forming a porous structure that bridges the apertures in the fabric. It is this bridged structure that determines the filtration efficiency for subsequent filtration periods. The graph also shows that the pressure drop after each cleaning cycle continues to rise until a steady state condition develops. Were this not to occur (broken lines), the pressure drop would continue to rise to the point where more power would be required to pull the gas through the system than the fan can produce. This would result in a reduction in flow rate, possible fabric damage and ultimately system shut down.

It follows from the above that, in steady state conditions, the amount of dust that is removed during cleaning is virtually equal to the amount that accumulates in the filtration phase. In reality a small, almost imperceptible increase in pressure drop may take place, resulting in a condition that will ultimately necessitate fabric removal. However, since this increase is typically less than 1 mm WG (water guage) per month, it is normally several months at least before this replacement becomes necessary.

13.2.4   Cleaning mechanisms

Fabric dust collectors are usually classified according to their cleaning mechanism, these being shake, reverse air and jet pulse. Whichever is employed it is important that a programme is devised to provide an optimum level of dust removal. In other words the cleaning should not be so excessive as to destroy the porous structure formed by the dust, which could lead to emission problems, but not so ineffective as to lead to an unacceptable pressure drop.

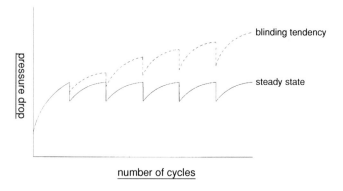

13.2   Resistance across filter medium.

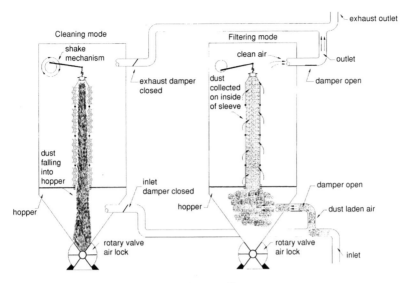

13.3   Shake collector.

13.2.4.1   Shake cleaning

Shake cleaning, as the name implies, involves switching off the exhaust fan and flexing the filter elements (or sleeves) with the aid of a shaking mechanism, either manually, as in traditional units, or automatically. In both cases the effect is to release the dust, which then falls into a hopper for collection and removal (Fig. 13.3). In this type of collector the filter sleeves, which may be up to 10 m in length, are suspended under controlled tension from the arm of a flexing mechanism which effects the cleaning action.

13.2.4.2   Reverse air cleaning

With this mechanism, cleaning is achieved again by switching off the exhaust fan but this time followed by reversing the airflow from outside to inside of the sleeves. There are two basic styles of reverse air collector. The first causes the sleeve to inflate during the collection phase and partially collapse during low pressure reverse air cleaning, whereas in the second, involving a higher cleaning pressure, the sleeves are prevented from total collapse by means of a number of metal rings inserted at strategic intervals along the length of the sleeve during fabrication. In some cases, reverse air cleaning may also be combined with a shake mechanism for enhanced performance.

13.2.4.3   Pulse jet cleaning

Compared with the mechanisms described so far, which normally involve dust collection on the inside of the sleeves, pulse jet collectors operate by collecting dust on the outside. On this occasion the sleeves, typically 3 m in length and 120–160 mm in diameter, are mounted on wire cages (Fig. 13.4). In operation, removal of the collected dust is effected by a short pulse of compressed air, approximately 8–14 litres in volume and 6 bar pressure, which is injected into a venturi tube located at the opening of the elements. This transmits a shock pulse that is sufficient to overcome the force of the exhaust fan and also to cause a rapid expansion of the filter sleeves. The dust is thus made to fall from the sleeves and to be collected in the hopper

13.4   Section of filter sleeve on cage.

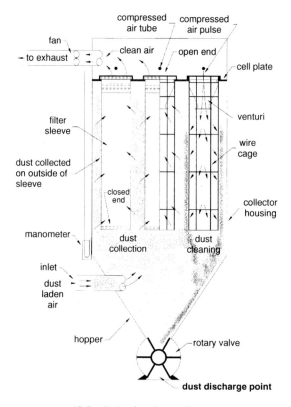

13.5   Pulse jet dust collector.

(Fig. 13.5). Of the three types of mechanism described, the pulse jet is the most widely used.

13.2.5   Fabric design or selection considerations
The primary factors which determine the selection of a fabric for a particular application may be summarised as:

- thermal and chemical conditions
- filtration requirements
- equipment considerations, and
- cost.

13.2.5.1 Thermal and chemical conditions

The thermal and chemical nature of the gas stream effectively determines which type of fibre is to be used. Table 13.1 identifies the more common types which are used in dust collection and also their basic limitations. For example, if the temperature of the gas stream is higher than can be sustained by the fibre, and cost considerations preclude the possibility of gas cooling prior to dust collection, then alternative means of collection – perhaps by means of ceramic elements – will have to be sought.

Depending on the duration of exposure, high temperatures may have several effects on the fibre, the most obvious of which are loss in tenacity due to oxidation and less effective cleaning due to cloth shrinkage.

From Table 13.1, the maximum operating temperature for each fibre may appear quite low, especially when compared with the respective melting points. Suffice it to say that, whilst the fibres may withstand short surges at 20–30 °C higher than the temperatures indicated, experience has shown that continuous operation above the temperatures listed will lead to a progressive reduction in tenacity.

The presence of moisture in the gas stream, which above 100 °C will be present in the form of superheated steam, will also cause rapid degradation of many fibres through hydrolysis, the rate of which is dependent on the actual gas temperature and its moisture content. Similarly, traces of acids in the gas stream can pose very serious risks to the filter fabric. Perhaps the most topical example is found in the combustion of fossil fuels. The sulphur that is present in the fuel oxidises in the combustion process to form SO_2 and in some cases SO_3 may also be liberated. The latter presents particular difficulties because, in the presence of moisture, sulphuric acid will be formed. Hence, if the temperature in the collector were allowed to fall below the acid dew point, which could be in excess of 150 °C, rapid degradation of the fibre could ensue. Polyaramid fibres are particularly sensitive to acid hydrolysis and, in situations where such an attack may occur, more hydrolysis-resistant fibres, such as produced from polyphenylenesulphide (PPS), would be preferred. On the debit side, PPS fibres cannot sustain continuous exposure to temperatures greater than 190 °C (and atmospheres with more than 15% oxygen) and, were this a major constraint, consideration would have to be given to more costly materials such as polytetrafluoroethylene (PTFE).

Because a high proportion of fabric dust collectors are not faced with such thermal or chemical constraints, the most commonly used fibre in dust collection is of polyester origin, this being capable of continuous operation at a reasonably high temperature (150 °C) and is also competitively priced. On the debit side, polyester is acutely sensitive to hydrolysis attack and, were this to pose a serious problem, fibres from the acrylic group would be the preferred choice.

13.2.5.2 Filtration requirements

Failure to collect dust particles efficiently will inevitably lead to atmospheric pollution, which will at least be undesirable if not positively harmful. It is important therefore that, in the first instance, the fabric is designed to capture the maximum number

Table 13.1 Chemical and abrasion resistance of fibres

Generic type	Examples	Max working temperature (°C)	Abrasion resistance	Acid resistance	Alkali resistance	Some damaging agents
Polyester	Dacron Trevira	150	VG	G	P	Quicklime, conc. mineral acids, steam hydrolysis
Polyaramid	Nomex Conex	200	VG	P	VG	Oxalic acid, mineral acids, acid salts
Polyimide	P84	260	VG	P	VG	Oxalic acid, mineral acids, acid salts
Cellulose	Cotton Viscose	100	G	P	VG	Copper sulphate, mineral acids, acid salts, bacteria
Silicate	Fibreglass	260	P	F	F	Calcium chloride, sodium chloride, strong alkalis
Homopolymer acrylic	Dolanit Zefran Ricem	140	G	G	F	Zinc and ferric chloride, ammonium sulphate, thiocyanates
Copolymer acrylic	Dralon Orlon	120	G	G	F	As above for homopolymer
Polypropylene	Moplefan (Trol)	90 (125)	G	E	E	Aluminium sulphate, oxidising agents, e.g. copper salts, nitric acid
PTFE	Teflon Rastex	260	F	E	E	Fluorine
Polyamide	Nylons	100	E	P	E	Calcium chloride, zinc chloride, mineral acids
Polypeptide	Wool	110	G	G	P	Alkalis, bacteria
Polyphenylene-Sulphide	Ryton Procon	190	G	VG	E	Strong oxidising agents
PEEK	Zyex	250	VG	G	G	Nitric acid

E = Excellent, VG = Very Good, G = Good, F = Fair, P = Poor, PEEK= polyetheretherketone.
a Can be improved by special finishing treatments.

of particles present. The particle size and size distribution will be of great importance to the media manufacturer since these will determine the construction of the fabric. If the particles are extremely fine this could lead to penetration into (and possibly through) the body of the fabric, plugging of the fabric pores, ineffective cleaning and a prematurely high pressure drop. The fabric would become 'blind'. The skill therefore will be to select or design a fabric, which will facilitate the formation of a suitable dust pore structure on or near the surface and will sustain an acceptable pressure drop over a long period.

The particles may also present a challenge according to their abrasive nature, this giving rise to internal abrasion that will be further aggravated by the flexing actions to which the sleeve will be subjected. Conventional textile abrasion test methods will be of marginal value in predicting performance unless a mechanism for introducing the actual dust being processed can be introduced.

The particles that are conveyed to the collector may also possess an electrostatic charge,[3] either preapplied or acquired en route that, if carried into the collection compartment, could accumulate with potentially explosive consequences. Such a case, involving white sugar dust handling systems, is the subject of a paper by Morden.[4] As static electricity is essentially a surface effect, were it likely that the accumulation of such a charge will pose a serious risk, then consideration would have to be given to constructing the filter fabric with antistatic properties, for example by means of a special surface treatment or through the inclusion of antistatic fibres such as stainless steel or carbon-coated polyester (epitropic). Provided the media are properly earthed, such inclusions will enable the charge to dissipate readily.

Assessment of a fabric's antistatic properties can be made relatively easily by measurement of surface resistivity (Ω) between two concentric rings placed on the surface of the fabric, each carrying a potential difference of 500 V.[5]

Conversely, by constructing the filter medium with a blend of fibres of widely contrasting triboelectric properties, it is claimed by a fibre manufacturer[6] that superior collection efficiency can be obtained. It is further claimed that, by virtue of this enhanced efficiency, a more open structure can be used with consequent advantages in respect of the reduced power consumption required to pull the dust-laden air through the collector. However, although this effect has been used to some advantage in clean air room filtration applications, considerably more research is necessary if the triboelectric effects in industrial dust collection are to be fully understood and exploited.

Yet another problem which may confront the engineer is the presence of very hot particles. Whether from a combustion, drying or other process, these particles have been known to be carried with the gas stream into the filtration compartment where they present a serious risk of fire. (In certain conditions even ostensibly nonflammable polyaramid fibres have been found to ignite.) Consequently, if adequate particle screening is not provided, the fabric may require a special flame-retardant treatment.

The above notwithstanding, arguably the most difficult conditions in dust collection arise from the presence of moisture in the gas stream or if the dust were of a sticky nature from previous processing. This situation will be aggravated if the fabric were subsequently allowed to dry out, resulting in the formation of nodes or agglomerations of dust particles and leading to an increased weight of dust cake and eventually a critical blinding situation. In such cases, it may be advisable for the fabric to be subjected to special hydrophobic or oleaphobic treatments as part of the finishing process.

13.2.5.3 Equipment considerations

Equipment considerations again focus on the cleaning mechanisms and in particular, the forces applied by them. In the case of shake collectors, the filter sleeves will be subjected to quite vigorous flexing, which could lead to the formation of creases and ultimately holes in the fabric through flex fatigue, a situation that, as stated previously, will be aggravated by the presence of abrasive particles in the gas stream. As a consequence, in addition to resisting stretch from the weight of the dust load, a filter fabric with superior flexibility – at least at the strategic flex points – will provide a longer life.

By comparison, in pulse jet collectors the fabric sleeve is mounted on a wire cage into which, at frequent intervals, a pulse of compressed air is injected. This causes the fabric to expand briefly in a lateral direction after which the force of the exhaust fan, coupled with the fabric's elastic recovery property, returns the element back to a snug fit on the cage. This action has been studied in some depth by Sievert and Loffler.[7] Critical factors here are the actual pulse force and frequency of cleaning, the design and condition of the cages and the 'fit' of the filter sleeve on the cage itself. Too tight results in inefficient cleaning and too slack may result in damage against the cage wires or possibly interference with adjacent elements. This aspect will be aggravated if the cleaning frequency is increased, as may occur with higher dust loading.

The style of filter will also determine the complexity of the sleeve design. Apart from knitted fabrics, which for this purpose are produced in tubular form, the chosen filter media will first have to be slit to an appropriate width, then formed into a tube. This may be achieved by sewing or, if the polymer is of a thermoplastic nature, by hot air welding, the latter having the advantages of both a higher production speed and obviating the need for sewing threads. In the case of high cost materials such as PPS, this could constitute a substantial saving.

In practice it is normal to manufacture the 'tubes' in quite long lengths, for example 100 m, after which the individual sleeves can be cut to ordered size in preparation for the next stage of fabrication. In the case of reverse air and shake collectors this may involve the fitting of anticollapse rings and possibly metal caps – attachments by which the filter sleeve can be suspended in the filter. Other reinforcements may also be included to enable the sleeve to withstand the effect of frequent flexing.

By comparison, filter sleeves in pulse jet collectors are located in an opening in a cell plate. In this respect they may be mounted in either a vertical or horizontal manner. Since dust is collected on the outside of these sleeves, the fixture at the location point is critical if by-pass of the filter and subsequent emission of dust into the atmosphere is to be avoided. Some possible gasketing arrangements are shown in Fig. 13.6.

13.2.5.4 Cost

In spite of all the design considerations and performance guarantees that are frequently required of the media manufacturer, this is still a highly competitive industry. As a consequence every effort is made to reduce media manufacturing costs, either by judicious sourcing of raw materials, or more efficient manufacturing (including fabrication) techniques.

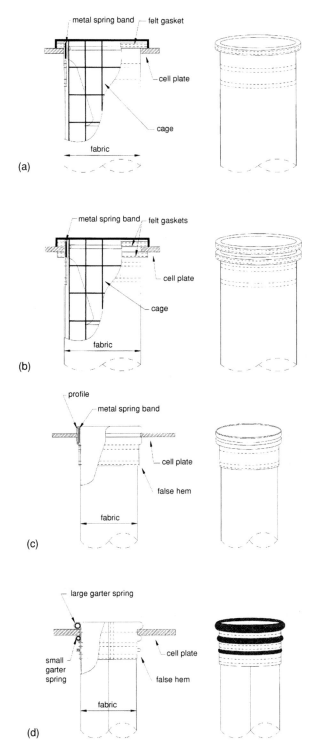

13.6   Filter sleeve location arrangements. (a) Single felt gasket, (b) double felt gasket, (c) spring band profile, (d) garter springs.

13.3   Fabric construction

Three basic types of construction are found in fabric dust collectors, viz., woven fabrics, needlefelts and knitted structures. The first two are produced in flat form and will require (i) slitting to appropriate width and (ii) converting into tubular sleeves, whereas knitted fabrics may be produced directly in tubular form.

13.3.1   Woven fabrics

Used predominantly in shake collectors, this class of filter fabric may comprise twisted continuous filament yarns, short staple-fibre yarns (cotton or woollen spinning system) or perhaps a combination of both. Weave patterns may be in the form of elementary twills, for example 2/1, 2/2 or 3/1, or perhaps simple satin designs, the latter providing greater flexibility and hence superior resistance to flex fatigue and a smoother surface for superior cake release. Woven fabric area densities are typically in the range 200–500 g m^{-2}.

Design requirements include resistance to stretch from the mass of the dust cake, resistance to flex fatigue from the shake cleaning mechanism, a surface that will facilitate efficient dust release and a construction that will effect maximum particle capture whilst at the same time providing minimum resistance to gas flow.

Depending on the choice of yarns, woven fabrics may present either a smooth continuous filament yarn surface to the gas stream, or a more bulky fibrous surface as provided by staple-fibre yarns. Whilst the former will provide superior cake release characteristics, the latter, by virtue of its greater number of pores, will permit higher filtration velocities, greater laminar flow and therefore a lower pressure drop across the fabric. By using a combination of continuous filament warp and staple-fibre weft yarns, preferably in a satin weave for a smoother surface and greater flexibility, an ideal compromise is possible. In this case, the filtration efficiency can be further enhanced by subjecting the weft side to a mechanical raising treatment.

13.3.2   Needlefelts

This type of construction, a cross-section view of which is shown in Fig. 13.7, is by far the most widespread in dust collection processes, providing an infinitely larger number of pores and facilitating considerably higher filtration velocities than woven fabrics.

In the majority of cases they are produced by needle punching a batt of fibre – a number of layers of carded fibre web formed by means of a cross-laying mechanism – on to both sides of a woven basecloth or scrim. This may be carried out in a continuous process or by attachment of a preformed and preneedled batt produced in a separate operation. After 'tacking' the fibre to the scrim, the assembly is consolidated in a secondary, more intensive needle-punching operation, usually with the aid of finer needles. This operation frequently addresses both sides of the felt simultaneously in a single 'double-punch' process.

The use of a woven scrim, whilst not employed in every case, provides the needlefelt with stability and the necessary tensile characteristics to withstand the stresses imposed by the predominantly pulse cleaning mechanism, whereas the batt provides the necessary filtration efficiency and also a measure of protection for the basecloth

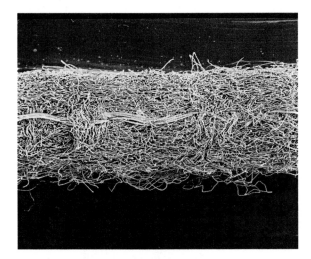

13.7   Scanning electron micrograph showing cross-section of needlefelt.

from abrasion caused by constant flexing against the cage wires. Depending on the tensile specification of the finished needlefelt, the area density of scrims is usually in the range 50–150 g m^{-2}.

Inevitably some damage to the scrim will occur in the needle punching operation, especially if it comprises continuous filament yarns. The design of the scrim is therefore frequently 'overengineered' to compensate for this and the damage may also be alleviated by judicious selection of (i) needle design, (ii) needle fineness, (iii) needle orientation, (iv) needle board pattern, and (v) needling programme, that is, punch rate and penetration.

The needles themselves, typically 75–90 mm in length, are mounted in a board, the arrangement or pattern of which is so designed as to provide a surface which is as uniform as possible and devoid of 'needle tracking lines'. Normally triangular in cross-section, the needles contain a series of barbs which are set into the corners. Typically nine barbs per needle, these are designed to engage the fibres on the downward stroke of the punching action yet emerge completely clean on the up-stroke. Hence the fibres become mechanically locked both to other fibres in the assembly and also to the woven scrim. The barbs may be regularly spaced over the length of the needle blade, or more closely spaced for more intensive needling and the production of a more dense structure. In another design, the barbs are located on only two of the three corners, this style being used where maximum protection to one of the scrim components is required.

The density of the needles in the needleboard, the frequency of needle punching, the style of needle and depth of penetration through the structure will all be influential in controlling the thickness and density of the final assembly, and also the strength retained by the scrim.[8]

The fibres, which form the batt, are normally in the range 1.66–3.33 decitex though trends to considerably finer 'microfibres' (e.g. less than 1 decitex) have gained some prominence. Whilst the latter will provide an even greater number of pores per unit area, and hence more efficient filtration, they will also require a much higher degree of carding, resulting in considerably reduced productivity. In another development, a similar construction can be achieved by means of so-called splittable fibres. Such fibres comprise a number of elements which are bonded together at the extrusion stage. However, as a result of the subsequent mechanical action of carding (or aqueous treatment in the case of water-soluble binders), the individual elements split from the parent structure to produce the appearance of a microfibre needlefelt.

Although most fibres utilised in dust collection are of circular cross-section, irregular, multilobal-shape fibres, as in the case of Lenzing's P84 and peanut shape fibres, as in DuPont's Nomex, are also possible. The latter are of particular value as they possess a higher surface area and hence facilitate the production of needlefelts with potentially superior particle collection capability. Some manufacturers have taken this a stage further by producing structures with a 'veneer' of high particle collection efficiency fibres on the surface whilst retaining coarser, less expensive fibres on the back.

Needlefelt area densities are typically in the range 300–640 g m^{-2}, lighter qualities being used in reverse air and shake collectors and heavier qualities in pulse jet collectors. The majority of needlefelts actually fall in the range 400–510 g m^{-2}, these facilitating generally higher filtration velocities. However, in the event that the dust is particularly abrasive, a longer life may be expected from felts in the 540–640 g m^{-2} range.

13.3.3   Knitted fabrics
Because they are capable of being produced in seamless tubular form, weft-knitted fabrics provide, in theory, an attractive and economic alternative to both woven and needled constructions. By inlaying appropriate yarns into the knitted structure, the elasticity which is normally associated with such fabrics can also be controlled and the same may be used to enhance the particle collection capability.[9] On the down side, in critical applications, the filtration efficiency will be inferior to a needlefelt construction and further problems are likely to be found in respect of the large number of sleeve diameters which the industry requires. Limitations are also inevitable in respect of the number of physical and chemical finishes which are frequently administered to both woven and needlefelt constructions.

13.4   Finishing treatments

These are designed essentially to improve (i) fabric stability, (ii) filtration collection efficiency, (iii) dust release, and (iv) resistance to damage from moisture and chemical agents. A number of finishing processes are employed to achieve these goals, for example heat setting, singeing, raising, calendering, 'special surface treatments' and chemical treatments.

13.4.1   Heat setting
Improved stability is essential in order to prevent shrinkage during use. Such shrinkage may be caused by the relaxation of tensions imposed on fibres and/or yarns during manufacture, or be due to the inherent shrinkage properties of the raw materials themselves.

The thermal conditions that are often found in a dust collector, will be conducive to fabric relaxation and, if not effectively addressed during manufacture, could lead to serious shrinkage problems during use. For example, in a pulse collector, lateral shrinkage could result in the fabric becoming too 'tight' on the supporting cage, leading to inefficient cleaning and ultimately an unacceptable pressure drop.

As heat is the primary cause of shrinkage, it is logical that fabric stability should be achieved by thermal means. Such an operation is normally referred to as heat setting, and may be carried out by surface contact techniques, 'through air' equipment, or by stentering, the latter two being preferred because they enable greater penetration of heat into the body of the structure. This is particularly relevant in the case of needlefelts because the scrim is to some extent insulated by the batt fibres. Whichever technique is employed, in order to ensure stability during use, the temperature in the heat setting operation will invariably be significantly higher than the maximum continuous operating temperature of the material in question. Furthermore, since complete fibre relaxation is a temperature–time related phenomenon, manufacturers will also process at speeds that are appropriate to achieve the desired effect.

In addition to stabilising the fabric, the heat setting process will also effect an increase in the density of the structure through increased fibre consolidation. This in turn will further assist in achieving a higher level of filtration efficiency.

13.4.2 Singeing

Filter fabrics, especially needlefelts, which are produced from short staple fibres, invariably possess surfaces with protruding fibre ends. Since such protrusions may inhibit cake release by clinging to the dust, it is common practice to remove them. This is achieved by singeing, a process in which the fabric is passed, at relatively high speed, over a naked gas flame or, in another technique, over a heated copper plate. The heat of the flame causes the fibres to contract to the surface of the fabric where, in the case of thermoplastic fibres, they form small hard polymer beads (Fig. 13.8). Singeing conditions (i.e. speed and gas pressure) will normally be adjusted according to polymer type and the intensity required by either the end-use application or the individual manufacturer's preference.

13.4.3 Raising

Whilst the singeing process is designed to denude the fabric of its protruding fibres, the raising process is designed actually to create a fibrous surface, normally on the outlet side of the filter sleeve, to enhance the fabric's dust collection capability. It follows therefore that this process is designed essentially for woven fabrics comprising staple-fibre yarns – at least in the weft direction. In operation the fabric is pulled over a series of rotating rollers termed 'pile' and 'counter pile', each of which is clothed with card wire and mounted concentrically on a large cylinder of approximately 1.5 m diameter. As the cylinder rotates, the pile rollers raise the fibres proud of the surface whereas the counter pile rollers stroke them into a more orderly fashion. Raised fabrics may comprise 100% staple-fibre yarns or a combination of multifilament and staple-fibre yarns, the latter being woven in satin style in which the face side is predominantly multifilament and the reverse side predominantly staple.

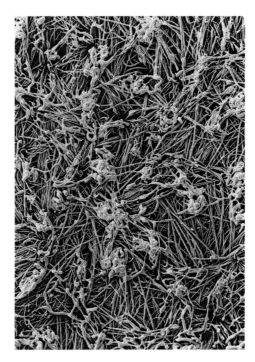

13.8   Scanning electron micrograph showing surface of (singed) needlefelt.

The smooth surface provided by the multifilaments will aid cake release whilst the raised staple yarns on the reverse side will enhance particle collection efficiency. A significant measure of width contraction invariably takes place during this operation and proper attention will have to be given to this when designing the fabric.

13.4.4   Calendering

The calendering operation fulfils two objectives, viz. to improve the fabric's surface smoothness and hence aid dust release, and to increase the fabric's filtration efficiency by regulation of its density and permeability. As a result of the latter, the yarns and fibres become more tightly packed, making it more difficult for particles to pass through or even into the body of the fabric.

Most calenders in the industry consist of at least two bowls, one manufactured from chrome-plated steel and the other from a more resilient material such as nylon or highly compressed cotton or wool fibres. The steel bowl is equipped with a heat source, for example gas, electric elements, superheated steam, or circulating hot oil. Thus, by varying the process temperature (usually according to polymer type), pressure and speed, the desired density and degree of surface polish can be achieved. In reality, rather than density, a more common control parameter is measurement of the fabric's air permeability, this normally being expressed in units of cfm (cubic feet per square foot per minute at $\frac{1}{2}$ inch water gauge) or litres per square decimetre per minute at 20 mm water gauge.[10]

The cotton or synthetic bowl may also posses a cambered profile in order to offset the deflection (bending) that occurs as the pressure is applied and that may otherwise lead to non-uniform calendering. Alternatively, since this camber only applies to the operating pressure for which it is designed, the generally preferred approach would be to employ a calender adopting a system such as developed by equipment manufacturers Kusters and Ramisch-Kleinwefer in which uniform pressure can be maintained across the full width of the fabric regardless of the applied force.

Although the calender is useful, not least in regulating permeability, it should not be regarded as a more economical substitute for reduced needling density or, in the case of woven fabrics, more economical thread spacing. Aggressive conditions in the filter may well negate the effect of the calendering operation before the fabric has become fully 'acclimatised' to the conditions. This is especially relevant where the fibres in the filter medium are of a particularly resilient nature, such as those in the acrylic family.

13.4.5   Chemical treatments
Chemical treatments are normally applied for one of two reasons, namely (i) to assist in dust release, especially where moist sticky dusts, possibly containing oil or water vapour are encountered, or (ii) to provide protection from chemically aggressive gases such as SO_2 and SO_3 referred to earlier. However, in the case of SO_3 it is possible that such chemical treatments, in the presence of moisture, will be less than 100% effective and, in such circumstances, a more chemically resistant fibre must be sought.

Other chemical treatments may also be employed for more specific purposes. For example, proprietary treatments, usually involving silicone or PTFE, enhance yarn-to-yarn or fibre-to-fibre 'lubricity' during pulse or flex cleaning and similarly, where flammability is a potential hazard, padding through commercially available flame-retardant compounds may be necessary.

13.4.6   Special surface treatments
This category of treatments is devoted to improving still further the fabric's filtration efficiency and cake release characteristics. In this respect there are basically two types of treatment, namely (i) attachment of a more efficient membrane, for example biaxially stretched PTFE (Fig. 13.9) in a lamination operation, and (ii) the application of a low-density microporous foam (Fig. 13.10). Both these treatments are designed to restrict the dust particles, as far as possible, to the surface of the fabric, thereby reducing the tendency for blinding. The PTFE membrane, comprising an extremely fine structure, is particularly effective in this respect. It may be laminated to the surface of the fabric either by special adhesives or, where appropriate, by flame bonding. Although highly efficient, the gossamer-like surface is rather delicate and care must be exercised when handling filter sleeves produced from such materials. In addition, as PTFE laminated fabrics are relatively expensive, their use is normally restricted to difficult applications, for example where the dust particles are extremely fine or of a particularly hazardous nature or where the interaction with a surface of this type shows unique advantages in respect of cake release.

By comparison, the foam treatment is achieved by (i) mechanically generating a low density latex foam, (ii) applying this foam to the fabric by the knife over roller

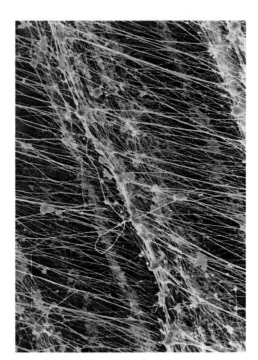

13.9   Scanning electron micrograph showing biaxially stretched PTFE membrane.

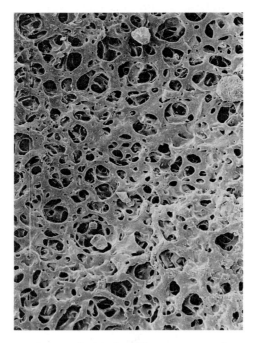

13.10   Scanning electron micrograph showing microporous coating on needlefelt substrate.

(or knife over air) technique, (iii) drying the foam at a modest temperature, (iv) crushing the foam to produce an open cell structure, and (v) curing the foam at a higher temperature to crosslink the chemical structure. Although the principal ingredient of the treatment is usually an aqueous-based acrylic latex, the precise formulation may comprise a variety of chemical agents to ensure the production of a fine, regular and stable pore structure and perhaps also to provide specific characteristics such as antistatic or hydrophobic properties. The actual density of the foam as applied to the material is also critical to a successful application, too high density leading to excessive wetting of the substrate and resulting in an unacceptable air permeability and too low density leading to inadequate penetration, poor mechanical bonding and hence the risk of delamination.

Acrylic foam-coated needlefelts produced in this manner are capable of continuous operation at temperatures up to approximately 120 °C. However, they are not normally resistant to hydrolytic conditions, these leading to collapse of the structure and hence, premature pressurisation. The latter notwithstanding, in view of the success of foam-coated structures operating in relatively 'safe conditions', the future will undoubtedly see more advanced products of this type, leading to structures that are both more efficient in particle capture and also capable of operation in more chemically and thermally challenging environments.

13.5   Solid–liquid separation

13.5.1   Introduction
Although there are several ways in which solid/liquid separation may be achieved (e.g. settling, floatation, hydrocyclones, evaporation, magnetic, electrostatic, gravity, centrifuge, vacuum, and pressure), the mechanisms that consume the largest volume of textile filter media and on which this section will concentrate, are those of pressure and vacuum.

In focusing on these mechanisms it will be appreciated that, apart from textile fabrics there are also many other forms of filter media. Some of the more common types and their relative collection efficiencies are listed in Table 13.2.

Table 13.2 Comparative particle collection efficiency for various media types

Media type	Approximate minimum particle size retained (μm)
Flat wedge screens	100
Woven wire	100
Sintered metal sheets	3
Ceramic elements	1
Porous plastic sheets	0.1
Yarn (cheese wound) cartridges	2
Compressed, fibre sheets	0.5
Filter aids (powders/fibres)	1
Membranes	0.1
Woven monofilaments	<10
Other woven fabrics	<5
Needlefelts	5
'Link' fabrics	200

13.5.2   Fabric design/selection considerations

There are many factors which confront the technologist when choosing or designing a fabric for a particular application. These may conveniently be grouped under the following general headings:

1   thermal and chemical conditions
2   filtration requirements
3   filtration equipment considerations and
4   cost.

13.5.2.1   Thermal and chemical conditions

Before the advent of synthetic materials, the only fibres available for industrial purposes were those of natural origin such as flax, wool and cotton. The last mentioned is still used in one or two applications even today; the tendency of this fibre to swell when wet facilitates the production of potentially highly efficient filter fabrics. On the other hand, the wide range of chemical conditions which prevail in industrial processes and, more significantly, the introduction of more chemically stable synthetic fibres, have effectively led to the demise of cotton in all but a few applications. But even synthetic fibres have their limitations. Polyamide – nylon 6.6 – arguably the first and most widely used true synthetic material is notoriously sensitive to strong acidic conditions and, conversely, polyester is similarly degraded by strong alkaline conditions.

By comparison, polypropylene is generally inert to both strong acids and alkalis and, primarily for these reasons, is the most widely used polymer in liquid filtration. On the down side, this material is limited by its relatively poor resistance to oxidising agents (Fig. 13.11) – nitric acid and heavy metal salts[11] fall into this category – and at temperatures above 90–95 °C stability problems may be encountered, especially if the filter fabric is also subjected to considerable stress. Resistance to organic solvents and mineral oils is also limited.

Some of the more common fibres (and their general properties) which are used in industrial filtration are listed in Table 13.3. Note that the maximum operating tem-

13.11   Scanning electron micrograph showing oxidation damaged polypropylene fibres.

Table 13.3 Fibres and their properties

Fibre type	Density (g cm^{-3})	Maximum operating temperature (°C)	Resistance to:		
			Acids	Alkalis	Oxidising agents
Polypropylene	0.91	95	E	E	P
Polyethylene	0.95	80	E	E	P
Polyester (PBT)	1.28	100	G	F	F
Polyester (PET)	1.38	100	G	P	F
Polyamide 6.6	1.14	110	P	VG	P
Polyamide 11	1.04	100	P	VG	P
Polyamide 12	1.02	100	P	VG	P
PVDC	1.70	75	E	VG	VG
PVDF	1.78	100	E	E	G
PTFE	2.10	120+	E	E	VG
PPS	1.37	120+	VG	E	F
PVC	1.37	75	E	E	F
PEEK	1.30	120+	G	G	F
Cotton	1.5	90	P	G	F

PBT = poly(butylene) terephthalate, PET = polyethylene terephthalate, PVDC = polyvinylidene chloride, PVDF = polyvinylidene fluoride, PVC = polyvinyl chloride.
E = excellent, VG = very good, G = good, F = fair, P = poor.

peratures shown in Table 13.3 are somewhat lower than in a previous table and reflect the influence of continuous exposure to aqueous conditions. However, in the absence of official published data these should only be used as a general comparative guide.

13.5.2.2   Filtration requirements

13.5.2.2.1   Filtrate clarity
The mechanisms by which particles are removed by fabric media may be identified as

1   Screening or straining: this is a simple mechanism in which particles are retained by the medium only as and when they are confronted with an aperture which is smaller than the particles themselves.
2   Depth filtration: in this mechanism the particles are captured through attachment to the fibres within the body of the filter medium, e.g. because of Van der Waal or electrostatic forces, even though they may be smaller than the apertures that are formed. This is particularly relevant to nonwoven media.
3   Cake filtration: this is undoubtedly the most widely encountered mechanism in industrial filtration and involves the accumulation of particles that 'bridge' together in a porous structure on the surface of the fabric. It follows from this that, once formed, the cake effectively becomes the filter medium with the fabric thereafter acting simply as a support. In cases where it is difficult for the particles to form a naturally porous cake, the use of a special precoat or body feed may be employed to assist in this task.

The various mechanisms outlined above are illustrated in Fig. 13.12. These have been the subject of many papers, the more significant of which have been eloquently summarised by Purchas.[12]

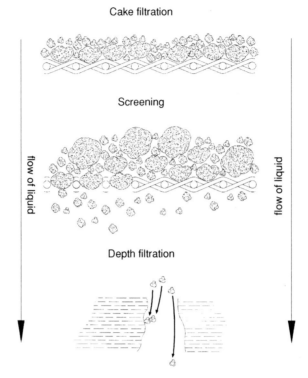

13.12   Solid/liquid filtration mechanisms.

Notwithstanding the fact that the filter fabric is used to effect the maximum separation of particles from liquids, absolute clarity is not always necessary. In certain gravity- or vacuum-assisted screening operations the filter fabric is simply designed to capture particles greater than a specific size and in other filtration systems a measure of solids in filtrate can be tolerated before cake filtration takes over and the necessary clarity is achieved. Recirculation of the slurry may also be possible in some applications until the same condition prevails. From this it will be appreciated that in some cases the solids are the more valuable component in the slurry whereas in others the process is concerned with clarification of the liquid, the solids thereafter being of little or no value.

13.5.2.2.2   Filtrate throughput
Although largely dictated by the equipment, restrictions to flow imposed by the unused filter fabric could pose serious pressure losses for a plant engineer and, in some applications, additional problems in forming a satisfactory filter cake.[13] In practice therefore, were it possible to tolerate the presence of a measure of solids in filtrate, some compromise is normally accommodated between throughput and clarity.

13.5.2.2.3   Low cake moisture content
As it is often necessary for filter cakes to be dried before moving to the next process and because drying by thermal means is energy intensive, it is important that as much liquid as possible is removed by mechanical means prior to the actual drying

operation. A similar situation applies in effluent treatment operations. If the processed effluent is transported to landfill sites, it is important to reduce the moisture content, first in order to meet local statutory regulations and second, because it is simply uneconomic to transport water. This also applies to mining ore concentrates that incur shipping costs as they too are transported, sometimes across oceans, for further processing.

As with filtrate throughput, whilst the choice of raw materials and the construction of the filter fabric will play a part in controlling cake moisture content, this aspect will again be governed largely by forces within the filter itself, for example membrane squeezing and cake drying by means of compressed air.

13.5.2.2.4   Resistance to blinding
Blinding is a term which is commonly applied to filter fabrics which, after normal cleaning operations, are so contaminated with embedded solids that the resistance to filtrate flow is unacceptably high. The blinding may be temporary or permanent; temporary in the sense that the cloth may be partly or completely rejuvenated by special laundering or in situ cleaning, for example with chemicals and/or high pressure hosing (Fig. 13.13 and Fig. 13.14), and permanent in the sense that the solids are irretrievably trapped within the body of the fabric, perhaps between fibres and filaments.

The compressible nature of the slurry, the shape and size of the particles and the possibility of crystal growth from the process itself are factors which will be addressed, particularly when selecting the fabric components. This will be discussed further in Section 6.

13.5.2.2.5   Good cake release
At the end of the filtration cycle the dewatered filter cake must be removed from the fabric in preparation for the next cycle. It is important that the cake is effectively discharged at this point since any delays will lead to extended filtration cycle times and therefore reduced process efficiency. This is particularly apt in filter press

13.13   Scanning electron micrograph showing used fabric before cleaning.

13.14   Scanning electron micrograph showing used fabric after cleaning.

operations where manual intervention may be necessary to remove sticky cakes. As a consequence, in addition to longer cycle times, the cost of the operator must also be considered. To some extent this topic may be linked to cake moisture content because, broadly speaking, wetter cakes will adhere more tenaciously to the cloth. This problem has been partly addressed by equipment manufacturers with the incorporation of high pressure wash jets and brush cleaning devices, and filter media producers also continue to pursue the development of fabrics that will facilitate the ultimate goal of perfect, unassisted cake release and hence the achievement of a fully automated operation.

13.5.2.2.6   Resistance to abrasive forces
The abrasive forces in this context arise from the shape and nature of the particles in the slurry. Materials with hard sharp quartz-like edges may lead to internal abrasion, the breakage of fibres and filaments and ultimately a weak point and possibly a pinhole in the fabric. Being the point of lowest resistance to flow, enlargement of such a hole then follows (Fig. 13.15) and eventually excessive solids in the filtrate ensues. The filter fabric should therefore be designed, as far as possible, to withstand the impact of such forces. This may be achieved by appropriate yarn and fabric construction, ideally manufactured from the toughest polymer consistent with the chemical conditions in the application.

13.5.2.2.7   Filter aids and body feed
In identifying filtration requirements, it is recognised that, in some cases, the filter fabric may require additional assistance, for example, by way of filter aids, body feeds or even filter papers. The use of filter aids, of which there are many types, is designed to precoat the fabric with a layer of powder, such as diatomaceous earth. This is carried out in order (i) to protect it from blinding, (ii) to assist in the collection of particularly fine particles, or (iii) to enable more efficient cake release. In special circumstances filter papers may also be used for similar reasons, especially where absolute clarity is essential. Body feeds, on the other hand, are added to the

13.15 Scanning electron micrograph showing mechanical damage from abrasive particles.

slurry to be filtered to enable the formation of a more porous cake than would otherwise be the case, thereby enhancing the rate of filtration flow.

13.5.2.3   Filtration equipment considerations

Having determined the preferred type of polymer and identified the various filtration requirements, of equal importance is the need to ensure that the fabric is capable of providing trouble-free performance on the equipment itself. In this respect it should provide (i) resistance to stretch, (ii) resistance to flex fatigue, and (iii) resistance to the abrasive forces that may be present on the filter itself.

13.5.2.3.1   Resistance to stretch

The propensity for stretch is evident on most types of filter and may arise as a result of cloth tensioning mechanisms, internal pressures or other forces such as the mass of the filter cake and the gravitational pull that it exerts on the fabric. In the case of filter belts, excessive force from the filter's tensioning mechanism may, in extreme cases, cause the belt to extend to the maximum length that the machine can handle. This in turn could lead to drive problems and hence the need to shorten the belt or even replace it. This will be discussed further in Section 13.9.1.

In similar manner, excessive stretch caused by the mass of filter cakes in filter press operations could result in the port holes in the fabric moving out of alignment with corresponding holes in the filter plate, thereby restricting flow of filtrate out of the press. Likewise, in other filtration systems such as pressure leaf filters, the same stretch could result in the formation of creases and ultimately mechanical damage to the fabric.

13.5.2.3.2   Resistance to flex fatigue

In addition to the overall dimensional changes to filter fabrics caused by sustained high tension, which will be aggravated at higher temperatures, corresponding changes to thread spacing may also be encountered, possibly leading to a more open structure and less efficient filtration. A similar type of thread disturbance has also

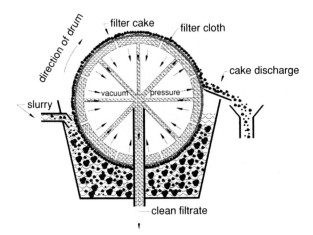

13.16   Rotary vacuum drum filter.

been observed on rotary vacuum drum and rotary vacuum disc filters, this time caused by flex fatigue.

In such systems, the filter fabric, which may be fabricated to envelop the filter element or simply caulked into its drainage surface, will operate in both vacuum and pressure modes (Fig. 13.16). During the initial phase, dewatering commences as the slurry is drawn by vacuum on to the surface of the immersed fabric and, as the equipment rotates, this continues until completion of approximately two-thirds of a revolution. At this point the vacuum is replaced with compressed air, which causes the fabric to expand. This in turn causes the dewatered filter cake to crack and fall from the fabric under force of gravity.

As previously mentioned, the constant flexing that the fabric receives in moving from vacuum to pressure can lead to a measure of fatigue and possible loss in filtration efficiency, a condition that is further aggravated by the presence of abrasive particles in the slurry.

13.5.2.3.3   Resistance to abrasive forces
Abrasive forces arising from the design and/or construction of the filter itself are found in many forms. In the case of filter belts, a potential reason for abrasion damage is due to the sustained, possibly excessive, pressure of the scraper blade which is engaged to ensure maximum cake removal at the discharge point. In addition to the general pattern of wear which is caused by this blade, local damage resulting from irregularities such as trapped particles, yarn knots or fabric creases will also be inflicted to the detriment of filtrate clarity. Once again the wear pattern will be intensified by the presence of abrasive slurries.

In addition to the scraper blade, abrasive damage and general cloth distortion can also be expected on belt filters from edge tracking or guiding mechanisms, especially if these are poorly maintained. In such cases, the damage can be alleviated by reinforcing the edge of the fabric for example by impregnation with resin treatments or perhaps with hot-melt polymers.

The surfaces against which the fabric will be expected to operate in filter press operations will also be influential in the choice of fabric construction. Whilst the introduction of advanced plastics has considerably reduced the damage that was

Table 13.4 Influence of yarn type on filtration properties

Order of merit	Maximum clarity	Maximum throughput	Low cake moisture content	Resistance to blinding	Ease of cake release	Abrasion resistance
1	Staple	Monofil	Monofil	Monofil	Monofil	Staple
2	Multifil	Multifil	Multifil	Multifil	Multifil	Multifil
3	Monofil	Staple	Staple	Staple	Staple	Monofil

previously inflicted on fabrics by rough, cast iron surfaces, there remain a large number of applications where the use of cast iron is still necessary.

In such circumstances, if the ideal fabric in purely filtration terms is incapable of withstanding abrasive forces of this nature, special fabrication techniques or the use of a backing cloth may be necessary. The latter, being of a more robust construction, will also be designed to facilitate the free flow of filtrate which passes through the primary filter cloth.

From the foregoing it will be evident that, from a technical point of view, the final choice of fabric may not be ideal in all respects. Therefore, as a general guide, Table 13.4 provides some direction about the types of yarn that are most suitable for a particular application.

13.5.2.4 Cost
In the majority of applications it can be shown that the cost of the filter fabric is a relatively small fraction of the total product cost. This notwithstanding, it is inevitable that in any application the filter fabric will, at some stage, have to be replaced. The onus of responsibility therefore rests with the cloth manufacturer to develop appropriate materials to provide maximum cost effective performance to ensure continuity of the operation for as long a period as possible.

13.6 Yarn types and fabric constructions

The technologist has basically four types of yarn to choose from when designing a filter fabric, namely monofilament, multifilament, fibrillated tape and staple-fibre yarns.

13.6.1 Monofilaments
Being manufactured from thermoplastic polymers, monofilament yarns are produced by extruding molten polymer chip through an orifice in a precision-engineered dye. On emergence from the extrusion point, the molten polymer is cooled, usually in a water bath, and drawn through a series of rollers to orientate the molecules and provide the monofilament with the desired stress–strain properties. The bath through which the monofil passes may also contain additives such as lubricants to assist in weaving, and antistatic agents to avoid shocks during high speed warping and also to alleviate the attraction of dust and 'fly'. The diameters of the monofilaments used range from 0.1 mm up to 1.0 mm, the smaller diameters being used mainly in applications involving filter presses, pressure leaf and candle filters, rotary vacuum disc and rotary vacuum drum filters, whereas the larger

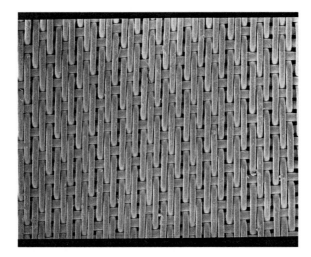

13.17   Scanning electron micrograph showing monofilament fabric, five-end satin weave.

diameters are used mainly in relatively coarse filtration applications involving heavy duty vacuum belt filters or multiroll filter presses. Although normally extruded in round cross-section, for special applications they may also be produced in flat or oval form.

The principal characteristics of monofilament fabrics (Fig. 13.17) may be summarised as (i) resistance to blinding, (ii) high filtrate throughput, and (iii) efficient cake release at the end of the filtration cycle. These characteristics are attributed to the smooth surface of the yarn and, in respect of cake release, weaving in a satin construction can further enhance this. On the down side, the apertures that are formed between adjacent threads and at the interlacing points in the weave (the only points where filtration can take place in monofilament fabrics) may prove to be too large for the separation of very fine particles such as dyestuffs and pigments, even though the warp threads may be quite densely packed. Fabrics containing over 110 threads per centimetre, each of 0.15 mm diameter, are not uncommon. Resistance to abrasive forces is also generally low with monofilament fabrics and some form of reinforcement may be necessary where this is likely to present difficulties.

For most filtration applications involving monofilaments, the majority of diameters used are in the range 0.15–0.35 mm, yielding fabric area densities between 180 and 450 g m^{-2}. Heavy-duty filter belt applications, on the other hand, usually employ diameters from 0.3–1.0 mm resulting in area densities from 500–1700 g m^{-2}.

13.6.2   Multifilaments

Although like monofilaments, multifilaments are also extruded through a precision-engineered dye, here the similarity ends; the dye on this occasion contains many more apertures of much smaller size. Moreover, the material to be extruded may again be in the form of a molten polymer or alternatively in the form of a solvent dope, the solvent evaporating on extrusion to be recovered for further use. Drawing of the threads is again carried out to orientate the molecules and develop the appropriate tenacity, this being typically of the order 5.5–6.5 centiN tex^{-1}.

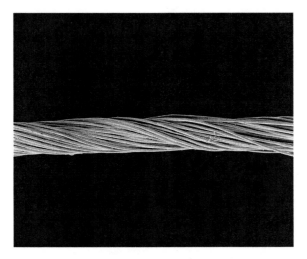

13.18   Scanning electron micrograph showing multifilament yarn.

In practice, manufacturers of multifilament yarns produce a number of standard linear densities that, for industrial filtration purposes, may range in fineness from 120 decitex to 2200 decitex, with individual filaments varying from 6–10 decitex. From this it can be shown that the diameter of such filaments will be of the order of 0.03 mm.

The filament assemblies may be held together by air intermingling, texturising or by twist (Fig. 13.18), the latter being preferred for warp purposes owing to the abrasive forces that will impact on the filaments – especially during weaving where the yarn is under considerable tension – and that may otherwise lead to filament breakage. Whilst weaving performance can be improved by suitable choice and addition of lubricant, determination of the optimum level of twist in the yarn will be critical to successful warping and weaving operations; too much twist presents handling difficulties in warping and too little results in yarn damage, inefficient weaving and substandard fabric.

Multifilament fabrics (Fig. 13.19) are characterised by their high strength and resistance to stretch, these properties being enhanced as the tenacity of the yarn increases. Multifilament yarns are also more flexible than monofilaments, a property which facilitates weaving of the tightest and most efficient of all woven fabrics. This is used to particular advantage when filtering fine particles ($<1 \mu m$) at very high filtration pressures, in some cases in excess of 100 bar.

In view of the tightness of fabric into which they are frequently woven, multifilament fabrics are generally inferior to monofilaments in respect of throughput and their resistance to blinding will be similarly reduced. This is due to the fact that, in addition to the filtration which takes place between adjacent threads, particles are also captured and possibly permanently trapped within the body of the threads themselves; this occurs despite the fact that the filaments may be tightly bound together by twist. The accumulation of such particles leads to swelling of the yarns, a reduction in pore size and a corresponding fall in filtrate throughput.

Fabric area densities in this category vary from as little as $100 \, g \, m^{-2}$ to around $1000 \, g \, m^{-2}$. The lighter fabrics, depending on the application, may require additional

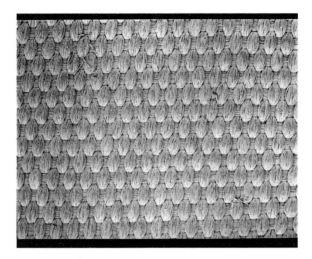

13.19   Scanning electron micrograph showing multifilament fabric.

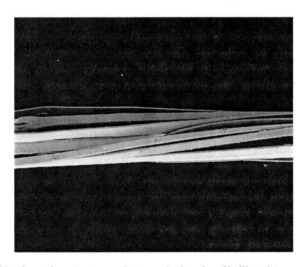

13.20   Scanning electron micrograph showing fibrillated tape yarn.

assistance in the form of a support or backing cloth. This is in order to avoid damage from abrasive filter plates or perhaps to avoid the fabric from being deformed into the indentations of the plate surface itself where it would impede escape of filtrate. Heavier fabrics, on the other hand, will be used, mainly unsupported, in more arduous, higher stress-related applications such as filter belts on vertical automatic filter presses.

13.6.3   Fibrillated tape ('split film') yarns
As the title suggests, these yarns are produced by taking a narrow width polypropylene film then splitting it into a number of components and binding these together by twist (Fig. 13.20). In this sense they may be seen as rather coarse multifilament

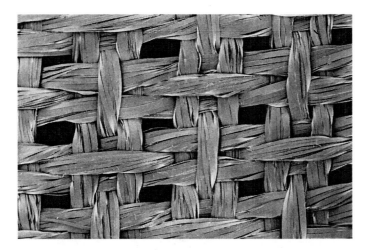

13.21   Scanning electron micrograph showing fibrillated tape fabric.

yarns. However, as they are considerably stiffer than the latter, they are not normally used in filter fabrics as such but rather in more open weave backing cloths. Therefore their function is to provide protection for the more delicate primary filter fabric from damaging surfaces, whilst at the same time permitting the free flow of filtrate from the filtration compartment. The use of a 'mock leno' weave (Fig. 13.21) is ideal in this respect. For the production of such fabrics, which are generally in the 400–600 g m^{-2} range, yarn linear densities of around 2200 decitex and higher are employed.

13.6.4   Staple-fibre yarns

The synthetic fibres which are used in these yarns are again produced by a continuous extrusion process, followed by conversion into a short staple length, which will facilitate processing on either rotor or cotton or woollen ring spinning systems. The cotton system tends to produce yarns that are rather lean in character whereas those from the woollen system are more bulky (Fig. 13.22). Similarly, for any given linear density, the cotton yarn tends to be stronger and less extensible than the woollen spun yarn, a feature that may be used to advantage when superior resistance to stretch is required. On the other hand, because of their bulk, higher flow rates may be expected in fabrics woven with woollen spun yarns (Fig. 13.23) and resistance to blinding from solid, non-compressible particles (as distinct from compressible slimes) will also be superior. Although difficult to substantiate, it is believed that this feature is related to the ease by which particles may enter and exit the bulky woollen spun structure.

In addition to particle collection efficiency, fabrics produced from woollen spun staple-fibre yarns are also characterised by their resistance to abrasive forces, such as may be found on rough, possibly chemically corroded cast iron filter plates. For this, and filtration purposes in general, the yarns are usually spun with 3.3 decitex fibres in relatively coarse linear densities, typically from 130–250 tex. Fabrics in this category are normally woven in area densities ranging from 350–800 g m^{-2}, the lighter and intermediate fabrics generally being used in pressure leaf and rotary vacuum drum filters and the heavier fabrics in filter presses.

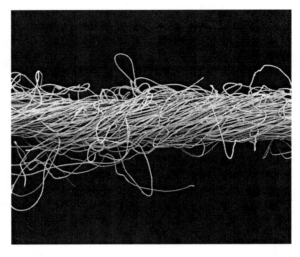

13.22   Scanning electron micrograph showing woollen ring-spun yarn.

13.23   Scanning electron micrograph showing woollen ring-spun fabric.

When woven in plain weave, maximum efficiency coupled with dimensional stability are usually the key operational requirements, whereas if woven in a twill weave, greater bulk and hence greater resistance to abrasive or compressive forces are usually the dominant factors.

13.6.5   Yarn combinations
By producing fabrics with different components in warp and weft it may be possible to create a structure that utilises the best features of each. The most popular combinations in this respect comprise multifilament warp and staple-fibre weft yarns (Fig. 13.24) and monofilament warp and multifilament weft yarns. In both cases the ratio of warp to weft threads is at least 2:1 and usually considerably higher. This facilitates the production of fabrics with a smooth warp-faced surface for efficient

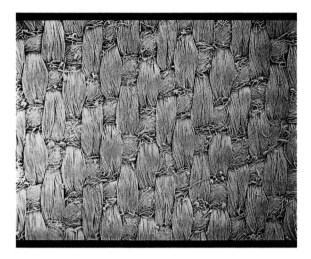

13.24 Scanning electron micrograph showing fabric woven with multifilament (warp) and woollen ring-spun (weft) yarns.

cake release and also higher warp tensile properties for greater resistance to stretch from the mass of heavy cakes. In the case of the multifilament and staple combination, the inclusion of a staple-fibre weft yarn provides scope for improved resistance to mechanical damage whilst maintaining a high particle collection efficiency and an acceptable throughput. Similarly, the inclusion of a multifilament weft yarn in a monofilament and multifilament fabric will lead to an improvement in filtration efficiency, especially if it is suitably texturised.

13.7   Fabric constructions and properties

13.7.1   Plain weave

This is the most basic weave of all woven structures that provides the framework for the tightest and most rigid of all single layer filter fabrics, see Fig. 13.19. Because of the sinusoidal path that the yarns follow, this weave is particularly suitable for flexible yarns of the multifilament and short staple-fibre types. The weave is also ideally suited to applications where thread displacement, due for example to high internal pressures, may otherwise be experienced.

13.7.2   Twill weaves

Usually produced in simple 2/2 or 2/1 style, twill weaves enable more weft threads per unit length to be crammed into the fabric than the preceding design (plain weave), as shown in Fig. 13.23. As a consequence, this facilitates the production of fabrics of higher area density and hence greater bulk, features which are particularly suited to woollen spun yarns. Twill weave fabrics are also marginally more flexible than plain weave fabrics, which may be advantageous when fabricating cloths of complex make up or indeed when fitting the cloths on the filter itself, for example, caulking into grooves.

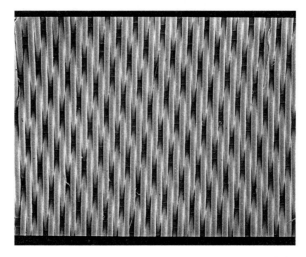

13.25   Scanning electron micrograph showing satin fabric (eight-end design).

13.7.3   Satin weaves

Both regular and irregular satin weaves are employed. The irregular weaves, such as the four-shaft construction, are frequently found in more densely sett high efficiency fabrics, often with two warp threads being woven as one (Fig. 13.24). Although maximum separation may be the principal requirement here, the combination of weave pattern and a double multifilament thread arrangement also creates a smooth surface for superior cake release. By comparison, the regular satin weaves such as the eight-shaft (Fig. 13.25) and 16-shaft constructions are usually employed where efficient cake release and throughput are of greater importance. From this it will be appreciated that the weaves with the longer floats are normally used in conjunction with monofilament yarns.

13.7.4   Duplex and semiduplex weaves

These weaves are frequently, though not exclusively, found in belt filters, either of the vacuum, continuous multiroll press, or of the vertical automatic pressure type. Owing to the interlacing pattern of the threads, it is possible to create fabrics with a measure of a solidity and stability that are ideally suited to filters of the types identified. On the debit side, the cost of weaving such high density fabrics tends to preclude their use in all but a limited number of niche applications.

13.7.5   Link fabrics

As shown in Fig. 13.26, link fabrics are produced by a novel technique in which polyester monofilaments are wound into spiral form then meshed with similar monofilaments, which are spiral wound in the opposite direction. The spirals are subsequently held together by a straight monofilament. By virtue of this form of construction it is possible to produce endless filter belts without the need for special joining techniques such as 'clipper' seams, which are often the weakest point in a filter belt.

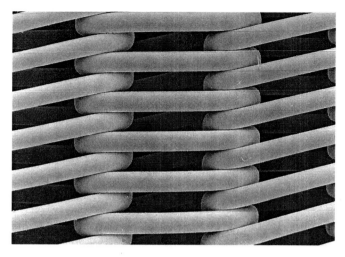

13.26 Scanning electron micrograph showing link construction from the surface (top view).

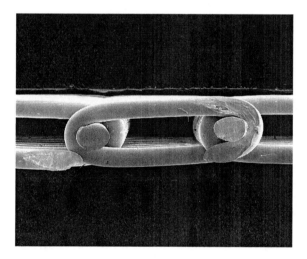

13.27 Scanning electron micrograph showing cross-section of link construction.

Because they are made from relatively coarse monofilaments of around 0.7 mm diameter, link constructions are generally of an open nature and are designed for the filtration of chemically flocculated sludges, these being relatively easy to separate but requiring efficient drainage. From the cross-sectional view (Fig. 13.27) it will also be seen that the monofilaments assume a 'race track' configuration relative to the direction of belt movement. This ensures that the wear pattern on the monofilaments is evenly distributed, that is, as distinct from certain woven fabrics where the warp threads can suffer local abrasion damage at the crown of the interlacing points between warp and weft. In the event that a more efficient link construction is required, additional monofilaments (or other threads) may be inserted as shown in Fig. 13.28 and Fig. 13.29.

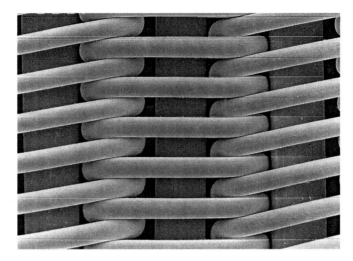

13.28   Scanning electron micrograph showing link construction with filler threads from the surface (top view).

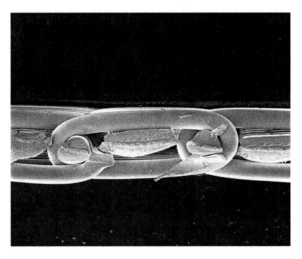

13.29   Scanning electron micrograph showing cross-section of link construction with filler threads.

These constructions, which are produced mainly with polyester monofilaments, are ideally suited to multiroll continuous pressure filters which combine both gravity and pressure filtration mechanisms. Such filters are used extensively in coal reclamation and effluent treatment operations.

13.7.6   Needlefelts
The construction of needlefelts has been described in general terms in the previous Section 13.2 on dust collection. For further reading on this subject the monograph by Purdy[14] provides an ideal introduction.

Although widely used in dust collection, needlefelts have found only limited use in liquid filtration because their thickness and density render them prone to blinding in many applications. One area where they have found some success, however, has been in the filtration of metal ore concentrates such as copper on horizontal vacuum belt filters. These applications tend to be very aggressive on the filter fabric, and hence a suitably designed and finished needlefelt is often more cost effective than a considerably more expensive woven fabric, especially if required in lengths of around 80 m and widths up to 6 m. The solids which are captured in such applications quickly form a cake on the surface and, should some penetration occur, as with woollen spun yarns, the bulky nature of the material provides scope for the particles to escape. For such arduous applications, needlefelts are generally in the area density range 800–1000 g m^{-2}.

13.8   Production equipment

13.8.1   Warping equipment

From the preceding information, it will be appreciated that in the production of woven filter fabrics, which are predominant in solid–liquid separation processes, there is a demand for a wide range of qualities. Because of this, coupled with the knowledge that the fabrics will be required in a variety of lengths and widths, the flexibility provided by section warping makes this the preferred warp preparation technique.

13.8.2   Weaving equipment

In the majority of cases, filter fabrics are woven on either flexible or rigid rapier looms, which require a smaller shed for weft insertion than more traditional shuttle looms. In this respect they generally inflict less damage on the warp sheet. Even so, because filter fabrics are frequently quite densely sett, looms with beat-up forces of the order of 15 kN m^{-1} in reed widths up to and in excess of 4 m may be necessary to achieve the required pick spacing. High weft thread densities also demand high warp tensions and these in turn impose substantial stresses on let-off, shedding and take-up mechanisms. As a consequence, only weaving machines that are adequately reinforced in these areas will be suitable for long term performance.

By comparison, heavy duty belt filters may require fabrics up to 8 m in width. For these purposes the warps usually consist of a series of precision wound 'minibeams' or spools which, after preparation, are mounted on a common let-off shaft on the weaving machine. The latter are, of necessity, extremely robust in construction, being similar in style to equipment normally employed in paper-machine fabric manufacture.

Although weft insertion on these heavy duty machines may also be by rapier, for the wider looms insertion by conventional shuttle or projectile shuttle is more common. Furthermore, with weft insertion rates approximately 66% lower than the narrower, more conventional weaving machines, productivity is not particularly high.

13.9   Finishing treatments

Finishing treatments for fabrics employed in liquid filtration applications are designed for three basic reasons, namely (i) to ensure dimensional stability during use, (ii) to modify the surface for more efficient cake release, and (iii) to regulate the permeability of the fabric for more efficient particle collection.

13.9.1   Dimensional stability treatments

As discussed in a previous section, in the production of woven filter fabrics, both fibres and yarns are subjected to considerable stress. Although in the majority of cases the applied forces, being within the material's elastic limit, are unlikely to result in permanent deformation, they will produce a degree of stretch which, with time, will recover. The application of heat will accelerate this recovery process and, similarly, the application of heat may also induce a measure of shrinkage, which is inherent in the fibre, or filament as received. This shrinkage, be it inherent in the fibre or due to a stress recovery phenomenon, may cause several problems during use. Examples include difficulties in actually fitting the cloth on to the filter, misalignment of holes in cloth and filter and, in extreme cases, partial by-pass of the filter cloth by unfiltered slurry. These difficulties will be further aggravated if the fabric is also subjected to a hot tumble-washing programme, which may be necessary in order to rejuvenate the material following temporary blinding.

As in the production of fabrics for dust collection applications, heat is again instrumental in inducing the necessary fabric stability, which, on this occasion, may be achieved through hot aqueous treatment, heat setting or a combination of both. In the case of aqueous treatments these may also include surfactants to remove unwanted fibre and yarn processing aids. Once again media manufacturers will be aware of the machine speeds and temperatures that will be necessary in these processes to achieve the maximum effect.

In overcoming the instability that may arise from fibre or yarn shrinkage, it is also important in liquid filtration to ensure that the material is equally stable to forces that may be applied either by the equipment itself, or indeed by the mass of the filter cake. In most cases this is achieved by selection of yarns of appropriate tenacity but, in the case of filter fabrics designed for use on belt filters, additional assistance is necessary. For such applications the fabrics are subjected to a thermal stretching operation that, in addition to increasing the fabric's initial modulus, also eradicates any tension variations that will have been introduced during yarn preparation or weaving and that may otherwise cause lateral tracking problems on the filter.

13.9.2   Surface modifications

Surface modifications include singeing, which has already been discussed in Section 13.4.2.

13.9.2.1   Special surface treatments

Although the surface of a fabric can be significantly enhanced by physical/thermal means such as singeing and calendering, the development of chemical coatings such as Madison Filter (formerly Scapa Filtration)'s Primapor (Fig. 13.30) has led to the

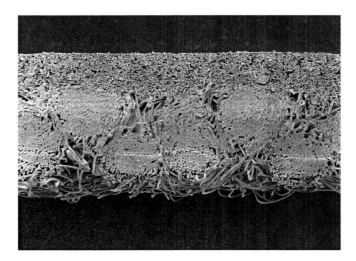

13.30 Scanning electron micrograph showing cross-section of Madison Filter's 'Primapor' microporous fabric.

production of still more efficient filter media. The use of stretched PTFE membranes in liquid filtration has also been reported although, it is suspected, the rather delicate nature of this material will restrict its use to niche applications.

As in dust collection, where surface coatings have been available for many years, the treatments are designed to present a microporous structure to the slurry which effectively restricts the penetration of particles to all but a few micrometres in depth. Consequently, a filter cake quickly forms on the surface of the coating and, by restricting the particles to the surface, the same cake can be easily discharged at the end of the filtration cycle. Unlike the coatings in dust collection however, the microporous structure in liquid applications has to withstand much higher pressures. Failure to do so will result in structural collapse and premature pressurisation of the filter.

It is predicted that the future will see much more development work in this area, targeted specifically at more efficient and more durable coatings both in terms of structural stability and also resistance to chemical and abrasive agents.

13.9.3 Permeability regulation

13.9.3.1 Calendering (see also Section 13.4.4)
The calendering operation is able both to modify the surface and also to regulate the fabric's permeability by means of heat and pressure. A third variable, namely the speed at which the fabric is processed, will also have a controlling influence on the effectiveness of the operation.

In the case of needlefelts, a reduction in pore size is achieved by compressing the fibres into a more dense structure (loads up to $300\,decaN\,m^{-1}$ may be necessary) and, by selection of the appropriate conditions, a more durable (Fig. 13.31) surface can also be obtained through partial fusion of the surface fibres. With woven fabrics, on the other hand, some deformation of the yarns may be necessary to achieve the optimum filtration properties. This is particularly graphic in the case of fabrics woven from monofilament yarns as shown in Fig. 13.32 and Fig. 13.33.

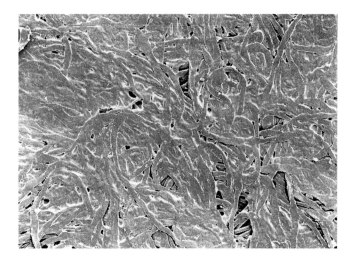

13.31 Scanning electron micrograph showing needlefelt with fused fibre surface.

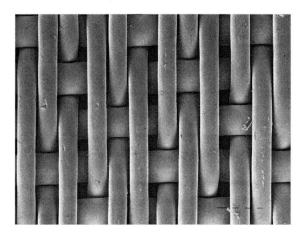

13.32 Scanning electron micrograph showing monofilament fabric before calendering.

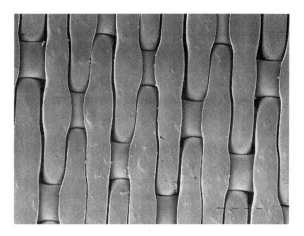

13.33 Scanning electron micrograph showing monofilament fabric after calendering.

13.9.3.2   Other techniques

Using the inherent shrinkage characteristics of fibres and yarns, it is possible to effect a reduction in permeability simply by application of heat alone. This has the effect of pulling the threads closer together, thus reducing the fabric's aperture/pore size and resulting in a tighter, more efficient fabric.

Detailed information from yarn and fibre suppliers on the reaction of their products to thermal conditions will be essential, (i) in order to obtain the desired effect, and (ii) in order to ensure the necessary finishing controls.

13.10   Fabric test procedures

13.10.1   General quality control tests

These are carried out in normal textile laboratories in order (i) to ensure that the materials under test have been manufactured in accordance with design specification, and (ii) to monitor any short, medium or long term trends. Such tests are concerned primarily with area density, fabric sett, yarn types and linear densities, fabric structure, air permeability, thickness and density (principally needlefelts), tensile properties and fabric stability.

The resistance to stretch is of particular interest with respect to tensile properties. From previous sections it will be appreciated that although filter fabrics are rarely subjected to forces that will result in tensile failure, they may suffer a degree of stretch that could have serious consequences. Resistance to stretch at relatively low loads (e.g. less than 100N per 5cm) is therefore of particular importance from a control point of view. Furthermore, since this phenomenon is temperature related, the ability to carry out such measurements at elevated temperatures is also a useful asset.

Shrinkage tests take one of several forms depending on whether the application is wet or dry. For dust collection applications, measurement of the fabric's free shrinkage in an air circulating oven is the standard practice, the time of exposure and temperature varying according to the specific test procedure.

By comparison, because it is not uncommon in liquid filtration applications for cloths to be removed from the filter and subjected to a laundering operation, a laboratory test programme has to be devised that will reproduce the mechanically induced shrinkage generated by an industrial tumble washing machine. Such action, by virtue of the mass of cloths involved, is inevitably more severe than a domestic machine.

Although test procedures exist for measuring the liquid permeability of fabrics (e.g. by measurement of the time for a specified volume of water to pass through the fabric), either under gravity (falling column) or at a specified vacuum, it is normally more convenient to quantify the permeability of fabrics by air techniques. A typical procedure is described in DIN 53887.[10]

Whichever technique is used, it is important to remember that, although permeability results are a useful pointer in characterising the efficiency of a fabric, they must not be viewed in isolation but rather in conjunction with other fabric parameters such as thickness (needlefelts), area density and threads per unit area (fabric sett).

13.10.2 Performance-related tests

Whilst the above procedures are ideal for routine quality control purposes, they provide very little guidance about the aperture size and hence the actual efficiency of the filter fabric in dealing with particles of known size. In the case of large mesh monofilament screening fabrics, it is possible to calculate the aperture size simply by means of thread diameters and thread spacing. With much tighter constructions on the other hand an alternative approach has to be taken.

Measurement of 'equivalent pore size' by a bubble point procedure[15] is perhaps the most well known and involves immersing the fabric in a suitable wetting fluid and then measuring the air pressure that is necessary to create a bubble on the surface. The pore size can be calculated from the relationship $r = 2T \times 10^5/\sigma Pg$, where r is the pore radius (μm), T is the surface tension of the fluid (mN m^{-1}), σ is the density of water at the temperature of test (g cm^{-3}), P is the bubble pressure (mm H$_2$O) and $g = 981$ cm s^{-2}.

An arguably more relevant approach to assessment of filtration efficiency is proposed by Barlow[16] in which a dilute suspension of flyash in glycerol is pumped through the fabric. By measuring the particle size distribution with the aid of a Coulter counter before and after passage through the fabric – and before the formation of a filter cake – a measure of its filtration efficiency can be obtained.

Information obtained from the above procedures, in combination with previous experience, will be of considerable value when selecting the appropriate fabric for a particular application.

With a small sample of slurry and a laboratory pressure vacuum leaf or piston press,[17] still further refinements can be made. (This of course presumes that the nature of the slurry will be a representative sample whose character will not change irreversibly on leaving the manufacturing plant.) From such tests speedy comparisons can be made on parameters such as throughput, filtrate clarity, cake moisture content and a subjective assessment of cake release. Note that medium/long term blinding will not normally be apparent from such procedures.

In dust collection, the number of laboratory test procedures are legion, most of these being designed to support a particular theory. A more practical procedure on the other hand is described by Barlow[16] and involves the construction of a pilot dust collector. The equipment houses four filter sleeves and is capable of operation in both reverse air and pulse cleaning modes with sufficient flexibility to change the rate of dust feed, velocity, cleaning frequency and pulse pressure. By virtue of its design construction the equipment is also capable of continuous operation for several days, if not weeks, which facilitates a better 'feel' for whether the media under test can cope with a particular situation.

A more convenient means of media comparison, whilst still retaining some of the above practical elements, is described by Anand, Lawton, Barlow and Hardman.[18] In this procedure a single sleeve of smaller size is mounted on a 'plastic cage' and located in a clear perspex cylindrical enclosure. A quantity of dust is added and this is then transformed into a cloud by means of compressed air injected at the base of the unit. The dust-laden air is drawn onto the surface of the sleeve and the subsequent pressure differential monitored by computer. Any particle emissions through the fabric are captured by an ultrafilter and measured gravimetrically. Once again the gas velocity, cleaning frequency, pulse pressure, and pulse duration can be adjusted and, by virtue of the clear Perspex cylinder, the effects of these can be visually observed.

Although both the above procedures provide useful comparative data, such information should only be used as a guide for media selection and not as a means of specifying filtration parameters or indeed actual filtration performance. Thermal and chemical conditions and the condition of the cages inside the filter, none of which can be simply reproduced in the laboratory, may well eclipse laboratory predictions.

References

1. C N DAVIES, *Air Filtration*, Academic Press, London, 1973.
2. E ROTHWELL, 'Fabric dust filtration', *The Chem. Engineer*, 1975 March 138.
3. N PLAKS, 'Fabric filtration with integral particle charging and collection in a combined electric and flow field', *J. Electrostatics*, 1988 **20** 247.
4. K MORDEN, 'Dust explosion hazards in white sugar handling systems', *Int. Sugar J.* 1994, **96**, issue number 142, Feb, 48.
5. BS6524: 1984 British Standard Method for Determination of the Surface Resistivity of a textile fabric, 1984.
6. A C HANDERMANN, *Basofil Filter Media – Efficiency Studies and an Asphalt Plant Baghouse Field Trial*, BASF Corporation, Fiber Products Division, Enka, N. Carolina, 1995.
7. J SIEVERT and F LOFFLER, 'Actions to which the filter Medium is subjected in Reverse Jet Bag Filters', *Zement – Kalk – Gips*, **3** 1986 no. 3 71–72.
8. A T PURDY, 'The structural mechanics of needlefelt filter media', *Second World Filtration Congress*, The Filtration Society, Uplands Press, London, 1979, 117–132.
9. S C ANAND and P J LAWTON, 'The development of knitted structures for filtration', *J. Textile Inst.*, 1991 **82**(3) 297.
10. DIN 53887 Determination of Air Permeability of Textile Fabrics, 1986.
11. S J HACZYCKI, 'The behaviour of polypropylene fibres in aggressive environments', *PhD Thesis*, University of Bradford, 1989.
12. D B PURCHAS, 'Practical applications of theory', *Solid/Liquid Separation Technology*, Uplands Press, Croydon, 1981, Chapter 10, 595–693.
13. A RUSHTON and P V R GRIFFITHS, in *Filtration, Principles and Practices Part I* ed C. Orr, Marcel Dekker, New York, 1977, 260.
14. A T PURDY, *Needle-Punching*, Monograph No. 3, The Textile Institute, Manchester, 1980.
15. BS3321: 1969, The Equivalent Pore Size of Fabrics (Bubble Pressure Test), BS Handbook 11, 1974.
16. G BARLOW, *Fabric Filter Medium Selection for Optimum Results*, Filtech Conference, Uplands Press, Croydon, 1981, 345.
17. D B PURCHAS (ed), *Solid/Liquid Separation Equipment Scale Up*, Uplands Press, Croydon, 1977.
18. S C ANAND, P J LAWTON, G BARLOW and E HARDMAN, *Application of Knitted Structures in Dust Filtration*, Filtration Society Meeting, Manchester, 15th May 1990.

14

Textiles in civil engineering. Part 1 – geotextiles

Peter R Rankilor

9 Blairgowrie Drive, West Tytherington, Macclesfield, Cheshire SK10 2UJ, UK

14.1 Introduction to geotextiles

Although skins, brushwood and straw–mud composites have been used to improve soft ground for many thousands of years, it is not realistic to refer to these as 'geotextiles'. The important factor that separates them from modern geotextiles is that they cannot be made with specific and consistent properties. When modern polymers were developed in the mid 20th century, it became possible to create textiles with designed forecastable performance and to produce them in large quantities with statistically consistent and repeatable properties. Once this was achieved, the science of geotextiles became possible. In essence, the difference between geotextiles and skins is their numerical or engineering capability.

In the early 1960s and 1970s, some pioneering engineers wondered if textiles could be used to control soils under difficult conditions. For example, very wet soils need draining and textiles were used to line drains, to prevent mud and silt from clogging up the drains. Similarly, engineers tried to use textiles beneath small access roads constructed over very soft wet soils. It was found that these textiles helped to increase the life and performance of roads. Also, early work was being undertaken in the laying of textiles on the coast to prevent erosion by wave action. A number of limited but historical publications were published.[1–2]

However, in those early days, it was not known exactly how these textiles performed their functions. How did they actually filter? How did a relatively weak textile apparently support heavy vehicles and improve road performance? This was a dangerous period for engineers, because it was quite possible that the experience-based employment of geotextiles could lead to their use in unsuitable constructions. It was likely that before long, an engineer would use textiles in a structure that was too large, too demanding or too stressful for the product; a significant failure could result. It was therefore vital that study and research should be undertaken to provide theories and preliminary design equations against which to test site results.

In 1977 Rankilor produced what was probably the first 'design' manual for a commercial product[3] and this was followed by a textbook written in 1980[4] which built on the extensive experience that had been amassed by this time. As is so typical of scientific development, many engineers were soon working worldwide on the development of geotextiles. Another significant textbook by Koerner and Welsh was published in 1980,[5] showing that work in the USA was at an advanced stage. The French, Japanese, Germans, Dutch and workers in other countries were equally active in the utilisation of textiles in civil engineering earthworks at that time.

During the last 20 years of the 20th century, the use of geotextiles spread geographically worldwide and in area terms their use increased almost exponentially. It is expected that their use will continue to increase into the 21st century unabated.

Once textiles were recognised as being numerically capable materials, engineers developed new types of textile and new composites to solve more difficult problems. Woven and nonwoven textiles were joined into composite products; nonwoven products were combined with plastic cores to form fin drains, and woven products were developed from stronger polymers such as polyester to extend the mechanical range of textiles and their uses in soil reinforcement. It is probable that the Dutch were the first to weave heavy steel wires into polypropylene textiles for incorporation into their major coastal land reclamation schemes. During the period 1984–85, Raz and Rankilor explored and developed the design and use of warp knitted fabrics for civil engineering ground uses.[6,7] Rankilor coined the term 'DSF' geotextiles – directionally structured fabric geotextiles; Raz specified the 'DOS' group within the main DSF range – directionally orientated structures.

Within a few years, more than six major manufacturers were producing warp knitted textiles for civil engineering earthworks. Currently, many are commercially available.

It can be considered that the 'first generation' of geotextiles were textiles that were being manufactured for other purposes (such as carpet or industrial sackings) but which were diverted and used for geotechnical purposes. The second generation of geotextiles became generated by manufacturers choosing specific textiles suitable for geotechnical purposes, but using conventional manufacturing techniques. The third generation textiles were actually designed and developed anew specifically for the purpose of geotechnical application – in particular DSF, DOS and composite products.

The development of geotextiles has always been an 'industry-led' science. Academic institutions have almost universally lagged well behind industry, with industrial designers acquiring experience at an ever-increasing rate. Currently, for example, in the USA, there are only a small number of universities teaching geotextile design as part of their main core programmes. In the UK, there are even fewer. Nonetheless, research publications from British academic institutions are of a high quality, showing specialised interests such as weathering,[8,9] filtration,[10] soil reinforcement[11,12] and computer applications.[13]

The establishment of the International Geotextile Society in 1978 led to a coordinated and coherent approach to international development of geotextile design and utilisation. The Society's four-yearly international symposium has been emulated by many other groups and countries, such that the rate of publication of papers is now very high, providing widespread exposure of developments to all interested engineers.

There are some interesting commercial aspects related to geotextiles that are specific to the industry. For example, availability must be considered in the light of the extreme size range of operations into which geotextiles are incorporated. About one-third of all geotextiles are used in small batches of three rolls or less, but a significantly large proportion are used in very large projects incorporating hundreds of thousands of square metres. Supply must therefore be available on call for one or two rolls from local stock and, simultaneously, must be available through agents or directly from the manufacturer in large quantities over a short space of time.

Delivery period is particularly onerous for textile suppliers. The majority of delivery requirements are of a standard industrial nature, but geotextile suppliers have to be able to supply large quantities within a short period for major engineering undertakings. This aspect has deterred many potential geotextile manufacturers from entering the field.

Price is also of interest, in that the cost of the polymer and manufacture can be irrelevant in certain cases. In civil engineering, a textile can be used to 'replace' a more conventional material such as sand in a granular filter. In this case, the cost of the product would be relevant and would be compared to the cost of the sand. Taken into account would be other marginal factors such as time saved in the laying of the textile as opposed to that of laying the sand. If the balance was in favour of the textile, then it might be adopted. However, in different circumstances the same textile might be worth considerably more as a sand replacement, for example, if sand were required to be placed under rapidly moving water or waves. In this case, if the textile could be placed where sand could not, then the comparison is not simply a matter of cost, but of the textile actually allowing construction to take place when the sand could not. Considerably more could be charged for a textile in these circumstances than in the former. Therefore, the cost of textiles is enhanced where they are sold and used as part of a 'system'.

Quality has to be controlled in much the same way as with other textiles – quality variation within the fabric and quality variations over time – but the implications of failure can be so much greater than with normal industrial products. If a major dam were to fail because the textile filter clogged, it would not just be a matter of apologising and replacing the filter with new product! The manufacturer does not take responsibility for the use of fabrics in the ground, but the design consultant does. He will not therefore be willing to certify the use of a textile if he is not satisfied that quality can be maintained at all levels of the process.

It is certainly necessary for modern-day geotextiles to be produced by manufacturers having ISO 9000 certification and it is ideal for this to include 9001, 9002 and 9003. The full range of these certifications covers the manufacturer's operation from raw material supplier through manufacture to storage and delivery.

14.2   Geosynthetics

In the field of civil engineering, membranes used in contact with, or within the soil, are known generically as 'geosynthetics'. This term encompasses permeable textiles, plastic grids, continuous fibres, staple fibres and impermeable membranes. Textiles were the first products in the field, extending gradually to include additional products, but have remained by far the most important of the range. Grids are formed from sheets of plastic that are punched and stretched; meshes are formed from melted extruded polymer; neither can be categorised as textiles. Geomembranes are continuous sheets of impermeable plastic and are not textiles. The more difficult areas of the geosynthetic range to categorise are those where discrete staple fibres or continuous filament fibres are mixed directly with soil. These are polymer textile fibres and therefore, as such, are included within the definition of geotextiles.

14.2.1 Geotextile types

Geotextiles basically fall into five categories – woven, heat-bonded nonwoven, needlepunched nonwoven, knitted and by fibre/soil mixing.

Woven fabrics are made on looms which impart a regular rectilinear construction to them, but which can vary in terms of the component fibres and the weave construction. They have a surprisingly wide range of applications and they are used in lighter weight form as soil separators, filters and erosion control textiles. In heavy weights, they are used for soil reinforcement in steep embankments and vertical soil walls; the heavier weight products also tend to be used for the support of embankments built over soft soils. The beneficial property of the woven structure in terms of reinforcement, is that stress can be absorbed by the warp and weft yarns and hence by fibres, without much mechanical elongation. This gives them a relatively high modulus or stiffness.

Heat-bonded nonwoven textiles are generally made from continuous filament fine fibres that have been laid randomly onto a moving belt and passed between heated roller systems. These fabrics acquire their coherence and strength from the partial melting of fibres between the hot rollers, resulting in the formation of a relatively thin sheet of textile.

Needlepunched nonwoven fabrics are made from blended webs of continuous or staple filaments that are passed through banks of multiple reciprocating barbed needles. The fabrics derive mechanical coherence from the entangling of fibres caused by the barbs on the reciprocating needles; these fabrics thus resemble wool felts.

In the case of needlepunched textiles, considerable thicknesses (up to more than 10 mm) and weights greater than $2000\,g\,m^{-2}$ can be achieved, whereas the heat-bonding process is limited in its efficacy as thickness increases. If sufficient heat is applied to melt the internal fibres of a thick fabric adequately, then the outer fibres will tend to be overheated and overmelted. Conversely, if appropriate heat is applied to the external fibres, then insufficient heat may be applied to the centre of the sheet, resulting in inadequate bonding and potential delamination in use.

Knitted fabrics, as used in the field of geotextiles, are restricted to warp-knitted textiles, generally specially produced for the purpose. Warp-knitting machines can produce fine filter fabrics, medium meshes and large diameter soil reinforcing grids. However, it is generally found that only the high strength end of the product range is cost effective, usually for soil reinforcement and embankment support functions.

14.2.2 The main geotextile fibre-forming polymers

The two most common fibre polymers used for the manufacture of geotextiles are polypropylene and polyethylene, but polyester is almost inevitably used when high strengths are required. There are other higher strength polymers available on the market, but geotextiles have to be produced in large quantities (some polymers are not available in large volumes) and economically (specialist polymers tend to be very expensive). On the overall balance of cost against performance, polyester is the present day optimum, while polypropylene and polyethylene vie for being the most chemically resistant.

Care must be taken when considering the properties of geotextile polymers that consideration is restricted to polymers as they are actually produced and used for geotextile manufacturing purposes; they are not in their chemically pure form. For example, raw polyethylene in its colourless translucent form is quite susceptible to light degradation. However, it is not used in this form in geotextiles, but usually contains carbon black as an ultraviolet (UV) light stabiliser. In this black form, it is arguably the most light-resistant polymer.

Also, it must be recognised that real in situ field testing of geotextile polymers is limited. Publications and authorities may quote accelerated laboratory results for xenon UV exposure, high temperature degradation testing, and similar, but these cannot take account of additional degradation factors such as biological attack, or synergistic reactions that may take place. The difficulty, therefore, arises that accelerated laboratory testing may well be pessimistic in one regard and optimistic in the other when used for ranking purposes.

Although polyamide is a common fibre-forming and textile material, nonetheless, it is rarely used in geotextiles, where its cost and overall performance render it inferior to polyester. Some woven materials, for example, have used polyamide in the weft direction, more as a 'fill', where its properties are not critical. Its main asset is its resistance to abrasion, but it displays softening when exposed to water, which appears to have made it unpopular for geosynthetic use. Polyvinylidene chloride fibre is used in Japan and in one or two products in the United States, but not in Europe.

14.3   Essential properties of geotextiles

The three main properties which are required and specified for a geotextile are its mechanical responses, filtration ability and chemical resistance. These are the properties that produce the required working effect. They are all developed from the combination of the physical form of the polymer fibres, their textile construction and the polymer chemical characteristics. For example, the mechanical response of a geotextile will depend upon the orientation and regularity of the fibres as well as the type of polymer from which it is made. Also, the chemical resistance of a geotextile will depend upon the size of the individual component fibres in the fabric, as well as their chemical composition – fine fibres with a large specific surface area are subject to more rapid chemical attack than coarse fibres of the same polymer.

Mechanical responses include the ability of a textile to perform work in a stressed environment and its ability to resist damage in an arduous environment. Usually the stressed environment is known in advance and the textile is selected on the basis of numerical criteria to cope with the expected imposed stresses and its ability to absorb those stresses over the proposed lifetime of the structure without straining more than a predetermined amount. Figure 14.1 compares the tensile behaviours of a range of geotextiles.

On the other hand, damage can be caused on site during the construction period (e.g. accidental tracking from vehicles) or in situ during use (e.g. punching through geotextiles by overlying angular stone). Clearly, in both cases, damage is caused by

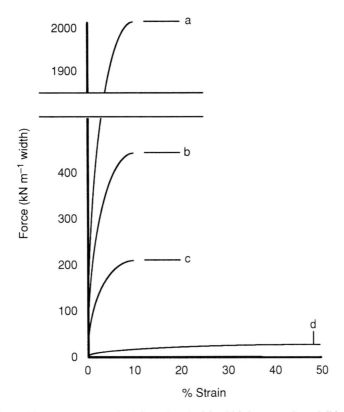

14.1 Typical ultimate stress–strain failure levels (a) of high strength and (b) of medium strength polyester woven geotextiles used for embankment support and soil reinforcement, (c) of geogrids and lower strength polyester woven geotextiles used for soil reinforcement and (d) of low strength, highly extensible nonwoven geotextiles used for separation and filtration. (c) represents the current maximum strength capacity of polyethylene geogrids.

an undesirable circumstance which is particularly difficult to remove by design. However, in the latter case, it is possible to perform advanced field testing and to allow appropriate safety factors in calculations.

The ability to perform work is fundamentally governed by the stiffness of the textile in tension and its ability to resist creep failure under any given load condition. The ability to resist damage is complex, clearly being a function of the fibre's ability to resist rupture and the construction of the fabric, which determines how stresses may be concentrated and relieved. In practical terms, geotextiles can be manufactured in a composite form, utilising the protective nature of one type of construction to reduce damage on a working element. For example, a thick non-woven fabric may be joined to a woven fabric; the woven textile performs the tensile work whilst the nonwoven acts as a damage protective cushion.

The filtration performance of a geotextile is governed by several factors. To understand this, it is essential to be aware that the function of the textile is not truly as a filter in the literal sense. In general, filters remove particles suspended in a fluid, for example, dust filters in air-conditioning units, or water filters, which are intended to remove impurities from suspension. Quite the opposite state of affairs exists with geotextile filters. The geotextile's function is to hold intact a freshly prepared soil

Table 14.1 Recommended time periods for maximum daylight exposure of geosynthetics. Beyond the limits shown damage may occur, depending upon sunlight intensity

	Temperate	Arctic	Desert	Tropical
April to Sept	8 Weeks	4 Weeks	2 Weeks	1 Week
Oct to March	12 Weeks	6 Weeks	2 Weeks	1 Week

surface, so that water may exude from the soil surface and through the textile without breaking down that surface. If water is allowed to flow between the textile and the soil interface, with particles in suspension, it will tend to clog up the textile which will fail in its function. In practice, it has been found that, in conjunction with a textile, the soil will tend to filter itself, provided that the integrity of its external surface is maintained. The actual process taking place is the passage of a liquid from a solid medium that is held intact by a permeable textile. The process is not one of restraining the passage of solids that are suspended within a liquid medium.

Geotextiles are rarely called upon to resist extremely aggressive chemical environments. Particular examples of where they are, however, include their use in the basal layers of chemical effluent containers or waste disposal sites. This can happen if and when leaks occur, permitting effluent to pass through the impermeable liner, or if the textiles have been incorporated directly in the leachate disposal system above the impermeable liner. Another example might be the use of textiles in contact with highly acidic peat soils, where in tropical countries, pH values down to 2 have been encountered. In industrialised countries where infrastructure developments are being constructed through highly polluted and contaminated areas, geotextiles can also come into contact with adverse environments.

Ultraviolet light will tend to cause damage to most polymers, but the inclusion of additives, in the form of antioxidant chemicals and carbon black powder, can considerably reduce this effect. The only time when a geotextile is going to be exposed to sunlight is during the construction period. It is generally considered that contracts should specify the minimum realistic period of exposure during site installation works. However, this will vary with time of year and latitude. In brief, it can be considered that exposure in UK and northern European type climates can be eight weeks in the summer and twelve in the winter. In tropical countries, however, exposure should be limited to seven days at any time of year before noticeable damage occurs. Table 14.1 lists typical maximum exposure periods.

14.3.1 Mechanical properties

The weight or area density of the fabric is an indicator of mechanical performance only within specific groups of textiles, but not between one type of construction and another. For example, within the overall range of needlepunched continuous filament polyester fabrics, weight will correlate with tensile stiffness. However, a woven fabric with a given area density will almost certainly be much stiffer than an equivalent weight needlepunched structure. Clearly the construction controls the performance. Therefore, it is impossible to use weight alone as a criterion in specifying textiles for civil engineering use. However, in combination with other

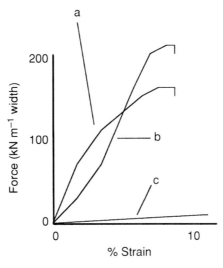

14.2   Different stress–strain curve shapes exhibited by the three main types of geosynthetic construction. (a) Geogrids absorb the imposed stresses immediately, giving a high initial modulus. Later, the curve flattens. (b) Woven fabrics exhibit initial straightening of warp fibres which produces a low initial modulus. Later the modulus increases as the straightened polymer fibres take the stress directly. (c) Nonwovens give a curvilinear curve, because extension is primarily resisted by straightening and realignment of the random fibre directions.

specified factors, weight is a useful indication of the kind of product required for a particular purpose.

The breaking strength of a standard width of fabric or 'ultimate strip tensile failure strength' is universally quoted in the manufacturers' literature to describe the 'strength' of their textiles. Again, this is of very limited use in terms of design. No designer actually uses the failure strength to develop a design. Rather, a strength at a given small strain level will be the design requirement. Therefore, the tensile resistance or modulus of the textile at say, 2%, 4%, and 6% strain is much more valuable. Ideally, continuous stress–strain curves should be provided for engineers, to enable them to design stress resisting structures properly.

Stress–strain curves, as shown in Fig. 14.1 and in Fig. 14.2 above, may well comprise a high strain sector, contributed by the textile structure straightening out, and a low strain sector, contributed by the straightened polymer taking the stress. Of course, the mechanical performance of the common geotextiles will be less as the ambient temperature rises. Because engineering sites are exposed to temperatures varying from −20 °C to 50 °C, this can have important consequences during installation and use.

Creep can cause the physical failure of a geotextile if it is held under too high a mechanical stress. It has been found that in practical terms, both polyester and polyethylene will stabilise against creep if stress levels can be maintained at a sufficiently low level. Although polypropylene does not seem to stabilise at any stress level, its creep rate is so low at small stresses that a 'no creep' condition may be considered to exist in practice.

The 'no creep' condition, measured as elongation, for any particular polymer

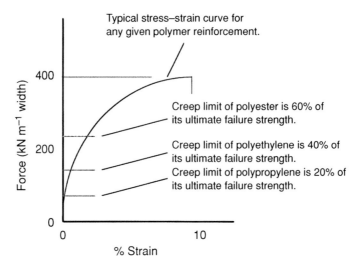

Typical stress–strain curve for
any given polymer reinforcement.

Force (kN m⁻¹ width)

400

Creep limit of polyester is 60% of
its ultimate failure strength.

200

Creep limit of polyethylene is 40% of
its ultimate failure strength.

Creep limit of polypropylene is 20% of
its ultimate failure strength.

0

0   10

% Strain

14.3   Approximate limits of creep resistance for different geosynthetic
polymer constructions.

textile is defined (usually as a percentage) with respect to the textile's ultimate load-carrying capability. For polyester, it is approximately 60%, for polyethylene about 40% and for polypropylene around 20%. Therefore, for example, a polyester fabric with an ultimate tensile strength of $100\,kN\,m^{-1}$ width cannot be loaded under a long term stress of more than $60\,kN\,m^{-1}$. The higher the level of imposed stress above this point, the more rapid will be the onset of creep failure. Figure 14.3 shows the safe loading limits for most commonly used geotextiles.

Wing tear, grab tear and puncture resistance tests may be valuable because they simulate on-site damage scenarios such as boulder dropping and direct over-running by machines. These tests are developed in standard form in a number of countries, with the standard geosynthetic test specification in the UK being BS 6906 which contains tests for:

1   tensile testing by means of a wide strip test
2   pore size testing by dry sieving
3   water flow testing normal to the plane of the textile
4   puncture resistance testing
5   creep testing
6   perforation susceptibility (cone) testing
7   water flow testing in the plane of the textile
8   testing of sand/geotextile frictional behaviour.

While not normally part of the mechanical requirements of a textile, the strength of joints between sheet edges is an important aspect of geotextile performance. When laying textiles on soft ground for supporting embankments, parallel sheets of textile have to be sewn together so that they do not separate under load. The strength of such sewn joints depends critically on the tensile strength of the sewing thread. Rarely will the sewn joint exceed 30% of the weft ultimate tensile strength. Research and field practice have shown that the strength of a sewn joint depends more upon the tenacity and tension of the sewing thread, the kind of sewing stitch

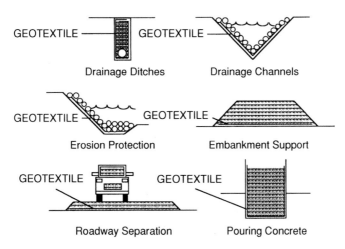

GEOTEXTILE ——— █   GEOTEXTILE ——— ◠◠◠

Drainage Ditches   Drainage Channels

GEOTEXTILE ◠◠◠   GEOTEXTILE ◠◠◠

Erosion Protection   Embankment Support

GEOTEXTILE   GEOTEXTILE

Roadway Separation   Pouring Concrete

14.4 Some different drainage and filtration applications for geotextiles in civil engineering.

and the kind of textile lap than the strength of the textile. An erroneous but common concept of joint 'efficiency' has developed which expresses the strength of a sewn seam as a percentage of the textile strength. In fact, relatively weak textiles can be sewn such that the joint is as strong as the textile, thus giving a 100% efficiency. The stronger the textile, the less is the relative strength of the sewn joint, leading to falling efficiencies with stronger fabrics. Thus it is reasonable to request a 75% efficient sewn joint if the textiles being joined are relatively weak, say 20 kN ultimate strength, but it would be impossible to achieve with a textile of say 600 kN ultimate strength. Unfortunately, it is the stronger textiles that tend to need to be joined, in order to support embankments and the like.

Adhesive joints, on the other hand, can be made using single-component adhesives whose setting is triggered by atmospheric moisture. These can be used to make joints which are as strong as the textile, even for high strength fabrics. Research is still needed on methods of application, but their use should become more widespread in the future.

Apart from tensile testing of joints, there is an urgent need to develop tests that give a meaningful description of the ways that textiles behave when stressed within a confining soil mass and additionally when stressed by a confining soil mass. The standard textile tests used in the past are not able to do this. Research work has been started along these lines but is so far insufficient to provide a basis for theoretical analysis.

14.3.2 Filtration properties

Filtration is one of the most important functions of textiles used in civil engineering earthworks. It is without doubt the largest application of textiles and includes their use in the lining of ditches, beneath roads, in waste disposal facilities, for building basement drainage and in many other ways (Fig. 14.4).

Of all the varied uses for geotextiles, only in a reinforced soil mass is there no beneficial filtration effect. In just about all other applications including drains, access

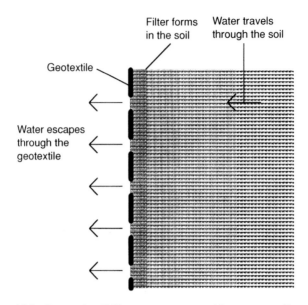

14.5   Internal soil filter zone generated by a geotextile.

roads, river defences, marine defences, embankment support and concrete pouring, the geotextile will play a primary or secondary filtering function.

The permeability of geotextiles can vary immensely, depending upon the construction of the fabric. Various national and international standards have been set up for the measurement of permeability that is required, most often at right angles to the plane of the textile (crossflow), but also along the plane of the textile (in-plane flow, called transmissivity). It is important in civil engineering earthworks that water should flow freely through the geotextile, thus preventing the build-up of unnecessary water pressure. The permeability coefficient is a number whose value describes the permeability of the material concerned, taking into account its dimension in the direction of flow; the units are rationalised in metres per second. Effectively the coefficient is a velocity, indicating the flow velocity of the water through the textile. Usually, this will be of the order of $0.001\,\mathrm{m\,s^{-1}}$. A commonly specified test measures a directly observed throughflow rate, which many feel is more practical than the permeability coefficient; this is the volume throughflow in litres per square metre per second at 100 mm head of pressure. Engineers also use a coefficient called the permittivity, which defines the theoretical permeability irrespective of the thickness of the fabric.

The filtration effect is achieved by placing the textile against the soil, in close contact, thus maintaining the physical integrity of the bare soil surface from which water is passing. Within the first few millimetres of soil, an internal filter is built up and after a short period of piping, stability should be achieved and filtration established (Fig. 14.5).

As previously discussed, filtration is normally achieved by making the soil filter itself, thus using a solid medium system, through which the liquid is flowing. There are, however, special cases where it is specifically required that the textile works in a slurry environment. Examples include tailing lagoons from mining operations and other industrial lagoons where water has to be cleared from slurries. Single textiles

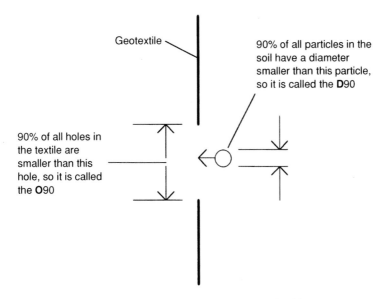

Geotextile

90% of all particles in the soil have a diameter smaller than this particle, so it is called the **D**90

90% of all holes in the textile are smaller than this hole, so it is called the **O**90

14.6 Relationship between O90 and D90.

do not work well under these conditions, but experimental work has suggested that double layers of different types of textile acting as a composite unit can improve the ability of the individual components to effect filtration without clogging.

The simplest combination reported is a smooth woven textile over a thick needlepunched nonwoven fabric placed so that the former is between the needlepunched component and the slurry. It appears that the woven fabric acts as a 'shield', protecting the nonwoven from the liquid and emulating a soil surface, thus permitting the nonwoven to function more effectively as a filter. The drainage effect of the underlying nonwoven also possibly acts to induce high hydraulic gradients which, reciprocally, assist the woven to function.

The procedure for matching a textile to the soil, in order to achieve stability under difficult hydraulic conditions, is to use a textile whose largest holes are equal in diameter to the largest particles of the soil (see Fig. 14.6 where O90 = D90). Where hydraulic conditions are less demanding, the diameter of the largest textile holes can be up to five times larger than the largest soil particles (O90 = 5D90). Particularly difficult hydraulic conditions exist in the soil (i) when under wave attack, (ii) where the soil is loosely packed (low bulk density), (iii) where the soil is of uniform particle size, or (iv) where the hydraulic gradients are high. Lack of these features defines undemanding conditions. Between the two extremes lies a continuum of variation which requires the engineer to use experience and judgment in the specification of the appropriate O90 size for any given application.

The largest hole sizes and largest particle sizes are assessed by consideration of the largest elements of the fabric and soil. Measuring the largest particles of a soil is achieved by passing the soil through standard sieves. In order to assess a realistic indication of the larger particle diameters, a notional size is adopted of the sieve size through which 90% of the soil passes. This dimension is known as the D90 by convention. Similarly, an indication of the largest holes in a textile is taken as the 90% of the biggest holes in the fabric, the O90.

Even under ideal conditions, if the O90 pore size is bigger than 5D90, then so-called piping will take place. The textile O90 pore size should be reduced from 5D90 towards D90 as the ground and hydraulic conditions deteriorate.

14.3.3 Chemical resistance

Although the chemical mechanisms involved in fibre degradation are complex,[8] there are four main agents of deterioration: organic, inorganic, light exposure and time change within the textile fibres.

Organic agents include attack by micro- and macrofaunas. This is not considered to be a major source of deterioration per se. Geotextiles may be damaged secondarily by animals, but not primarily. For example, few animals will eat them specifically, but in limited instances, when the textile is buried in the ground, it may be destroyed by animals burrowing through. Microorganisms may damage the textiles by living on or within the fibres and producing detrimental by-products. Possibly the most demanding environment for geotextiles is in the surf zone of the sea where oxygenated water permits the breeding of micro- and macroorganisms and where moving water provides a demanding physical stress.

Inorganic attack is generally restricted to extreme pH environments. Under most practical conditions, geotextile polymers are effectively inert. There are particular instances, such as polyester being attacked by pH levels greater than 11 (e.g. the byproducts of setting cement), but these are rare and identifiable.

Geotextiles can fail in their filtration function by virtue of organisms multiplying and blocking the pores, or by chemical precipitation from saturated mineral waters blocking the pores. In particular, water egressing from old mine workings can be heavily saturated with iron oxide which can rapidly block filters, whether textile or granular.

Ultraviolet light will deteriorate geotextile fibres if exposed for significant periods of time, but laboratory testing has shown that fibres will deteriorate on their own with time, even if stored under dry dark cool conditions in a laboratory. Therefore, time itself is a damaging agent as a consequence of ambient temperature and thermal degradation, which will deteriorate a geotextile by an unknown amount.

14.4 Conclusions

Geotextiles are part of a wider group of civil engineering membranes called geosynthetics. They are extremely diverse in their construction and appearance. However, they are generally made from a limited number of polymers (polypropylene, polyethylene and polyester), and are mostly of five basic types: woven, heat-bonded, needlepunched, knitted and direct soil mixed fibres.

The physical properties of this diverse group of products vary accordingly, with ultimate strengths reaching up to $2000 \, kN \, m^{-1}$, but commonly between 10 and $200 \, kN \, m^{-1}$. Ultimate strains vary up to more than 100%, but the usable range for engineers is generally between 3 and 10%. Similarly, the filtration potential and permeability of different geotextiles vary enormously.

Geotextiles are used in civil engineering earthworks to reinforce vertical and steep banks of soil, to construct firm bases for temporary and permanent roads and highways, to line ground drains, so that the soil filters itself and prevents soil from filling up the drainpipes and to prevent erosion behind rock and stone facing on river banks and the coast. They have been developed since the mid 1970s, but the advent of knitted and composite fabrics has led to a revival in attempts to improve textile construction in a designed fashion. Better physical properties can be achieved by using more than one fabric and by utilising the best features of each.

References

1. H A AGERSCHOU, 'Synthetic material filters in coastal protection', *J. Amer. Soc. Civil Engineers (Waterways and Harbours Division)*, 1961 **87**(No.WW1) (February) 111–124.
2. H J M OGINK, *Investigations on the Hydraulic Characteristics of Synthetic Fabrics*, Delft Hydraulics Laboratory, Publication No.146, May 1975.
3. P R RANKILOR, *Designing with Terram*, a Technical Design Manual published by ICI Fibres Terram Division, 1977.
4. P R RANKILOR, *Membranes in Ground Engineering*, John Wiley & Sons, Chichester, UK, 1980.
5. R M KOERNER and J P WELSH, *Construction and Geotechnical Engineering Using Synthetic Fabrics*, John Wiley & Sons, New York, 1980.
6. P R RANKILOR, 'The fundamental definition and classification of warp knitted DSF geotextiles for civil engineering and geotechnical end-uses', International Geotextiles Society's Fourth International Conference, on *Geotextiles, Geomembranes and Related Product*, ed. G Den Hoedt, held in The Hague, The Netherlands, 28th May–1st June 1990, Balkema, Rotterdam, 1990, 819–823.
7. S RAZ, *The Mechanical Tear Properties of Warp Knitted D.O.S. Fabrics*, Karl Mayer & Sons, Frankfurt, Germany, 1989.
8. A R HORROCKS, 'The durability of geotextiles', 1992, EUROTEX, Bolton Institute of Higher Education, and *Degradation of Polymers in Geomembranes and Geotextiles*, eds. S H Hamidi, M B Amin and A G Maadhah, Marcel Dekker, London & New York, 1992, 433–505.
9. P R RANKILOR, 'The weathering of fourteen different geotextiles in temperate, tropical, desert and permafrost conditions', PhD Thesis, 1989, University of Salford, UK, 1989.
10. S CORBET and J KING, Editors, *Geotextiles in Filtration and Drainage*, Proceedings Geofad Symposium, Cambridge University Sept.92, Thomas Telford, London, 1993.
11. A MCGOWN and K Z ANDRAWES, 'The load-strain-time behaviour of Tensar geogrids', *Polymer Grid Reinforcement in Civil Engineering*, Thomas Telford, London, 1984, 11–30.
12. R A JEWELL, 'Revised design charts for steep reinforced slopes', *Proceedings Symposium on Reinforced Embankments: Theory and Practice in the British Isles*, Thomas Telford, London, 1989.

14

Textiles in civil engineering. Part 2 – natural fibre geotextiles

Martin Pritchard, Robert W Sarsby and Subhash C Anand

Department of the Built Environment, Faculty of Technology, Bolton Institute, Deane Road, Bolton BL3 5AB, UK

14.5   Introduction

Processes for the selection, specification, production and utilisation of synthetic geotextiles are well established in developed countries. In many ground engineering situations, for example temporary haul roads, basal reinforcement, consolidation drains, and so on, geotextiles are only required to function for a limited time period whereas suitable synthetic materials often have a long life. Hence, the user is paying for something which is surplus to requirement. Also, conventional geotextiles are usually prohibitively expensive for developing countries. However, many of these countries have copious supplies of cheap indigenous vegetable fibres (such as jute, sisal and coir) and textile industries capable of replicating common geotextile forms. Although, there are numerous animal and mineral natural fibres available, these lack the required properties essential for geotextiles, particularly when the emphasis of use is on reinforcing geotextiles.

Synthetic geotextiles not only are alien to the ground, but have other adverse problems associated with them, in that some synthetic products are made from petroleum-based solutions. As a result of the finite nature of oil, the oil crisis in 1973, the conflict with Kuwait and Iraq in 1991, and the potentially political volatile state of some of the world's other oil producing countries, both the cost and the public awareness of using oil-based products have considerably increased. Natural fibre products of vegetable origin will be much more environmentally friendly than their synthetic equivalents and the fibres themselves are a renewable resource and biodegradable.

14.6   Development of natural materials as geotextiles

The exploitation of natural fibres in construction can be traced back to the 5th and 4th millennia BC as described in the Bible (Exodus 5, v 6–9) wherein dwellings were

14.7   Woven mat and plaited rope reeds used as reinforcement in the Ziggurat at Dur Kurigatzu.

formed from mud/clay bricks reinforced with reeds or straw. Two of the earliest surviving examples of material strengthening by natural fibres are the ziggurat in the ancient city of Dur-Kurigatzu (now known as Agar-Quf) and the Great Wall of China.[1] The Babylonians 3000 years ago constructed this ziggurat using reeds in the form of woven mats and plaited ropes as reinforcement (Fig. 14.7). The Great Wall of China, completed circa 200 BC, utilised tamarisk branches to reinforce mixtures of clay and gravel.[1,2] These types of construction however, are more comparable to reinforced concrete than today's reinforced earth techniques, because of the rigid way in which stress was transferred to the tensile elements and the 'cemented' nature of the fill.

Preconceived ideas over the low apparent tensile strength of natural materials and the perception that they have a short working life when in contact with soil limited their uses, especially for strengthening soil, in geotechnical engineering at this early stage. Also, the lack of reliable methods of joining individual textile components to form tensile fabrics presented a major limitation to their usage.

The first use of a textile fabric structure for geotechnical engineering was in 1926, when the Highways Department in South Carolina USA[3] undertook a series of tests using woven cotton fabrics as a simple type of geotextile/geomembrane, to help reduce cracking, ravelling and failures in roads construction. The basic system of construction was to place the cotton fabric on the previously primed earth base and to cover it with hot asphalt; this however made the fabric perform more like a geomembrane than a geotextile. Although published results were favourable, especially for a fabric that had been in service for nine years, further widescale development of this fabric as a geotextile did not take place. This was probably due to the high extensibility and degradable nature of this particular natural fibre together with the advent of chemical fibres.

The earliest example of jute woven fabric geotextiles for subgrade support was in the construction of a highway in Aberdeen in the 1930s.[4] The British Army also used a special machine to lay canvas or fascines over beaches and dunes for the invasion of Normandy in 1944.[5]

For thousands of years the textile industry has been spinning fibres to make yarns which in turn can be woven into fabrics. Up until the mid 1930s, these fibres were all naturally occurring, either vegetable or animal. At the beginning of this century the use of natural polymers based on cellulose was discovered, and this was quickly followed by production of chemical or synthetic products made from petroleum-based solutions.

The use of chemical fibre-based geotextiles in ground engineering started to develop in the late 1950s, the earliest two references being (i) a permeable woven fabric employed underneath concrete block revetments for erosion control in Florida[6] and (ii) in the Netherlands in 1956, where Dutch engineers commenced testing geotextiles formed from hand-woven nylon strips, for the 'Delta Works Scheme'.[7]

In the early 1960s, the excess capacity of synthetic products caused the manufacturers to develop additional outlets such as synthetic geotextiles for the construction industry. The manufacturers refined their products to suit the requirements of the engineer, rather than the engineer using the available materials to perform the requisite functions, because to a certain extent, fibre fineness and cross-sectional area can be modified to determine satisfactory tensile properties in terms of modulus, work of rupture, creep, relaxation, breaking force and extension. This led to the prolific production of synthetic materials for use in the geotextile industry. These synthetic geotextiles have monopolised the market irrespective of the cost both in economical and ecological terms. This put severe pressure on the manufacturers of ropes and cordage made from natural fibres, almost to the point of their extinction. In 1973, three fundamental applications were identified for the use of geotextiles, namely, reinforcement, separation and filtration,[8] with drainage applications (fluid transmission) also being a significant area. During the 1990s over 800 million m^2 of synthetic geotextiles have been produced worldwide,[9] making it the largest and fastest growing market in the industrial/technical fabrics industry.[10]

Although natural fibres have always been available, no one visualised their potential as a form of geotextile until synthetic fibres enabled diverse use and applications of geotextiles to emerge. Manufacturers are now attempting to produce synthetic fibres which will mimic the properties of natural fibres, but at a greater expense.

14.7   Natural fibres

The general properties of chemical fibres compared to natural fibres still tend to fall into distinct categories. Natural fibres possess high strength, modulus and moisture uptake and low elongation and elasticity. Regenerated cellulose fibres have low strength and modulus, high elongation and moisture uptake and poor elasticity. Synthetic fibres have high strength, modulus and elongation with a reasonable amount of elasticity and relatively low moisture uptake.

Natural fibres can be of vegetable, animal or mineral origin. Vegetable fibres have the greatest potential for use in geotextiles because of their superior engineering properties, for example animal fibres have a lower strength and modulus and higher

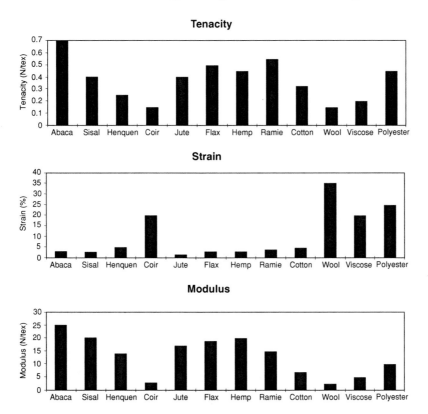

14.8 Typical strength, elongation and modulus values of natural fibres relative to those of synthetic fibres.

elongation than vegetable fibres. Mineral fibres are very expensive, brittle and lack strength and flexibility. Figure 14.8 shows typical strength, elongation and modulus values of natural fibres relative to those of synthetic fibres.

The pertinent factor for a geotextile, especially for reinforcement, is that it must possess a high tensile strength. It is known that the best way of obtaining this criterion is in the form of fibres which have a high ratio of molecular orientation. This is achieved naturally by vegetable fibres, but for synthetic polymers the molecules have to be artificially orientated by a process known as stretching or drawing, thus an increase in price is incurred. Hence nature provides ideal fibres to be used in geotextiles. In strength terms vegetable fibres compare very well with chemical fibres, in that the tenacity for cotton is in the region of $0.35\,\mathrm{N\,tex^{-1}}$ and for flax, abaca and sisal it is between 0.4–$0.6\,\mathrm{N\,tex^{-1}}$ when dry, increasing when wet to the strength of high tenacity chemical fibres – the tenacity of ordinary chemical fibres is around $0.4\,\mathrm{N\,tex^{-1}}$ (polyester). The Institute Textile de France showed (prior to 1988) that individual flax fibres (separated from their stems within the laboratory, using a process that does not weaken them) have a strength of $2 \times 10^{6}\,\mathrm{kN\,m^{-1}}$ and modulus of $80 \times 10^{6}\,\mathrm{kN\,m^{-1}}$, that is, of the same order as Kevlar,[11] a chemically modified polyamide, with exceptionally high strength compared to other synthetic fibres.

Natural fibre plants may be cultivated mainly for their fibre end-use (e.g. jute, sisal and abaca), but vegetable fibres are often a byproduct of food/crop production. Flax fibre can be extracted from the linseed plant. Also, hemp fibre is extracted for paper pulp or textile use whilst the soft inner core of the stem is used for livestock bedding. The cultivation of flax and hemp fibre allows farmers to grow the fibre crops on set-aside land (land out of food production as part of a European Union policy to decrease surpluses) which would otherwise be standing idle.

Nature provides plants with bundles of fibres interconnected together by natural gums and resins to form a load-bearing infrastructure. These fibres are pliable, have good resistance to damage by abrasion and can resist both heat and sunlight to a much greater extent than most synthetic fibres. Some fibres can also withstand the hostile nature of the marine environment. However, all natural fibres will biodegrade in the long term as a result of the action of the microorganisms. In certain situations this biodegradation may be advantageous.

Vegetable fibres contain a basic constituent, cellulose, which has the elements of an empirical formula $(C_6H_{10}O_5)_n$. They can be classed morphologically, that is according to the part of the plant from which they are obtained:

1   *Bast* or *phloem fibres* (often designated as soft fibres) are enclosed in the inner bast tissue or bark of the stem of the dicotyledonous plants, helping to hold the plant erect. Retting is employed to free the fibres from the cellular and woody tissues, i.e. the plant stalks are rotted away from the fibres. Examples of the most common of these are flax, hemp and jute.
2   *Leaf fibres* (often designated as hard fibres) run hawser-like within the leaves of monocotyledonous plants. These fibres are part of the fibrovascular system of the leaves. The fibres are extracted by scraping the pulp from the fibres with a knife either manually or mechanically. Examples of these are abaca and sisal.
3   *Seed and fruit fibres* are produced by the plant, not to give structural support, but to serve as protection for the seed and fruit that are the most vulnerable parts of the plant normally attacked by predators. Examples of these are coir and cotton. With coir fibre, the coconut is dehusked then retted, enabling the fibre to be extracted.

A natural fibre normally has a small cross-sectional area, but has a long length. This length is naturally formed by shorter fibres (often referred to as the cell length) joined together by a natural substance, such as gum or resin (the exception to this is the fibre from the seeds of the plant, vis-à-vis cotton and kapok where the length of the fibre is the ultimate fibre length).

Of the 1000 to 2000 fibre-yielding plants throughout the world,[12] there are some 15–25 plants that satisfy the criteria for commercial fibre exploitation although a number of these are only farmed on a small scale. These main fibres are flax, hemp, jute, kenaf, nettle, ramie, roselle, sunn, urena (bast fibres) abaca, banana, cantala, date palm, henequen, New Zealand flax, pineapple, sisal, (leaf fibre) coir, cotton and kapok (seed/fruit fibres). Figure 14.9 indicates the principal centres of fibre production. The main factors affecting the production/extraction of vegetable fibres are:

1   The quantity of the fibre yield from the plant must be adequate to make fibre extraction a viable proposition.
2   There must be a practical and economical procedure for extracting the fibres, without causing damage to them, if they are to be of any value as a textile material.

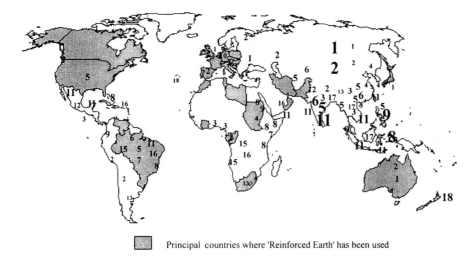

Principal countries where 'Reinforced Earth' has been used

14.9 Principal centres of fibre production. 1 Flax, 2 hemp, 3 sunn, 4 ramie, 5 jute, 6 kenaf, 7 roselle, 8 sisal, 9 abaca, 10 nettle, 11 coir, 12 cantala, 13 henequen, 14 kapok, 15 urena, 16 pineapple, 17 banana, 18 New Zealand flax. The size of the numbers indicates the most important countries for the production of fibre.

3 The pertinent properties of the fibre must be equivalent or superior to the existing chemical fibres used for the same given purpose in terms of both end production and machinability.
4 The annual yield of the fibre must be 'repeatable' and sufficiently large, i.e. if a plant has a high yield of fibre, say only every five years, then its marketability declines. Consideration must also be given to the time of harvest, i.e. late harvest yields lower quality fibres.
5 Whether there is a demand for the fibre properties on the market.
6 If there are problems of plant diseases and insect attack – protection from which has seen major improvements in the 20th century.

Advantages of developing such indigenous geotextiles would be:

• robust fibre
• environmental friendliness
• low unit cost
• strength/durability of some natural fibres, which are superior to chemical products
• reinforcing material is on the doorstep of developing countries
• increase in demand for the grower, therefore more money entering the country
• good drapability
• biodegradability
• additional use of byproducts or new use for waste.

14.7.1 Vegetable fibre properties
When selecting the most suitable vegetable fibres for geotextiles, consideration must be given to the general properties of available natural fibres in terms of strength,

elongation, flexibility, durability, availability, variability and their production forms, from the civil engineering and textile aspects. Also, factors affecting the economics of fibre cultivation and extraction on a large commercial scale should be taken into account. Allowing for the above factors, six vegetable fibres have been selected as the most promising to form geotextiles: flax, hemp, jute, abaca, sisal and coir (not in the order of priority). A generalised description of these plants/fibres is given in Tables 14.2 to 14.7, with typical values of their physical, mechanical, chemical and morphological characteristics shown in Tables 14.8 to 14.11.

Hemp and flax can be cultivated in the climatic conditions experienced in temperate countries such as the UK. Hemp does not require any pesticide treatment whilst growing. Both hemp and flax are very similar types of plant and are grown/cultivated in virtually identical conditions, producing almost similar properties in terms of fibre. However, hemp requires a licence from the Home Office for its cultivation, which imposes disadvantages compared to flax. Jute has emerged from its infancy in geotechnical engineering and has found a potential market in the erosion control industry, but may lack durability for other end-uses.

Strength properties of abaca may be superior to those of sisal, but the overall properties/economics of sisal may just outweigh those of abaca, that is, abaca is only cultivated in two countries throughout the world, with a production of less than one-fifth of sisal fibre. However, with leaf fibres retting has to be conducted within 48 hours of harvesting because otherwise the plant juices become gummy, and therefore fibre extraction is more difficult and unclean fibre is produced.

In certain categories, coir does not perform to the same standards as other fibres (i.e. low strength and high elongation), but general factors related to coir overshadow most of the other fibres for specific applications. The energy required to break the coir fibres is by far the highest of all the vegetable fibres, indicating its ability to withstand sudden shocks/pulls. Also, it is one of the best fibres in terms of retention of strength properties and biodegradation rates (in both water and sea water).

Further prioritisation of these six vegetable fibres will ultimately depend on the utilisation/end application of the geotextile.

14.8   Applications for natural geotextiles

The use of geotextiles for short-term/temporary applications to strengthen soil has a particular niche in geotechnical engineering. Geotextiles are used extensively in developed countries to combat numerous geotechnical engineering problems safely, efficiently and economically. They have several functions which can be performed individually or simultaneously, but this versatility relies upon the structure, physical, mechanical and hydraulic properties of the geotextile. Details of the general properties required to perform the functions of the geotextiles for various applications are given in Table 14.12.

14.8.1   Soil reinforcement
Soil is comparatively strong in compression, but very weak in tension. Therefore, if a tensile inclusion (geotextile) is added to the soil and forms intimate contact with it, a composite material can be formed which has superior engineering

Table 14.2   General description of flax plant/fibre

Fibre names and family	Flax	(Liniaceae)
Genus and species	*Linum usitatissimum*	(Bast fibre)
Plant type – harvesting	Annual plant, stem diameter 16–32 mm, stem length 0.9–1.2 m. Harvested after 90 days of growth when stems are green-yellow. 30 bundles of fibre in stem, each bundle contains 10–14 individual (ultimate) fibres. Low input crop fits well on a rotation scheme (6–7 years)[13]	
Countries of cultivation	Russia (80%), China, Egypt, Turkey, Philippines, Malaysia, Sri Lanka, Japan, New Zealand, UK, Poland, France, Belgium, Netherlands, USA, Canada, Argentina, West Indies, Japan and Taiwan	
Environmental – climate requirements	90% of the world's production is grown between 49° to 53°N, but can be cultivated between 22° to 65°N and 30° to 45°S	
Soil type	Rich deep loams, slightly acidic[13]	
Components of yield	One quarter of stem consists of fibre. Stem is pulled out of ground not cut, therefore longer fibres are obtained. 5–7 tonnes of flax per hectare, of which 15–20% can be extracted as long fibre, 8–10% as short fibre or tow, 5–10% seed, 45–50% woody core or shives.[13] There has been little flax produced in the UK because until recently there was no processing industry, however there is now an EC subsidy for seed flax to produce linseed oil[13]	
Uses	Linen, twines, ropes fishing nets, bags, canvas & tents. Tow fibre; high grade paper, i.e. cigarette paper and banknotes. Linseed oil and linseed flax fibre	
World annual production (tonnes)	830 000[14] 3rd most important fibre in terms of cash and acreage. Ranks 4th for the total fibre production (1st cotton, 2nd jute, 3rd kenaf)	
£ per tonne	Long line 800–2000, tow 300–700[15]	
Fibre extraction – retting	Retting affects the colour, 85% by dew retting (3–7 wks) fibre grey in colour, producing cheaper better fibre by less labour intensive way than dam, tank and chemical retting. After retting, fibre is broken away from the stems and combed	
Effects from water, sea water, pests, etc.	Fibre strength increases when wet. Pests; flea, beetles and thrips, however in general flax is not very vulnerable[13]	
Cross-section bundles	Roundish elongated irregular[16]	
Ultimate fibre	Nodes at many points, cell wall thick and polygonal in cross-section. Cell long and transparent	
Longitudinal view	Cross-marking nodes and fissures[16]	
Fibre cell ends	Ends taper to a point or round[17]	
Properties compared to other fibres	Physical and chemical properties are superior to cotton	
General fibre detail, colour, etc.	Yellowish-white, soft and lustrous in appearance. High degree of rigidity and resists bending. Russian flax weak but very fine	
General	Inextensible fibre, more elongation obtained when dry. One of the highest tensile strength and modulus of elasticity of the natural vegetable fibres. Density same as polymers, thus used as alternative reinforcement to glass, aramid and carbon in composites. Good conductor of heat and can be cottonised	

Table 14.3   Generalised description of hemp plant/fibre

Fibre names and family	Hemp (Moraceae)
Genus and species	*Cannabis sativa* (Bast fibre)
Plant type – harvesting	Annual plant, stem diameter 4–20 mm, stem length 4.5–5 m. Harvested after 90 days
Countries of cultivation	Russia, Italy, China, Yugoslavia, Romania, Hungry, Poland, France, Netherlands, UK and Australia
Environmental – climate requirements	Annual rainfall >700 mm mild climate with high humidity
Soil type	Best results from deep, medium heavy loams well-drained and high in organic matter. Poor results from mucky or peat soils and should not be grown on the same soil yearly
Components of yield	Not hard to grow. Hemcore Ltd in 1994 grew 2000 acreage in East Anglia
Uses	Ropes, marine cordage, ships sails, carpets, rugs, paper, livestock bedding and drugs
World annual production (tonnes)	214 000[14] Ranks 6th in importance of vegetable fibre
£ per tonne	300–500[15]
Fibre extraction – retting	Same process as flax, at 15–20°C retting takes 10 to 15 days. Separation of the fibre from the straw can be carried out mechanically; this is commercially known as green hemp
Effects from water, sea water, pests, etc.	Not weakened or quickly rotted by water or salt water. No pesticide protection required for growth
Cross-section bundles	Similar to flax[16]
Ultimate fibre	Similar to flax, polygonal in cross-section
Longitudinal view	Similar to flax[16]
Fibre cell ends	Rounded tips, ends of cell are blunt
Properties compared to other fibres	Stronger, more durable, stiffer and more rigid and coarser than most vegetable fibres
General fibre detail, colour, etc.	Harsh, stiff and strong fine white lustrous and brittle. Suitable for weaving of coarse fabric
General	30 varieties, narcotic drug terrahydrocannabinol (THC), in some countries cultivation illegal (cultivated now <0.3% THC thus no narcotic value). Lacks flexibility and elasticity, i.e. brittle fibre. One hectare of hemp produces as much pulp as 4 acres of forest. Can be cottonised, i.e. up to 50% hemp, does not spin easily but produces useful yarns

characteristics to soil alone. Load on the soil produces expansion. Thus, under load at the interface between the soil and reinforcement (assuming no slippage occurs, i.e. there is sufficient shear strength at the soil/fabric interface) these two materials must experience the same extension, producing a tensile load in each of the reinforcing elements that in turn is redistributed in the soil as an internal confining stress. Thus the reinforcement acts to prevent lateral movement because of the lateral shear stress developed (Fig. 14.10). Hence, there is an inbuilt additional lateral confining stress that prevents displacement. This method of reinforcing the soil can be extended to slopes and embankment stabilisation. The following exam-

Table 14.4   Generalised description of jute plant/fibre

Fibre names and family	Jute (Tiliaceae)
Genus and species	*Corchorus capsularus* and *Corchorus olitorius* (Bast fibre)
Plant type – harvesting	Annual plant, stem diameter 20 mm, stem length 2.5–3.5 m. Harvested after 90 days, small pod stage best fibre yield
Countries of cultivation	India, Bangladesh, China, Thailand, Nepal, Indonesia, Burma, Brazil, Vietnam, Taiwan, Africa, Asia and Central and South America
Environmental – climate requirements	Annual rainfall >1800 mm required >500 mm during the growing season, high humidity between 70–90%, temperature between 70–100 °F, i.e. hot damp climates
Soil type	Rich loam soils produce best results, well-drained soils obtain reasonable results, with rocky – sandy soils producing poor results
Components of yield	Easily cultivated and harvested. Line sowing increases yield by 25–50% and reduces cost of cultivation by 25%
Uses	Ropes, bags, sacks, cloths. Erosion control applications; geojute, soil-saver, anti-wash, etc.
World annual production (tonnes)	2 300 000[14] 2nd most important fibre in terms of cash and acreage
£ per tonne	300–500[15]
Fibre extraction – retting	Same process as flax. Late harvest requires prolonged retting. 1–5% oil & water emulsion is added to soften the fibre for spinning into yarns
Effects from water, sea water, pests, etc.	Fibre deteriorates rapidly when exposed to moisture. Plant; damage by excessive: heat, drought, rainfall and floods. Pests; semilooper, mite, hairy, caterpillar and apion
Cross-section bundles	Varying size roundish or elongated[16]
Ultimate fibre	Sharply polygonal, rounded (5–6 sides) corners; wall thickness varies
Longitudinal view	Fissures and cross marking are unlikely. Lumen varies in size along each fibre
Fibre cell ends	Round tips partly pointed and tapered
Properties compared to other fibres	Not as strong as hemp and flax nor as durable
General fibre detail, colour, etc.	White, yellow, red or grey; silt like and easy to spin. Difficult to bleach and can never be made pure white owing to its lack of strength. If kept dry will last indefinitely, if not will deteriorate in time
General	Holds 5 times its weight of water. Cheap and used in great quantities, high initial modulus, but very little recoverable/ elasticity (woody fibre); exhibiting brittle fracture, having small extension at break. Poor tensile strength, good luster (silky), high lignin content. Individual fibres vary greatly in strength owing to irregularities in the thickness of the cell wall

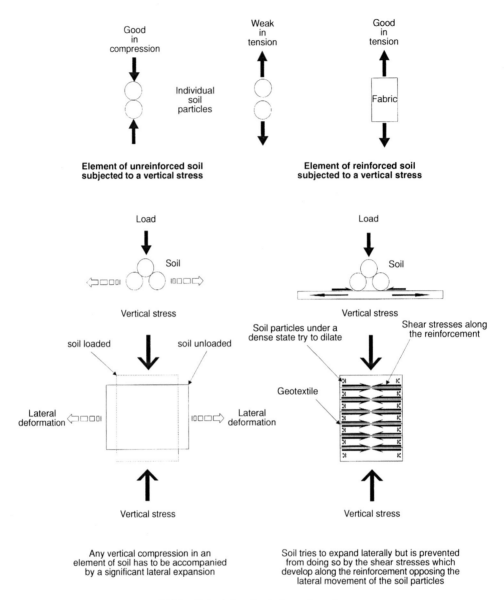

14.10   Principle of reinforced earth.

ples illustrate typical applications where geotextiles are employed to strengthen soil
for a limited amount of time.

14.8.1.1   Long-term embankments

Many developing countries have engineering situations where geotextiles could
be employed to great benefit, for example hillside stabilisation, embankment
and flood bank strengthening and construction over soft ground. Such countries
often have copious, renewable supplies of natural fibres. Labour is also abundant in
these developing countries, therefore it is more desirable to construct inexpensive
short-term projects, monitor and assess their stability periodically and rebuild them

Table 14.5 Generalised description of abaca plant/fibre

Fibre names and family	Abaca or Manila hemp (Musaceace)
Genus and species	*Musa textilis* (leaf fibre)
Plant type – harvesting	Perennial plant, 12–30 stems per plant, leaves 2–4 m, stem diameter 130–300 mm, stems length 7.5 m. Stem contains 90% water/sap with 2–5% fibre the rest soft cellular tissue. Plant life 10–20 years without replanting, fertilisation or rotation, thus impoverishes the soil. Productivity life 7–8 years, harvest 3 stalks every 4–5 months
Countries of cultivation	Philippines (85%) & Ecuador (15%)
Environmental – climate requirements	Warm climate, shade, abundant moisture and good drainage. Altitude <900 m heavy rainfall (2500–3000 mm per year) uniformly distributed throughout the year and high humidity are most advantageous. Too much heat causes excessive evaporation from the leaves, thus damages them and the fibres. Also immersion in water injures the plant
Soil type	Needs little cultivation, best grown in very fertile and well-drained soils
Components of yield	Fibre obtained from the stem of the leaves not the expanded portion of the leaf. After efflorescence plant dies. Yield up to 1 tonne of dry fibre per acre. Maximum production between 4–8 years. 100 kg of fresh leaves produce 1–3 kg of fibre
Uses	Marine cordage (naturally buoyant), fishing nets, mission ropes, well-drilling cables, paper and tea bags
World annual production (tonnes)	70 000[14]
£ per tonne	680–1150
Fibre extraction – retting	Within 48 hours if not the plant juices become gummy thus fibre extraction is more difficult and unclean fibre is produced also waste water is acidic. Fibre extracted by separating the ribbons (tuxies) of the fibre from the layers of pulp by a knife to remove the residual pulp then hung to dry (this process can be carried out by machines)
Effects from water, sea water, pests, etc.	Good water-resisting properties, hydroscopic, not affected by salt water. Pests; brown aphids, corm weevil, slug caterpillar
Cross-section bundles	Roundish, slightly indented or round to elliptical[16]
Ultimate fibre	Cells are uniform, smooth and regular surface, thus poor interlock, i.e. hard to make into a yarn. Polygonal, slightly rounded corners[16]
Longitudinal view	Smooth cross-markings rare[16]
Fibre cell ends	Thin smooth walls and sharp or pointed ends tapered. Cell diameter 3–4 times thicker than the cell wall. Cylindrical, long and regular in width
Properties compared to other fibres	Superior to flax, better than hemp for marine ropes and hawers
General fibre detail, colour, etc.	Cream and glossy, stiff and tenacious; even texture, very light weight
General	Strong and sufficiently flexible to provide a degree of give when used in ropes where strength, durability and flexibility are essential. There are 4 groups of fibre yielded from this plant, depending on where the leaves have come from; (i) Outside sheaths (*Primera baba*) dark brown/light purple & green strips (i.e. exposure to sun) Grade 4–5. (ii) Next to outside (*segunda baba*) light green and purple, Grade 3. (iii) Middle (Media), Grade 2. (iv) Innermost (Ubod), Grade 1. The grade also depends on knife type to extract the fibre

Table 14.6 Generalised description of sisal plant/fibre

Fibre names and family	Sisal
Genus and species	*Agave sisalana* (Leaf fibres)
Plant type – harvesting	Perennial plant, leaves 1–2 m long each containing about 1000 fibres
Countries of cultivation	Central America, Mexico, Brazil, Philippines, India, Florida, Africa, Venezuela, Tanzania, Kenya, Madagascar, Mozambique, Angola and Ethiopia
Environmental – climate requirements	If rainfall is erratic growth is spasmodic, thus low annual yield. Temperature between 27–32 °C (<16 °C), frost damages leaves, optimum rainfall 1200–1800 mm, but can withstand droughts, when other plants would perish, requires substantial amounts of strong sunlight
Soil type	Grows on dry, porous, rocky, not too acidic or low in nutrients free draining soils. Hardy plant can grow in mimimum rainfall 250–375 mm per year. Waterlogging and salinity are fatal to sisal
Components of yield	If the leaves are in the shade poor quality fibre is produced. Also cold, frost and hail can damage the leaves (fibre). There are spines at the tips of leaves. The leaves are harvested after 2–4 years of growth and then at intervals, after efflorescence plant dies, 45 kg of leaves produce approximately 2 kg of long and tow fibre
Uses	Twines, ropes, rugs, sacking, carpets, cordage and agricultural. Tow (waste product) used for upholstery
World annual production (tonnes)	378 000[14]
£ per tonne	450–1100[15]
Fibre extraction – retting	Within 48 hours if not the plant juices become gummy, thus fibre extraction is more difficult and unclean fibre is produced. Machines are used which scrape the pulpy material from the fibre, after washing, the fibre is dried and bleached in the sun, or oven-dried
Effects from water, sea water, pests, etc.	It was once believed that sisal deteriorated rapidly in salt water; experience has shown that this is not the case. Sisal is widely used for marine ropes.
Cross-section bundles	(i) Crescent to horse-shoe often split.[16] (ii) Few or no hemi-concentric bundles with cavities.[16] (iii) Round ellipt[16]
Ultimate fibre	Polygonal wall, thick to medium.[16] Stiff in texture, wide central cavity (may be wider than the cell wall), marked towards the middle
Longitudinal view	Smooth[16]
Fibre cell ends	Same thickness as abaca, but half as long. Rounded tips, seldom forked – pointed[16]
Properties compared to other fibres	Shorter, coarser and not quite as strong as abaca. Also lower breaking load and tends to break suddenly without warning. Can be spun as fine as jute. Sisal can be grown under a wider range of conditions then henequen
General fibre detail, colour, etc.	Light yellow in colour, smooth, straight, very long and strong fibre. Number of different types of cells inside a sisal plant; normal fibre cell straight, stiff, cylindrical and often striated
General	Blooms once in its lifetime then dies. Cheap, stiff, inflexible, high strength and good lustre. Sisal fibre is equivalent hand or machine stripped. Dark bluish-green leaves, having a waxy surface to reduce water loss

Table 14.7 Generalised description of coir plant/fibre

Fibre names and family	Coir	(Coconut fibre)
Genus and species	*Cocos nucifera*	(Seed/fruit fibre)
Plant type – harvesting	Perennial plant 70–100 nuts per year, fruit picked every alternate month throughout the year. Best crop between May & June, economic life 60 years. Two types of coir; brown and white. Brown coir obtained from slightly ripened nuts. White coir obtained from immature nuts (green coconuts) fibre being finer and lighter in colour	
Countries of cultivation	India (22%), Indonesia (20%), Sri Lanka (9%), Thailand, Malaysia, Brazil, Philippines, Mexico, Kenya, Tanzania, Asia, Africa, Kerala State, Latin America and throughout the Pacific regions	
Environmental – climate requirements	20°N to 20°S latitude, planted below an altitude of 300 m. Temperature 27–32 °C, diurnal variations <7 °C, rainfall between 1000–2500 mm, >2000 hours of sunshine i.e. high humidity and plenty of sunlight	
Soil type	Wide range of soils. Best results are from well-drained, fertile alluvial and volcanic soils	
Components of yield	Husk to nut ratio, size of nuts, fibre quality, huskability, pests and diseases. Harvesting: men climb trees, from ground, or use a knife on the end of a bamboo pole, monkeys (*Macacus nemestrima*) also climb trees to collect the nuts	
Uses	Known as the tree of life, because source of many raw materials; leaves used for roofs and mats, trunks for furniture, coconut meat for food, soap and cooking oil, roots for dyes and traditional medicines, husk for ropes, cordage and sailcloths; in marine environments	
World annual production (tonnes)	100 000[14]	
£ per tonne	200–800[15]	
Fibre extraction – retting	Retting pits (brown fibre up to 9 months, white fibre 2–6 weeks). Dehusked manually or mechanically (brown fibre only)	
Effects from water, sea water, pests, etc.	Coir is resistant to degradation by sea water, endures sudden pulls, that would snap the otherwise much stronger ropes, made from hemp or other hard fibres	
Cross-section bundles	Round mostly, with cavities, hemi-concentrical bundles[16]	
Ultimate fibre	Polygonal to round, also oblong walls, medium thickness. Round and elliptical in cross-section[16]	
Longitudinal view	Smooth[16]	
Fibre cell ends	Blunt or rounded[16]	
Properties compared to other fibres	Mature brown coir fibre contains more lignin and less cellulose than fibres such as flax and cotton	
General fibre detail, colour, etc.	Reddish-brown strong, elastic filaments of different lengths, thicker in middle and tapers gradually towards the ends. Naturally coarse, suitable for use in sea water, high lignin content makes it resistant to weathering	
General	Extremely abrasive and rot resistant (high % of lignin) under wet and dry conditions and retains a high percentage in tensile strength. Surface covered with pores, but relatively waterproof, being the main natural fibre resistant to damage by salt water	

Table 14.8 Typical values of chemical, mechanical, morphological and physical characteristics of vegetable fibres

	Chemical composition of plant fibres						
Fibre type	Cellulose (%)	Hemi-cellulose (%)	Pectin (%)	Lignin (%)	Water-soluble (%)	Fat and Wax (%)	Moisture (%)
Flax	64.1	16.7	1.8	2.0	3.9	1.5	10.0
Jute	64.4	12.0	0.2	11.8	1.1	0.5	10.0
Hemp	67.0	16.1	0.8	3.3	2.1	0.7	10.0
Sisal	65.8	12.0	0.8	9.9	1.2	0.3	10.0
Abaca	63.2	19.6	0.5	5.1	1.4	0.2	10.0
Coir	35–45	1.25–2.5		30–46		1.3–1.8	20

Figures in Tables 14.8 to 14.11 are obtained from reference sources Lewin and Pearce,[17] McGovern,[18] van Dam,[19] and Mandal.[20]

Table 14.9 Mechanical parameters from stress–strain for vegetable fibres

Fibre type	Tensile ($kN\,m^{-2} \times 10^6$)	Tenacity ($N\,tex^{-1}$)	Initial modulus ($N\,tex^{-1}$)	Extension at break (%)	Work of rupture ($N\,tex^{-1}$)
Flax	0.9	0.54–0.57	17.85–18.05	1.6–3	0.0069–0.0095
Jute	0.2–0.5	0.41–0.52	19.75	1.7	0.005
Hemp	0.3–0.4	0.47–0.6	17.95–21.68	2.0–2.6	0.0039–0.0058
Sisal	0.1–0.8	0.36–0.44	25.21	1.9–4.5	0.0043
Abaca	1.0	0.35–0.67	17.17	2.5–3	0.0077
Coir	0.1–0.2	0.18	4.22	16	0.0157

Table 14.10 Morphological plant fibre characteristics

Fibre type	Long length (mm)	Diameter (mm)	Fineness (Denier)	Cell length (mm)	Cell diameter (um)
Flax	200–1400	0.04–0.62	1.7–18	4–77	5–76
Jute	1500–3600	0.03–0.14	13–27	0.8–6	5–25
Hemp	1000–3000	0.16	3–20	5–55	10–51
Sisal	600–1000	0.1–0.46	9–406	0.8–8	7–47
Abaca	1000–2000	0.01–0.28	38–400	3–12	6–46
Coir	150–350	0.1–0.45		0.3–1.0	15–24

Table 14.11 Physical plant fibre characteristics

Fibre type	Specific gravity (%)	Specific heat ($cal\,g^{-1}\,°C^{-1}$)	Moisture regain (%) 65% RH 20°C	Absorption (%)	Volume swelling (%)	Specific heat ($cal\,g^{-1}\,°C^{-1}$)	Porosity (%)	Apparent density ($g\,cm^{-3}$)	True density ($g\,cm^{-3}$)
Flax	1.54		12	7	30		10.7	1.38	1.54
Jute	1.5	0.324	13.8	10–12.5	45	0.324	14–15	1.23	1.44
Hemp	1.48	0.323	12	8		0.323			1.5
Sisal	1.2–1.45	0.317	14	11	40	0.317	17	1.2	1.45
Abaca	1.48		14	9.5			17–21	1.2	1.45
Coir	1.15–1.33			10					1.15

Table 14.12 Functional requirements for geotextiles

Geotextiles functions	Tensile strength	Elongation	Chemical resistance	Biodegradability	Flexibility	Friction properties	Interlock	Tear resistance	Penetration	Puncture resistance
Reinforcement	iii	iii	ii-iii	iii	i	iii	iii	i	i	i
Filtration	i-ii	i-ii	iii	iii	i-ii	i-ii	iii	iii	i	ii-ii
Separation	ii	iii	iii	iii	iii	i	ii	iii	iii	ii-ii
Drainage	na	i-ii	iii	iii	i-ii	na	ii	ii-iii	iii	iii
Erosion control	ii	ii-iii	i	iii	iii	ii	i	ii	ii	i-ii

Geotextiles functions	Creep	Permeability	Resistance to flow	Properties of soil	Water	Burial	UV light	Climate	Quality assurance & control	Costs
Reinforcement	iii	na-i	i	iii	iii	iii	ii	na	iii	iii
Filtration	na	ii-iii	i	ii	iii	iii	na	ii	iii	iii
Separation	na	ii-iii	i	na	na	iii	na	i	ii	iii
Drainage	na	iii	i	na	iii	iii	na	iii	iii	iii
Erosion control	na	ii	iii	na	iii	na	iii	iii	i	iii

iii = Highly important, ii = important, i = moderately important, na = not applicable.

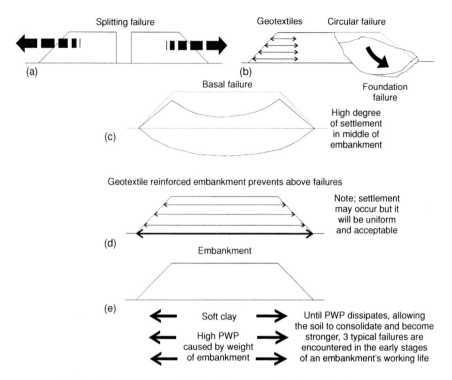

14.11   Short-term applications for geotextiles in embankments.

after a number of years if necessary (i.e. when the natural material has lost sufficient strength owing to the degradation process that it can no longer withstand the applied tensile forces). Furthermore, this procedure enriches the soil thereby improving growing conditions without introducing harmful residues. Although, it is not suggested, these natural geotextiles would be a universal panacea; they would have a significant impact on the economy of developing countries.

14.8.1.2   Short-term embankments
Geotextiles provide an invaluable solution to the problem of constructing embankments over soft compressible ground where water fills the pores between the soil particles under the embankment. The load from the embankment fill increases the tendency for the embankment to fail. Figure 14.11(a) to (c) illustrates three typical modes of failure that may be encountered (splitting, circular and basal) caused because the underlying soft soil does not have sufficient strength to resist the applied shear stresses (water has no shear strength). The use of geotextiles at vertical increments in an embankment and/or at the bottom of it, between the underlying soft soil and embankment fill (Fig. 14.11(d)), would provide extra lateral forces that either prevent the embankment from splitting or introduce a moment to resist rotation. Compression of the soft soil beneath the embankment will occur, but this will be uniform, which is acceptable. The embankment loading increases the water pressure in the pores in the underlying ground, especially at the centre of the embankment, whilst the pore water pressure (PWP) in the soil at and preceding the

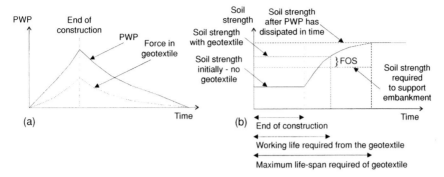

14.12   Stabilising force to be provided by the geotextile will diminish with time.
FOS = factor of safety.

extremities of the embankment is low in comparison (Fig. 14.11(e)). Thus, there is a pressure gradient set-up and water migrates from beneath the embankment sideways so that the PWP falls. Stability of the embankment will improve in time (1–2 years) as the excess PWP from the underlying soft soil dissipates (Fig. 14.11(e) and Fig. 14.12(a)). Hence its strength will increase and the stabilising force that has to be provided by the geotextile will diminish with time as shown in Fig. 14.12(b). This decrease (in the required stabilising force) can be designed to correspond to the rate of deterioration of the vegetable fibre geotextile. If necessary the rate of dissipation of the excess PWP can be enhanced by the use of consolidation drains.

14.8.1.3   Specialist areas – short-term
The armed forces often have to construct temporary roads/structures very quickly when they are dealing with confrontations. Also, these structures must be capable of being demolished if the soldiers have to retreat. By employing indigenous vegetable fibre materials as reinforcing geotextiles, the additional costs associated with the long life of synthetic geotextiles are not incurred. Decommissioning the reinforced structure is a low cost procedure – the structure can be destroyed by machinery or explosives and the natural geotextiles left to rot in the soil or set on fire, without leaving any resources for the enemy to exploit.

14.8.2   Drainage (fluid transmission)
Normally the strength of soil is determined by its water content; as the water content decreases its strength increases and vice versa. A geotextile can convey fluids or gases within the plane of the geotextile to an egress point.

14.8.2.1   Consolidation/basal drains
The drainage system allows dissipation of excess pore water pressure, thus consolidation can take place and the soil strength is increased. The rate of dissipation of excess pore water pressure can be enhanced by using temporary drains in the soil so that the drainage path is reduced (Fig. 14.13). This type of drain is only required to perform for a limited time period, until consolidation has taken place (Fig. 14.14).

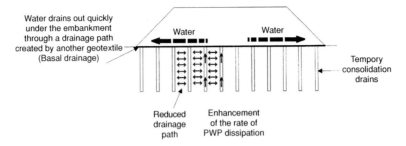

14.13   Temporary consolidation drains.

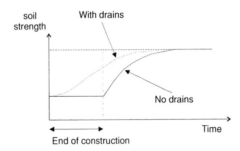

14.14   Comparison of time and strength of soil with and without consolidation drains.

14.8.3   Filtration

A geotextile acts as a filter by permitting the flow of liquid and gases, but preventing the passage of soil particles which can cause settlement due to loss of ground. The pore size within the geotextile is selected to avoid blocking, blinding and clogging.

Ground drains are used to prevent/intercept water flow, normally to reduce the risk of a rise in pore water pressure. Typically these drains are vertically sided trenches, lined with a geotextile and then filled with coarse gravel. Initial loss of soil particles will be high adjacent to the geotextile. This causes a zone (in the remaining soil particles) to bridge over the pores in the geotextile and retain smaller particles, which in turn retain even smaller particles. Thus a natural graded filter is formed which will prevent additional washout of fine particles, after which the geotextile becomes more-or-less redundant. If the geotextile was not used to encapsulate the coarse granular drainage material, too much wash-through of particles would occur and this would either cause the drain to block or cavities to develop and lead ultimately to subsidence.

14.8.4   Separation

A geotextile acts as a separator by preventing the intermixing of coarse and fine soil materials whilst allowing the free flow of water across the geotextile. For instance, when a geotextile is placed between the subsoil and the granular sub-base of an unpaved road, it prevents the aggregate from being punched down into the soil during initial compaction and subsequently from the dynamic loading of vehicle axles. An example of a short-term use of a geotextile is in a temporary haul road that is formed during the construction of the permanent works, where it is only required to function for a limited amount of time before being removed. The temporary haul road is dug up and disposed of. A geotextile made from natural fibres, such as jute, coir, and so on, would be more suitable for such applications, because it would be biodegradable and hence more environmentally friendly.

14.8.5   Erosion control/absorption

A rapidly developing area for geotextiles is in the erosion control industry where they are employed for short-term effects. This usage differs from the other applications of geotextiles in that they are laid on the surface and not buried in the soil. The main aim is to control erosion whilst helping to establish vegetation which will control erosion naturally. The geotextile is then surplus to requirements and can degrade, enriching the soil. Geotextiles can reduce runoff, retain soil particles and protect soil which has not been vegetated, from the sun, rain and wind. They can also be used to suppress weeds around newly planted trees. Erosion control can be applied to riverbanks and coastlines to prevent undermining by the ebb and flow of the tide or just by wave motion.

14.9   Engineering properties of geotextiles

The physical and mechanical properties of soil are virtually unaffected by the environment over substantial periods. The natural fibre geotextiles could be used where the life of the fabrics is designed to be short. The definition of a short-term timescale varies from site to site and application to application. It depends ultimately on a number of factors, such as the size of the job, the construction period, the time of the year (weather), and so on. However, from the wealth of accumulated knowledge, a conservative design life expectancy of the geotextiles may be made for each given end-use. Applications exist where geotextiles are only required to perform for a few days after laying (drainage/filtration) or have to last up to a hundred years (reinforced earth abutments). The design life of natural fibre fabrics will be dictated by the type of fibre and the conditions to which they will be subjected. However, design lives of a few months to 4–5 years should be achieved for natural fibre geotextiles used in non-extreme situations, particularly since the need for the geotextile declines with the passage of time.

Natural materials such as timber have been used in the construction industry for a long time. However, the use of timber is limited because it is only used as a block, that is, the individual components are not utilised. With natural fibres the stalks/stems can be stripped away to leave just the fibre which can be adapted to suit many different purposes in numerous forms and shapes with a wide range of properties. The key to developing geotextiles from natural fibres is the concept of designing by function, that is, to identify the functions and characteristics required to overcome a given problem and then manufacture the product accordingly. Provided the function can be satisfied technically and economically, these can compete with synthetic materials and in some situations they will have superior performance to their artificial counterparts.

14.10   Present state and uses of vegetable fibre geotextiles

The major use of vegetable fibre geotextiles is in the erosion control industry. Jute is readily biodegradable and ideally suited for the initial establishment of vegetation that in turn provides a natural erosion prevention facility. By the time natural vegetation has become well established the jute has started to rot/break down and disappear (6–12 months), without polluting the land.

Bangladesh, China, India and Thailand produce and sell jute geotextiles for erosion control. These are coarse mats, with open mesh woven structures made from 100% jute yarn produced on traditional jute machines. The jute geotextiles are laid on the surface of the slopes, where the weight and drapability of the mats encourage close contact with the soil. Between 1960 and 1980 a number of studies conducted by universities and highway departments demonstrated the effectiveness of jute geotextiles for surface erosion control.[21] Typical properties of a jute geotextiles[22] are:

- Pore size: 11 mm by 18 mm
- Open area ratio: 60–65%
- Water permeability: >500 litres $m^{-2} s^{-1}$ (100 mm head)
- Water absorption: 485%
- Breaking strength:
  warp 7.5 kN m^{-1}
  weft 5.2 kN m^{-1}

Some research has been directed towards reducing the degradation rate of jute, which can be made almost rot-proof by treating the fabric with a mixture of oxides and hydroxides of cobalt and manganese with copper pyroborate. Even after 21 days exposure in multiple-biological culture tanks, jute which had been subjected to this treatment had retained 96% of its original tensile strength. In soil incubation tests, the chemically treated jute had a 13-fold increase in lifetime over untreated jute.[23] Tests have been conducted on phenol formaldehyde-treated polypropylene–jute blended fabrics buried in soil to assess their susceptibility to microbial attack compared to untreated samples. It was concluded that the treated jute could withstand microbial attack more effectively than the untreated jute.[24] However, treated jute loses some of its 'environmental friendliness'.

There has been no substantial research on the engineering properties of vegetable fibres for soil strengthening or on the development of new and novel geotextile structures made from vegetable fibres for exploiting the beneficial properties of the fibre, fabric and ground for short-term or temporary applications.

14.11   Performance of natural fibre geotextiles for soil strengthening

An area which may offer the most new and upcoming potential for the use of veg-etable fibres as geotextiles is to strengthen soils, as demonstrated by Sarsby et al.[25] in 1992. Hence, the remainder of this chapter is devoted to the use of vegetable fibres for this specific application.

Factors affecting the suitability of vegetable fibres for reinforcing geotextiles can be identified as: durability, tensile properties, creep behaviour, manufacturing feasibility and soil/geotextile interaction. To be accepted these materials must satisfy/fulfil all of the above criteria to some degree. The aim of this section is not to 'design' for a specific limited application, but to determine whether acceptable balances of properties may be established. To achieve this, comparisons are made between different vegetable fibre yarns for long-term stability, that is, for biodegra-dation and creep. Also, nine different vegetable fibre geotextiles are compared with two synthetic products in terms of fabric stress–strain and shearing interactive properties.

14.11.1   Long-term stability of natural fibre geotextiles

A geotextile should show the ability to maintain the requisite properties over the selected design life. One of the reasons for using vegetable fibre geotextiles is that they biodegrade when they have served their working life, but they must be sufficiently durable in different and aggressive ground conditions to last the prescribed duration. Only purely environmental deterioration will be considered, no damage to the geotextile caused by installation will be taken into account. The effects of biodegradation and creep will be considered for four vegetable fibre yarns which are particularly suitable for soil reinforcement: flax, abaca, sisal and coir.

14.11.1.1   Durability/biodegradation rates
There are numerous factors which combine together to influence the rate of de-terioration of vegetable fibres. However, to demonstrate simply the differences in the rates of deterioration, the change in strength and elongation of the four veg-etable fibre yarns (fully immersed in water) is shown in Fig. 14.15.

The values shown are the average of five samples, tested after every three months. The samples were removed from the water and tested immediately, in other words the wet strength is given. This was chosen as representing the conditions most likely to be found in the ground. The original conditioned tex values were used at each testing stage to determine the yarns' tenacity. The initial strength of abaca and sisal (both leaf fibres) yarns increases by approximately 4% and 9%, respectively, when wet. However, with flax and coir (bast and seed/fruit fibres, respectively) there is a reduction in strength by 31% and 18%, respectively. This is in contrast to the earlier reference to the strength properties of flax increasing when wet. The reduction in strength could be accounted for by the yarn structure, rather than the fibre prop-erties themselves. For the flax, abaca and sisal yarns there is a steady reduction in tenacity with time, with slight variations in strain. In contrast, the variations in the strength of coir yarn are minor, the difference between the groups of readings prob-ably resulting from variations in the natural product itself. Figure 14.16 shows the

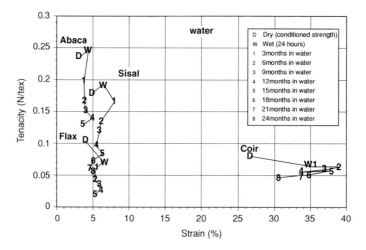

14.15   Effect of the deterioration process on the stress–strain properties of vegetable fibre yarns.

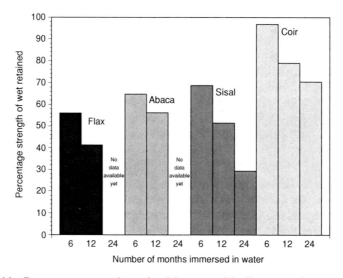

14.16   Percentage strength retained for vegetable fibre yarns in water for 6, 12 and 24 months.

percentage of the 24 hours wet strength retained for 6, 12 and 24 months. It can be seen that coir has retained by far the highest amount of strength. This was also true for coir and jute ropes which were immersed in pulverised fuel ash (PFA) for 10 and 36 months[25] – the reduction in strength for the coir was 38% and 47%, respectively whereas for jute it was 75% and 100%.

14.11.1.2   Creep
Creep and stress relaxation of geotextiles are prime factors in serviceability failure over the fabric's design life. Creep is a time-dependant increase in strain under

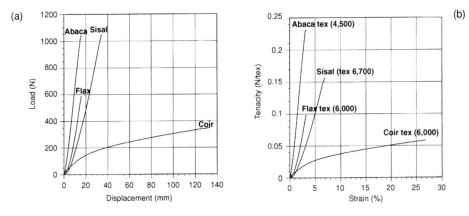

14.17   (a) Load–displacement and (b) stress–strain properties of the four vegetable fibre yarns.

constant load (e.g. reinforced walls), whereas stress relaxation is the reduction in tensile stress with time when subjected to constant strain (e.g. basal embankment reinforcement).

The main variables influencing creep for vegetable fibres geotextiles could be related to:

1   The fibre cell structure (e.g. abaca contains spiral molecules which are in a parallel configuration to each other, producing low extension).
2   Yarn type (e.g. between adjacent flax fibres cohesion is present, however with sisal no cohesion is present, the fibres are held together by twist only).
3   Fabric structure forms (e.g. crimp in woven structures).

Laboratory tests have been carried out in which the variables were load and time, with temperature and relative humidity being kept constant at 20 °C and 65%, respectively. Uniform loads of 40%, 20% and 10% of the maximum load (representing factors of safety of 2.5, 5 and 10) were applied to the four different vegetable fibre yarns and a gauge length of 500 mm was monitored.

Figure 14.17 illustrates typical short term load/extension curves at constant strain for flax, abaca, sisal and coir yarns, with the values of total strain and creep strain given in Table 14.13. Total strain includes the initial strain the sample undergoes when the load is applied plus the creep strain (this latter is the increase in change in length due to the passage of time, after the initial elongation).

14.12   Geotextile structure forms

Table 14.14 indicates the eleven different types of geotextile structure and fibre type together with their standard properties.

The creation of reinforcing geotextiles made from vegetable fibres introduces new manufacturing restraints, compared with the use of synthetic fibres and structures on existing textile machines. Numbers 1 to 5 of these structures have been designed, developed and produced in the Textile Centre at Bolton Institute from novel structure runs with selected natural fibres, namely flax, sisal and coir, to enable

Table 14.13 Total strain and creep strain of vegetable fibre yarns

Type of yarn	Max. load (kN)	Max. strain (%)	40% load (kN)	Strain at 40% load	20% load (kN)	Strain at 20% load	10% load (kN)	Strain at 10% load
Sisal	1.05	6.90	0.42	3.50	0.21	2.30	0.11	1.50
Abaca	1.04	3.19	0.42	1.40	0.21	0.80	0.10	0.50
Coir	0.35	26.71	0.14	3.70	0.07	1.50	0.04	0.70
Flax	0.68	4.02	0.27	2.30	0.14	1.50	0.07	0.90

Total strain for 10 min

Type of yarn	% of Max. load		
	40	20	10
Sisal	4.6	2.7	1.5
Abaca	1.8	1.3	0.6
Coir	5.1	2.0	1.7
Flax	2.4	1.5	0.9

Creep strain for 10 min

Type of yarn	% of Max. load		
	40	20	10
Sisal	1.1	0.4	0.0
Abaca	0.4	0.5	0.1
Coir	1.4	0.5	1.0
Flax	0.1	0.0	0.0

Total strain for 100 min

Type of yarn	% of Max. load		
	40	20	10
Sisal	4.6	2.8	1.6
Abaca	1.9	1.4	0.7
Coir	6.0	2.3	1.8
Flax	2.6	1.5	1.0

Creep strain for 100 min

Type of yarn	% of Max. load		
	40	20	10
Sisal	1.1	0.5	0.1
Abaca	0.5	0.6	0.2
Coir	2.3	0.8	1.1
Flax	0.3	0.0	0.1

Total strain for 1000 min

Type of yarn	% of Max. load		
	40	20	10
Sisal	4.8	3.0	1.8
Abaca	2.0	1.4	0.7
Coir	6.9	2.7	2.0
Flax	2.8	1.5	1.1

Creep strain for 1000 min

Type of yarn	% of Max. load		
	40	20	10
Sisal	1.3	0.7	0.3
Abaca	0.6	0.6	0.2
Coir	3.2	1.2	1.3
Flax	0.5	0.0	0.2

Total strain for 10 000 min

Type of yarn	% of Max. load		
	40	20	10
Sisal	5.0	3.2	2.0
Abaca	2.1	1.5	0.8
Coir	7.9	3.1	2.1
Flax	2.9	1.6	1.2

Creep strain for 10 000 min

Type of yarn	% of Max. load		
	40	20	10
Sisal	1.5	0.9	0.5
Abaca	0.7	0.7	0.3
Coir	4.2	1.6	1.4
Flax	0.6	0.1	0.3

Total strain for 100 000 min

Type of yarn	% of Max. load		
	40	20	10
Sisal	5.2	3.3	2.1
Abaca	2.2	1.5	0.8
Coir	8.8	3.4	2.3
Flax	3.1	1.7	1.2

Creep strain for 100 000 min

Type of yarn	% of Max. load		
	40	20	10
Sisal	–	1.0	0.6
Abaca	–	0.7	0.3
Coir	5.1	1.9	1.6
Flax	–	0.2	0.3

Table 14.14 Standard properties of vegetable fibres and commercially available geotextiles

Average of 5 fabric samples for all test results shown	Disp. at max. load (mm)	Load at max. (kN)	% Strain at max. load	Stress at max. load (MPa) (N mm⁻²)	Load/Width at max. load (kN m⁻¹)	Modulus (kN m⁻¹)	Toughness (MPa) (N mm⁻²)	Mass (g m⁻²)	Thickness (mm) weight 100 g
1 Knitted flax sisal inlay (strength direction)	16.35	10.33	8.18	38.98	206.60	4657.64	3.85	1753.23	5.3
Knitted flax sisal inlay (x-strength direction)	80.04	1.03	40.02	3.74	20.57	93.02	0.50		
2 Knitted grid flax sisal (strength direction)	14.88	7.88	7.44	32.63	143.58	2647.04	3.88	1613.81	4.4
Knitted grid flax sisal (x-strength direction)	97.76	1.09	48.88	4.35	19.15	84.26	0.47		
3 Plain weave sisal warp flax weft (warp direction)	19.28	8.99	9.64	49.94	179.80	2604.24	4.34	1289.95	3.6
Plain weave sisal warp flax weft (weft direction)	58.07	0.22	29.04	1.22	4.40	50.65	0.06		
4 Plain weave sisal warp coir weft (warp direction)	32.68	5.65	16.34	14.86	113.00	683.24	1.59	1895.48	7.6
Plain weave sisal warp coir weft (weft direction)	51.70	1.32	25.85	3.48	26.42	256.73	0.73		
5 6 × 1 woven weft rib sisal warp coir weft (warp direction)	16.69	8.53	8.35	14.10	170.60	2947.56	0.96	3051.75	12.1
6 × 1 woven weft rib sisal warp coir weft (weft direction)	68.16	5.58	34.08	9.23	111.70	710.27	2.87		
6 Plain weave coir geotextile (warp direction)	56.23	0.99	28.12	2.47	19.74	114.56	0.50	1110.99	8.0
Plain weave coir geotextile (weft direction)	44.01	0.89	22.00	2.23	17.86	142.40	0.41		
7 Knotted coir geotextile (long direction)	105.90	0.92	52.95	5.93	18.38	56.11	1.87	605.37	3.1
Knotted coir geotextile (width direction)	389.60	0.33	194.80	2.12	6.56	12.80	0.90		
8 Nonwoven hemp (machine direction)	112.70	0.11	56.37	0.48	2.15	2.48	0.16	683.16	4.5
Nonwoven hemp (x-machine direction)	85.47	0.17	42.74	0.76	3.43	4.14	0.22		
9 Nonwoven coir latex (machine direction)	12.26	0.20	6.13	0.74	4.07	107.58	0.07	1018.24	5.5
Nonwoven coir latex (x-machine direction)	11.18	0.15	5.59	0.54	2.95	72.05	0.05		
10 Plain weave synthetic polyester (warp direction)	16.35	2.07	8.17	51.62	41.30	768.72	2.94	432.09	0.8
Plain weave synthetic polyester (weft direction)	19.62	2.30	9.81	57.50	46.00	669.36	3.52		
11 Synthetic warp knitted polyester (warp direction)	55.78	2.32	27.89	27.31	46.42	446.42	4.18	430.13	1.7
Synthetic warp knitted polyester (weft direction)	102.87	0.11	51.43	1.31	2.23	70.64	0.26		

the creation of the most suitable compositions of fabrics. They have been created with the fundamental properties required to form geotextiles to reinforce soil, in that they have been designed to provide:

1   The highest possible strength in one direction, combined with ease of handling and laying on site
2   Soil particle interlock with the fabric to such an extent that the soil/fabric interface exhibits greater shearing resistance than the surrounding soil, i.e. the soil/fabric coefficient of interaction (α) is greater than one
3   A degree of protection to the high strength yarns during installation
4   A tensile strength in the range of $100–200\,kN\,m^{-1}$.
5   Ease of manufacture on conventional textile machines.

Numbers 1 and 2 are the most novel structures developed, being of weft knitting origin. The knitted structure is formed from a flax yarn (tex $\approx$ 400) encapsulating high strength sisal yarns (tex $\approx$ 6700). Knitted flax and inlay sisal yarns can be substituted by other natural fibres yarns.

The knitted flax/sisal inlay number 1 (Fig. 14.18) has as many straight inlay yarns as possible in one direction which gives the geotextile its high strength, without introducing crimp into these yarns. Thus a fabric is produced which has low extensibility compared with conventional woven structures. The knitted loops hold the inlay yarn in a parallel configuration during transportation and laying on site; under site conditions it would be impractical to lay numerous individual sisal yarns straight onto the ground. The knitted loops also provide some protection for the sisal inlay yarns during installation/backfilling. The most advantageous use of the knitted loops in this structure is that they form exactly the same surface on both sides of the fabric and the sand is in contact not only with the knitted loops but with the inlay yarns as well. Thus the shear stress from the sand is transmitted directly to both the inlay yarns and the knitted skeleton.

With the grid flax/sisal geotextile number 2 (Fig. 14.18), at predetermined intervals needles were omitted and the sisal inlay yarn left out, to produce large apertures in the geotextile. This is similar in form to the Tensar Geogrid (commercial polymer grids designed for soil reinforcement), which allows large gravel particles to penetrate into the structure thereby 'locking' the gravel in this zone and forcing it to shear against the gravel above and below the geotextile, rather than just relying on the surface characteristics.

Structures 3 to 5 employed traditional woven patterns, but exploited combinations of different types of yarn and thickness to produce advantageous fabric properties for reinforcing geotextiles.

The plain weave sisal warp/flax weft geotextile number 3 (Fig. 14.18) allows the maximum possible number of the high strength sisal yarns to be laid in one direction, whilst the flax weft yarns hold the sisal yarns together during transportation and laying on site. By only using very thin weft yarns compared to the warp yarns no crimp is introduced in these warp yarns. This structure is not as stable as the knitted structures and the flax weft yarns offer no protection for the sisal warp yarns during installation.

The plain weave sisal warp/coir weft geotextile number 4 (Fig. 14.18) provides the sisal strength yarn in one direction whilst using the coir weft yarn to form ridges in the structures caused by its coarseness, thus creating abutments which the soil has to shear around. By using a thick weft yarn, crimp is introduced into the warp

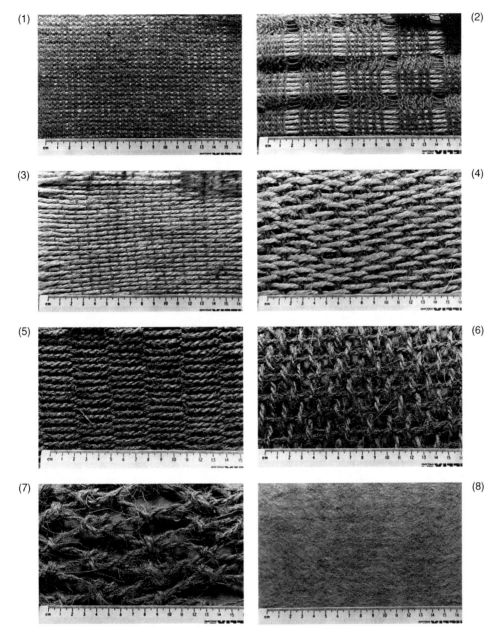

14.18 Photographs of different fabric structures used for tensile and shear interactive tests. (1) Knitted flax/sisal inlay, (2) knitted grid flax/sisal, (3) plain weave sisal warp/flax weft, (4) plain weave sisal warp/coir weft, (5) 6 × 1 woven weft rib sisal warp/coir weft, (6) plain weave coir warp/coir weft, (7) knotted coir grid, (8) nonwoven hemp, (9) nonwoven coir latex, (10) plain weave synthetic polyester, (11) synthetic warp-knitted polyester.

yarn and this in turn creates a more extensible geotextile, as well as providing no protection for the sisal strength yarns.

The woven 6 × 1 weft rib geotextile number 5 (Fig. 14.18) was designed to provide the ultimate protection for the high strength sisal yarns but without introducing any

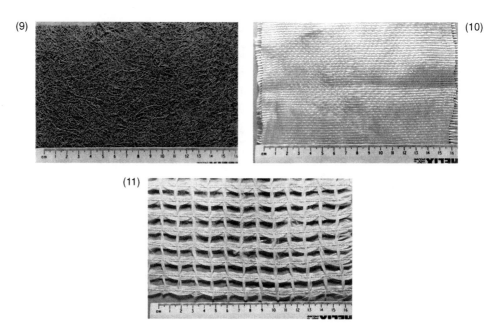

14.18   *Continued.*

crimp. However, this structure has comparatively lower productivity because of the high weft cover factor and thus it is more costly.

Numbers 6 to 11 are all commercially available geotextile products, with 6 to 9 being of a natural fibre origin. The coir knotted geotextile (Fig. 14.18) was chosen to study the effect of larger particle interlock with the fabric and large abutments formed by the knots. This geotextile was obtained from India (Aspinwall & Co. Ltd.) where the knots are produced by hand. The nonwoven samples 8 and 9 (Fig. 14.18) were obtained from Thulica AB, Sweden, for a comparison with the knitted and woven natural fibre structures. However, geotextiles 10 and 11 are of a synthetic origin from the midrange of synthetic products commercially available. These were used for a direct comparison with the natural fibre geotextiles using exactly the same tests and procedures. Both of these synthetic geotextiles were made of polyester, number 10 was a plain weave structure and number 11 was a warp-knitted grid (Fig. 14.18).

14.13   Frictional resistance of geotextiles

The frictional shearing resistance at the interface between the soil and the geotextile is of paramount importance since it enables the geotextile to resist pull-out failure and allows tensile forces to be carried by the soil/geotextile composite. The resistance offered by the fabric structure can be attributed to the surface roughness characteristics of the geotextile (soil sliding) and the ability of the soil to penetrate the fabric, that is, the aperture size of the geotextile in relation to the particle size of the soil, which affects bond and bearing resistance (Fig. 14.19). Bond resistance is created when soil particles interlock with the geotextile and permit these 'locked'

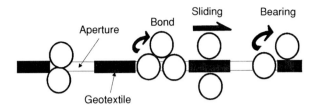

14.19   Forms of shearing resistance; sliding, bond and bearing.

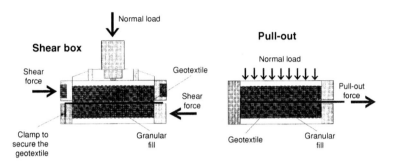

14.20   Laboratory tests to determine the frictional resistance of a geotextile.

particles in the apertures to shear against ambient soil in close vicinity above and below the geotextile surface, whereas bearing resistance, which can only really be assessed by pull-out tests, is the effect of soil having to shear around abutments in the geotextiles, or at the end of the apertures, in the direction of shear. This mode of resistance is very similar to that encountered in reinforced anchors and is determined by relating the pull-out force to the sum of projected area of the transverse members in the geotextile.

The efficiency of geotextiles in developing shearing resistance at the soil–fabric interface is indicated by the coefficient of interaction (α) defined as the ratio of the friction coefficient between soil and fabric ($\tan \delta$) and the friction coefficient for soil sliding on soil ($\tan \phi$). There are two conventional laboratory tests to determine the frictional resistance of a geotextile; the direct shear box and the pull-out test (Fig. 14.20). The main distinction between these tests is that in the direct shear box test, the soil is strained against the fabric, whereas in the pull-out test, strain is applied to the fabric thereby mobilising different degrees of shearing resistance along the fabric corresponding to a relative position of the fabric from the applied load and the extensibility of the fabric.

14.13.1   Performance of vegetable fibre geotextiles during shear

The stress–strain response and volumetric behaviour for all the geotextiles in both sand and gravel are typical of a densely packed granular dilating medium. Figure 14.21 illustrates typical curves that should be expected, relating the physical properties of the geotextiles to that of the fill. Initial volumetric compression (A) would occur to a higher degree than in plain soil as a result of the soil bedding in the geotextile. At relatively small strains, the stress level would increase rapidly more-or-

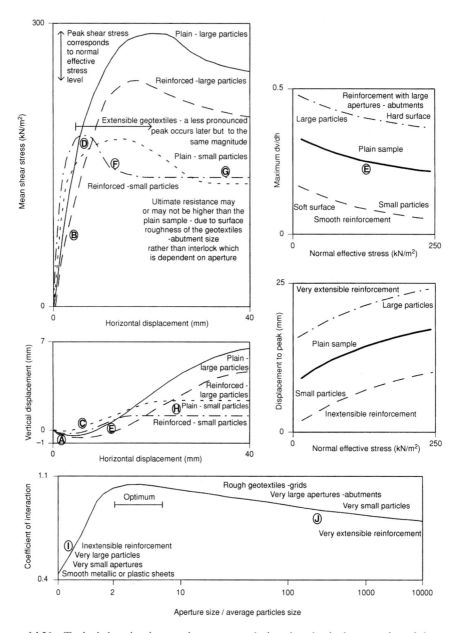

14.21   Typical shearing interactive curves, relating the physical properties of the geotextiles to that of the fill. Values indicated on these proposed charts are only shown as an estimate of the range of typical values which may lie within, for dense sand and gravel fills.

less linearly with strain (B). The stress increase will be at a higher rate than in plain soil if the geotextile limits the movement between adjacent soil particles caused by soil interlocking with the fabric. Volumetric expansion will develop at the same time, that is the soil will be dilating (C). At maximum shearing resistance the stress–strain response should produce a well-defined peak in the shear stress (D), more pronounced than in plain soil, because of the 'locked' nature of the soil particles. This should correspond to the rate of maximum volume change (E), which is likely to be greater in reinforced samples than in plain soil, particularly if introduction of the geotextile produces 'abutments' around which the soil has to shear. At this stage all the available shearing resistance, under the given vertical pressure, has been mobilised and the shear stress at the soil–fabric interface is equal to the shear strength. This stage is followed by a reduction in shearing resistance, as particle interlocking is 'released', (F) towards the final state (G), where constant volume is maintained (H). Thus a thin rupture zone of the soil at critical density is produced. By increasing the particle size in the direct shear box the behaviour will be modified slightly, because the nature ratio of the soil–fabric contact will be reduced. The opportunity for movement between ambient soil particles will be reduced as will the soil–fabric interlock as the size of the apertures in the geotextile approach the diameter of the particles (I). With very small particles or large apertures the converse will apply (J), in that the ratio of particles to aperture size will be large and thus will permit additional freedom of movement between sand grains in the shearing zone. Furthermore, the use of larger particles will produce a less rapid stress/strain response, i.e. considerably more horizontal shear displacement is needed, more effort is therefore required, to enable a gravel particle to ride over another gravel particle than it would for corresponding sand particles. Therefore no constant volume shearing zones will be expected in a sample with large particles. The extensibility of the fabric is of paramount importance for producing different degrees of soil strain. The geotextile is required to strain sufficiently to permit maximum soil strength to be mobilised, but not to the extent that serviceability failure occurs.

14.13.2   Coefficient of interaction

Values of the shearing angle and coefficient of interaction, α, of the geotextiles sheared in sand and gravel are shown, together with a summary of their stress–strain values, in Table 14.15. The results for the nonwoven samples were not as favourable as the other geotextiles, for tensile strength and shearing interactive properties, indicating that these structures are not as suitable for soil reinforcement. Some of the α values are more than 1 for the sand, indicating that by introducing the geotextile in the sand it actually strengthens the ambient sand. This could possibly be due to the surface texture of some of these geotextiles, because the sand grains can interlock with the fabric and reduce movement. This scenario can be described as if sand were sheared against sandpaper producing a higher frictional resistance than shearing sand against sand. As a result of sand shearing against sand, the sand grains above and below the failure plane are free to move, but with the sandpaper the sand grains grains are unable to move. In a practical situation if α is more than 1, the failure surface would just be pushed up away from the geotextile into the region of sand against sand. The fabric structure can be further assessed by applying a flow rule analysis to the soil/fabric interface data, as demonstrated by Pritchard,[26] to enable an assessment of whether a higher shearing resistance was developed from the surface roughness characteristics of the geotextile (smoothness of the fabric) or as a result of interlock, that is, from a higher dilational component (the effect from the apertures and abutments in the fabric).

Table 14.15 Shearing interactive values of vegetable fibre geotextiles compared to two synthetic geotextiles

	Geotextiles	(kN m⁻¹)	% Strain at max.	θ'max Sand	∝ for θ'max	θ'r Sand	∝ for θ'r	θ'max Gravel	∝ for θ'max
	Fill vs Fill			40.5°	1.00	33.1°	1.00	54.7°	1.00
1	Knitted flax sisal inlay	207	8	40.9°	1.01	33.0°	1.00	50.5°	0.86
2	Knitted grid flax sisal	144	7	38.8°	0.94	32.5°	0.98	50.9°	0.87
3	Plain weave sisal warp flax weft	180	10	40.0°	0.98	32.4°	0.97	49.8°	0.84
4	Plain weave sisal warp coir weft	113	16	42.1°	1.06	33.1°	1.00	53.4°	0.95
5	6 × 1 Woven weft rib sisal coir	170	8	42.0°	1.05	33.2°	1.00	50.9°	0.87
6	Plain weave coir geotextile	20	28	41.9°	1.05	33.1°	1.00	51.2°	0.88
7	Knotted coir geotextile	18	53	43.5°	1.11	36.7°	1.21	51.8°	0.90
8	Nonwoven hemp	2	56	39.3°	0.96	34.8°	1.07	44.6°	0.70
9	Nonwoven coir latex	4	6	34.7°	0.81	–	–	36.4°	0.52
10	Plain weave synthetic (polyester)	41	8	40.4°	1.00	31.9°	0.95	46.6°	0.75
11	Warp knitted grid synthetic (polyester)	46	28	38.4°	0.93	31.8°	0.95	51.3°	0.88

Tests conditions are: 300 × 300 mm shear box. Fill: Leighton Buzzard sand and limestone gravel (average particle diameter 0.8 mm and 6 mm, respectively). Nominal normal stress; 200, 150, 100 and 50 kN m⁻². Nominal unit weight of 96% and 94% of the maximum nominal dry unit weight for the sand and gravel, respectively. Accuracy of ±0.01 Mg m⁻³ from the mean dry density in subsequent shear box tests. Leading side of the bottom half of the shear box had the geotextile clamped to it. θ'max = maximum shear angle.

14.14   Conclusions

Vegetable fibre geotextiles offer environmentally friendly, sustainable, cost effective, geotechnical solutions to many ground engineering problems, in both developed and less developed countries. The main area where they have been employed is in the erosion control industry, but new and novel structures are being produced which exploit advantageous fabric/ground interaction properties. One of the main areas, with the largest potential for development, is to use these natural products temporarily to strengthen the ground, during and just after construction, until the soil consolidates and becomes stronger. These reinforcing geotextiles then biodegrade leaving no alien residue in the ground.

From the extensive research conducted on vegetable fibres, the six most promising fibres for geotextiles are flax, hemp, jute (bast), sisal, abaca (leaf) and coir (seed/fruit). These can be refined down to the four most suitable fibres, flax, abaca, sisal and coir, when taking into account the relevant properties required for soil reinforcement.

It has been shown that coir yarns are far more durable than any of the other vegetable fibre yarns when tested in water. Also, the coir rope exhibited excellent durability qualities compared to that of the jute rope when subjected to a hostile environment of PFA. However, the coir yarn exhibited significantly higher creep rates than flax, abaca and sisal at increased load levels.

Vegetable fibre geotextiles have been found to have superior properties to the mid-range reinforcing synthetic geotextiles for soil reinforcement, when considering tensile strength (between $100–200\,kN\,m^{-1}$) and frictional resistance (α approximately 1). The high degree of frictional resistance of the vegetable fibre geotextiles probably develops from both the coarseness of the natural yarns and the novel structure forms.

Finally it must be pointed out that the success of synthetic geotextiles is due to excess manufacturing capacity and the large amount of research and development that has been carried out in relation to their production, properties and application and not simply because they are superior to fabrics made from natural fibres.

14.15   Relevant British Standards

BS 2576: 1986 Determination of breaking strength and elongation (strip method) of woven fabrics.

BS 6906: Part 1: 1987 Determination of the tensile properties using a wide width strip.

BS 6906: Part 2: 1987 Determination of the apparent pore size distribution by dry sieving.

BS 6906: Part 3: 1987 Determination of the water flow normal to the plane of the geotextile under a constant head.

BS 6906: Part 4: 1987 Determination of the puncture resistance (CBR (California bearing ratio) puncture test).

BS 6906: Part 5: 1987 Determination of creep.

BS 6906: Part 6: 1987 Determination of resistance to perforation (cone drop test).

BS 6906: Part 7: 1987 Determination of in-plane flow.

BS 6906: Part 8: 1987 Determination of geotextile frictional behaviour by direct shear.

References

1. C J F P JONES, *Earth Reinforcement and Soil Structures*, London, Thomas Telford, 1996.
2. Department of Transport (UK), *Reinforced Earth Retaining Walls for Embankments including Abutments*, Technical Memorandum, 1977.
3. W K BECKMAN and W H MILLS, 'Cotton Fabric Reinforced Roads', *Eng. News Record*, 1957 **115**(14) 453–455.
4. P RANKILOR, 'Designing textiles into the ground', *Textile Horizons*, 1990 March (2) 14–15.
5. J C THOMSON, 'The role of natural fibres in geotextile engineering', 1st Indian Geotextiles Conference, on *Reinforced Soil and Geotextiles*, 1988, 25–29.
6. J N MANDAL, 'Potential of geotextiles', *The Indian Textile J.*, 1995 May 14–19.
7. J N MANDAL, 'The role of technical development of natural fibre geotextiles', *Man-made Textiles India*, 1988 Oct **XXX1** (10) 446.
8. N W M JOHN, *Geotextiles*, London, Blackie, 1987, 5.
9. ANON, 'Aspinwall launches geotextiles project based on jute/coir' *Asian Textile J.*, 1995 Feb **3**(4) 49–51.
10. M HOMAN, 'Geotextile market update', *Market Report*, 1994, 37.
11. E LEFLAIVE, 'The use of natural fibres in geotextile engineering' 1st Indian Geotextiles Conference, on *Reinforced Soil and Geotextiles*, 1988, 81–84.
12. R H KIRBY, *Vegetable Fibres*, Leonard Hill, London, 1963.
13. S RIDDLESTONE, 'Flax (*Linum usitatissimum*)', Leaflet, Fibre project *Bioregional Development Group*, Sutton Ecology Centre, Honeywood Walk, Carshalton, Surrey 1993, Oct.
14. J BOLTON, 'The Case for Plant Fibres', MAFF Conference *UK – Grown Non Wood Fibres: Meeting the Needs of Industry*, Queen Elizabeth II Conference Centre, 1997, Feb 21st, 1–14.
15. H GILBERTSON, 'UK sources for plant fibre material and existing technology to process them', MAFF Conference *UK – Grown Non Wood Fibres: Meeting the Needs of Industry*, Queen Elizabeth II Conference Centre, 1997, Feb 21st, 1–11.
16. B LUNIAK, *The Identification of Textile Fibres*, Pitman, London, 1953, 124.
17. S B BATRA, 'Other long vegetable fibres', *Handbook of Fibre Science and Technology*: *Volume 4 Fibre Chemistry*, eds M Lewin and E Pearce, Dekker, Chapter 9, 1985.
18. J N MCGOVERN, 'Fibres vegetable', *Polymers – Fibre and Textiles*, A Compendium, 1990, 412–430.
19. E G VAN DAM, 'Environmental advantages of application of jute fibres in innovative industrial products market potentials for plant lignocellulose fibres', *International Consultation on Jute and the Environment*, The Hague, Netherlands 1993, 26–29th Oct, 1–8.
20. J N MANDAL, 'Role of natural materials in geotextiles engineering', *Man-made Textiles India*, 1989 **32**(4) April 151–157.
21. S R RANGANATHAN, 'Jute geotextiles in soil erosion control', International Erosion Control Association 23rd Annual Conference and Trade Exposition, Reno, Nevada, USA, February, 1986, 1–19.
22. S R RANGANATHAN, 'Development and potential of jute geotextiles', *Geotextiles and Geomembranes*, 1994 **13** 421–433.
23. S N PANDEY, *Potential for Use of Natural Fibres in Civil Engineering*, Jute Technological Research Laboratory, 400–410.
24. M K TALUKDAR, A K MAJUMDAR, C R DEBNATH and A MAJUMDAR, 'A study of jute and polypropylene needle punched nonwoven fabrics for geotextiles', 1st Indian Geotextiles Conference *Reinforced Soil and Geotextiles*, 1988, 3–7.
25. R W SARSBY, M ALI, R DE ALWIS, J H KHAFFAF and J MCDOUGALL, 'The use of natural and low cost materials in soil reinforcement', International Conference *Nonwoven*, New Delhi, India, Textile Institute (North India Section), 1992 Dec, 297–310.
26. M PRITCHARD, 'Vegetable geotextiles for soil', *Institution of Civil Engineers Cooling Prize File*, 1997.

15

Medical textiles

Alistair J Rigby and Subhash C Anand

Faculty of Technology, Bolton Institute, Deane Road, Bolton BL3 5AB, UK

15.1 Introduction

An important and growing part of the textile industry is the medical and related healthcare and hygiene sector. The extent of the growth is due to constant improvements and innovations in both textile technology and medical procedures. The aim of this chapter is to highlight the specific medical and surgical applications for which textile materials are currently used. A variety of products and their properties that make them suitable for these applications will be discussed.

Textile materials and products that have been engineered to meet particular needs, are suitable for any medical and surgical application where a combination of strength, flexibility, and sometimes moisture and air permeability are required. Materials used include monofilament and multifilament yarns, woven, knitted, and nonwoven fabrics, and composite structures. The number of applications are huge and diverse, ranging from a single thread suture to the complex composite structures for bone replacement, and from the simple cleaning wipe to advanced barrier fabrics used in operating rooms. These materials can be categorised into four separate and specialised areas of application as follows:

- **Nonimplantable materials** – wound dressings, bandages, plasters, etc.
- **Extracorporeal devices** – artificial kidney, liver, and lung
- **Implantable materials** – sutures, vascular grafts, artificial ligaments, artificial joints, etc.
- **Healthcare/hygiene products** – bedding, clothing, surgical gowns, cloths, wipes, etc.

The majority of the healthcare products manufactured worldwide are disposable, while the remainder can be reused. According to a survey in the USA during the decade 1980–1990, the growth of medical textile products occurred at a compound annual rate of 11%. It is estimated that the annual growth was around 10% during 1991–2000. In western Europe the usage of nonwoven medical products between 1970 and 1994 rose from 3000 tonnes to 19 700 tonnes[1] (Fig. 15.1). The medical

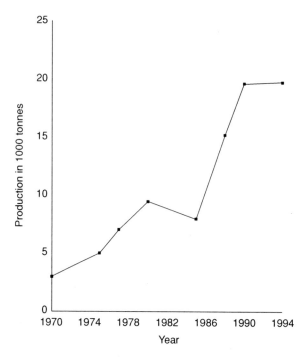

15.1   Nonwoven medical products in western Europe.

product sales of textile-based items in the USA amounted to $11.3 billion in 1980 and $32.1 billion in 1990. This figure is expected to have reached a staggering $76 billion by the year 2000.[2] The US market for disposable healthcare products alone was estimated to rise from $1.5 billion in 1990 to $2.6 billion in 1999[3] (Fig. 15.2). In Europe, medical textiles already have a 10% share of the technical textiles market, with 100000 tonnes of fibre, a growth rate of 3–4% per year and a market of US$7 billion.[32]

Although textile materials have been widely adopted in medical and surgical applications for many years, new uses are still being found. Research utilising new and existing fibres and fabric-forming techniques has led to the advancement of medical and surgical textiles. At the forefront of these developments are the fibre manufacturers who produce a variety of fibres whose properties govern the product and the ultimate application, whether the requirement is absorbency, tenacity, flexibility, softness, or biodegradability.[4] A number of reviews concerning textile materials for medical applications have also been reported elsewhere.[5–7]

15.2   Fibres used

15.2.1   Commodity fibres
Fibres used in medicine and surgery may be classified depending on whether the materials from which they are made are natural or synthetic, biodegradable or nonbiodegradable. All fibres used in medical applications must be non-toxic, non-allergenic non-carcinogenic, and be able to be sterilised without imparting any change in the physical or chemical characteristics.

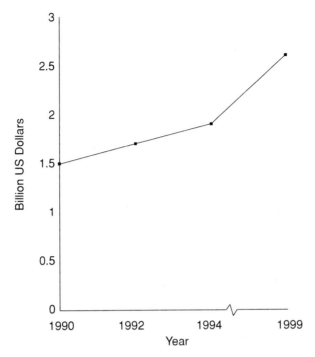

15.2 Disposable healthcare products in the USA.

Commonly used natural fibres are cotton and silk but also included are the regenerated cellulosic fibres (viscose rayon); these are widely used in nonimplantable materials and healthcare/hygiene products. A wide variety of products and specific applications utilise the unique characteristics that synthetic fibres exhibit. Commonly used synthetic materials include polyester, polyamide, polytetrafluoroethylene (PTFE), polypropylene, carbon, glass, and so on. The second classification relates to the extent of fibre biodegradability. Biodegradable fibres are those which are absorbed by the body within 2–3 months after implantation and include cotton, viscose rayon, polyamide, polyurethane, collagen, and alginate. Fibres that are slowly absorbed within the body and take more than 6 months to degrade are considered nonbiodegradable and include polyester (e.g. Dacron), polypropylene, PTFE and carbon.[8]

15.2.2   Speciality fibres

A variety of natural polymers such as collagen, alginate, chitin, chitosan, and so on, have been found to be essential materials for modern wound dressings.[9] Collagen, which is obtained from bovine skin, is a protein available either in fibre or hydrogel (gelatin) form. Collagen fibres, used as sutures, are as strong as silk and are biodegradable. The transparent hydrogel that is formed when collagen is crosslinked in 5–10% aqueous solution, has a high oxygen permeability and can be processed into soft contact lenses.[10] Calcium alginate fibres are produced from seaweed of the type Laminariae.[11] The fibres possess healing properties, which have proved to be effective in the treatment of a wide variety of wounds, and dressings comprising calcium alginate are non-toxic, biodegradable and haemostatic.[12] Chitin, a polysaccharide that is obtained from crab and shrimp shells, has excellent antithrombogenic characteristics, and can be absorbed by the body and promote healing. Chitin nonwoven fabrics used as artificial skin adhere to the body stimulating new skin formation which accelerates the healing rate and reduces pain. Treatment of chitin with alkali yields chitosan that can be spun into filaments of similar strength to viscose rayon. Chitosan is now being developed for slow drug-release membranes.[10] Other fibres that have been developed include polycaprolactone (PCL) and polypropiolactone (PPL), which can be mixed with cellulosic fibres to produce highly flexible and inexpensive biodegradable nonwovens.[13] Melt spun fibres made from lactic acid have similar strength and heat properties as nylon and are also biodegradable.[14] Microbiocidal compositions that inhibit the growth of microorganisms can be applied on to natural fibres as coatings or incorporated directly into artificial fibres.[15]

15.3   Non-implantable materials

15.3.1   Introduction

These materials are used for external applications on the body and may or may not make contact with skin. Table 15.1 illustrates the range of textile materials employed within this category, the fibres used, and the principal method of manufacture.

Table 15.1   Non-implantable materials

Product application	Fibre type	Manufacture system
Woundcare		
absorbent pad	Cotton, viscose	Nonwoven
wound contact layer	Silk, polyamide, viscose, polyethylene	Knitted, woven, nonwoven
base material	Viscose, plastic film	Nonwoven, woven
Bandages		
simple inelastic/elastic	Cotton, viscose, polyamide, elastomeric yarns	Woven, knitted, nonwoven
light support	Cotton, viscose, elastomeric yarns	Woven, knitted, nonwoven
compression	Cotton, polyamide, elastomeric yarns	Woven, knitted
orthopaedic	Cotton, viscose, polyester polypropylene, polyurethane foam	Woven, nonwoven
Plasters	Viscose, plastic film, cotton, polyester, glass, polypropylene	Knitted, woven, nonwoven
Gauzes	Cotton, viscose	Woven, nonwoven
Lint	Cotton	Woven
Wadding	Viscose, cotton linters, wood pulp	Nonwoven

15.3.2 Wound care

A number of wound dressing types are available for a variety of medical and surgical applications (Fig. 15.3). The functions of these materials are to provide protection against infection, absorb blood and exudate, promote healing and, in some instances, apply medication to the wound. Common wound dressings are composite materials consisting of an absorbent layer held between a wound contact layer and a flexible base material. The absorbent pad absorbs blood or liquids and provides a cushioning effect to protect the wound. The wound contact layer should prevent adherence of the dressing to the wound and be easily removed without disturbing new tissue growth. The base materials are normally coated with an acrylic adhesive to provide the means by which the dressing is applied to the wound.[16] Developments in coating technology have led to pressure sensitive adhesive coatings that contribute to wound dressing performance by becoming tacky at room temperature but remain dry and solvent free. The use of collagen, alginate, and chitin fibres has proved successful in many medical and surgical applications because they contribute significantly to the healing process. When alginate fibres are used for wound contact layers the interaction between the alginate and the exuding wound

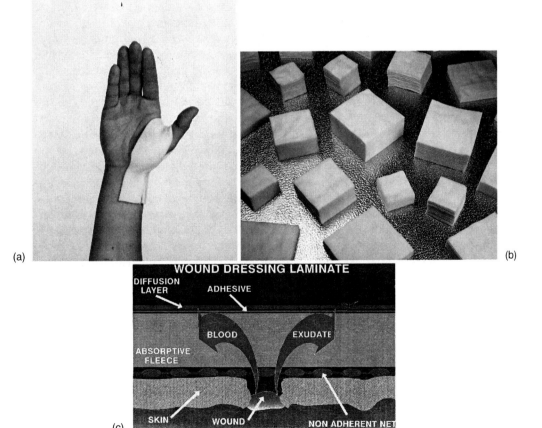

(a)

(b)

(c)

15.3 Wound dressings. (a) and (b) wound dressings, (c) wound dressing concept.

creates a sodium calcium alginate gel. The gel is hydrophilic, permeable to oxygen, impermeable to bacteria, and contributes to the formation of new tissue.[17]

Other textile materials used for wound dressing applications include gauze, lint, and wadding. Gauze is an open weave, absorbent fabric that when coated with paraffin wax is used for the treatment of burns and scalds. In surgical applications gauze serves as an absorbent material when used in pad form (swabs); yarns containing barium sulphate are incorporated so that the swab is X-ray detectable.[18] Lint is a plain weave cotton fabric that is used as a protective dressing for first-aid and mild burn applications.[19] Wadding is a highly absorbent material that is covered with a nonwoven fabric to prevent wound adhesion or fibre loss.[18]

15.3.3   Bandages

Bandages are designed to perform a whole variety of specific functions depending upon the final medical requirement. They can be woven, knitted, or nonwoven and are either elastic or non-elastic. The most common application for bandages is to hold dressings in place over wounds. Such bandages include lightweight knitted or simple open weave fabrics made from cotton or viscose that are cut into strips then scoured, bleached, and sterilised. Elasticated yarns are incorporated into the fabric structure to impart support and conforming characteristics. Knitted bandages can be produced in tubular form in varying diameters on either warp or weft knitting machines. Woven light support bandages are used in the management of sprains or strains and the elasticated properties are obtained by weaving cotton crepe yarns that have a high twist content. Similar properties can also be achieved by weaving two warps together, one beam under a normal tension and the other under a high tension. When applied under sufficient tension, the stretch and recovery properties of the bandage provides support for the sprained limb.[18,20] Compression bandages are used for the treatment and prevention of deep vein thrombosis, leg ulceration, and varicose veins and are designed to exert a required amount of compression on the leg when applied at a constant tension. Compression bandages are classified by the amount of compression they can exert at the ankle and include extra-high, high, moderate, and light compression and can be either woven and contain cotton and elastomeric yarns or warp and weft knitted in both tubular or fully fashioned forms. Orthopaedic cushion bandages are used under plaster casts and compression bandages to provide padding and prevent discomfort. Nonwoven orthopaedic cushion bandages may be produced from either polyurethane foams, polyester, or polypropylene fibres and contain blends of natural or other synthetic fibres. Nonwoven bandages are lightly needle-punched to maintain bulk and loft. A development in cushion bandage materials includes a fully engineered needlepunched structure which possesses superior cushion properties compared with existing materials.[21]

A selection of bandages and non-implantable materials products are shown in Fig. 15.4 and 15.5.

15.4   Extracorporeal devices

Extracorporeal devices are mechanical organs that are used for blood purification and include the artificial kidney (dialyser), the artificial liver, and the mechanical lung. The function and performance of these devices benefit from fibre and textile

(a)

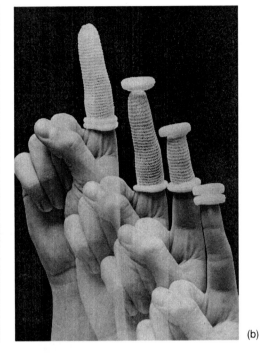

(b)

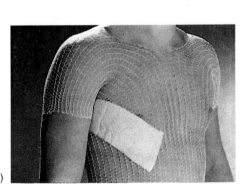

(c)

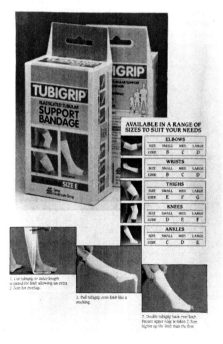

(d)

15.4 Different types of bandages and their application. (a) Elasticated flat bandage, (b) tubular finger bandages, (c) tubular elasticated net garment, (d) tubular support bandages, (e) and (f) orthopaedic casting bandage, (g) pressure gloves, (h) pressure garment, (i) hip spica, (j) lumbar/abdominal support, (k) anti-embolism stockings.

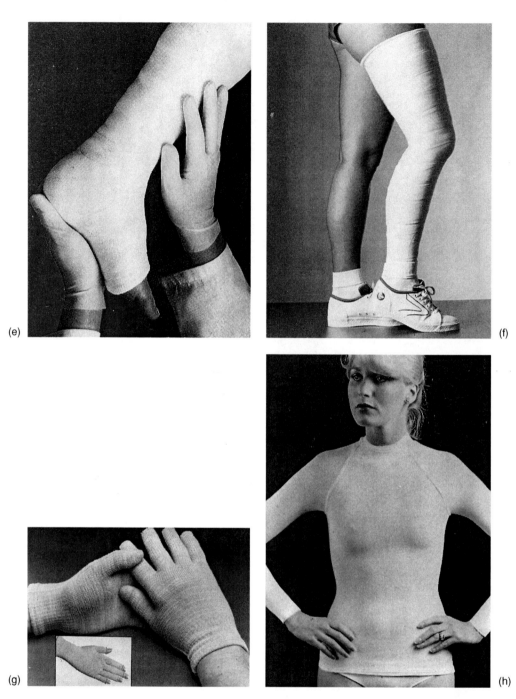

(e)

(f)

(g)

(h)

15.4 *Continued.*

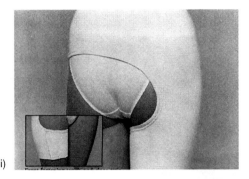
(i)

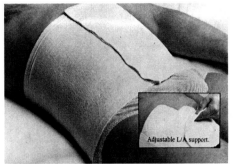
(j)

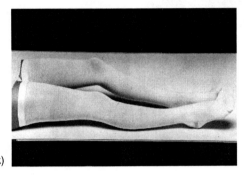
(k)

15.4 *Continued.*

technology. Table 15.2 illustrates the function of each device and the materials used in their manufacture.

The function of the artificial kidney is achieved by circulating the blood through a membrane, which may be either a flat sheet or a bundle of hollow regenerated cellulose fibres in the form of cellophane that retain the unwanted waste materials.[10,22] Multilayer filters composed of numerous layers of needlepunched fabrics with varying densities may also be used and are designed rapidly and efficiently to remove the waste materials.[23] The artificial liver utilises hollow fibres or membranes similar to those used for the artificial kidney to perform their function.[10] The microporous membranes of the mechanical lung possess high permeability to gases but low permeability to liquids and functions in the same manner as the natural lung allowing oxygen to come into contact with the patient's blood.[10,22]

15.5   Implantable materials

15.5.1   Introduction

These materials are used in effecting repair to the body whether it be wound closure (sutures) or replacement surgery (vascular grafts, artificial ligaments, etc.). Table 15.3 illustrates the range of specific products employed within this category with the type of materials and methods of manufacture. Biocompatibility is of prime

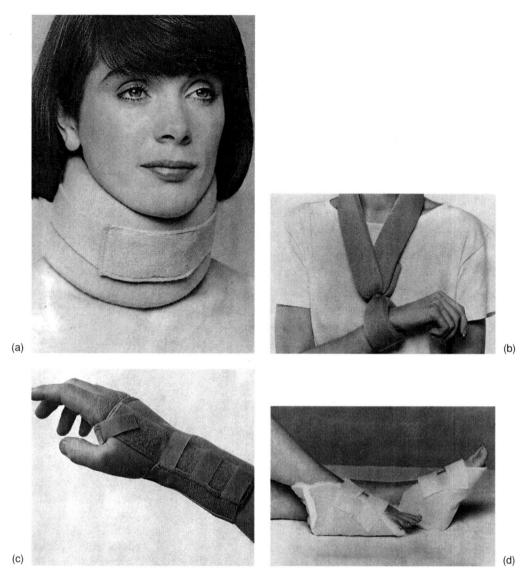

(a)

(b)

(c)

(d)

15.5 Miscellaneous surgical hosiery and other products made from non-implantable materials. (a) Cervical collar, (b) foam padded arm sling, (c) adjustable wrist brace, (d) anti-decubitus boots.

importance if the textile material is to be accepted by the body and four key factors will determine how the body reacts to the implant. These are as follows:

1   The most important factor is porosity which determines the rate at which human tissue will grow and encapsulate the implant.
2   Small circular fibres are better encapsulated with human tissue than larger fibres with irregular cross-sections.
3   Toxic substances must not be released by the fibre polymer, and the fibres should be free from surface contaminants such as lubricants and sizing agents.

Table 15.2   Extracorporeal devices

Product application	Fibre type	Function
Artificial kidney	Hollow viscose, hollow polyester	Remove waste products from patients blood
Artificial liver	Hollow viscose	Separate and dispose patients plasma, and supply fresh plasma
Mechanical lung	Hollow polypropylene, hollow silicone, silicone membrane	Remove carbon dioxide from patients blood and supply fresh blood

Table 15.3   Implantable materials

Product application	Fibre type	Manufacture system
Sutures		
biodegradable	Collagen, polylactide, polyglycolide	Monofilament, braided
non-biodegradable	Polyamide, polyester, PTFE, polypropylene, steel	Monofilament, braided
Soft-tissue implants		
artificial tendon	PTFE, polyester, polyamide, silk, polyethylene	Woven, braided
artificial ligament	Polyester, carbon	Braided
artificial cartilage	Low density polyethylene	Nonwoven
artificial skin	Chitin	
eye contact lenses/artificial cornea	Polymethyl methacrylate, silicone, collagen	
Orthopaedic implants		
artificial joints/bones	Silicone, polyacetal, polyethylene	
Cardiovascular implants		
vascular grafts	Polyester, PTFE	Knitted, woven
heart valves	Polyester	Woven, knitted

4   The properties of the polymer will influence the success of the implantation in terms of its biodegradability.

Polyamide is the most reactive material losing its overall strength after only two years as a result of biodegradation. PTFE is the least reactive with polypropylene and polyester in between.[24]

15.5.2   Sutures
Sutures for wound closure are either monofilament or multifilament threads that are categorised as either biodegradable or nonbiodegradable. Biodegradable sutures are used mainly for internal wound closures and nonbiodegradable sutures are used to close exposed wounds and are removed when the wound is sufficiently healed.

15.5.3 Soft-tissue implants

The strength and flexibility characteristics of textile materials make them particularly suitable for soft-tissue implants. A number of surgical applications utilise these
characteristics for the replacement of tendons, ligaments, and cartilage in both
reconstructive and corrective surgery. Artificial tendons are woven or braided
porous meshes or tapes surrounded by a silicone sheath. During implantation the
natural tendon can be looped through the artificial tendon and then sutured to itself
in order to connect the muscle to the bone. Textile materials used to replace
damaged knee ligaments (anterior cruciate ligaments) should not only possess biocompatibility properties but must also have the physical characteristics needed for
such a demanding application (Fig. 15.6). Braided polyester artificial ligaments are

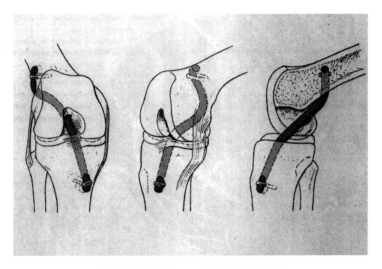

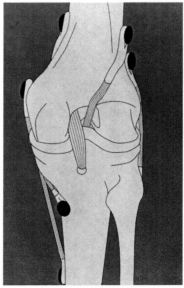

15.6   Anterior cruciate ligament prostheses.

strong and exhibit resistance to creep from cyclic loads. Braided composite materials containing carbon and polyester filaments have also been found to be particularly suitable for knee ligament replacement. There are two types of cartilage found within the body, each performing different tasks. Hyaline cartilage is hard and dense and found where rigidity is needed, in contrast, elastic cartilage is more flexible and provides protective cushioning.[25] Low density polyethylene is used to replace facial, nose, ear, and throat cartilage; the material is particularly suitable for this application because it resembles natural cartilage in many ways.[22] Carbon fibre-reinforced composite structures are used to resurface the defective areas of articular cartilage within synovial joints (knee, etc.) as a result of osteoarthritis.[26]

15.5.4   Orthopaedic implants

Orthopaedic implants are those materials that are used for hard tissue applications to replace bones and joints. Also included in this category are fixation plates that are implanted to stabilise fractured bones. Fibre-reinforced composite materials may be designed with the required high structural strength and biocompatibility properties needed for these applications and are now replacing metal implants for artificial joints and bones. To promote tissue ingrowth around the implant a non-woven mat made from graphite and PTFE (e.g. Teflon) is used, which acts as an interface between the implant and the adjacent hard and soft tissue.[27] Composite structures composed of poly(D, L-lactide urethane) and reinforced with polyglycolic acid have excellent physical properties. The composite can be formed into shape during surgery at a temperature of 60 °C and is used for both hard and soft tissue applications.[28] Braided surgical cables composed of steel filaments ranging from 13–130 μm are used to stabilise fractured bones or to secure orthopaedic implants to the skeleton.[29]

15.5.5   Cardiovascular implants

Vascular grafts are used in surgery to replace damaged thick arteries or veins 6 mm, 8 mm, or 1 cm in diameter.[10] Commercially available vascular grafts are produced from polyester (e.g. Dacron) or PTFE (e.g. Teflon) with either woven or knitted structures (Fig. 15.7). Straight or branched grafts are possible by using either weft or warp knitting technology.[18] Polyester vascular grafts can be heat set into a crimped configuration that improves the handling characteristics. During implantation the surgeon can bend and adjust the length of the graft, which, owing to the crimp, allows the graft to retain its circular cross-section.[18,24] Knitted vascular grafts have a porous structure which allows the graft to become encapsulated with new tissue but the porosity can be disadvantageous since blood leakage (haemorrhage) can occur through the interstices directly after implantation. This effect can be reduced by using woven grafts but the lower porosity of these grafts hinders tissue ingrowth; in addition, woven grafts are also generally stiffer than the knitted equivalents.[30]

In an attempt to reduce the risk of haemorrhage, knitted grafts have been developed with internal and external velour surfaces in order to fill the interstices of the graft. Another method is to seal or preclot the graft with the patient's blood during implantation. This is a time-consuming process and its effectiveness is dependent upon the patient's blood chemistry and the skill of the surgeon.[31] Presealed grafts

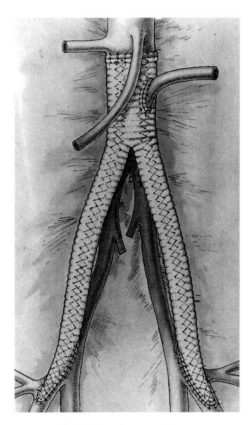

15.7   Vascular prosthesis.

have zero porosity when implanted but become porous allowing tissue ingrowth to occur. The graft is impregnated with either collagen or gelatin that, after a period of 14 days, degrades to allow tissue encapsulation.[30,31] Artificial blood vessels with an inner diameter of 1.5 mm have been developed using porous PTFE tubes. The tube consists of an inner layer of collagen and heparin to prevent blood clot formation and an outer biocompatible layer of collagen with the tube itself providing strength.[10] Artificial heart valves, which are caged ball valves with metal struts, are covered with polyester (e.g. Dacron) fabrics in order to provide a means of suturing the valve to the surrounding tissue.[27]

15.6   Healthcare/hygiene products

Healthcare and hygiene products are an important sector in the field of medicine and surgery. The range of products available is vast but typically they are used either in the operating theatre or on the hospital ward for the hygiene, care, and safety of staff and patients. Table 15.4 illustrates the range of products used in this category and includes the fibre materials used and the method of manufacture.

Textile materials used in the operating theatre include surgeon's gowns, caps and masks, patient drapes, and cover cloths of various sizes (Fig. 15.8). It is essential that the environment of the operating theatre is clean and a strict control of infection is

Table 15.4   Healthcare/hygiene products

Product application	Fibre type	Manufacture system
Surgical clothing		
gowns	Cotton, polyester, polypropylene	Nonwoven, woven
caps	Viscose	Nonwoven
masks	Viscose, polyester, glass	Nonwoven
Surgical covers		
drapes	Polyester, polyethylene	Nonwoven, woven
cloths	Polyester, polyethylene	Nonwoven, woven
Bedding		
blankets	Cotton, polyester	Woven, knitted
sheets	Cotton	Woven
pillowcases	Cotton	Woven
Clothing		
uniforms	Cotton, polyester	Woven
protective clothing	Polyester, polypropylene	Nonwoven
Incontinence diaper/sheet		
coverstock	Polyester, polypropylene	Nonwoven
absorbent layer	Wood fluff, superabsorbents	Nonwoven
outer layer	Polyethylene	Nonwoven
Cloths/wipes	Viscose	Nonwoven
Surgical hosiery	Polyamide, polyester, cotton elastomeric yarns	Knitted

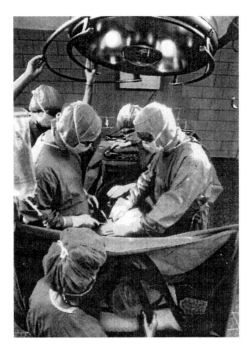

15.8   Surgical garments.

maintained. A possible source of infection to the patient is the pollutant particles shed by the nursing staff, which carry bacteria. Surgical gowns should act as a barrier to prevent the release of pollutant particles into the air. Traditionally, surgical gowns are woven cotton goods that not only allow the release of particles from the surgeon but are also a source of contamination generating high levels of dust (lint). Disposable nonwoven surgical gowns have been adopted to prevent these sources of contamination to the patient and are often composite materials comprising nonwoven and polyethylene films for example.[16]

The need for a reusable surgical gown that meets the necessary criteria has resulted in the application of fabric technology adopted for clean room environments, particularly those used for semiconductor manufacture. Surgical masks consist of a very fine middle layer of extra fine glass fibres or synthetic microfibres covered on both sides by either an acrylic bonded parallel-laid or wet-laid nonwoven. The application requirements of such masks demand that they have a high filter capacity, high level of air permeability, are lightweight and non-allergenic. Disposable surgical caps are usually parallel-laid or spun-laid nonwoven materials based on cellulosic fibres. Operating room disposable products and clothing are increasingly being produced from hydroentangled nonwovens. Surgical drapes and cover cloths are used in the operating theatre either to cover the patient (drapes) or to cover working areas around the patient (cover cloths).

Nonwoven materials are used extensively for drapes and cover cloths and are composed of films backed on either one or both sides with nonwoven fabrics. The film is completely impermeable to bacteria while the nonwoven backing is highly absorbent to both body perspiration and secretions from the wound. Hydrophobic finishes may also be applied to the material in order to achieve the required bacteria barrier characteristics. Developments in surgical drapes has led to the use of loop-raised warp-knitted polyester fabrics that are laminated back to back and contain microporous PTFE films in the middle for permeability, comfort and resistance to microbiological contaminants.

The second category of textile materials used for healthcare and hygiene products are those commonly used on hospital wards for the care and hygiene of the patient and includes bedding, clothing, mattress covers, incontinence products, cloths and wipes. Traditional woollen blankets have been replaced with cotton leno woven blankets to reduce the risk of cross-infection and are made from soft-spun twofold yarns which possess the desirable thermal qualities, are durable and can be easily washed and sterilised.[20] Clothing products, which include articles worn by both nursing staff and patients, have no specific requirements other than comfort and durability and are therefore made from conventional fabrics. In isolation wards and intensive care units, disposable protective clothing is worn to minimise crossinfection. These articles are made from composite fabrics that consist of tissue reinforced with a polyester or polypropylene spun-laid web.[16]

Incontinence products for the patient are available in both diaper and flat sheet forms with the latter used as bedding. The disposable diaper is a composite article consisting of an inner covering layer (coverstock), an absorbent layer, and an outer layer. The inner covering layer is either a longitudinally orientated polyester web treated with a hydrophilic finish, or a spun-laid polypropylene nonwoven material. A number of weft- and warp-knitted pile or fleece fabrics composed of polyester are also used as part of a composite material which includes foam as well as PVC sheets for use as incontinence mats. Cloths and wipes are made from tissue paper or nonwoven bonded fabrics, which may be soaked with an antiseptic finish. The cloth or wipe may be used to clean wounds or the skin prior to wound dressing application, or to treat rashes or burns.[26]

Surgical hosiery with graduated compression characteristics is used for a number of purposes, ranging from a light support for the limb, to the treatment of venous disorders. Knee and elbow caps, which are normally shaped during knitting on circular machines and may also contain elastomeric threads, are worn for support and compression during physically active sports, or for protection.

15.7 Conclusions

Textile materials are very important in all aspects of medicine and surgery and the range and extent of applications to which these materials are used is a reflection of their enormous versatility. Products utilised for medical or surgical applications may at first sight seem to be either extremely simple or complex items. In reality, however, in-depth research is required to engineer a textile for even the simplest cleaning wipe in order to meet the stringent performance specifications. New developments continue to exploit the range of fibres and fabric-forming techniques which are available. Advances in fibre science have resulted in a new breed of wound dressing which contribute to the healing process. Advanced composite materials containing combinations of fibres and fabrics have been developed for applications where biocompatibility and strength are required. It is predicted that composite materials will continue to have a greater impact in this sector owing to the large number of characteristics and performance criteria required from these materials. Nonwovens are utilised in every area of medical and surgical textiles. Shorter production cycles, higher flexibility and versatility, and lower production costs are some of the reasons for the popularity of nonwovens in medical textiles.

References

1. P BOTTCHER, *Int. Textile Bull. – Nonwovens/Industrial Textiles*, 1995 **4** 4.
2. G S KWATRA, *Indian Textile J.*, 1992 **102** 18–21.
3. J RUPP, *International Textile Bulletin – Nonwovens/Industrial Textiles*, 1995 **4** 24.
4. K SHAMASH, *Textile Month*, 1989, December 15–16.
5. S C ANAND, *Textile Technol. Int.*, 1994, 220–223.
6. A J RIGBY and S C ANAND, *Technical Textiles Int.*, 1996 **5**(7) 22–28.
7. A J RIGBY and S C ANAND, *Technical Textiles Int.*, 1996 **5**(8) 24–29.
8. M D TELI and N VED V VERMA, *Man-made Textiles India*, 1989 **32** 232–236.
9. Y QIN, C AGBOH, X WANG and D K GILDING, in *Proceedings of Medical Textiles 96 Conference*, Bolton Institute, UK, Woodhead Publishing, Cambridge, 1997, pp. 15–20.
10. T HONGU and G O PHILLIPS, in *New Fibres*, Ellis Horwood, UK, 1990, p. 145.
11. R W MONCRIEFF, in *Man-made Fibres*, 6th edn, Newnes–Butterworths, London, 1975, p. 300.
12. J G COOK, in *Handbook of Textile Fibres – Man-made Fibres*, 5th edn, Merrow, Co. Durham, 1984, p. 153.
13. ANON, *Medical Textiles*, 1993, October 1–2.
14. ANON, *JTN*, 1994, November (No.480) 14.
15. ANON, *Medical Textiles*, 1991, October 1–2.
16. M KRULL, *Nonwoven Bonded Fabrics*, eds J Lunenschloss and W Albrecht, Ellis Horwood, UK, 1985, pp. 399–403.
17. ANON, *Medical Textiles*, 1991, February 1–2.
18. D G B THOMAS, *Textiles*, 1975 **4** 7–12.
19. S Y CHITRE and C SAHA, *Sasmira Technical Digest*, 1987 **20** 12–20.

20. H M TAYLOR, *Textiles*, 1983, **12** 77–82.
21. A J RIGBY, S C ANAND, M MIRAFTAB and G COLLYER, in *Proceedings of Medical Textiles 96 Conference*, Bolton Institute, UK, Woodhead Publishing, Cambridge, 1997, pp. 35–41.
22. P R CHATTERJI, *J. Sc. Ind. Res.*, April **46**, 1987, pp. 14–16.
23. L BERGMANN, *Technical Textiles Int.*, 1992, September 14–16.
24. R W SNYDER, in *High-tech Fibrous Materials*, eds T L Vigo and A F Turbak, ACS Symposium Series 457, 1991, pp. 124–131.
25. C R SCHNEIDERMAN, in *Basic Anatomy and Physiology in Speech and Hearing*, College Hill Press, California, 1984, p. 3.
26. J BROWN, in *Proceedings of Industrial, Technical and Engineering Textiles Conference*, The Textile Institute, Manchester, Paper 8, 1988.
27. A S HOFFMAN, in *Fibre Science*, ed. M Lewin, Applied Polymer Symposium 31, John Wiley & Sons, New York, 1977, p. 324.
28. R F STOREY, J S WIGGINS and A D PUCKETT, *Polymer Reprints*, 1992, **33**(2) 452–453.
29. ANON, *Medical Textiles*, 1994, April 5.
30. ANON, *Medical Textiles*, 1991, February 5–6.
31. ANON, *Medical Textiles*, 1989, September 1–5.
32. S C ANAND, *Techtextil North America*, Atlanta, USA, Conference Proceedings, 24 March, 2000, p. 3.

16

Textiles in defence*

Richard A Scott

Defence Clothing and Textiles Agency, Science and Technology Division, Flagstaff Road, Colchester, Essex CO2 7SS, UK

16.1 Introduction

To be prepared for War is one of the most effectual means of preserving Peace
(George Washington, 1790)[1]

Defence forces on land, sea, or air throughout the world are heavily reliant on technical textiles of all types – whether woven, knitted, nonwoven, coated, laminated, or other composite forms. Technical textiles offer invaluable properties for military land forces in particular, who are required to move, live, survive and fight in hostile environments. They have to carry or wear all the necessities for comfort and survival and thus need the most lightweight, compact, durable, and high performance personal clothing and equipment. The life-critical requirements for protecting individuals from both environmental and battlefield threats have ensured that the major nations of the world expend significant resources in developing and providing the most advanced technical textiles for military use.

16.2 Historical background

Military textile science is not new, and one of the earliest documented studies can probably be credited to Count Rumford, or Benjamin Thompson. Rumford was an American army colonel and scientist who issued a paper in 1792 entitled 'Philosophical Transactions', which reported on the importance of internally trapped air in a range of textile fabrics to the thermal insulation provided by those fabrics.[2] He was awarded the Copley Medal for his paper, as the significance of his discovery was recognised immediately.

16.2.1 Pre-Twentieth century

Up until the end of the 19th century military land battles were fought at close quarters by individual engagements. Military uniforms were designed to be bright, shiny
and colourful, both for regimental identification and to intimidate the enemy.
'Danger' colours such as scarlet were widely used, and uniforms carried embellish-
ments such as large epaulettes to increase the apparent width of the shoulders.
Tall headwear made from animal furs (bearskin caps), feathers (ostrich), or carrying tall plumes were worn to increase the apparent height of troops. The materials
used were all of natural origin, based upon wool and goat hairs, cotton, silk, flax,
leather, horsehair, pig bristle, furs from bears, seals, tigers, and leopards, and feathers from birds such as chickens, peacocks and ostriches. Such uniforms were
heavy, uncomfortable, and impractical in the field, incurring irreparable damage in
a short time.

16.2.2 The twentieth century

At around the turn of the 20th century advances in technology and science provided
more lethal long-range weapons with improved sighting. Visual detection equipment
became more sophisticated at about this time. These combined effects caused
rapid changes in military strategy and tactics, as engagements could be made at a
distance. It now became important to hide troops and equipment by blending in
with the background. The British Forces adopted khaki coloured uniforms (khaki
meaning dung or dust in Persian and Urdu).[3] The first khaki drill (or KD) made
from cotton twill or drill entered service for tropical use in 1902, although it had
been adopted in the South African Boer War before that time.[4] This cotton drill was
found to give insufficient protection from the elements in temperate climates, so
that wool worsted serge (twill fabric) uniforms were issued in the khaki or brown
colours.

At this time all non-clothing textile items such as tents, shelters, covers, nets,
load-carriage items and sleeping systems were made from natural fibres based upon
wool, cotton, flax, jute, hemp, sisal, and kapok. Those used for screens, covers and
tents were heavy, cumbersome, and prone to degradation by insects, moisture and
biological organisms.

The natural environment has always been as big a threat to military forces as
enemy action. History provides many instances where the weather defeated armies,
navies, and in recent times, air forces. Examples include the Napoleonic wars, World
Wars I and II, the Korean War, and more recently, the Falklands War, where countless numbers of forces incurred casualties due to extremes of cold, wet, or hot
climates.

In the 1930s the UK War Office became increasingly aware of the need for new
and more rational combat dress to meet the needs of mechanisation on land, sea,
and in the air. This was to provide better protection, comfort, and practicality.[5]

During World War II advances in textile fibres, fabrics, and treatments saw
notable landmarks such as the use of the new fibre 'Nylon' for light strong
parachute canopies, and the development of Ventile® cotton fabric for aircrew survival clothing for those who had to ditch into the cold North Sea. Ventile was the
first waterproof/water vapour-permeable fabric – invented by scientists at the
Shirley Institute (now the British Textile Technology Group). It was based upon low-
twist Sea Island cotton yarns in a very tightly woven construction. A very efficient

stearamido derivative water-repellent finish (Velan® by ICI) improved the waterproofness of this technical fabric, which is still widely used today by many air forces.

The well-known worsted serge 'battledress' uniform was introduced in 1939. Prototypes of this were made at the Garment Development Section, Royal Dockyard, Woolwich, London. The first formal specification was designated E/1037 of 28th October 1938, issued by the Chief Inspector of Stores and Clothing (CISC).[5]

The 'Denison smock', in a lightweight windproof cotton gaberdine fabric, and bearing rudimentary camouflage patterning, was introduced for airborne paratroopers in 1941. Captain Denison served with a special camouflage unit commanded by Oliver Messel, an eminent theatrical stage designer!

Armoured fighting vehicle crews were issued with a one-piece coverall in black cotton denim. This led to the development of a sand-coloured version in 1944 for desert use.[5]

Woollen serge as a material for field uniforms became obsolete when the United States army introduced the 'layered' combat clothing concept in 1943. The British and other Allies followed this lead, introducing their own layered system as a winter uniform in the Korean War of the 1950s to avoid weather-related casualties.[6]

The next great landmark in combat dress appeared in 1970, when the olive green (OG) 100% cotton satin drill fabric appeared. This was followed in 1972 by the first four-colour disruptively patterned material (DPM) for temperate woodland camouflage. The UK was one of the first forces to introduce such a printed material for combat forces.

From the 1960s to the present day the military textiles, clothing and equipment of all major nations have become ever more sophisticated and diverse. They now utilise the most advanced textile fibres, fabrics and constructions available. It has now been recognised that, no matter how sophisticated weapon systems and equipment become, they ultimately depend for their effectiveness on a human operator to make the final decisions. This has led to significant increases in the reliance on scientific and technical solutions to solve the perennial problems associated with protection of the individual from environmental and battlefield threats, with the need to maintain comfort, survivability and mobility of fighting forces.

16.3   Criteria for modern military textile materials

The main functional criteria for military textiles are dealt with here under a range of headings. These include the physical, environmental, camouflage, specific battlefield threats, flames, heat and flash, and the economic considerations (Tables 16.1 to 16.6).

16.4   Incompatibilities in military materials systems

The functional performance requirements for military textiles are manifold and complex, as indicated in Section 16.3. This complexity inevitably results in serious incompatibilities. It is the attempt to solve these many incompatibilities which occupies the efforts of scientists and technologists.

Table 16.1   Physical requirements for military textiles

Property	Comments
Light weight and } Low bulk	Items have to be carried by individuals or vehicles with minimal space available
High durability and } Dimensional stability }	Must operate reliably in adverse conditions for long periods of time without maintenance.
Cleanable	
Good handle and drape	Comfortable
Low noise emission	Tactically quiet – no rustle or swish
Antistatic	To avoid incendive or explosive sparks

Table 16.2   Environmental requirements

Property	Comments
Water-repellent,	For exterior materials exposed to cold/wet weather
Waterproof,	" "
Windproof and	" "
Snow-shedding	" "
Thermally insulating	For cold climates
Water vapour permeable	For clothing and personal equipment (tents etc.)
Rot-resistant	For tents, covers, nets etc.
UV light resistant	For environments with strong sunlight
Air permeable.	For hot tropical climates
Biodegradable	If discarded or buried

Table 16.3   Camouflage, concealment and deception requirements

Property	Comments
Visual spectrum	Exposed materials match visual colours, texture and appearance of natural backgrounds
Ultraviolet	To match optical properties of snow and ice
Near infrared	To match reflectance of background when viewed by image intensifiers and low light television
Far infrared	To minimise the heat signature emitted by humans and hot equipment. Detection by thermal imagers
Acoustic emissions	Rustle and swish noises emitted by certain textile materials Detected by aural means, unattended ground sensors and microphones
Radar spectrum	Detection of movement by Doppler radar

Table 16.4   Requirements for flame, heat and flash protection

Property	Comments
Flame retardance	Of outer layers exposed to flames and heat
Heat resistance	Avoid heat shrinkage and degradation
Melt resistance	For textiles in contact with the skin
Low smoke emission	To allow escape in confined spaces
Low toxicity	Of combustion products in confined spaces such as ships, submarines, buildings, vehicles

Table 16.5   Specific battlefield hazards

Hazard	Comments
Ballistic fragments	From bombs, grenades, shells, warheads
Low velocity bullets	From hand guns, pistols, etc.
High velocity bullets	From small-arms rifled weapons from 5.56 mm up to 12.7 mm calibre
Flechettes	Small, sharp, needle shaped projectiles
Chemical warfare agents	Including blood agents, nerve agents, vessicants
Biological agents	Bacteria, toxins, viruses
Nuclear radiation	Alpha, beta and gamma radiation
Directed energy weapons (DEW)	Includes laser rangefinders and target designators

Table 16.6   Economic considerations

Property	Comments
Easy-care	Smart, non-iron, easily cleanable
Minimal maintenance	Maintenance facilities not available in the field
Long storage life	War stocks need to be stored for 10–20 years.
Repairable	Repairable by individuals or HQ workshops
Decontaminable or disposable	Against nuclear, biological or chemical contamination
Readily available	From competitive tendering in industry against a standard or specification
Minimal cost	Bought by taxpayers and other public funding

Figure 16.1 shows the relationships between the properties of textile material systems. A cross between two properties highlights a particular problem.[7]

If we start the interpretation of Fig. 16.1 at the box marked 'Low Weight and Bulk', we can see that it is difficult to produce high durability textiles at low weight and good thermal insulation at low bulk. Ballistic protection requires the use of heavy, bulky, relatively inflexible materials. Commodity artificial fibres such as nylon and polyester are widely used, but suffer from their lack of flame retardance. Good snow-shedding properties are provided by flat continuous filament fabrics which are not readily available in flame-retardant forms. Continuous filament fabrics tend to be noisy, producing a characteristic 'swishing' or 'rustling' noise when rubbed or crumpled.

If we provide materials which offer ballistic protection against bullets and bomb fragments, the ballistic packs also offer high thermal insulation, which causes heat stress in the wearer. If we consider the thermophysiological effects further, we encounter the classic problem of providing waterproof fabrics which are permeable to water vapour and air. If water vapour permeability is limited, the problem of activity related heat stress occurs. Conversely, water vapour and air permeable fabrics do not readily provide barriers to chemical warfare agents.

Air-permeable fabrics which are ideal in hot tropical climates, allow biting insects such as mosquitos to penetrate the fabrics. Moreover, many insects are attracted to the printed visual camouflage colours, which include green, khaki and brown.

Camouflage properties are required to cover a wide range of the electromagnetic spectrum, including ultraviolet (<400 nm), visible range (400–800 nm), the near-infrared (NIR) (750–1200 nm), and the far-infrared (FIR) (2600–14000 nm =

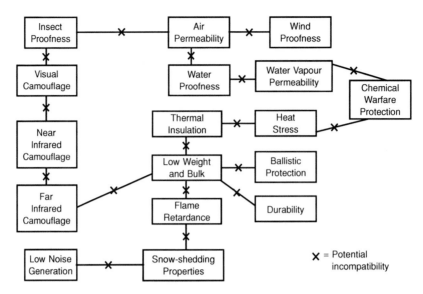

16.1   Incompatibilities in combat materials systems.

2.6–14 µm). Whilst it is relatively easy to to print a wide range of textile fibre types with colour fast dyes of the correct visual shades, it is more difficult to achieve NIR and FIR cover on the same fabric. Artificial fibres such as nylon, polyester, aramids, modacrylics and polyolefins cause particular problems. These requirements are discussed further in Section 16.10.

Since the mid 1970s, research and development effort has resulted in improved knowledge, new products and processes which can go some way towards solving and eliminating these incompatibilities, many of which are discussed later in this chapter.

16.5   Textiles for environmental protection

Military forces have to be prepared to operate in all parts of the globe from arctic, through temperate, to jungle and desert areas. As such they experience the widest range of climatic conditions possible, encountering rain, snow, fog, wind, lightning, sunlight, and dust. They have to survive the attendant heat, cold, wetness, UV light, windchill and other discomforts on land, sea, and in the air. Defence Standard 00-35[8] defines the worldwide climatic conditions in which men, women, equipment and weapons have to operate effectively.

The environment is considered to have the highest priority where protection of the individual is considered. Whether forces are operating at headquarters, during training, on internal security, or peace-keeping duties, or involved in full scale war, the environment is ever present. The battlefield threats – whilst probably much more life threatening – occur at much less frequent intervals.

16.5.1   Underwear materials
Textile materials used for next-to-skin clothing are primarily worn for hygiene reasons. The thermal insulation properties tend to be less important than the tactile properties and the way the material handles moisture (mainly perspiration) in order

to remove it from the skin. Tactile properties are associated with fit, flexibility, roughness, and dermatitic skin reactions.[9] A significant proportion of the population has a true allergic reaction to untreated scaly wool fabrics. Military combat underwear fabrics used by many nations, including the UK, need to be made from non-thermoplastic fibres to minimise contact melt/burn injuries (see Section 16.11 on flame and heat protective materials).

The perspiration and handling properties of knitted underwear materials are extremely critical for mobile land forces such as infantry soldiers, marines and special forces. Their activities range from rapid movement on foot carrying heavy loads, to total immobilisation for long periods when lying in ambush or on covert reconnaissance operations in rural areas. Unlike their civilian counterparts outdoors, military forces cannot choose the level of activity, or wait for better conditions before venturing out. This makes it all the more important to stay dry and comfortable. Sweat-wetted clothing is, at least, uncomfortable, but in the worst situations the loss in dry thermal insulation and the wind chill effect on wet skin and clothing can rapidly lead to hypothermia in cold/wet conditions.

Individuals can characterise the sweat content of a fabric in contact with the skin using a subjective scale of wetness, where 1 is 'dry', 2 and 3 are 'damp', and 4 is 'wet'.[10] Modern laboratory methods allow us to measure the capacity of underwear to handle pulses of sweat from the body.[11] This 'buffering capacity' is measured using a sweating guarded hot plate in accordance with ISO 11092 (The Hohenstein Skin Model Apparatus).[12] The test simulates a condition where the garment is lying on the wearer's wet skin. The passage of water through a sample of the material is measured at intervals. It gives an indication of the water that has passed from the plate into the environment, and also that which has been absorbed from the plate into the sample. The buffering index (Kf) has values between 0 (no water transported) and 1 (all water transported). Values above 0.7 are indicative of 'good' performance. Table 16.7 shows the results for a range of underwear materials[15] based upon special high performance polyester fabrics, and blends with cotton, compared with 100% cotton rib – the UK in-service arctic underwear.[13]

The results show that a wide range of fabrics possess very similar buffering indices when exposed to large amounts of sweat. Values above 0.7 indicate that a fabric will have good wicking and drying properties. The best fabrics in these tests were blends of hollow polyester and cotton in a two-sided (bicomponent) double jersey construction. The 100% cotton military fabric, purported to be a poor fabric

Table 16.7 Sweat buffering indices for a range of underwear fabrics

Underwear fabric	Buffering index (Kf)	Ranking
100% Cotton 1 × 1 rib (olive) (**13**)	0.644	5 =
100% Hollow polyester 1 × 1 rib (olive)	0.641	5 =
100% Quadralobal polyester 1 × 1 rib (olive)	0.720	4
70% Hollow polyester/30% cotton, 2-sided rib	0.731	3
67% Hollow polyester/33% cotton double jersey	0.765	1 =
64% Hollow polyester /36% cotton double jersey	0.764	1 =
72% Quadralobal polyester/28% cotton two-sided rib	0.645	5 =
63% Quadralobal polyester/37% cotton double jersey	0.635	8

for performance underwear, actually performed better than a blend of a special quadralobal polyester and cotton, and equally as well as some of the high performance 100% polyester fabrics specifically developed for sports underwear. The main advantage of non-absorbent synthetic fibres is that they dry more rapidly on the body than cotton fabrics and minimise the the cold 'cling' sensation. These laboratory results are augmented by carefully controlled wear trials and physiological trials with human subjects.[14] The differences in performance are shown to be marginal when worn by highly active humans in outdoor situations.

16.6   Thermal insulation materials

Military forces of many nations need to survive and fight in the most extreme conditions known on earth. The cold/wet regions tend to cause the most severe problems, as it is necessary to provide and maintain dry thermal insulation materials.

The cold/dry areas, including the arctic, antarctic, and mountainous regions require the carriage and use of clothing, sleeping bags, and other personal equipment which possess high levels of thermal insulation. Military forces are prone to sacrificing thermal comfort for light weight and low bulk items. The Royal Marines unofficial motto 'travel light, freeze at night' bears out this assertion.[16] Significant effort has been expended by military research establishments to solve this incompatibility.

Any fibrous material will offer some resistance to the transmission of heat, because of the air enclosed between and on the surface of the fibres. What really determines the efficiency of the fibrous insulator is the ratio of fibre to air, and the way in which the fibres are arranged in the system. An efficient insulator will be composed of about 10–20% of fibre and 80–90% of air, the fibre merely acting as a large surface area medium to trap still air.[17] There is a secondary effect that is governed by the diameter of the fibres. Large numbers of fine fibres trap more still air, owing to the high specific surface area. However, fine fibres give a dense felt-like batting. There is a compromise between fineness and flexural rigidity which gives the fibre the ability to maintain a degree of 'loft', resilience and recovery from compression which is essential for clothing and sleeping bags. Finer fibre battings are more suitable for insulated footwear and handwear, where low thickness is an important factor.

16.6.1   Insulation efficiency

The insulation efficiency of military clothing and equipment is critical, as we endeavour to achieve the highest insulation value at the lowest weight and thickness. Figure 16.2 shows the warmth/thickness ratios in Tog per centimetre for a range of woven, knitted, pile and quilted textile assemblies.[18] The Tog is the SI unit of thermal insulation, measured on a 'Togmeter'.[19] The definition of the Tog unit is: 1 Tog = $m^2 K/10$ Watts.

Figure 16.2 shows that the warmth/thickness values only vary over a small range, although there is an increase in the value for microfibre products such as Thinsulate®.

If we take the same materials and measure them on a warmth to weight basis ($Tog\,m^2\,kg^{-1}$) we can see from Fig. 16.3 that there is now a significant difference in

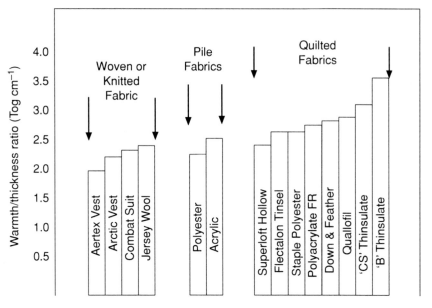

16.2   Warmth/thickness efficiency ratio of textile materials.

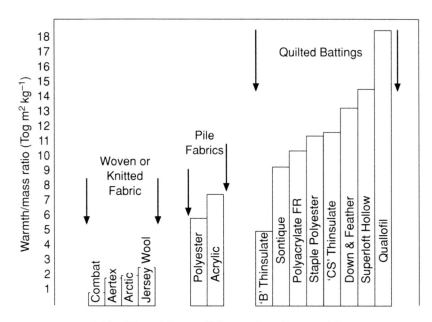

16.3   Warmth/mass efficiency of textile materials.

efficiency. Woven and knitted fabrics offer poor insulation for their mass. The pile fabrics are intermediate in efficiency, but the quilted battings are the most efficient. Hollow fibres and down fillings are 13 to 17 times more efficient than a polyester/cotton woven fabric if insulation needs to be carried by the individual.

16.6.2   Effect of moisture on insulation

Any fibrous, porous insulation material is adversely affected by the presence of moisture, whether this is perspiration or rain. Replacing air of low thermal conductivity by water of high conductivity is the primary cause. Moreover, fibrous materials, particularly pile fabrics or quilted battings, have a high affinity for wicking and entrapping large amounts of moisture. Figure 16.4 shows the dramatic effect on a range of quilted battings. The presence of 10–20% by weight of moisture is sufficient to cause up to 50% loss in the dry insulation value.[20] All military personnel are trained to look after their arctic clothing according to the following phrases:

- Keep it **C**lean
- Do not **O**verheat
- Wear it in **L**ayers
- Keep it **D**ry

At regular intervals reflective metallised insulation materials appear on the market which claim to offer improved insulation performance in clothing and sleeping bags, by reflecting back body heat, and being unaffected by moisture. Unfortunately, these claims do not stand close scrutiny, as the reflective component is not used in a way which offers any advantages for active humans operating outdoors on Earth.[17] Such materials operate by reflecting radiant energy, and work well in the vacuum of space, or where large temperature differences occur. On Earth, the major modes of heat loss outdoors are convection and conduction.

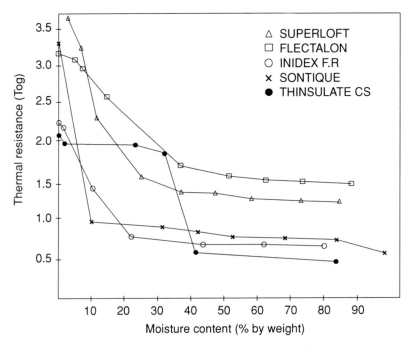

16.4   Loss of thermal insulation in wet battings.

16.7 Water vapour permeable/waterproof materials

One of the basic incompatibilities in technical textiles is that associated with providing waterproof materials which allow free passage of water vapour (perspiration). Without this facility, physiological problems can occur when impermeable clothing is worn by highly active soldiers, marines, and special forces.[21] Table 16.8 shows the consequences of such situations.

In the most extreme war operations individuals cannot choose either the climatic conditions or the intensity of their activities. This can result in injury or death due to hypothermia or hyperthermia.

Over the twenty years since around 1980, appreciable effort has been expended by polymer and textile manufacturers to solve this problem. There are now on the market a wide range of woven, coated, or laminated fabrics which are waterproof and water vapour permeable. Many national forces are issued with combat clothing and equipment which provide these properties.

Table 16.8 Effects of wearing impermeable clothing in different conditions

Conditions	Activity	Consequences
1. Cold/wet climate	Medium activity	Discomfort
2. Cold/wet climate in sweat-wetted clothing	High activity followed by low activity	Hypothermia (cold stress)
3. Hot/moist climate and wearing protective clothing	High activity	Hyperthermia (heat stress)

16.7.1 Types of water vapour permeable barrier fabrics
There are three main categories of materials of this type:[22]

1 **High density woven fabrics** – are typified by Ventile cotton fabric. There are also a range of fabrics based on woven microfibre polyester of Japanese origin such as Teijin Ellettes®, Unitika Gymstar®, and Kanebo Savina®. Ventile was originally developed for military use during World War II, and is still widely used by military and civilian forces.[23]

2 **Microporous coatings and films** – are widely available in many variants. Such membranes are typified by having microporous voids of pore sizes from 0.1–5 μm. The most well-known product, Gore-Tex®, utilises a microporous polytetrafluoroethylene (PTFE) membrane. There are also a range of products based upon polyurethane chemistry, with tradenames such as Cyclone®, Entrant®, and Aquatex®. Other products are based upon microporous acrylic, (Gelman Tufferyn®), and polyolefin (Celguard®). In some cases these membranes or coatings incorporate a top coat of a hydrophilic polymer to resist contamination of the pores by sweat residues, and penetration by low surface tension liquids.[22]

3 **Hydrophilic solid coatings and films** – in contrast to microporous films, the hydrophilic products are continuous pore/free solid films. As such they have a high resistance to ingress of liquids. Diffusion of water vapour is achieved by the incorporation of hydrophilic functional groups into the polymer such as -O-, CO-, -OH, or $-NH_2$ in a block copolymer. These can form reversible hydrogen bonds with the water molecules, which diffuse through the film by a stepwise action along the molecular chains.[24]

Many products are based upon segmented polyurethanes with polyethylene oxide adducts, and have trade names such as Witcoflex Staycool®. This particular polymer was originally developed for military use with research funding from the Defence Clothing and Textiles agency (formerly the Stores and Clothing Research and Development establishment, Colchester, Essex)

Other European market products are based upon a modified type of polyester into which polyether groups have been introduced.[25] The film laminate Sympatex® is typical of this class of textile.

16.7.2   Relative performance of vapour permeable barrier textiles

Work carried out by the UK Defence Clothing and Textiles Agency on a wide range of products[22] has enabled general comparisons of performance to be made. Table 16. 9 compares the main properties in terms of star ratings.

Physiological trials using instrumented human subjects wearing identical garments made from a range of materials have been made.[26] The results show that the differences in vapour permeability between materials are much smaller when worn in garment form than the laboratory test figures would indicate.

Table 16.9   Comparison of performance of water vapour permeable fabrics

Type of barrier	Water vapour permeability	Liquid proofness	Cost	Comments
PTFE laminates	*****	*****	High	Market leader, versatile, expensive
Microporous polyurethanes	** to *****	*** to *****	Medium to high	Widely used, reasonable durability
Hydrophilic polyurethanes and polyesters	** to ***	*** to *****	Low to medium	Cheap, widely available, some durability problems
High density woven fabrics	*****	*	Medium to high	Ventile is expensive, waterproofness low
Impermeable coatings	–	** to *****	Low to medium	Uncomfortable

* = Poor, ** = low, *** = medium, ***** = good, ***** = excellent.

16.7.3   Military usage of waterproof/vapour permeable textiles

Table 16.10 shows the range of military items specified by the UK MOD.

16.8   Military combat clothing systems

Current combat clothing systems are based upon the layer principle, where each layer performs a specific function in the Combat Soldier 95 (the combat clothing system worn by UK Forces which entered service in 1995) assembly. Details of the

Table 16.10   Military items using vapour permeable barrier fabrics

Water vapour permeable barrier	End item usage	Material specification
PTFE Laminates	Waterproof suits, army, Royal Marines and Royal Air Force, camouflaged;	UK/SC/5444 PS/13/95
	MOD police anorak, black;	UK/SC/4978
	Arctic mittens;	UK/SC/4778
	Insock, boot liners;	PS/04/96
	Cover, sleeping bag, olive;	UK/SC/4978
	Tent, one man	UK/SC/4960
Microporous	Suit, waterproof, aerial erectors	UK/SC/5070
Polyurethane and	Suit, foul weather, Royal Navy	PS/15/95
hydrophilic polyurethane	Gaiter, snow, general service	UK/SC/5535
Ventile	Coverall, immersion, aircrew, RAF;	
high density	Jacket, windproof, aircraft carrier deck;	
woven cotton	Coveralls, swimmer canoeist	

Table 16.11   Combat Soldier 95 clothing layers

Layer	Material	Specification
Underwear	100% Cotton knitted 1 × 1 rib, olive	UK/SC/4919
Norwegian shirt	100% Cotton, knitted plush terry loop pile, olive	UK/SC/5282
Lightweight combat suit	Cloth, twill, cotton/ polyester, camouflaged DPM, near IRR camouflaged	UK/SC/5300
Windproof field jacket	Cloth, gaberdine, 100% cotton with nylon rip-stop, water-repellent, near IRR, DPM	UK/SC/5394
Fleece pile jacket	Cloth, knitted, polyester, fleece pile, double-faced	UK/SC/5412
Waterproof rain suit	Cloth, laminated, nylon/PTFE/nylon, waterproof/water vapour permeable, DPM, near IRR camouflaged	PS/13/95

composition of the textiles in each layer are given in Table 16.11. This is the basic fighting system to which can be added other special protective layers, including a ballistic protection system comprising body armour and helmet, a nuclear, biological and chemical (NBC) oversuit, and a snow camouflage oversuit.

16.8.1   Thermal and water vapour resistance data for combat clothing systems

The Combat Soldier 95 layered system has been evaluated using a sweating guarded hotplate apparatus (Hohenstein Skin Model) conforming with ISO 11092:1993.[12] Both the thermal resistance (Rct) and the water vapour resistance (Ret) have been measured and are reported here in Table 16.12. The water vapour permeability index (imt) is defined as $S \times Rct/Ret$, where $S = 60\,Pa\,W^{-1}$. The imt has values between 0 and 1.

The thermal resistance (Rct) and vapour resistance (Ret) values for each layer are additive, which gives an indication of the total value for the clothing assembly, excluding air gaps, which can add significantly to both values.[27]

Table 16.12 Rct and Ret values for Combat Soldier 95 clothing system

Textile layer	Rct ($m^2 K W^{-1}$)	Ret ($m^2 Pa W^{-1}$)	imt
Cotton underwear	0.03	5.1	0.3
Norwegian shirt	0.05	8.6	0.3
Polyester fleece	0.13	13.4	0.6
Lightweight combat suit	0.01	4.3	0.2
Windproof field jacket	0.005	4.8	0.1
'Breathable' rain suit	0.003	11.2	0.01
Total =	0.228	47.4	–

To gain an insight into the effect of vapour resistance (Ret) on a comfort rating system, we can compare these figures with the requirements laid down in the European Standard EN 343:1996 (Clothing for protection against foul weather). This puts the vapour resistance of clothing layers into three categories or classes, which are used in the CE marking of personal protective equipment, as follows:

- **Class 1 materials** have Ret values greater than $150 m^2 Pa W^{-1}$, and are considered to be impermeable, i.e. they offer no perceivable comfort to the wearer.
- **Class 2 materials** have Ret values between 20 and $150 m^2 Pa W^{-1}$, and are rated as medium performance, offering some breathable performance. The majority of products on the market fit into this category.
- **Class 3 materials** have Ret values less than $20 m^2 Pa W^{-1}$ and have the best performance in terms of 'breathability'.

From Table 16.12 we can see in general that all the materials have class 3 performance, although the total clothing assembly would be a class 2 overall.

16.8.2 Vapour permeability of footwear

Leather military footwear for cold/wet climates can be fitted with a waterproof/vapour permeable liner or 'sock'. Its main purpose is to improve the waterproofness of leather boots. Tests have been carried out[27] using a sweating/guarded hot-plate to measure the Ret value of the leather, the liner, and the complete assembly:

- Sock liner: $23.9 m^2 Pa W^{-1}$
- Boot leather: $80.2 m^2 Pa W^{-1}$
- Combined boot + liner: $113.4 m^2 Pa W^{-1}$

Thus, the leather is seen to be the determining factor here, its high resistance is then increased markedly when worn with a liner. The Ret value for the combination is approaching the level at which sweat condensation inside the boot becomes a problem.

16.8.3 Vapour permeability of sleeping bags

The heat and moisture transport properties of fibrous battings for temperate weight sleeping bags have been measured[27] and appear in Table 16.13. Five variants based on polyester fibres are shown.

Table 16.13   Performance of polyester fibre fillings for sleeping bags

Sleeping bag filling type	Density $(g\,m^{-2})$	Water vapour resistance $(m^2\,Pa\,W^{-1})$	Thermal resistance $(m^2\,K\,W^{-1})$
Polyester fibre	175	48.1	0.45
Polyester fibre	200	53.4	0.51
Polyester 4 hole fibre	200	54.5	0.52
Poly synthetic down	285	45.7	0.31
Mixed denier poly	300	49.6	0.39

The four-hole hollow fibre product has been specified for UK military sleeping bags, and is the most thermally efficient for its weight. Note the approximate relationship between Rct and Ret – as one increases so does the other. The Ret values for such battings are high, even though they are open fibrous structures.

16.9   Camouflage concealment and deception

The word camouflage comes from the French word 'camoufler' (to disguise) and was first introduced by the French during World War I to define the concealment of objects and people by the imitation of their physical surroundings, in order to survive. There are earlier examples of the use of camouflage by skirmishing infantry from the 1750s to 1800s, followed by the use of khaki colouring after 1850 in India. In essence, effective camouflage must break up the object's contours, and minimise contrasts between the object and the environment.

Observation in the visual region, either by the eye, or by photography, remains the primary means of military surveillance and target acquisition.[28] However, modern battlefield surveillance devices may operate in one or more wavebands of the electromagnetic spectrum, including the ultraviolet (UV), Near Infra Red (NIR), Far Infra Red (FIR), and millimetric or centimetric radar wavebands. Figure 16.5 shows the relevant parts of the spectrum. A basic objective is that observation and detection should occur at as long a range as possible, and it should be a passive process, that is, it should not be itself detectable. Shining a torch to illuminate an enemy object is likely to meet with unwanted retaliatory action.[29] A notable feature of warfare is that advances in technology proceed in discrete stages. As soon as one threat is countered by technology, another more complex threat emerges. Camouflage research is a good example of this, as new threats in different parts of the spectrum are developed and then subsequently defeated.

Textiles are widely used as the camouflage medium, in the form of light flexible nets, covers, garnishing and clothing items.

16.9.1   Ultraviolet waveband

Only in the snow covered environment is UV observation of military importance. The threat is mainly from photographic systems which use quartz optics and blue/UV sensitive film emulsions. Developments have seen the use of CCD video camera systems which can now operate in this short wavelength region. Snow has a uniform high reflectance at all visible wavelengths, that is, it appears white, but it also continues to have a high reflectance in the UV region. The spectral curves for

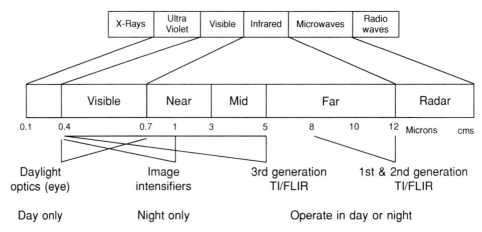

16.5   Electromagnetic spectrum.
TI/FLIR = Thermal imaging/forward looking infrared.

light, heavy, and melting snows vary somewhat, as the texture and crystal structures are different.

The detection problem occurs with white textiles or coatings, as the titanium dioxide pigment which is commonly used as a low-cost widely available treatment for artificial fibres is visually white, but has low reflectance in the UV. Luckily, other pigments such as barium sulphate are suitable and can be incorporated into textile coatings. Figure 16.6 shows the NATO standard reference curve for the snow camouflage UVR colour.[30] Materials must match this curve to achieve good camouflage. It is interesting to note that the reflectance of snow is between 80 and 98% in the UV and visible bands.

Lightweight nylon or polyester filament fabrics coated with a pigmented acrylic coating are widely used for covers, nets and clothing. The coated fabric is cut or incised into textured shapes or blocks to mimic the snow-laden background. Figure 16.7 shows a typical vehicle concealment net in use.

16.9.2   Visible waveband

In this range we are trying to mimic natural or even artificial backgrounds, not just in terms of colour, but also patterns, gloss and texture. Colour can be measured in terms of tri-stimulus coordinates using a spectrophotometer in the laboratory. Camouflage is one of the unique areas where textile coloration is used for a functional purpose, rather than for aesthetic purposes.

If we consider the vegetated temperate environment as an example, can we define an average or standard background against which to develop woodland camouflage? A tree or bush, for instance, will have a different appearance during different parts of the day as the quality of illumination changes. The leaves and bark also change appearance throughout the seasons of the year, deciduous vegetation

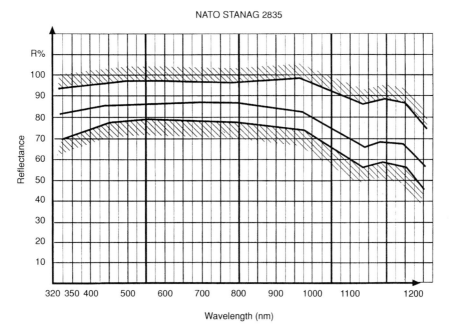

16.6 NATO standard reference curves for snow camouflage.

16.7 Snow camouflage net in use.

showing the widest variation of colour, texture, and appearance from summer to winter. Any measurements or standards which we develop are only modestly accurate, and we have to select colours and patterns which, on average, perform the best. This is still done very empirically, with much trialling of prototypes in the field using direct observation or photographic assessment.

In practice, each military nation has adopted its own visual colours and patterns. Colours often include khaki, green, brown and black, with additional colours such as olive, yellow, orange, pink, grey, beige, and sand to extend use to other urban, rural and desert backgrounds. The UK Disruptively Patterned Material (DPM) printed for clothing uses the first four colours in carefully calculated areas to mimic temperate woodland areas. There is also a two coloured brown and beige pattern for desert use.

Most textile fibres can be dyed to match the visual shades of a standard pattern. Nets, garnishing and covers for vehicle windscreens, machinery and large weapons are often made from lightweight polyurethane or acrylic-coated nylon which is pigmented to give the appropriate visual colours. Figure 16.8 shows soldiers wearing

16.8   Soldiers wearing camouflage-patterned clothing.

clothing in the UK DPM print.[31] This is most effective when viewed with bushes or trees in the immediate vicinity, which is where soldiers tend to conceal themselves. It is not so effective in open grassland, although troops will enhance the effectiveness by covering themselves in freshly cut branches and other vegetation.

16.9.3   Visual decoys

Textile materials are widely used to fabricate and simulate the outline of high value military targets such as aircraft, tanks, missile launchers, and other vehicles. These decoys vary in their complexity depending on the source of the potential attack. If surveillance and target acquisition is at short range, and with sufficient time to study detail, then the decoy has to be a realistic three-dimensional copy of the genuine item. Inflatable decoys made from neoprene or hypalon-coated nylon fabrics have been used to mimic armoured fighting vehicles (AFV), missile launcher/tracker modules, artillery, and other vulnerable equipment. These are cheap and easy to transport and deploy. If target surveillance is at long range and with short acquisition times, as in the case of high speed aerial attack, then the decoys can be a simple two-dimensional representation of the target. As long as it approximates to the shape and size of the original, and casts a shadow that authenticates it, decoys made from fabricated textile materials on a simple supporting frame are adequate for the purpose. Figure 16.9 shows a textile structure developed to mimic the tornado multirole front line aircraft.

The tactical advantages of decoys are obvious: they confuse the enemy into believing that opposing forces are larger than in reality. They may also cause the enemy to release expensive weaponry and ordnance at worthless targets, wasting valuable mission effort and exposing themselves to the risk of retaliation from 'real' weapons.

16.9   Textile decoy of a tornado aircraft.

16.9.4   Near infrared camouflage

The NIR region of the spectrum covers the wavelength range from 0.7–2.0 µm, although current camouflage requirements concentrate on the 0.7–1.2 µm range. In this region objects are still 'seen' by reflection. The military camouflage threat is posed by imaging devices which amplify low levels of light, including moonlight and starlight, which go under the generic name of image intensifiers. These can be in the form of monoculars, binoculars, or low-light television systems. The earliest image intensifiers were developed during Word War II[32,33] after many nations had solved the problem of avoiding visual detection by the eye. Modern image intensifiers use microchannel plates (MCP) technology, and gallium arsenide photocathodes. They are now smaller, lighter and more capable than earlier systems, and hence more readily usable. They tend to operate in the range from 0.7–1.0 µm. Modern infrared photographic 'false colour' films tend to work in the range 0.7–1.3 µm[34] but are only useful for photographing installations which are unlikely to move during the time taken to process the film.

The attribute which is required by camouflage to degrade the threat is related to the reflectance spectrum of leaves, bark, branches, and grasses in the NIR. Figure 16.10 shows the reflectance spectrum of natural objects, including leaves.[35] Note the maximum reflectance at 0.55 µm in the visual range which gives rise to the green colour. As we pass into the NIR there is a dramatic rise in reflectance between 0.7 and 0.8 µm, up to about 40% reflectance. This 'chlorophyll rise' or 'edge' has to be matched by the dyes and pigments used in the camouflage textiles. This is a complex problem, as few dyes, coatings, and pigments exhibit this behaviour in the NIR. Moreover, the reflectance of vegetation varies widely: deciduous tree leaves have relatively high infrared reflectance compared with coniferous needles. There is also the change in reflectance with the seasons, as detailed in Section 16.10.2. The overall NIR reflectance in winter tends to be much lower than in summer.

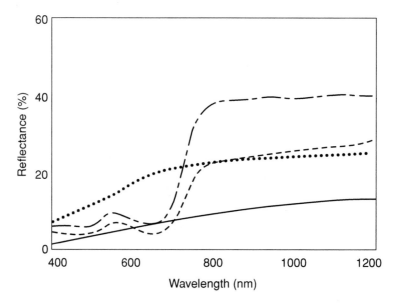

16.10   Typical reflectance curves of natural objects. — -—, Lime tree leaf; … , dry sand; ---, silver fir needle; —, soil. (Reproduced from Ciba-Geigy[35] with permission.)

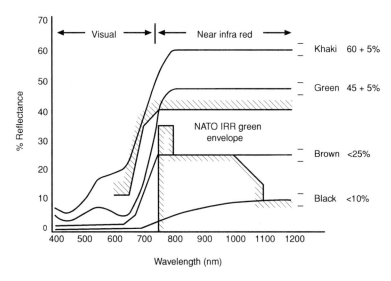

16.11   Reflectance curves for four-colour disruptively patterned textiles.

Figure 16.11 shows the spectral reflectance curves for a four-colour disruptively patterned printed textile using khaki, (60 ± 5%), green (45 ± 5%), brown (<25%) and black (<10%) colourants. Note that each colour has to meet a specified reflectance value. Moreover, the overall reflectance values, integrated with the area of each colour in the print, have to fall within the envelope for NATO IRR Green[36] in accordance with the equation: $(0.16 \times \text{black}) + (0.35 \times \text{brown}) + (0.34 \times \text{green}) + (0.15 \times \text{khaki}) = \text{NATO near infrared green envelope}$, which is superimposed onto Fig. 16.11.

Similar requirements are laid down for camouflage for use in desert regions. The UK uses a disruptive pattern of two colours, brown and beige. The brown must achieve a NIR reflectance value of 45 ± 5%, and the beige a value of 65 ± 5% between 1.0–1.2 μm.

16.9.5   Dyes for near infrared camouflage

Cellulosic fibres and blends thereof have been successfully dyed with a selected range of vat dyes which have large conjugated systems of aromatic rings. These have met NATO requirements for many years.[35] Other fibres such as wool, and synthetic fibres such as nylon, polyester, aramids (Nomex®, Conex® or Kermel ®), and poly-olefins have proved more difficult, since these fibres are dyed with small molecules which have either little or no absorption, and thus high reflectance, in the NIR region.

Many vat dyes have been specifically developed for the express purpose of NIR camouflage, and many patents appear in the literature.[37] They tend to be based upon large anthraquinone–benzanthrone–acridine polycyclic ring systems. They possess very high light, rub, and wash fastness on cellulosic materials, as well as resistance to chemical agents. Many sulphur dyes also exhibit NIR control, but have poor light and wash fastness. They are precluded from military use because of their corrosive interaction with materials used for bullet and shell casings, and detonator compositions. Until recently, it was necessary to incorporate strongly IR absorbing pigments, such as carbon black, which can be melt spun into polyester fibres such as Rhone Poulenc 'grey' polyester, which contains about 0.01% by weight of carbon. Finely divided carbon can also be mixed into printing pastes with suitable binders and applied to textiles. Such processes are difficult to control in production, and subsequent washing can remove some of the carefully metered carbon during use, causing changes in the NIR reflectance values.

Advances in dye application chemistry, funded by the UK MOD, now offer the possibility of NIR camouflage on a wider range of synthetic substrates, including nylon, polyester, aramids, polyolefins, and polyurethane elastane fibres.[38] Such treatments confer desirable properties of high rub, wash and light fastness on military textiles. Pigment printing of textiles using azoic colorants or isoindolinone residues has been reported.[37] Green and black pigments can be screen printed onto textiles in synthetic binders.

Clearly the requirements of NIR camouflage will change as advances in surveillance technology are made, particularly as observation of the battlefield at longer wavelengths, from 1.0–2.0 μm, may be possible in the future.

16.9.6   Thermal infrared camouflage waveband

The thermal or far infrared (FIR) wavebands are, militarily, defined as being from 3–5 μm, and 8–14 μm. In these two bands or 'windows' the atmosphere is sufficiently transparent to allow long range surveillance and target acquisition. Objects are detected by the heat energy they emit or reflect.

Thermal imagers have been around for many years. Early applications were in the medical field in laboratories, but more rugged and compact military systems are now available which can detect vehicles at ranges of several kilometres, and fixed facilities such as storage depots and airfields at ranges of tens of kilometres.

The relationships between energy emitted, emissivities, wavelengths, and temperatures are covered by mathematical relationships derived by Planck, Wien, and Stefan. In simplified terms these are:

$$\text{Wien;} \qquad \lambda_{max} T = \text{a constant} \tag{1}$$

where λ is the wavelength and T is absolute temperature.

$$\text{Stefan;} \qquad E = \eta \sigma T^4 \tag{2}$$

where η is the emissivity and σ is a constant.

Planck's equation relates the spectral radiant emittance of energy to wavelengths at various absolute temperatures. To simplify the mathematics, if we consider two typical military target temperatures, say 33 °C or 306 K for a human body, and 427 °C or 700 K for a typical aircraft or vehicle exhaust, we can consult Planckian curves, which give the following results:

- At 306 K the maximum emittance of radiation is at about 10 μm.
- At 700 K the maximum emittance of radiation is at about 3 μm.

Thus, at higher temperatures the emittance is at shorter wavelengths and vice versa. Therefore we need sensors to cover the range of targets adequately in both windows, 3–5 μm and 8–14 μm.

Stefan's law states that the amount of emitted radiation is proportional to the fourth power of the absolute temperature T, and the emissivity η of the material in question. Therefore, there are two things that we can do to reduce the thermal signature of targets, reduce the temperature and the emissivity of the target.

1 **Reduce the temperature of the target** – vehicles need to be designed so that hot exhaust systems are cooled by air or liquids, by insulating the hot components, or by rerouting the hot piping so that it is covered and not visible. This all adds to the cost and complexity of military vehicles.

   With human targets we can lower the thermal signature by wearing more insulated clothing, putting on covers, or increasing the external surface area using fur or pile-type structures. Unfortunately, this adds to the thermal discomfort of the individual in all but the coldest climates. Additionally, humans are reluctant to wear insulative coverings on the face, one of the most thermally highlighted parts of the body.

2 **Reduce the emissivity of the target** – emissivity is a measure of how efficiently an object radiates its energy. It has a scale of values from 1.0 for a perfect emitter, to 0.0 for materials which emit no energy at all. The list below shows typical emissivities of a range of common materials:[29]

   - Textile fabrics: 0.92–0.98
   - Sandy soil: 0.91–0.93
   - Old snow: 0.98
   - Concrete: 0.94–0.97
   - Hardwood: 0.90
   - White paint: 0.91
   - Black paint: 0.88
   - Stainless steel: 0.12
   - Aluminium: 0.04–0.09.

   Most surfaces are good emitters, except those which are shiny and metallic. Therefore, we can lower the emissivity of the target by using a shiny reflective cover, although this will obviously interfere with visual camouflage (see Section 16.4).

Practical thermal camouflage-screening materials tend to be complex laminates which include a textile fabric support in woven slit film polyolefin carrying a film of aluminium or other shiny metal foil. The foil is covered by a dull green coloured coating, which has the correct visual and NIR characteristics. The green coating is formulated so that it allows the thermal imager to 'look' through it at the underlying metal layer. Such materials have been in service for some years and are specified by UK MOD.[39] The thermal screen is used in conjunction with a standard green/brown incised camouflage net. In this form it avoids other complex thermal reactions caused by solar radiation warming the material, or reflection of radiant energy from the 'cold' ($-50\,°C$) sky, which the imager 'sees' as a negative contrast. The thermal screen is bulky, stiff, and impermeable to sweat vapour, which precludes its use in clothing.

Current work includes studies to provide comfortable, practical thermal camouflage materials for clothing. Further speculative research is examining the feasibility of smart, adaptive camouflage using thermochromic,[40] photochromic, or electrochromic dyes, along with phase change materials. These could provide 'chameleon' type camouflage over a wide range of the spectrum.

16.10   Flame-retardant, heat protective textiles

There is a unique difference between civilian and military fire events. The majority of civilian fires are accidental events, whereas the majority of military fires are deliberate, planned events specifically intended to destroy equipment and installations, or to maim and kill human life.

Military textile materials are often the first materials to ignite. These propagate small fires leading rapidly to large conflagrations. The threat is such that defence forces have paid particular attention to the use of flame-retardant textiles for many applications. These specifically include:

- **Protective clothing** – for firefighters, bomb disposal (explosive ordnance disposal, EOD) crews, nuclear, biological and chemical (NBC) protection, AFV tank crews, naval forces aboard ships and submarines, aircrew, and special forces such as SAS (Special Air Service), SBS (Special Boat Service), and US navy seals.
- **Equipment** – such as tents, shelters, vehicle covers, and bedding.

16.10.1   Military flame and heat threat

The threats to humans and equipment are as follows:

1. open flames from burning textiles, wood, vegetation, furnishings and fuels
2. radiant weapon flash – whether conventional or nuclear weapons
3. exploding penetrating munitions, especially incendiary devices
4. conducted or convected heat, including contact with hot objects
5. toxic fumes generated in confined spaces
6. smoke which hinders escape in confined spaces, and can damage other equipment
7. molten, dripping polymers, which can injure clothed humans and spread fires in furnishings and interior fittings.

16.10.2   Severity of the military threat

Taking each of the main threats in turn will allow a worst-case threat envelope to be constructed. Table 16.14 gives details of the severity of the threat to tank (AFV) crewmen, coupled with the attendant exposure time limits to ensure survival.[41]

The worst-case envelope must therefore encompass fluxes up to $600\,kW\,m^{-2}$ normally. The nuclear thermal pulse situation is complicated, because we assume

Table 16.14   Severity of the military heat and flame threat

Threat source	Typical heat flux ($kW\,m^{-2}$)	Survival time (s)
Burning fuels	~150	7–12 s for no injury
Exploding munitions	~200	<5 s
Penetrating warheads	~500–560	<0.3 s
Nuclear thermal pulse inside closed vehicle	~600–1300	<0.1 s

that an enclosed vehicle is not in a region above its nuclear blast survivability levels. The thermal pulse may not be instantly significant, but the total heat absorbed by the mass of a vehicle may produce heating effects which cause fires inside the vehicle.

16.10.3   Criteria for protection of the individual

We must consider the following criteria to protect forces exposed to the threats listed in 16.10.1:

1. Prevent the outer clothing and equipment catching fire by the use of flame-retardant, self-extinguishing textiles. The material should still be intact and have a residual strength not less than 25% of the original. It should not shrink more than 10% after the attack.[42]
2. Prevent conducted or radiated heat reaching the skin by providing several layers of thermal insulation or air gaps.
3. Minimise the evolution of toxic fumes and smoke in confined spaces by careful choice of materials. This is mainly a hazard posed by clothing and textiles in bulk storage. The submarine environment is a particularly hazardous problem, as it relies on a closed cycle air conditioning system. Some toxic fumes may not be scrubbed out by the air purification system.
4. Prevent clothing in contact with the skin melting, by avoiding thermoplastic fibres such as nylon, polyester, polyolefins, and polyvinylidene chloride (PVDC).

16.10.4   Toxic fumes and smoke

All fires cause oxygen depletion in the immediate area of the fire, and deaths can occur if the oxygen content falls from the normal 21% down to below 6%.

All organic fuels produce carbon monoxide (CO), especially in smouldering fires where complete oxidation of the fuel does not occur. A survey[43] involving almost 5000 fatalities showed that the vast majority of the deaths were attributable to carbon monoxide poisoning. Moreover, the lethal concentrations of CO were much lower than previously believed. Another study[44] concluded that carbon monoxide yields in big fires are almost independent of the chemical composition of the materials burning.

The stable product of all combustion processes and developing fires is carbon dioxide (CO_2), an asphyxiant. It plays a major part in the complex effects which toxic products have on human organisms.

Textile fibres which contain nitrogen, such as wool, nylon, modacrylics, and aramids will produce volatile cyanide compounds to a lesser or greater extent. It has been confirmed that only 180 ppm in the atmosphere will cause death after 10 min. Whether the concentrations available from such fibres is high enough to be a significant threat in real fires is a subject for continuing debate.

Other toxic species from military textile materials include halogenated compounds from polyvinyl chloride (PVC) and neoprene-coated fabrics and PVDC fibres. A range of very toxic oxy-fluoro compounds can be released from PTFE laminates or coatings, and acrolein (an irritant) from cellulosic or polyolefin fibres. Finally, antimony compounds are used in conjunction with halogens to confer flame-retardation properties in fibres, finishes, and coatings. It is somewhat ironic that these two species confer flame-retardant properties, but at the expense of increasing the levels of toxicants in the atmosphere. Textiles used in submarines and ships are required to meet low toxicity and smoke qualification standards.

16.10.5 Thermoplastic melt hazard

There have been documented situations where forces have experienced the adverse effects of molten fibre polymer sticking to the skin of the wearer in fire and flash situations. This can cause more severe injuries in certain specific cases.

Table 16.15 shows that thermoplastic fibres have melting points as low as 105 °C and if used in underwear can shrink onto the skin prior to melting. The most commonly used fibres today are polyester ($Tm = 255$ °C) and nylon ($Tm = 250$ °C), often used in blends with cotton or other fibres.

There is a justifiable argument that the melt hazard is an academic problem, since if anyone is caught in open flames and their underwear reaches temperatures of 250 °C or more, the individual would already be severely injured by primary heat source burns. However, if we consider weapon flash burns, the situation is different, in that large amounts of energy are delivered to the clothing in a fraction of a second. There are multiple effects from a melt burn event, as follows:

1. There may be little or no 'pain alarm time' in which the individual has time to register the pain and move away from the heat source.
2. Latent heat, which is taken in when the fabric melts, is released again on resolidification. This causes more heat to be pumped into a localised area of skin.
3. Molten polymer residues shrink and stick to the skin, causing additional difficulties when medical help attempts to remove the remains of the clothing.
4. Polymer degradation products may enter broken skin wounds and circulate in the blood stream.

Research work which attempted to simulate melt burns from a range of polyester/cotton fabrics[45] concluded that there is enough energy in one molten drop of a polyester-rich blend with cotton to cause skin burns, if it were to fall on unprotected skin. Burns occur from all blends containing more than 35% polyester. The report concluded that the cotton component in the blend can absorb some of the molten polymer, and that the problem can be avoided if blends containing less than 35% polyester are utilised.

Table 16.15 Thermoplastic textile fibres

Fibre type	Trade names	Melting point (Tm°C)
Polyester	Terylene, Dacron, Trevira, Thermastat, Coolmax, Patagonia	255
Polypropylene	Meraklon, Leolene, Ulstron	150
Polyamide	Nylon 6, Nylon 6–6, Tactel	250
Poly Vinylidene Chloride	Damart Thermolactyl, Rhovyl	Shrinks 95 Melts 105
Modacrylic	Teklan, SEF, Velicren	175
Spandex (Elastic Fibres)	Lycra, Vyrene	250

The melt hazard issue is still a cause for much debate, especially in its inferences for infantry and marines operating in cold climates. Some nations ignore this potential problem, whilst others, including the UK, observe the risk in certain special situations for all aircrew, tank crew, and all naval action clothing. The UK has recently relaxed the restrictions on the use of thermoplastic textiles in certain cold weather operations.

16.10.6 Flame-retardant textiles in military use

Although the range of flame-retardant products is large, the actual number of types used by military forces is quite small. Table 16.16 shows those which are used and the applications. The most widely used of these is Proban®-treated cotton, a tetrakis hydroxymethyl phosphonium hydroxide product, bound to the fibre and cured in ammonia. Its advantage is its wide availability and low cost. It provides a finish which is resistant to many (careful) launderings, and gives good protection with low thermal shrinkage in a fire. Its disadvantages are that it liberates fumes and smoke when activated, the treatment can weaken the fabric or spoil its handle, and it must not be laundered using soap and hard water, as these can leave flammable residues in the fabric.

The use of Proban in naval action dress (shirt and trousers) is in a blend with 25% polyester, which improves appearance and durability. The Royal Navy action coverall is a two-layer Proban cotton garment in antiflash white, and it is worn in conjunction with Proban-treated white knitted headover and gloves during high alert action states on board ship.

The meta-aramid fibres possess good physical durability, low toxicity and low smoke evolution properties. However, their high cost limits their use to the special

Table 16.16 Flame-retardant textiles in military use

Fibre/fabric type	Treatment type	Cost	Military uses
Proban cotton	Chemical additive	Relatively cheap	Navy action dress Navy action coverall Anti-flash hood and gloves Air maintenance coverall Welder's coverall
Aramid	Inherent fibre property	Expensive	Tank crew coverall Aircrew coverall Bomb disposal suit Submariner's clothing Arctic tent liners
Zirpro wool	Chemical additive	Medium/high	Navy firefighters RAF firefighters Foundry workers
Modacrylic	Inherent fibre property	Medium/low	Nuclear, biological, and chemical clothing Tent liners
Flame-retardant viscose	Chemical additive	Medium	In blends with aramid fibres only

end-uses listed in Table 16.16. They are available in a wide range of fabric types, invariably in blends with para-aramids, or flame-retardant viscose.

Wool treated with colourless hexafluoro-titanium or -zirconium complexes (Zirpro®) treatments are used for certain heavy firefighter's clothing fabrics, such as the navy 'Fearnought' coverall, and the RAF ground crew coverall. These are typically heavy felted-type fabrics of weights in excess of $1000 \, \text{g m}^{-2}$, which provide good thermal insulation properties for high risk duties.

All UK fighting forces in navy, army and airforce would have to go to full scale war wearing a two-layer oversuit with boots and gloves to protect them from nuclear, biological, and chemical (NBC) warfare threats. The outer fabric is currently a woven twill with a nylon warp and modacrylic weft. The modacrylic component provides a limited degree of flame and flash protection (see Section 16.12).

The general service military tentage material currently consists of a polyester/cotton core-spun base fabric which is coated with a mixture of PVC and PVDC resins with antimony oxide as a flame inhibitor.[46] It also contains pentachlorophenyl laurate (PCPL) as a rot-proofing agent, although this is in the process of change, owing to the adverse effects of PCPL on the environment. Future tentage materials may be made from wholly synthetic polyester-coated textiles which do not require rot proofing. Coatings made from specially formulated PVC, polyurethane, or silicone polymers may be used. There are currently a range of neoprene and hypalon rubber-coated nylon and polyester fabrics which are used for flame-retardant covers, inflatable decoys, and shelters.

Finally, the exotic polybenzimidazole (PBI) fibre has been used in US aircrew clothing and UK military firefighters have recently been equipped with clothing made from PBI Gold fibre.

16.11 Ballistic protective materials

Most military casualties which are due to high speed ballistic projectiles are not caused by bullets. The main threat is from fragmenting devices. In combat, this means, in particular, grenades, mortars, artillery shells, mines, and improvised explosive devices (IEDs) used by terrorists. Table 16.17 shows statistics for casualties in general war, including World War II, Korea, Vietnam, Israel, and the Falklands conflicts.[47]

The main cause of injury to civilians (including police officers) has been bullets. These can be classed as 'low velocity' bullets fired from hand guns (revolvers, pistols) at close range. 'High velocity' weapons, such as rifles and machine guns tend to be used at longer ranges. Generally speaking, the velocity itself is less important than the kinetic energy, bullet shape, or composition of the bullet.[47] In terms of lethality, however, bullets are more likely to kill than bomb fragments, which will tend to

Table 16.17 Cause of ballistic casualties in general war

Cause of casualty	Percentage
Fragments	59
Bullets	19
Other	22

inflict several wounds, ranging in severity, depending on the source and distance of the blast. There may also be casualties from the secondary effects of bombs, including collapsing buildings, exploding aircraft, sinking ships, and flying debris.

16.11.1   Levels of protection

Providing the appropriate level of protection for an individual is rarely a problem. The limiting factors governing protection are related to the weight, bulk, rigidity and thermophysiological discomfort caused by body armour. Given these restrictions, it is apparent that textile structures should be prime candidates to provide the low weight, flexibility, and comfort properties required. Textile body armours may give protection against fragments and low velocity bullets, but not against other threats such as high velocity bullets of, typically, 5.56 mm, 7.62 mm and even 12.7 mm calibre. Textile armours are also defeated by flechettes, which are small, sharp, needle-shaped objects, disseminated in large numbers by exploding warheads or shells. In the case of these high speed projectiles we have to resort to using shaped plates made from metals, composites or ceramics. These are placed over the vital organs such as the heart. Figure 16.12 shows the reduction in casualties which result from wearing various levels of body armour and helmets. It is clear that the more a person wears, the better are the chances of avoiding injury. There does, however, seem to be a law of diminishing returns operating here owing to the bulk and weight factors mentioned earlier. For all the reasons stated here, ballistic protection of the active individual is always a compromise.

To illustrate the compromises that have to be made, the lightest fragment protective combat body armour (CBA), covering the minimum area of the body might

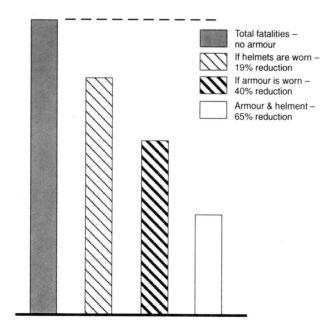

Total fatalities –
no armour

If helmets are worn –
19% reduction

If armour is worn –
40% reduction

Armour & helment –
65% reduction

16.12   Estimated reduction in casualties resulting from wearing body armour (troops standing in the open, threatened by mortar bomb.)

16.13 Soldier wearing combat body armour.

weigh 2.5–3.5 kg (see Fig. 16.13). If we then provide additional protection against high velocity bullets, using rigid plates and increasing the area of torso coverage, the weight might conceivably reach 13–15 kg, or about one-fifth of the weight of an average fit adult, and this does not include helmets, visors, and leg protection![48] The ultimate clothing system for whole body and head protection is the EOD suit, which is shown in Fig. 16.14.

16.11.2   Textile materials for ballistic protection

Ballistic protection involves arresting the flight of projectiles in as short a distance as possible. This requires the use of high modulus textile fibres, that is those having very high strength and low elasticity. The low elasticity prevents indentation of the body and subsequent bruising and trauma caused by the protective pack after impact. Woven textiles are by far the most commonly used form, although non-woven felts are also available.

16.14   Explosive ordnance disposal clothing.

One of the earliest materials used was woven silk, and work done in the USA has examined the use of genetically engineered spiders silk to provide protection. High modulus fibres based on aliphatic nylon 6-6 (ballistic nylon), have a high degree of crystallinity and low elongation, and are widely used in body armours and as the textile reinforcement in composite helmets.[49]

Since the 1970s a range of aromatic polyamide fibres have been developed (para-aramids). These are typically based on poly-para benzamide, or poly-para phenylene terephthalamide. Fibres with tradenames such as Kevlar® (Du Pont) and Twaron® (Enka) are available in a wide range of decitexes and finishes.

A range of ultra high modulus polyethylene (UHMPE) fibres have been developed. They are typically gel spun polyethylene (GSPE) fibres, with tradenames such as Dyneema® (DSM) and Spectra® (Allied Signal). Fraglight® (DSM) is a needle felt fabric having chopped, randomly laid GSPE fibres. These GSPE fibres have the lowest density of all the ballistic fibres at about $0.97 \, g \, ml^{-1}$. The main disadvantage of these fibres is their relatively low melting point at about 150 °C. Research

Table 16.18   Comparison of ballistic textile performance against steel wire

Property	Steel wire	Ballistic nylon	Kevlar 129	Dyneema SK60
Tensile strength (MPa)	4000	2100	3400	2700
Modulus (MPa)	18	4.5	93	89
Elongation (%)	1.1	19.0	3.5	3.5
Density (g ml^{-1})	7.86	1.14	1.44	0.97

work on the formation of composite materials for helmets using these polyethylene fibres has indicated that excellent ballistic performance was possible with significant reductions in areal density of about 45% compared with ballistic nylon.[50,51] Table 16.18 is a comparison of the properties of these synthetic fibres with steel wire.

It is clear that these specialist textile fibres offer the great advantages of low density and high tenacity compared with steel wire. Para-aramids and polyethylene fibres have demonstrated the vast improvements in performance which are possible with these fibrous polymers.

16.11.3   Fabric types and compositions

The majority of ballistic fabrics are of a coarse loose plain-woven construction. Continuous multifilament yarns with the minimum of producer twist tend to give the best results. The loose woven construction produces a light flexible fabric ideal for shaped clothing panels. However, with a loose sett there is a high probability of a projectile sliding between the individual filaments. In addition, a certain amount of bulk is necessary, as ballistic resistance increases with overall areal density. This necessitates the use of many layers, typically between 5 and 20, to produce a ballistic pack which will perform adequately. Figure 16.15 shows the inner layer of such a pack at the point where a bullet or fragment has been arrested. Each body armour layer is allowed to move independently, the pack is secured by stitching quilting lines or squares to maintain a degree of flexibility. This allows the wearer to bend, turn, and make arm movements. It is necessary to seal the ballistic vest inside a waterproof and light-tight cover, as the presence of moisture and UV light can reduce the ballistic performance.

16.11.4   Ballistic testing and evaluation

Material packs are tested in instrumented firing ranges. It is necessary to fire a projectile of standardised weight and size, and at a range of velocities, which are aimed at the pack. Using these fragment-simulating projectiles in a series of test firings enables a measure known as the V_{50} for each material pack and projectile to be made. The V_{50} is the velocity (in m s^{-1}) at which there is an expected probability of penetration of 0.5, that is, 50% go through and 50% do not. This can be used as a quality control measure. We also need to know the V_0, which is the highest velocity at which no penetration occurs at all. This is sometimes known as the V_c or critical velocity. The V_c is thought by some to be a more practical measure, since the objective of the armour is to stop all projectiles reaching the wearers body!

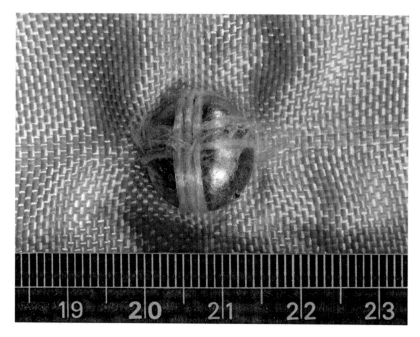

16.15   Ballistic pack showing arrested bullet.

In order to judge how effective a protective armour is likely to be in combat, we have to use a model which simulates a combat situation. The initial information fed into the model includes the V_0 for several sizes of fragment against a particular armour, together with the area of body coverage. We then use data about real weapons ranged against unprotected versus protected individuals. This casualty reduction analysis enables us to predict the real effectiveness of the armour in reducing casualties and fatalities.

16.12   Biological and chemical warfare protection

Biological and chemical warfare is a constant world threat. The toxic agents used are relatively easy to produce and their effects are emotionally and lethally horrific to the general population. They are weapons of insidious mass destruction. The fact that they have not been used in recent conflicts may be due, in part, to the difficulty of delivering and disseminating such weapons onto specific chosen targets. It is imperative to avoid adverse meteorological effects, such as wind blowing the agents back onto the delivering force! There is also the deterrent effect, as the use of such weapons may invoke massive escalating retaliation with other means of mass destruction, such as nuclear weapons.

The types of classic agent which might be used are outlined in Table 16.5. Perhaps one of the most common is mustard agent, which attacks both moist skin, tissues and the respiratory system, causing severe blistering, swelling and burns. Normal mustard agent is bis-(2-chloroethyl) sulphide, which was first produced in 1822, but first used in the later stages of World War I.

The nerve agents were first developed in the 1930s by German chemists,[52] and are so called because they affect the transmission of nerve impulses in the nervous system. They are all organophosphorus compounds such as phosphonofluoridates and phosphorylcyanides which are rapidly absorbed by the skin and respiratory system, although they are primarily respiratory hazards. They were given names such as Tabun (GA), Sarin (GB), and Soman (GD). Agent VX was developed by the USA in the late 1950s, and is one of the most toxic and persistent agents known.[53,54]

The borderline between biological and chemical agents has become less clear over the years as developments in biotechnology have multiplied the types of agent now possible. Classical biological agents would include bacteria, viruses, and rickettsia, but we can now include genetically modified forms of these, and add other toxins, peptides, and bioregulators.

The primary and essential devices for protection of the individual are ori-nasal or full-face respirators which are designed to filter out and deactivate the toxic species. However, mustard agents attack the skin, and nerve agents can be absorbed by damaged skin and at pressure points such as fingers, knees and elbows, necessitating the use of full body protective clothing, Most current clothing systems use activated carbon on a textile substrate to absorb the agent vapour. Activated carbon can be used in the form of a finely divided powder coating, small beads, or in carbon fibre fabric form. This form of carbon has a highly developed pore structure and a high surface area, enabling the adsorption of a wide spectrum of toxic gases. Those with boiling points greater than 60 °C are readily physically adsorbed on the charcoal, but vapours boiling at lower temperatures must be chemically removed by impregnants supported on the carbon.[55]

In practice most nations carry the activated charcoal on an air-permeable non-woven support, on a foam-backed textile, or in a laminate consisting of two textile fabrics sandwiching a charcoal layer. The UK uses a nonwoven, multifibre fabric onto which is sprayed charcoal in a carrier/binder.[56] This fabric is treated with an oil- and water-repellent finish. The charcoal layer is used in conjunction with an outer woven twill fabric consisting of a nylon warp and modacrylic weft, and carries a water-repellent finish. This layer wicks and spreads the agent to attempt to evaporate as much as possible before it transfers to the charcoal layer underneath (see also Section 16.10.6).

NBC protective clothing is currently worn over existing combat clothing, and is cumbersome and uncomfortable to wear in active situations. Much development work is devoted to attempting to reduce the thermophysiological load on the wearer, by reducing the number or bulk of layers in the clothing system whilst maintaining a high level of protection.

References

1. GEORGE WASHINGTON, *Speech to American Congress*, 8th January 1790.
2. C S BROWN, 'Benjamin Thompson – Count Rumford', *MIT Press*, 1995, p. 128.
3. M CHAPPELL, *The British Soldier in the 20th Century, Part 1, Service Dress 1902–1940*, Wessex Military Publishing, Hathersleigh, Devon, 1987, p. 3.
4. M CHAPPELL, *The British Soldier in the 20th Century, Part 6. Tropical Uniforms*, Wessex Military Publishing, Hathersleigh, Devon, 1988, p. 3.
5. M G BURNS, 'British combat dress since 1945', *Arms and Armour Press*, Classell, London, 1992, pp. 6, 7, 15.
6. M CHAPPELL, *The British Soldier in the 20th Century. Part 5. Battledress 1939–1960*, Wessex Military Publishing, Hathersleigh, Devon, 1987, pp. 8–16.
7. G T HOLMES, Paper presented at 8th Commonwealth Conference on Clothing and General Stores, Department National Defence, Canada, 1965.

8. Defence Standard 00-35, Issue 2, *Environmental Handbook for Defence Materials*, Chap. 1-01, Table 2, MOD Directorate of Standards 6, March 1996, p. 5.

9. R F GOLDMAN, 'Biomedical effects of underwear', *Handbook on Clothing*, ed. L Vangaard, Chap. 10, NATO Research Study Group 7, Natick, Massachussetts, USA, 1988.

10. N R S HOLLIES, 'Comfort characteristics of next-to-skin garments', Proceedings of International Seminar on Textiles for Comfort, Shirley Institute (BTTG), Manchester, 1971.

11. K H UMBACH, *Clothing Physiology – Tasks and Problems*, Vortrag VIII, UNIDO-Chemiefaserkurs, Wien, 1981.

12. ISO 11092. (DIN 54101). Measurement of stationary thermal and water vapour resistance by means of a thermo-regulatory model of human skin, 1993.

13. Specification UK/SC/4919, *Vest and Drawers, Cold Weather, Olive*, UK MOD, DCTA, Didcot, Oxon, November 1986.

14. C E MILLARD, D M KELM and A J ROBERTS, *An Assessment of Various Types of Cold Weather Underwear*, Centre for Human Sciences Report DRA/CHS/PHYS/95/045, Oct. 1995.

15. A HOBART and S HARROW, *Comparison of Wicking Underwear Materials*, Defence Clothing and Textiles Agency Report, January 1994.

16. M F HAISMAN, 'Physiological aspects of protective clothing for military personnel', *Clothing Comfort*, eds. N R S Hollies and R F Goldman, Ann Arbor Science, Michigan, 1977, Chap. 2, p. 5.

17. C COOPER, 'Textiles as protection against extreme winter weather', Textiles, 1979 **8**(3) 72–83.

18. R A SCOTT, *Efficiency of Thermal Insulation Materials*, unpublished work, Defence Clothing and Textiles Agency (DCTA), Colchester, UK, 1985.

19. B.S 4745: 1971 'Measurement of Thermal Insulation using a Guarded Hotplate Apparatus'.

20. R A SCOTT, 'Effect of moisture on thermal insulation materials', unpublished work, DCTA, Colchester, 1985.

21. N A GASPAR, 'Technical problems associated with protective clothing for military use', Paper presented at 36th International Man-made Fibres Conference, Osterreichisches Chemiefaser Institut, Dornbirn, Austria, 17–19th September 1997.

22. R A SCOTT, 'Coated and laminated fabrics', in *Chemistry of the Textile Industry*, ed. C M Carr, Blackie Academic and Professional, 1995, pp. 234–243.

23. C HIGENBOTTAM, *Water Vapour Permeability of Aircrew clothing*, MOD, DERA Report PLSD/CHS5/TR96/089 of December 1996.

24. G R LOMAX, 'Hydrophilic polyurethane coatings', J. Coated Fabrics, 1990 **20** 88–107.

25. M DRINKMANN, 'Structure and processing of Sympatex® laminates', J. Coated Fabrics, 1992 **21** 199–211.

26. N C GRAY and C E MILLARD, *Moisture Vapour Permeable Garments – a Physiological Assessment*, DERA, Farnborough Report PLSD/CHS5/CR96/010 April 1996.

27. D CONGALTON, 'Thermal and water vapour resistance of combat clothing', paper presented at *International Soldier Systems Conference*, Colchester, UK, ed. R A Scott, MOD, Defence Clothing and Textiles Agency, October 1997, 389–404.

28. Courtaulds Ltd, *Courtaulds Countersurveillance System*, Brochure FSM 92616c/9, 1992.

29. A F VICKERS, *Camouflage – What the Eye Can't See*, MOD, DCTA Lecture to Physics Department, Durham University, November 1996.

30. NATO Standardisation Agreement (STANAG) 2835, 'NATO UVR White Colour – Reference Curve and Limit Curves/Colorimetric Tolerances and Characteristics', edition 2, September 1995.

31. T NEWARK, Q NEWARK and J F BORSARELLO, *Brasseys Book of Camouflage*, Brasseys, London, 1996.

32. Y A IVANOV and B V TYAPIN, *Infra-red Technology in Military Matters*, English translation AD 610765, Air Force Systems Command, Wright – Patterson Air Force Base, Ohio, USA, published by Brasseys, London, 1965.

33. M A RICHARDSON, I C LUCKRAFT, R S PICTON, A L RODGERS and R F POWELL, '*Surveillance and Target Acquisition Systems*', Brasseys Battlefield Weapons and Systems, Technology Vol. VIII, Brasseys, London, 1998.

34. J FABIAN, H NAKAZUMI and M MATSUOKA, 'Infra red photography', *Chem. Rev.* 1992 **92** 1197.

35. Ciba-Geigy Ltd., *Dyes for Infra-Red Camouflage*, CIBA, Basle, Switzerland, 1972.

36. NATO STANAG 2338 'NATO Green Camouflage Spectral Curves – Visual and Near Infra-Red Requirements', now published as Defence Standard 00-23, Ops. 5, Glasgow, Oct. 1980.

37. S M BURKINSHAW, G HALLAS and A D TOWNS, 'Infra-red camouflage', Rev. Prog. Colouration, 1996 **26** 47–53.

38. S M BURKINSHAW, P J BROWN, 'Dyed materials', US Patent 5,607,483 dated 4th March 1997, Secretary of State for Defence.

39. Specification UK/SC/5154, *Cloth, Camouflage, Low Emissivity*, MOD, DCTA, UK, October 1992.

40. D AITKEN, S M BURKINSHAW, J GRIFFITHS and A D TOWNS, 'Textile applications of thermochromic systems', Rev. Prog. Textile Colouration, 1996 **26** 1–9.

41. NATO AC/301 Group on Standardisation of Material and Engineering Practices (B5), Allied Combat Clothing Publication no. 2, 12 November 1992.

42. S F ELTON, 'UK research into protection from flames and intense heat for military personnel', Fire and Materials, 1996 **20** 293–295.

43. S M DEBANNE, M M HIRSCHLER and G L NELSON, 'The importance of carbon monoxide in the toxicity of fire atmospheres', *Fire Hazard and Fire Risk Assessment*, ASTM STP 1150, ed. M M Hirschler, Amer. Soc. Testing & Materials, Philadelphia, 1992, 9–23.

44. M M HIRSCHLER, S M DEBANNE, J B LARSON and G L NELSON, 'Carbon monoxide and human lethality', *Fire and Non Fire Studies*, Elsevier, London, 1993.

45. R A J STAPLES, *The Melt/Burn Hazard from Cotton/Polyester Materials*, S&TD Tech. Memo 96/10, MOD, DCTA, Colchester, October 1996.

46. Specification UK/SC/4436, *Cloth, Duck, Cotton and Polyester, NATO Green, IRR, Flame, Water and Rot Resistant*, UK, MOD, Nov. 1991.

47. L TOBIN, *Military and Civilian Protective Clothing*, MOD, DCTA lecture given to RMCS Wound Ballistics course, January 1994.

48. A L MARSDEN, *Current UK Body Armour and Helmets*, MOD, DCTA lecture 1994.

49. R G SHEPHARD, *The Use of Polymers in Personal Ballistic Protection*, MOD, DCTA lecture, 28 November 1986.

50. A G ANDREOPOULOS and P A TARANTILI, 'Corona treatment yields major advantages', *Textile Month*, February 1997, 30–31.

51. S S MORYE, P J HINE, R A DUCKETT, D J CARR and I M WARD, *A Preliminary Study of the Properties of Gel Spun Polyethylene Fibre-based Composites*, UK, MOD, DCTA lecture, Personal Armour Symposium, Colchester, Essex, Sept. 1996.

52. A HAY, 'At war with chemistry', *New Scientist*, 22 March 1984, 12–18.

53. U IVERSSON, H NILSSON and J SANTESSON (eds.), *Briefing Book on Chemical Weapons*, No. 16, Forsvarets Forskningsanstalt (FOA), Sundbyberg, Sweden, 1992.

54. A HAY, S MURPHY, J PERRY-ROBINSON and S ROSE, 'The poison clouds hanging over Europe', *New Scientist*, 11 March 1982, 630–635.

55. B TUCKER, 'Protection and Decontamination in Chemical and Biological Warfare', *J. Defence Sci. (DERA)*, HMSO, 1997 **2**(3) 280–283.

56. Specification UK/SC/3346G, *Cloth, Bonded, Multi-Fibre Anti-gas*, MOD, DCTA, QPS, Didcot, Oxon., Sept. 1982.

17

Textiles for survival

David A Holmes

Faculty of Technology, Department of Textiles, Bolton Institute, Deane Road, Bolton BL3 5AB, UK

17.1 Introduction

The question concerning textile use in this chapter is, surviving what? The main emphasis here is on the preservation of human life. The clothing itself provides the protection rather than an individual textile material, but textile fabric is the critical element in all protective clothing and other protective textile products. As the safety barrier between the wearer and the source of potential injury, it is the characteristics of the fabric that will determine the degree of injury suffered by the victim of an accident.

There has been a large increase in the hazards to which humans are exposed as a result of developments in technology in the workplace and on the battlefield, for example. The need to protect against these agencies is paralleled by the desire to increase protection against natural forces and elements. The dangers are often so specialised that no single type of clothing will be adequate for work outside the normal routine. During the 1980s and 1990s extensive research has been carried out to develop protective clothing for various civilian and military occupations.[1] One such investigation was carried out by the United States Navy Clothing and Textile Research Facility to determine future clothing requirements for sailors exposed to potential and actual hazardous environments.[2] The results indicated that a series of protective clothing ensembles is required for a variety of potential hazards. Woven, knitted and nonwoven fabrics have been designed to suit specific requirements.[1]

In order to be successful, designers need to work closely with quality assurance and production personnel as well as potential customers and users from the earliest stages of development.[3] The types of protective garment specifically mentioned in the literature are:

- tents
- helmets
- gloves (for hand and arm protection)
- sleeping bags
- survival bags and suits

- fire-protective clothing
- heat-resistant garments
- turnout coats
- ballistic-resistant vests
- biological and chemical protective clothing
- blast-proof vests
- antiflash hoods and gloves
- molten metal protective clothing
- flotation vests
- military protective apparel including antihypothermia suits and ducted warm air garments
- submarine survival suits
- immersion suits and dive skins
- life rafts
- diapers
- antiexposure overalls
- arctic survival suits
- ropes and harnesses.

The types of occupation and activities for which protective garments and other products are made specifically mentioned in the literature are:

- police
- security guards
- mountaineering
- caving
- climbing
- skiing
- aircrew (both military and civil)
- soldiers
- sailors
- submariners
- foundry and glass workers
- firefighters
- water sports
- winter sports
- commercial fishing and diving
- offshore oil and gas rig workers
- healthcare
- racing drivers
- astronauts
- coal mining
- cold store workers.

All clothing and other textile products provides some protection. It is a matter of timescale which decides the degree and type of protection required. Hazards to be survived can be divided into two main categories:

- Accidents: these involve short term exposure to extreme conditions.
- Exposure to hazardous environments: this involves long term exposure to milder conditions than those normally associated with accidents or disasters.

Accident protection includes protection from:[2,4-6]

- fire
- explosions including smoke and toxic fumes
- attack by weapons of various types, e.g. ballistic projectiles, nuclear, chemical, biological
- drowning
- hypothermia
- molten metal
- chemical reagents
- toxic vapours.

Long term protection includes protection from:[2,4]

- foul weather
- extreme cold
- rain
- wind
- chemical reagents
- nuclear reagents
- high temperatures
- molten metal splashes
- microbes and dust.

There is obviously no sharp dividing line between short term and long term exposure to hazards and some hazards could fall into either category.

David Rigby Associates has used information from trade sources to estimate the European protective clothing market.[7] The data published focuses on high performance products such as those used by firefighters and other public utilities, the military and medical personnel. It excludes garments for sporting applications and foul weather clothing. The overall total market is over 200 million square metres of fabric. Of this, an increasing proportion is being provided by nonwovens, which was estimated to account for about 60% in 1998. The European protective clothing market is expanding at an attractive rate but most of the expansion is for applications excluding military and public utilities. In rapidly expanding applications such as the medical field, nonwovens are taking more of the market traditionally provided by woven and knitted fabrics as they are better able to match the performance and cost requirements of the customers. It is considered that suppliers of high performance fibres will find greater opportunities in the future in the developing countries as their requirements for protective textiles mirrors the increase in population. Table 17.1 shows the European consumption of fabric in protective clothing during 1996.[7]

17.2   Short term (accident) survival

17.2.1   Drowning and extreme low temperatures

Hypothermia is a condition which is known as the 'killer of the unprepared' and occurs when the heat lost from the body exceeds that gained through food, exercise and external sources. The risk increases with exertion or exposure to wet and

Table 17.1 European consumption of fabric in protective clothing during 1996 ($10^6 \, m^2$)[7]

Product function	End-use	Public utilities	Military	Medical	Industry, construction, agriculture	Total
Flame-retardant, high temperature	Woven/knit	5	2	–	15	22
	Nonwoven	–	–	–	–	–
	Total	5	2	–	15	22
Dust and particle barrier	Woven/knit	–	–	12	22	34
	Nonwoven	–	–	62	10	72
	Total	–	–	74	32	106
Gas and chemical	Woven/knit	1	1	–	4	6
	Nonwoven	3	–	–	47	50
	Total	4	1	–	51	56
Nuclear, biological, chemical	Woven/knit	–	2	–	–	2
	Nonwoven	–	2	–	–	2
	Total	–	4	–	–	4
Extreme cold	Woven/knit	–	1	–	2	3
	Nonwoven	–	–	–	–	–
	Total	–	1	–	2	3
High visibility	Woven/knit	11	1	–	3	15
	Nonwoven	–	–	–	–	–
	Total	11	1	—	3	15
Totals	Woven/knit	17	7	12	46	82
	Nonwoven	3	2	62	57	124
	Total	20	9	74	103	206

wind conditions. It is a major cause of death in areas where there is a severe climate, such as in Alaska.[8] Over 100 'man overboard' incidents resulting in 30 deaths occurred between 1972 and 1984 in the North Sea oil and gas industry. Immersion times were usually less than 5 min but could be as long as 10 min. In the North Sea the mean sea temperature is below 10 °C for nine months of the year and rarely exceeds 15 °C for the remaining three months. Initial and short term survival is, therefore, the main consideration.[9] Various strategies have been developed for preventing hypothermia, including the use of flotation and thermal protection devices. The United States Navy uses workwear coveralls which provide buoyancy and thermal insulation in case of accidental and emergency immersions in cold water. These provide survival times of 70–85 min in agitated water.[10]

All 140 members of the International Convention for the Safety of Life at Sea (SOCAS) require a thermal protection aid (TPA) to be carried on board vessels as standard equipment in case of shipwreck and the thermal protection required is from cold to prevent hypothermia. The spun bonded polyolefin fibre fabric, Tyvek®, made by Du Pont, when aluminised and made into survival suits[11] and survival bags,[12] satisfies the SOLAS criteria. These suits can also be used in Arctic emergencies.[11] Thermal insulation overalls made from Tyvek are also carried by many Merchant Navy ships and by several airlines flying the polar route in case the aircraft is forced down onto the Arctic ice.[13]

The use of personal water craft for sport is growing at a rate of 40% per year, therefore the demand for specially designed flotation vests has also increased.

Approximately 150 000 vests are sold each year.[14] Du Pont claim that their high tenacity yarns can be used in such products and maintain their strength and ultra-violet protection after prolonged exposure to sunlight and ultraviolet radiation. One of the leading manufacturers claims that their flotation products made from such yarn far exceed the United States Coast Guard Standards for protection and durability. Aesthetics are also important for this market and fabrics made from these nylon yarns can be dyed to create products with vivid colours and elaborate designs.[14]

17.2.2 Ballistic protection

Textile fibres are being used very effectively to protect against fragmenting munitions. They are able to absorb large amounts of energy as a consequence of their high tenacity, high modulus of elasticity and low density. Work is being carried out to establish the best fibres and the best constructions. The most commonly used fibres are currently glass fibre, nylon 6.6 and aramid fibre (e.g. Kevlar® and Twaron®). In addition to applications such as flak jackets, these fibres are being used in helmets and, in conjunction with ceramic inserts, provide armour sufficient to stop high velocity rifle bullets.[15]

Blast-proof vests are most frequently made from aramid fibres, such as Kevlar® (DuPont) and Twaron® (AKZO) and Dyneema® (DSM) high tenacity polyethylene fibre. Different fabric constructions are required for protection against low velocity and high velocity ammunition. Yarns made from aramid fibres have the best resilience to ballistic impact owing to their outstanding elasticity and elongation properties. In addition to ballistic resistance it is also important to know the amount of energy that the zone receiving the impact can absorb by way of deformation.

Most traditional ballistic armour used for bullet-resistant vests relies on multiple layers of woven fabric. The number of layers dictates the degree of protection. Neoprene coating or resination are also commonly used.[16]

In general it has been found that plain square weaves are most effective in ballistic protection. Knitting, of course, could offer considerable advantages in terms of cost and in the production of the final design of a contoured armour, but it has not proved successful, probably because of the high degree of interlocking of the yarns that occurs in the knitting process and resulting fabric with too low an initial modulus.

High performance polyolefin fibres such as Dyneema polyethylene are used to make needle punched nonwoven fabric.[17] This is claimed to provide outstanding ballistic protection and outstanding protection against sharp shrapnel fragments by absorbing projectile energy by deformation rather than fibre breakage as is the case with woven and unidirectional fabrics. It is claimed that the fabric is such a light weight, low density and thin construction that ballistic protection vests are hardly noticeable during normal military service. However, considerable care is needed to optimise the felt structure. Ideally the felt needs to have a high degree of entanglement of long staple fibres but with a minimum degree of needling. Excessive needling can produce too much fibre alignment through the structure, which aids the projectile penetration. Felts with very low mass per unit area are probably the most effective materials for ballistic protection, but, as the mass increases, woven textiles are superior to felts in performance.[16]

17.2.3 Protection from fire

The most obvious occupation requiring protection from heat and flame is firefighting. However, the 1992 Survival Conference held at Leeds University[4] noted that there is a growing need for protective apparel for other occupations, such as police and security guards.

Exposure to heat and flames is one of the major potential hazards that offshore oil and gas rig workers face. Protection from elements other than fire is also required and fabrics have been developed which combine fire-resistant characteristics with water and oil repellency (Dale Antiflame, http://www.offshore-technology.com, July, 1997).

Simulated mine explosions, involving coal dust and methane, endure for 2.2–2.6 s and reach maximum heat flux levels from 130–330 kW m^{-2}. Values from 3–10 s have been quoted for escape through aircraft or vehicle crash fuel spills with heat flux intensities peaking between 167 and 226 kW m^{-2}. The projected time to second degree burns at a heat flux of 330 kW m^{-2} is only 0.07 s. The introduction of a material only 0.5 mm thick increases the protection time significantly to longer than the flashover or explosion time. The danger, however, lies with the parts of the body not covered by clothing, confirmed by statistics showing that 75% of all firefighters' burn injuries in the USA are to the hands and face.[18]

Heat and flame-resistant textiles are used extensively to provide protection from fire and to do so need to prevent flammability, heat conduction, melting and toxic fume emissions.[6] Some of these textiles are made from conventional fibres that are inherently flame retardant (i.e. wool) or fibres that have been flame-retardant finished (principally cellulosic fibres). The newer high performance fibres are also used either alone or in blends. Some of these blends can be a complex mixture of high performance and conventional fibres with a large number of components.[19]

Firesafe Products Corporation has patented a fabric meeting the above criteria. It is woven from inherently non-combustible glass fibre and coated with a series of proprietary water-based polymers. The fabric does not ignite, melt, drip, rot, shrink or stretch and is noted for its low level and toxicity of smoke emissions in a fire. Several versions of the fabric are made for use in furniture barriers, fire and smoke curtains in ships and cargo wraps in aircraft.[20]

Securitex (Turnout Gear Selection, http://www.securitex.com, July, 1997) consider that there is a misconception in the widely held belief that the outer type of shell fibre is the critical factor in determining whether or not a firefighter is injured during flashover conditions and that the use of high performance fibres namely Kevlar and PBI (polybenzimidazole) provide better protection than fabrics made from other fibres. Their tests show that these fibres provide no more thermal protection than any other fabric of equivalent weight. After 10 s exposure to flashover conditions the outer shell was charred and broken up. Under less than flashover conditions, for example, compression bars, it is the type of moisture barrier and water absorption characteristics of the thermal barrier, not the type of outer shell fabric that are the critical factors influencing the type of burn injury.

17.3 Long term survival

Protection from heat, flame, molten metal splashes, severe cold and frost, radiation sources and so on is a prime requirement for both civil and defence applications. The conditions influencing demand depend upon specific environmental hazards, the degree of protection, the level of comfort, the durability of the garments, aesthetics and sociological factors, such as legislation, consumer awareness of possible hazards and so on.[1]

17.3.1   Extreme weather conditions

Events in the industrial field dictate that workers will be required to work and otherwise function in colder and colder temperatures, in weather conditions which hitherto would have been a sufficient excuse to 'down tools and head for the tea hut'. We are asked to provide means for personnel to work and function efficiently in temperatures well below −30 °C in wind, rain and snow, and in the case of military personnel, other hazards. Decisions have been made to send men to drill for oil under ice caps, and to fight in conditions when breathing out is accompanied by icicle formation. These situations are now attracting growing attention among various sectors of the textile and apparel industries.[21] There is also a need for more adequate weatherproof clothing for people who work in less extreme conditions, such as petrol station attendants, surveyors and engineers. Often they use conventionally accepted garments which can be more expensive but far less effective than ones specifically designed for the purpose.[3]

The design of protective military apparel for operating in extreme climates can be complex because of variations in conditions. Requirements for the components in protective apparel sometimes conflict and these demands stimulate interdisciplinary research for new textile materials, equipment and technologies. These disciplines include textile engineering, industrial engineering and design, apparel design, textile science and physiology. Current protection requirements are for normal, combat and emergency survival operations in both peacetime and war. One of the most important current problems is designing apparel that is effective and comfortable.[22]

Price is often synonymous with quality yet this is not always the case in practice. In many cases there is insufficient knowledge of the requirements for high technical performance so that even specialists find difficulty in making judgements on clothing for outdoor pursuits. The design and manufacture of the garment is of considerable importance, particularly the method of seaming for waterproof garments.[23]

Submarine suits must protect the wearer against drowning and hypothermia for long periods under severe weather conditions. The Swedish Division of Naval Medicine of the Swedish Defence Research Establishment has found that survival suits could maintain body temperature for up to 20 hours in cold water simulating winter conditions.[24] Effective survival suits may include a life raft and even diaper material for urine collection. In addition, thermal insulation and buoyancy of the suits are very important. Thermal insulation can be partly provided by an aluminized inner coverall worn over the uniform.[24] It was also concluded that survival suits should be developed consisting of a double layered suit with a life raft, a single layered suit with extra buoyancy and a life raft, or a modified double layered suit with extra buoyancy.[25]

Helly-Hansen, the Norwegian company which specialises in foul weather and survival gear for commercial fishermen, claims a 52% share of the world market. They accept the layering principle and prefer three layers, namely:

1   an inner layer with good skin contact, not too absorbent
2   an insulating layer trapping large volumes of still air and helping transportation of moisture away from the skin and
3   a wind/ water barrier layer.

Tests have shown that it is not the fibre which is important for insulation value of a fabric, but the construction of the fabric, for example, knitted versus woven, thickness, resistance to compression, weight and so on. This and the design of the garment are key factors and the ability to close the garment at the neck, wrist and ankle are important.[26,27]

Other manufacturers produce multilayer insulation systems for use in clothing for severe weather. Northern Outfitters (Superior Technology Means Superior Performance, http://www.northern.com, July, 1997) use their VÆTREX (Vapour Attenuating and Expelling Thermal Retaining insulation for EXtreme cold weather clothing) to make what they claim to be the warmest clothing and boots in the world. VÆTREX uses special open cell polyurethane foam as the principal insulating medium which has permanent loft and allows the expulsion of perspiration. This is sandwiched between two fabric layers, the outer one which deflects the wind and stabilizes the air in the insulation and the inner one which allows moisture vapour transfer. The construction of VÆTREX is shown in Fig. 17.1.

The hollow viscose fibre, Viloft® (Courtaulds) has been mixed with polyester to give a high bulk, low density material for thermal underwear. It gives high water permeability and water absorbency combined with resilience, strength and shape retention properties of the synthetic fibre. All these properties are essential in thermal underwear and both laboratory and field tests indicated Viloft/polyester had a substantial market potential.[15]

The development of Thinsulate® (3M) has been described.[15] This combines polyester staple with polypropylene microfibre and has undergone extensive tests both in the laboratory and in the field. Examples of the latter include use by postmen, ski centres and survival posts in the northern USA and in underwear for US Navy divers. Excellent results have also been recorded on the recent British winter expedition to Everest.[15]

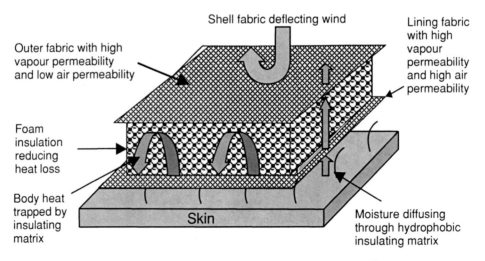

Shell fabric deflecting wind

Lining fabric with high vapour permeability and high air permeability

Outer fabric with high vapour permeability and low air permeability

Foam insulation reducing heat loss

Body heat trapped by insulating matrix

Skin

Moisture diffusing through hydrophobic insulating matrix

17.1   Construction of VÆTREX insulation system.[30]

Metallised coatings are frequently used to improve thermal insulation by reflecting heat radiated by the body.[28] An aluminium reflecting surface is very efficient in the part of the spectrum in which the body gives out radiation. It can reflect 95% of the radiant heat back into the body and in addition acts instantly giving fast warm-up in cases of acute suffering from cold. Flectalon is a filling of metallised and shredded plastic film for use in apparel and applications such as survival blankets. The product, besides its ability to reflect thermal radiation, allows diffusion of moisture and retains its reflecting properties when wet and when compressed. Trials have been successfully carried out by coastal rescue and cave rescue organisations, in mountaineering situations, and also for the protection of new born babies. Metallised polyvinyl chloride (PVC) can be used when flame resistance is also important.[15]

Sommer Alibert (UK) Ltd exhibited a novel composite fabric for insulation purposes which consists of a needled acrylic wadding, polyethylene film and aluminium foil, called Sommerflex for use in lightweight and windproof interlinings for anoraks, sleeping bags, gloves, mittens, continental quilts and mountaineering wear.[21]

Temperature-regulating fabric is made from cotton to which poly(ethylene glycol) has been chemically fixed. At high temperatures the fabric absorbs heat as the additive changes to a high energy solid form. At low temperatures the reverse process takes place. Such fabrics have potential in thermal protective clothing such as skiwear.[29] One such product is Outlast™ (Gateway Technologies, Inc). They claim that ski gloves containing Outlast fabric maintain the skin at a higher temperature than conventional gloves ten times thicker by utilizing the energy conserved during exercise. These thinner gloves allow better dexterity.[29,30]

Choice of appropriate fabric is not the only consideration when designing survival wear. The correct design of the suit is very important, particularly with regard to water ingress. Good suits allow less than 5 g of water ingress but many suits allow up to 1 litre. Even this amount of ingressed water would contribute about 50% of the heat loss from the body.[9]

17.3.2   High temperatures and associated hazards
Table 17.2 illustrates occupations where protection from heat and flame are important.[1] In the occupations listed in the table the human skin has to be protected from the following hazards:[1]

- flames (convective heat)
- contact heat
- radiant heat
- sparks and drops of molten metal
- hot gases and vapours.

The main factors that influence burn injury are:[18]

1   the incident heat flux intensity and the way it varies during exposure
2   the duration of exposure (including the time it takes for the temperature of the garment to fall below that which causes injury after the source is removed)
3   the total insulation between source and skin, including outerwear, underwear, and the air gaps between them and the skin
4   the extent of degradation of the garment materials during exposure and the subsequent rearrangement of the clothing/ air insulation

Table 17.2 Hazardous occupations requiring protection against heat and flame[1]

Industry	Flame	Thermal contact	Radiated heat
Foundry (steel manufacture, metal casting, forging, glass manufacture)	*	**	**
Engineering (welding, cutting boiler work)	*	**	*
Oil, gas and chemicals	*	0	0
Munitions and pyrotechnics	0	0	0
Aviation and space	*	0	0
Military	**	*	*
Firefighters	**	*	*

** Very important, * Important, 0 little importance.

5   condensation on the skin of any vapour or pyrolysis products released as the temperature of the fabric rises.

These factors may not be adequately considered when performance specifications are set for materials.

The most serious garment failure for the wearer is hole formation. When the fabric remains intact, its heat flow properties do not change greatly even when the component fibres are degraded, because heat transfer is by conduction and radiation through air in the structure and by conduction through the fibres (which is relatively small). Only when fibres melt or coalesce and displace the air, or when they bubble and form an insulating char, are heat flow properties substantially altered.

Shrinkage or expansion in the plane of the fabric does not substantially change the thermal insulation of the fabric itself. However, the spacing between fabric and skin or between garment layers may alter, with a consequent change in overall insulation. For example if the outer layer shrinks and pulls the garment on to the body, the total insulation is reduced and the heat flow increases.

17.3.2.1   Fibres suitable for protective clothing
These fibres can be divided into two classes:[31]

1   inherently flame-retardant fibres, such as aramids, modacrylic, polybenzimidazole (PBI), semi-carbon (oxidised acrylic) and phenolic (novaloid) in which flame retardancy is introduced during the fibre-forming stage, and
2   chemically modified fibres and fabrics, for example, flame-retardant cotton, wool and synthetics where conventional fibres, yarns or fabrics are after treated.

Flame retardants that are incorporated into fibres or applied as finishes may be classified into three major groups:[31]

1   Flame retardants, based primarily on phosphorus and frequently combined with other products. Phosphorus containing agencies usually operate in the solid phase, frequently with nitrogen showing synergistic effects;
2   Halogenated species (chlorine- or bromine-containing), which are active in the gaseous phase, and in many cases, are applied together with antimony compounds in order to obtain synergistic effects;
3   Compounds, such as alumina hydrate or boron compounds that provide endothermic dehydration reactions, remove heat and aid overall performance of a phosphorous or halogen-based retardant formulation.

Flame-retardant workwear has been available for many years. Inherently flame-retardant fabrics are produced from aramid (e.g. Nomex®), modacrylic (e.g. Velicren FR®, Montefibre), flame-retardant polyester (e.g. Trevira CS®, Hoechst), flame-retardant viscose (e.g. Visil®, Soteri) and other speciality fibres. Some of these fibres have one of a multitude of problems, such as high cost, thermoplastic properties, difficulties in weaving and dyeing, and poor shrinkage properties, which have prevented them from gaining universal acceptance in all industries.

Durable flame-retardant treated fabrics are composed of natural fibres, such as cotton or wool that have been chemically treated with fire-retardant agents that chemically interact with them or are bound to fibre surfaces.[37]

PBI is claimed to offer improved thermal and flame resistance, durability, chemical resistance, dimensional stability and comfort in comparison with other high performance fibres. Wearer trials have shown that PBI fibre exhibits comfort ratings equivalent to those of 100% cotton. Although PBI is expensive the outstanding combination of thermal and chemical resistance and comfort makes it an ideal fibre for protective clothing applications where a high degree of protection is required, such as firefighting suits, escape suits for astronauts, and aircraft furnishing barrier fabrics for the aircraft industry. PBI is easily processed on all conventional textile equipment and can be readily formed into woven, knit and nonwoven fabrics. PBI's excellent dimensional stability and nonembrittlement characteristics allow fabrics to maintain their integrity even after exposure to extreme conditions. By blending PBI with other high performance fibres, the design engineer can usually improve the performance of currently available protective apparel. PBI offers improved flammability resistance, durability, softness and retained strength after exposure to heat sufficient to damage other flame- and heat-resistant fibres. A 40/60 PBI/ aramid blend ratio has been determined as being optimal for overall fabric performance.[31,32]

Oxidised acrylic fibres have excellent heat resistance and heat stability. They do not burn in air, do not melt and have excellent resistance to molten metal splashes. There is no afterglow and the fabrics remain flexible after exposure to flame. They are ideal where exposure to naked flame is required, are resistant to most common acids and strong alkalis, are very durable and are said to be comfortable to wear. Universal Carbon Fibres markets an anti-riot suit made from oxidised acrylic fibre specifically designed for police and paramilitary forces. The suit is designed to provide protection against both flame and acid and to permit maximum freedom of movement of body and limbs.[34] For lower specification protective clothing, modacrylic fibres are being used successfully. Fibres with improved thermal stability up to 190 °C are available and these can be used alone or in blends for protective clothing.[31] In defence wear-life trials, it has been found that blends of modacrylic and wool can substantially reduce initial cost, reduce maintenance and improve the wear life performance of flame-retardant sweaters.

Wool is regarded as a safe fibre from the point of view of flammability. It may be ignited if subjected to a sufficiently powerful heat source, but will not usually support flame and continues to burn or smoulder for only a short time after the heat source is removed. Wool is particularly advantageous because it has a high ignition temperature, relatively high limiting oxygen index, low heat of combustion, low flame temperature and the material does not drip.[3,15] For foundry workers, who are at risk from being splashed by molten metals such as steel, cast iron, copper, aluminium, zinc, lead, tin and brass, small scale tests using molten iron and copper, verified by large scale tests, showed that wool fabric finished with Zirpro® (IWS) flame retardant offered the best protection. Untreated cotton was found to be second best. Fabric made from glass, asbestos and aramid fibres was considered to be unsuitable for protection against most metals. Molten metal tends to stick to most fabrics allowing time for the skin underneath to reach a high temperature, whereas it runs off wool fabric.[33] Aramid fibres soften and/or melt at around 316 °C and this causes trapping of the molten metal and subsequent excessive heat transfer. Untreated cotton offers good protection against molten aluminium, but the application of some organophosphorous-based flame-retardant compounds makes molten aluminium adhere to the fabric, with excessive heat transfer. This is not the case with some flame retardants based on organobromine compounds. To prevent metal from penetrating the fabric at the point of impact, a relatively heavy and tightly constructed outer fabric is required. In this case, contrary to the case with exposure to the flames, a multilayer approach to garment design is not suitable for protection against molten metal hazards.[34]

Benisek et al.[37] have studied the influence of fibre type and fabric construction on protection against molten iron and aluminium splashes. They made the following recommendations for optimum protection:

1   The fibre should not be thermoplastic and should have a low thermal conductivity.
2   The fibre should preferably form a char, which acts as an efficient insulator against heat from molten metal.
3   With increasing weight and density, the fabric should withstand increasing weights of molten aluminium; and
4   Ideally the fabric surface should be smooth to prevent trapping of metal.

Zirpro finished wool meets the above requirements. Decabromodiphenyl oxide/antimony oxide-acrylic resin finished cotton fabrics (Caliban, White Chemical) have also been found to be suitable for workers in the aluminium industry.[31]

Table 17.3 shows the thermal characteristics of protective clothing fabrics made from different types of natural and synthetic fibre.[31]

17.3.2.2   Use of fibre blends in protective clothing
Some aramid fabrics shrink and break open under intense heat, so a fabric blend known as Nomex III® (Du Pont) has been developed by blending regular Nomex with 5% Kevlar®, which itself has much higher resistance to disintegration. Aramid fibres are expensive but cheaper products can be made by blending them with flame-retardant viscose and flame-retardant wool.[31]

Table 17.3 Thermal characteristics of clothing fabrics[34]

Property	Wool	Cotton	FR Cotton	Nomex III	Kevlar
Mass ($g\,m^{-2}$)	240	320	315	250	250
Reaction to thermal exposure	Ignites	Ignites	FR Degrades		Degrades
Degradation temperature (°C)	260	340	320	430	430
Energy to cause thermal failure ($kJ\,m^{-2}$)	437	504	418	749	667
Energy per fabric mass ($kJ\,m^{-2}/g\,m^{-2}$) × 100	1.44	1.25	1.06	2.36	2.13
Mode of failure	Break open	Ignition	Tar deposition		Heat transfer

In France, Kermel®, a fibre with similar chemical structure and performance to Nomex is especially used for firefighters.[31] Its field of use is widening to include both military and civilian occupations in which the risk of fire is higher than usual. Like aramids it has a high price but this may be offset by blending with 10–15% viscose, which also eliminates static electricity generation. Blends of 25–50% Kermel with flame-retardant viscose offer a price advantage and resistance to UV radiation. Blending with 30–40% wool produces more comfortable woven fabrics with enhanced drape. By using blends it is possible to produce garments that are comfortable enough for the wearer to forget that they are wearing protective clothing. In the metal industry where protection from molten metal is needed and the life of a garment is extremely limited, a 50/50 blend gives very good results but a 65/35 Kermel/flame-retardant viscose blend would be preferred. The characteristics of Kermel blends are shown in Table 17.4.[31] Panox® (RK Textiles) is an oxidised acrylic fibre which has been blended with other fibres. Panox/wool blends are suitable for flying suits. In conjunction with aramid fibres, they can be used for military tank crews, where high resistance to abrasion is required. However, fabrics made from the black oxidised acrylic fibre have high thermal conductivity and are non-reflecting. Hence it is essential to have suitable underwear to protect the skin. For this purpose, a 60/40 Panox /modacrylic fibre double jersey fabric and a 60/40 wool/Panox core fabric have been devised. To prevent transfer of radiant heat, Panotex fabrics (containing Panox fibres from Universal Carbon Fibres) generally need to be metallized. An aluminized oxidised acrylic fabric is suitable for fire proximity work but not for fire entry. In some cases, the heat conduction of oxidised acrylic fabric can be an advantage in the construction of covers for aircraft seats. A fabric with Zirpro-treated wool face and Panox back will spread the heat from a localised ignition source and delay ignition of the underlying combustion-modified foam. Another advantage of proofed oxidised acrylic outer fabric is that it sheds burning petrol and can withstand several applications of napalm.[31]

The integrity and flexibility of specific flame-retardant viscose Durvil®, Nomex®, PBI fibre and wool and their blends has been studied. The results are shown in Table 17.5.[34] In all cases except PBI and PBI/viscose, the fabrics are hard and brittle so

Table 17.4 Effect of fabric mass, fibre, and blend ratio on limiting oxygen index[34]

Composition	Mass (g m^{-2})	LOI
100% Kermel	250	32.8
100% Kermel	190	31.3
100% FR Viscose	250	29.4
100% FR Viscose	145	28.7
50/50% Kermel/ Viscose	255	32.1
50/50% Kermel/ Viscose	205	29.9

Table 17.5 Thermal convective testing of different fibres and fibre blends[34]

	Time to 2nd degree burn (s)	Exposure energy (J cm^{-2})
Durvil	6.5	54.2
Nomex	8.9	73.4
80/20% Durvil/Nomex	4.8	40.3
PBI	7.6	63.4
80/20% Durvil/PBI	6.3	52.9
Wool	10.5	87.8
65/35% Durvil/wool	6.4	53.4
FR Cotton	3.8	31.5

they crack severely and break apart on relatively low flexing. The PBI and PBI/flame-retardant viscose blends can withstand repeated flexing with no effect on fabric integrity. Fabrics of 100% flame-retardant viscose are recommended for overalls and outerwear for military suits, whereas 40/60 viscose wool blends are found to be more suitable for firefighting uniforms. Karvin® (DuPont), a blend of 5% Kevlar, 30% Nomex and 65% Lenzing flame-retardant viscose has been designed for the production of flame-resistant protective clothing. Fabrics made from Karvin have an optimum combination of wear comfort, protection and durability.[31]

Using Dref friction spinning technology, special fibres such as aramid, polyimide, phenol, carbon and preoxidised and other flame-retardant fibres can be simply and economically processed, and special yarn constructions can be created by means of layer techniques.[35] These yarns incorporate the inherently flame-retardant melamine resin fibre, Basofil®.

Multicomponent yarns with cores (e.g. glass filament, metal wire), sheathed with flame-retardant fibre material are increasingly being sought. In the high price sector, which is only implemented when major protection requirements exist, para-aramid or preoxidised stretch broken slivers (up to 40% of yarn as core) are utilised to replace more expensive materials.

The Basofil product description includes the following:

Table 17.6   Characteristics of Basofil fabrics with aluminium coating[38]

Area weight ($g m^{-2}$) (without coating)	400	580
Fabric construction	2/2 Twill	Plain weave
Tensile strength (decaN/5 cm)	Warp 250 Weft 150	Warp 135 Weft 50
Convective heat according to EN 367 (s)	7.2	9.4
Radiated heat ($40 kW m^{-2}$) according to EN 366 (s)	150	115
Contact heat 300 °C according to EN 702 (s)	4.4	8.2
Limited flame spread EN 531	Index 3	Index 3

- flame-resistant, temperature-resistant melamine resin, staple fibres
- LOI 31–33%
- 4% moisture regain
- continuous service temperature: approx 200 °C
- hot air shrinkage, 1 hour at 200 °C, <2%
- coatable and dyeable.

Currently, 300 000 tonnes of coarse range Dref yarns are produced annually of which about 15–20% are technical yarns for the protective clothing sector.

Mechanical characteristics are determined by the choice of suitable cores, and heat and fire protection by the sheath material. Owing to this clear functional allocation, the individual components can be matched for the optimisation of the overall system. Two fabrics have proved especially advantageous for medium and heavy-weight fire and heat protection. For additional protection against extreme radiated heat, the outside of the fabric can be coated with aluminium. One fabric has a yarn consisting of a glass core and a sheath comprised of a blend of 80% Basofil and 20% p-aramid. The other fabric has a yarn consisting of a glass core and 100% Basofil sheath. It is claimed that the fabrics have resistance to convective and contact heat twice that of standard fabrics. Typical uses for the fabric are found in foundries and the metal production industry. Table 17.6 shows the characteristics of fabrics made from yarns incorporating Basofil fibre.[35]

17.3.2.3   Fabric constructions for protection
The optimum properties required of fabric intended for protection against heat and flame have been enumerated as:[3]

1   High level of flame retardance: must not contribute to wearers injury
2   Fabric integrity: maintains a barrier to prevent direct exposure to the hazard
3   Low shrinkage: maintains insulating air layer
4   Good thermal insulation: reduces heat transfer to give adequate time for escape before burn damage occurs

5   Easy cleanability and fastness of flame resistance: elimination of flammable contamination (e.g. oily soil) without adverse effect on flame retardance and garment properties
6   Wearer acceptance: lightweight and comfortable
7   Oil repellency: protection from flammable contamination, such as oils and solvents.

The influence of fabric construction and garment manufacture on flammability and thermal protection has been studied extensively.[31] Fabric construction and weight per unit area play an important role in determining suitability for different applications. Different fabric weights have been recommended for thermal protection under various working conditions. For a hot environment in which the fire hazard is principally a direct flame, a lightweight tightly woven construction, such as 150–250 g m^{-2} flame-retardant cotton sateen would be most suitable. For full fire-fighting installations, a flame-retardant cotton drill of about 250–320 g m^{-2} is recommended. For work in which the garment is exposed to a continuous shower of sparks and hot fragments as well as a risk of direct flame a heavier fabric is required and a raised twill or velveteen of about 320–400 g m^{-2} in flame-retardant cotton could be chosen. With molten metal splashes protection of the wearer from heat flux is important and fabric densities of up to 900 g m^{-2} are found useful.

For heat hazards of long duration, protection from conductive heat is required. Heat flow through clothing reaches a steady state and fabric thickness and density are the major considerations, since the insulation depends primarily on the air trapped between the fibres and yarns. Reducing the fabric density for a given thickness increases thermal insulation down to a minimum level of density below which air movement in the fabric increases and reduces insulation. For short duration hazards, increasing the fabric weight increases the heat capacity of the material, which increases protection.[31] In this case of protection against radiant heat, aluminised fabrics are essential. Clean reflective surfaces are very effective in providing heat protection, but aluminised surfaces lose much of their effectiveness when dirty.[31]

Table 17.7 shows the effect of two heat sources on various types of fabric.[19]

Woven and nonwoven fabrics of different masses made from aramid and PBI fibre have been compared[34] and the results are shown in Table 17.8. The original

Table 17.7 Comparison of radiant and convective heat sources[19]

Fabric	Thickness (mm)	Burn threshold (s) at 80 kW m^{-2}	
		Radiant	Convective
Aluminised glass	0.53	>30	2.6
FR Cotton	0.72	2.2	2.4
Aramid	0.97	2.7	3.1
Zirpro wool	1.16	3.1	4.1
Wool melton	3.64	6.7	8.8
Aramid + cotton interlock	1.77	5.0	5.5
Zirpro wool + cotton interlock	1.96	4.3	6.8
Bare skin		0.5	0.5

Table 17.8 Effect of construction on thermal-protective performance (TPP)[34]

Fibre type	Construction	Mass (g m^{-2})	TPP	TPP/mass[a]
PBI	woven	272	17.6	2.2
	nonwoven	296	28.4	3.3
PBI/Kevlar	woven	245	16.2	2.2
	nonwoven	282	26.0	3.1
Nomex	woven	255	16.4	2.2
	nonwoven	238	19.8	2.8

[a] At $84 \, kW \, m^{-2} \, s^{-1}$ 50/50 radiant/convective heat exposure.

Table 17.9 Effect of fabric constructional parameters on protection for PBI fabric[34]

Weave (twill)	Mass (g m^{-2})	Thickness (mm)	Temperature rise (°C/3 s)	Blister protection (s)
2/1	99	0.19	22.1	2.6
2/1	160	0.29	20.0	3.0
2/1	211	0.39	17.7	3.5
3/3	167	0.31	18.1	3.4

fabric constructional data were published in imperial units and they have been converted to SI units for consistency. Woven fabrics were designed as the outer shell material in firefighters' turnout coats, and the needlefelt nonwoven fabrics could be considered for use as a backing or thermal liner in thermally protective apparel. This work shows that nonwoven fabrics provide consistently better thermal protection than woven fabrics of equivalent mass per unit area.

Table 17.9 shows the effect of fabric constructional parameters on the protection provided by fabric made from PBI fibre.[31] Again the original fabric constructional data were published in imperial units.

There are conflicting requirements of protection and comfort in protective clothing. Fabric thickness is a major factor in determining the protection afforded against radiant and convective heat, but at the same time it impedes removal of metabolic heat from the body by conduction and sweat evaporation. Hence it is necessary to have a suitable garment design to enable body heat to be dissipated.[31] Gore-Tex microporous PTFE (polytetrafluoroethylene) film (Gore Associates) has been developed and used in producing waterproof and windproof fabrics with moisture vapour permeability to provide comfort to the wearer. For this purpose, a three-layer Nomex III/ Gore-Tex/modacrylic fabric has been found to be extremely good.[31]

The high performance aramid fibres such as Kevlar and Nomex have been and are made into both woven and nonwoven fabrics for protection against chemical, thermal and other hazards. As the relatively high cost of Nomex garments precludes their use in limited wear applications, Du Pont has developed Nomex spun-laced fabrics for low cost protective apparel for wear over regular work clothing. Spun-laced technology has also enabled the development of lighter weight turnout coats for firefighters. In the 1990s, Du Pont focused its research and development efforts on improving comfort, dyeing technology and moisture absorption and transmission of Kevlar and Nomex fibres, creating new fibres, reducing fibre linear densities and making permanently antistatic Nomex commercially available.[36]

Cotton textile garments were considered by the US National Aeronautics and Space Administration (NASA) to be suitable for protective apparel for space shuttles on the basis of skin sensitivity, comfort, electrical sensitivity and so on. Chemically treated flame-resistant cotton fabrics for space shuttle apparel were made from a two-ply sateen ($244\,g\,m^{-2}$), a weft sateen ($153\,g\,m^{-2}$), and a two-ply, mercerised, knitted single jersey fabric ($187\,g\,m^{-2}$). A space suit designed for NASA has been developed by combining new technology in fabric moulding with shuttle weaving. The tubular fabric, which is woven on an X-2 Draper shuttle loom from polyester continuous filament yarns is coated and moulded into specific shapes.[31]

A recommendation has been made for the use of an aluminised fabric as a fire blocking layer to encase polyurethane foam in aircraft seating. DuPont has developed lightweight multilayered spunlaced Nomex/Kevlar structures as fire blocking layers in aircraft seat upholstery. Nomex provides fire resistance whereas Kevlar provides added strength.[31]

17.3.2.4 Finishes for heat and flame resistance

The application of various finishes to cellulose fibres has been reviewed by Horrocks.[37] The commercially most successful durable finishes are the N-methylol dialkyl phosphonopropionamides (e.g. Pyrovatex®, Ciba; TFRI®, Albright and Wilson) and tetrakis(hydroxymethyl) phosphonium salt condensates (e.g. Proban®, Albright and Wilson).

To fulfil stringent requirements the natural flame-retardant properties of wool can be enhanced by various flame-retardant finishes. Titanium and zirconium complexes are very effective flame retardants for wool and this has led to the development of the IWS Zirpro finish. The Zirpro finish produces an intumescent char, which is beneficial for protective clothing where thermal insulation is a required property of a burning textile. A multipurpose finish incorporating Zirpro and a fluorocarbon in a single bath application makes wool flame retardant as well as oil-, water- and acid-repellent. This is extremely useful for end-uses where the protective clothing could become accidentally or deliberately contaminated with flammables, such as grease and oil and petrol, such as police uniforms where high moisture vapour permeability with low heat transfer and adequate durability are also important. Wool fabric finished with Zirpro flame retardant and a permanent fluorocarbon oil-repellent finish has demonstrated satisfactory performance under laboratory conditions. This combination of finishes is considered to give better overall performance.[3,25] With increasing environmental awareness, the use of such heavy metal-based finishes has been questioned.

Shirts containing 100% cotton, flame-retardant cotton, flame-retardant wool and Nomex aramid fibre have been evaluated for their protective and wear life performance. The greatest protection was provided by flame-retardant cotton and wool fabrics. Nomex fibre fabric gave less protection and untreated cotton the least.[31]

Flame-retardant finishes for synthetic fibres have been developed. Ideally, these should either promote char formation by reducing the thermoplasticity or enhance melt dripping so that the drips can extinguish away from the ignition flame. For protective clothing, char-forming finishes would be desirable. Flame-retardant finishes for nylon 6 or 6.6 do not seem to have had any commercial success.[31,36] Flame retardancy can be imparted to acrylic fibres by incorporating halogen or phosphorus-containing additives.[34] although this never happens in practice because modacrylics yield equivalent and acceptable performance levels.

17.3.2.5   Garment construction

Fireighters fighting a room fire can be exposed to up to $12.5\,kW\,m^{-2}$ and up to $300\,°C$ temperature for a few minutes. Heat exposure in a fire consists primarily of radiation, but convective and conductive heat (if, for example, molten metal or hot paint falls on a garment) may also be encountered. Under any of these conditions, the garments should not ignite; they should remain intact, that is, not shrink, melt or form brittle chars, and must provide as much insulation against heat as is consistent with not diminishing the wearer's ability to perform their duties. Several garment characteristics are important for protecting the wearer from pain and burn injury.[38] The major protective property is thermal resistance, which is approximately proportional to fabric thickness. Moisture content reduces this resistance. The resistance can be reduced by high temperature, especially if this causes the fibres to shrink, melt or decompose. Curvature decreases the resistance requiring more thickness to protect fingers than large body areas. As stated previously, clean reflective surfaces are very effective in providing heat protection. Surface temperatures of fabrics exposed to radiation are reduced to about half in still air by the use of aluminised surfaces.

Moisture present in a heat protective garment cools the garment, but it may also reduce its thermal resistance and increase the heat stored in it. If the garment gets hot enough, steam may form inside and cause burn injury. In the USA most firefighters turnout coats contain a vapour barrier either on the outside or between the outer shell and inner liner. This prevents moisture and many corrosive liquids from penetrating to the inside, but, on the other hand, it interferes with the escape of moisture from perspiration and increases the heat stress. Some European fire departments omit vapour barriers.

Thermal protective clothing should meet the following requirements:[31]

1   flame resistance (must not continue to burn)
2   integrity (garment should remain intact, i.e. not shrink, melt or form brittle chars which may break open and expose the wearer)
3   insulation (garments must retard heat transfer in order to provide time for the wearer to take evasive action; during combustion they must not deposit tar or other conductive liquids) and
4   liquid repellency (to avoid penetration of oils, solvents, water and other liquids).

The requirements for US firefighting bunker gear are (Turnout Gear Selection, http://www.seritex.com, July, 1997)

1   Those affecting garment life:
   • tear and abrasion resistance
   • resistance to UV degradation (for strength and appearance)
   • thermal damage tolerance (influencing ability to be reused after exposure to high temperatures)
   • resistance to molten metal splatter and burning embers
   • cleanability.

2   Those affecting firefighter safety:
  • ice shedding ability
  • water absorption on the fire ground
  • weight and suppleness
  • mobility
  • visibility
  • thermal protection
  • breathability to water vapour.

Bunker gear is made from a three-layer system consisting of (Turnout Gear Selection, http://www.seritex.com, July, 1997) an outer shell, a moisture barrier and a thermal barrier. The outer shell is the first line of defence for the firefighter. It provides flame resistance, thermal resistance and mechanical resistance to cuts, snags, tears and abrasion. There are a variety of outer shell fabrics available each with advantages and disadvantages. Most fabrics have a fibre content of Nomex and Kevlar, a twill or ripstop woven construction and an area density of 200–250 g m^{-2}.

The moisture barrier is the second line of defence. Its principal function is to increase firefighter comfort and protection by preventing fire ground liquids from reaching the skin. It also provides some burn protection due to its insulation value and ability to block the passage of hot gas and steam. The moisture barrier consists of a film or coating applied to a textile substrate. The substrate is either woven, usually ripstop, or spunlaced nonwoven both with aramid fibre content. The coatings and films can be either breathable or non-breathable.

Breathable films are usually microporous PTFE such as GoreTex and breathable coatings are either microporous or hydrophilic polyurethane. Breathable moisture barriers permit the escape of body perspiration and reduce the incidence of heat stress which is a major cause of death amongst firefighters. The thermal liner blocks the transfer of heat from the firefighting environment to the body of the wearer. It usually consists of a spunlaced, nonwoven felt or batting quilted or laminated to a woven lining fabric. The felt or batting is made from Nomex or Kevlar or a mixture of these two fibres. One manufacturer uses a closed cell foam made from PVC and nitrile rubber to reduce water absorption and increase drying time. The woven lining fabric has usually been made from spun Nomex. Continuous filament Nomex has been used to increase mobility and make it easier to take the garments on and off.

Benisek et al. have also described using flame-retardant finished wool in a multilayer approach to garment design.[34] Tightly woven outerwear fabric with a high integrity against flames, and a bulky low density, thick, knitted innerwear fabric, both made from char forming fibres such as Zirpro wool, offer additional insulation against flame exposure, associated with the air trapped in the knitted fabric. As condensation of moisture in the fabric has the effect of increasing both its thermal capacity and its thermal conductivity, it can be seen why a wet outer and a dry inner fabric offer the best protection. In this context, it is surprising to find that garments designed to encourage moisture condensation in the inner layers, by using vapour impermeable barriers, are officially approved for use in fire hazards.

Tables 17.10 to 17.12 show the protection provided by a range of multilayer garments.[31,38]

Table 17.10 shows the effect of total garment mass and thickness for garments containing various fibres and consisting of different numbers of layers.[31] Table 17.11

Table 17.10 Effect of garment mass and thickness on protection index[34]

Garment layer			Total mass $(g\,m^{-2})$	Total thickness (mm)	Protection index
Outside	Intermediate	Inside			
100% Kermel 250 g m⁻²	None	None	250	0.8	15.5
100% Kermel 250 g m⁻²	None	Kermel knitted underwear 330 g m⁻²	580	3.4	20.5
100% Kermel 250 g m⁻²	Kermel knitted underwear 300 g m⁻²	Kermel knitted pantihose 200 g m⁻²	780	4.8	35
100% Kermel 250 g m⁻²	Kermel knitted underwear 330 g m⁻²	Cotton knitted underwear 320 g m⁻²	900	5.5	43
100% Kermel 250 g m⁻²	Kermel knitted underwear 330 g m⁻²	Wool sweater 650 g m⁻²	1230	7.9	58

Table 17.11 Heat protection characteristics of combinations of shell fabrics, vapour barriers and thermal barriers ($84\,kW\,m^{-2}$ flame exposure)[34]

Shell fabric fibre content	Mass $(g\,m^{-2})$	Vapour barrier	Thermal barrier	Time to burn injury (s)	Total heat after 60 s $(kJ\,m^{-2})$
Aramid	245	NCN	QN	28	925
		CPCN	QN	31	900
		CPCN	NPN	40	780
		CPCN	FW	38	740
		GN	QN	32	900
		NCNPN		39	745
Coated Aramid	320	none	QN	34	850
Novaloid/Aramid	340	NCNPN		30	980
Aramid blend	255	NCN	QN	30	940
		CPCN	QN	50	590
FR Cotton	440	CPCN	QN	38	855

NCN = Neoprene-coated aramid fabric, 255 g m⁻²; CPCN = Coated aramid pyjama check fabric, 265 g m⁻²; GN = Gore-Tex-coated nylon base fabric; NCNPN = Neoprene-coated aramid needlepunched fabric, 735 g m⁻²; QN = Aramid quilt, consisting of aramid fibre batting with aramid pyjama check fabric attached, 245 g m⁻²; NPN = Aramid needlepunched fabric, 245 g m⁻²; FW = Wool felt, 430 g m⁻².

shows the range of heat protection properties during an $84\,kW\,m^{-2}$ exposure of US firefighters' turnout coats.[34] The inner thermal barrier is generally a batting or a needlepunched construction with fabric linings on the inside or both sides. The shell fabrics are of aramid fibre, an aramid/novaloid blend, or flame-retardant cotton.

Table 17.12   TPP ratings for protective clothing[42]

Ensembles	Total weight (g m^{-2})	Total thickness (mm)	Thermal load (% of 84 kW m^{-2})	TTP rating
Nomex III shell; Neoprene polycotton vapour barrier; Nomex quilt liner	766	5.4	50/50 convective/ radiant	38.6
Nomex III shell; Neoprene polycotton vapour barrier; Nomex quilt liner	899	5.6	100 radiant	42.5
Nomex III shell; Gore-Tex vapour barrier; Nomex quilt liner	719	5.4	50/50	46.5
Nomex III shell; Gore-Tex vapour barrier; Nomex quilt liner	726	5.8	100	44.7
Aluminized Nomex I shell	302	0.4	100	67.7
Aluminized Kevlar shell	346	0.5	100	78

Table 17.12 lists representative thermal protective performance (TPP) values for some materials used in structural and proximity clothing.[39] The first two ensembles show the effect of different vapour barriers on the overall thermal performance. The last two materials show the dramatic effect on radiant heat protection of aluminizing the outer shell material.

It has been suggested that standard firefighter's turnout gear or coat should be a multilayer construction with a durable fire-retardant outer shell fabric.[31] The outer shell may or may not be air and water permeable. The turnout coat may or may not incorporate a detachable vapour barrier or insulative liner. The vapour barrier is intended to protect the firefighter from steam and harmful chemicals, but on the other hand, it interferes with the escape of moisture from perspiration and increases the heat stress, which may result in subsequent hazards to health and safety. Hence an inner liner vapour barrier, mostly used in firefighters turnout coats in the United States, is excluded by some European fire departments. Table 17.13 compares the construction of tunics used in a number of European countries up to 1990.[31] In an attempt to produce a normalised EU standard for protective clothing, the current standards relate to clothing performance in general and flame and heat penetration in particular.

The excellent flame protection properties of woven outerwear fabric with bulky knitted underwear has been demonstrated and this combination has formed the basis of many clothing assemblies for protection against heat and flame hazards, for example the racing driver's garment assembly described in Table 17.14.[31] A glass fibre matrix has been used in wool fabric to give integrity to the char. Glass yarn was incorporated into the wool at the spinning stage. The yarn was then plied with all wool yarn and woven alternately with all wool yarn in both warp and weft direc-

Table 17.13 European firefighters tunics[34]

Country	Tunic				
	Mass (kg)	Thickness (mm)	Number of layers	Vapour barrier fibre	Outer fabric content
Austria	1.3	5.7	3	no	Wool/FR viscose/ FR polyester
	0.9	0.7	1	no	Zirpro wool
Belgium	1.9	5.1	3	yes	Coated aramid
Denmark	1.6	2.5	2	no	wool
Finland	2.5	2.3	2	yes	Wool/acrylic
France	2.1	2.3	2	no	Leather
Germany	1.0	1.05	1	no	Zirpro wool
Holland	2.3	3.9	1	no	Wool
	2.1	2.2	3	no	Wool/nylon
	2.2	4.9	3	yes	Aramid
Norway	1.8	1.8	1	no	Wool
Sweden	1.3	5.8	2	no	Zirpro wool
	1.8	2.2	2	no	Wool/FR viscose
	2.2	2.6	2	no	Wool/FR viscose
UK	1.9	4.5	3	no	Wool/nylon
	1.7	3.2	4	no	Aramid

Table 17.14 Specification for racing drivers' garment assembly[34]

Outer Layer

Fibre content	85/15 wool/glass	
Structure	2/2 twill	
Mass ($g\,m^{-2}$)	350	
Sett (threads/cm)		
warp	17.3	
weft	15.7	
Yarn linear density	R80 tex/2, all wool	Alternately in warp and weft
	R94 tex/2, 70/30 wool/glass	

Inner Layer

Fibre content	100% wool
Structure	2×2 rib with successive tucks
Mass ($g\,m^{-2}$)	175
Loop length (cm)	0.6
Yarn linear density	R44 tex/2

tions. Thermal underwear was also developed to offer increased bulk and thickness by incorporating a series of tuck stitches into the 2×2 rib construction.

17.3.3 Chemical, microbiological and radiation hazards

Today's user of chemical protective clothing is faced with a formidable task when selecting appropriate clothing. Among many factors that must be considered are cost, construction, style, availability, mode of use (disposable versus reusable). However, the most important factor is the effectiveness of the clothing as a barrier to the chemicals of interest.[40]

Both routine and emergency chemical handling may result in direct exposure to toxic chemicals. Examples of such situations include the following:[41]

1   handling of liquid chemicals during manufacture
2   maintenance and quality control activities for chemical processes
3   acid baths and other treatments in electronics manufacture
4   application of pesticide and agricultural chemicals
5   chemical waste handling
6   emergency chemical response and
7   equipment leaks or failures.

The degree of protection is important, particularly in the chemical industry, in which protective clothing is worn to prevent exposure to chemicals during production, distribution, storage and use. These chemicals may be gases, liquids or solids but, whatever their state, working with them safely means knowing how protective the clothing is or, more specifically, how resistant it is to chemical permeation.[42]

The selection of an appropriate level of protective clothing to be used in any particular situation should be based on a number of objective and subjective factors, including:[41]

• potential effects of skin contact with the chemical (for example, corrosiveness, toxicity, physical damage, allergic reaction)
• exposure period (time of contact)
• body zone of potential contact (for example, hands, feet, arms, legs, face, chest, back)
• permeability or penetration potential of the protective garment (breakthrough time and steady state rate)
• characteristics of potential contact (for example, splash, immersion)
• additive or synergistic effects of other routes of exposure (that is, inhalation and ingestion)
• physical properties required of the protective garment (for example, flexibility, puncture and abrasion resistance, thermal protection)
• cost (that is, based on single or multiple use and acceptable exposure).

Millions of farm workers and pesticide mixers across the world risk contamination by pesticides. Since dermal absorption is the significant route of pesticide entry, any barrier that can be placed between the worker and the chemical to reduce dermal contact can reduce exposure. With the exception of properly filtered air systems in enclosed cabs, the only significant type of barrier available to applicators is protective clothing.[43]

It has been reported[44] that nonwoven fabrics, in general, appear to perform better than most woven fabrics. However, heavy woven fabrics of twill construction, such as denim, have performed quite well in the limited studies completed to date. There is a direct correlation between fabric weight and thickness with pesticide penetration. Fabrics made from synthetic fibres show more wicking of pesticide onto the skin than fabrics containing cotton. Cotton-containing fabric with a durable press finish showed the lowest rate of absorbency and wicking. Fluorocarbon soil-repellent finishes have been found to be excellent barrier finishes against pesticides. Gore-Tex has been found to possess the most effective combination of barrier and thermal comfort properties.

Factors such as fibre content, yarn and fabric geometry and functional textile finish determine the response of a textile to contamination from liquid pesticides.[45] Fluorocarbon polymers alter the surface properties of fabrics so that oil as well as moisture has less tendency to wet the fabric surfaces and wicking is reduced. Liquid soil is partially inhibited from wetting, wicking, or penetrating the fabric. An undergarment layer offers better protection than does a single layer of clothing. The contamination of the second layer is generally less than 1% of the contamination of the outer garment layer; thus, the pesticide is not available for dermal absorption. A tee-shirt undergarment is recommended over other fabrics studied. Spun bonded olefin fabric offers similar protection to the fluorocarbon finish. The use of the disposable spun bonded olefin garments or fluorocarbon finish applied to non-durable-press work clothing has also been recommended. Theoretically, the greatest protection is produced by the use of disposable olefin garments worn during mixing, handling, and application of pesticides, in addition to fluorocarbon finish on the usual work clothing.

Several investigations have been carried out to determine the effectiveness of various types of fabric for protection against pesticides.[46–48] In one investigation two fabrics commonly worn by the agricultural workforce and five potential protective fabrics served as the test fabrics.[46] A 100% cotton 500 g m^{-2} denim with Sanforcet finish and a 100% cotton woven chambray shirting weight fabric represented typical clothing used for pesticide application. The five potential protective fabrics included three laminate variations, an uncoated spun bonded 100% olefin, and a 65% polyester/35% cotton 193 g m^{-2} coated poplin. Laminate 1 consisted of a microporous membrane of PTFE laminated between a face fabric of 100% nylon ripstop and an inner layer of nylon tricot. Laminate 2 consisted of a microporous membrane of PTFE laminated between an outer fabric of 100% polyester taffeta and an inner fabric of polyester tricot. Laminate 3 was a two-layer laminate with the microporous membrane serving as the face fabric and laminated to a polyester/cotton woven fabric backing. Table 17.15, which lists Duncan's multiple range test results, shows that the fabrics can be divided into three significantly different groupings. The fabrics offering the greatest protection included the three laminate variations and the spun bonded olefin. The second group of fabrics included the 100% cotton denim, the polyester/cotton coated poplin and the 100% spun bonded polyolefin. The fabric offering the least protection was the 100% chambray shirting weight fabric.

Table 17.15 Duncan's multiple range test results for disintegrations per minute (DPM) by fabric[49]

Variable	Mean DPM	Number	Duncan's grouping
Chambray	53 708	36	A
Coated poplin	10 505	36	B
Denim	8 407	36	B
Spunbonded polyolefin	5 914	36	B, C
Laminate 3	52	36	C
Laminate 2	42	36	C
Laminate 1	11	36	C

Lloyd describes four types of fabric selected for suitability for protection from pesticides but does not report any scientific study.[47] In this work, material A was a conventional woven fabric (polyester/cotton) used in work wear for many purposes but relying essentially on absorption for protection against liquids. Material B was a conventional woven fabric (nylon) that had been 'proofed' with a silicone, and thereby was similar to many products used in rainwear with the potential for 'shedding' spray liquids. Material C comprised a three-layer laminate of a micro-porous PTFE membrane sandwiched between two layers of nylon fabrics (woven and knitted). This product had the potential for resistance to penetration by spray liquids along with improved comfort to the wearer. Material D comprised a woven nylon fabric coated with neoprene. This product is virtually impermeable to air and is used commonly in the construction of general purpose chemical protective clothing.

An evaluation of aerosol spray penetrability of finished and unfinished woven and nonwoven fabrics has also been carried out.[48] The results presented in Table 17.16

Table 17.16   Aerosol protection performance of different types of fabric[51]

Test fabric	Water	Aerosol spray test performance	
		Water/surfactant	Cottonseed oil/ surfactant
Category I			
Spun-lace polyester, Scotchgard	pass	pass	pass
Spun-lace polyester, commercially finished	pass	pass	pass
Spun-bonded olefin, Scotchgard	pass	pass	pass
Spun-bonded/melt-bonded ($90\,g\,m^{-2}$), Scotchgard	pass	pass	pass
Spun-bonded/melt-bonded ($50\,g\,m^{-2}$), Scotchgard	pass	pass	pass
Category II			
Commercially finished (Scotchgard) cotton	pass	pass	fail
Cotton, Quarpel	pass	pass	fail
Cotton/polyester, 35/65 Quarpel finished	pass	pass	fail
Category III			
Spun-bonded olefin	pass	fail	fail
Spun-bonded/melt-bonded ($90\,g\,m^{-2}$)	pass	fail	fail
Spun-bonded/melt-bonded ($50\,g\,m^{-2}$)	pass	fail	fail
Spun-bonded/melt-bonded ($50\,g\,m^{-2}$) commercially finished	pass	fail	fail
Category IV			
Denim, unwashed	fail	fail	fail
Denim, 1 wash	fail	fail	fail
Denim, 3 washes	fail	fail	fail
Cotton drill	fail	fail	fail
Spun-lace polyester, unfinished	fail	fail	fail

indicate that the presence of a fluorocarbon finish increases the resistance to aerosol penetration. The woven fabrics tested failed to meet the criterion of being resistant to oil-based spray penetration. The finished nonwoven fabrics, with the exception of the commercially finished spun bonded/melt bonded, passed the spray tests using both the oil-based and water-based solutions. Comparisons of the comfort properties of air permeability, water vapour permeability, and density indicate that the spun lace class of nonwoven fabrics was very similar to the woven fabrics. The spun laced fabrics are denser or more 'clothlike' than many of the spun bonded fabrics.

An alternative need for chemical and microbiological protection is within the hygiene and medical fields. Textiles and fibrous materials may be subjected to various finishing techniques to afford protection of the user against bacteria, yeast, dermotaphytic fungi and other related microorganisms for aesthetic, hygienic or medical purposes.[49] A range of acrylic fibres many of which are phenol derivatives, have been chemically modified in order to fix bacteriostats. Courtaulds have developed a range of bactericidal acrylic fibres marketed under the brand name Courtek M. These fibres contain any of several widely accepted bactericides permanently bound to the fibre at a range of concentrations chosen for specific requirements. Bioactive acrylic polymers are highly effective in both fibre and nonwoven fabric forms. They are for use in the medical, sanitary, personal hygiene, and general patient care areas.

Protection against airborne radioactive particles is a problem in the nuclear industry. Microfilament yarns are densely woven to produce fabrics with maximum pore size of 20–30 μm compared with the 75–300 μm pore size of typical cotton and polyester cotton fabrics for use in the nuclear industry. Incorporation of filaments with a carbon core greatly reduces the attraction of radioactive particulates by static electricity on to the fabric during wear (Protech 2000; The First Protective Clothing Specifically Designed From Fiber to Finished Garment, http://www.wlst.com, July, 1997).

The spun bonded polyolefin fibre fabric Tyvek® while featuring extensively in all types of protective clothing, is claimed to be superior to other fabrics in the nuclear industry in preventing penetration and holding water borne contamination, dry particulate, tritiated water and tritium gas.[50] The fabric can be used uncoated or

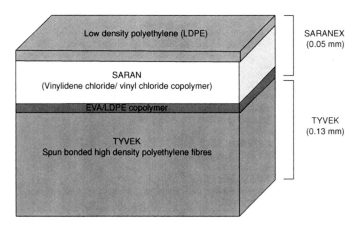

17.2   Saranex-coated Tyvek fabric.

laminated with Saranex® (Dow Chemical Co.) depending on the degree of protection required. Saranex-coated Tyvek is a complex fabric. The outer layer is Saranex 23, a coextruded multilayered film 0.05 mm thick. It has an outside layer of high density polyethylene, an inner layer of Saran, a copolymer of vinylidene chloride and vinyl chloride, and the other outside layer of ethyl vinyl acetate (EVA) / low density polyethylene copolymer, which is used for bonding to the Tyvek. Figure 17.2 shows the construction of Saranex-coated Tyvek fabric.[50]

Garments for very specialised applications have to provide protection from a number of agencies, for example military garments for protection against nuclear, biological and chemical weapons (NBC suits). The fabric used to make these garments consists of two layers. The inner layer, usually of nonwoven construction, is impregnated with activated carbon to adsorb the agent particles thus preventing them from reaching the wearers skin. The outer layer is a woven fabric of relatively high porosity to allow the agents to penetrate to the inner layer and to provide mechanical strength, heat and flame-resistance and some weather protection. The heat and flame resistance is achieved by using inherently flame-resistant synthetic fibres.

17.4   Conclusions

As society becomes more safety conscious and has to survive in more arduous conditions in order to provide raw materials, energy and to push the frontiers of knowledge further, there is a need to provide a safe working environment. The modern textile industry plays a part in providing this environment by developing and supplying sophisticated clothing and other products. The degree of sophistication and specialisation is increasing and many products are very highly specified requiring a complex combination of properties. Much has been gained by developing traditional technologies such as, spinning, weaving, knitting and finishing. However the newer technologies appear to be developing at a faster rate. For example, nonwoven technology when it first appeared was seen only as a low cost method of producing fabric for unsophisticated products. Nonwoven fabric is now used extensively in survival products by combining the appropriate fibre content and appropriate method of production with other materials, such as chemical finishes, laminates and coatings.

References

1. P A BAJAJ and A K SENGUPTA, *Protective Clothing, Textile Prog.*, 1992 **22**(2/3/4) 1.
2. F S ANDRUK, *Protective Clothing for the United States Navy in the 1980s*, Navy Clothing and Textile Research Facility – Natick, Series No. AD-A956 145, 1980.
3. ANON, *Int. Dyer Textile Printer*, 1979 May 389.
4. I HOLME, *Apparel Int.*, 1992 **21**(5) May 31.
5. I HOLME, *Indian Textile J.*, 1992 **102**(9) June 134.
6. M MASRI, *Tech. Textiles Int.*, 1992 June 13.
7. B DAVIES, *Tech. Textiles Int.*, 1998 March 15.
8. J DZUGAN, 'Hypothermia', *Fisheries Safety and Survival Series*, Alaska Sea Grant Coll. Program – Fairbanks, Series No PB92-157937, 1992.
9. I HOLME, *Textile Month*, June 1988 38.
10. J GIBLO, *Thermal Performance of Navy Anti-exposure Coverall to Different Water Exposure Conditions*, Navy Clothing and Textile Res. Facility – Natick, Series No. ADA293950, August, 1993.

11. ANON, Allgemeiner Vliestoff-Report, 1987 **15**(2) 136(E137).
12. ANON, *High Performance Textiles*, 1991 March 10.
13. ANON, *Nonwovens Rep. Int.*, 1987 (194) May 10.
14. ANON, *Du Pont Magazine*, 1996 **90**(2) March/April 18.
15. ANON, *Textile Month*, 1981 June 23.
16. P A BAJAJ and A K SENGUPTA, *Protective Clothing, Textile Prog.*, 1992 **22**(2/3/4) 65.
17. ANON, *Vliesstoff Nonwoven Int.*, 1994 (6–7) 179.
18. B V HOLCOMBE and B N HOSCHKE, *Performance of Protective Clothing*, eds. R L Barker and G C Coletta, ASTM Special Technical Publication 900, Philadelphia, 1986, p. 327.
19. W G BURCKEL, *Heat Resistant Fiber Blends For Protective Garments*, US Patent Office, Patent No 1 486 997, 1974.
20. S ALEXANDER, *Tech. Textiles Int.*, 1995 **4**(6) July/August 12.
21. T POLFREYMAN, *Textile Month*, 1980 May 19.
22. L A SHANLEY et al, *Clothing Textiles Res. J.*, 1993 **11**(3) 55.
23. I HOLME, *Textile Month*, 1987 September 112.
24. A LARSSON et al, *Test of Modified Submarine Escape And Immersion Suit*, National Defence Research Establishment, Sundyberg, Sweden, Series No BP92-125624, 1991.
25. M GENNSER et al, *Survival Suit for Submarine Personnel*, Swedish Defence Material Administration-Stockholm, Series No PB 93-204634, March, 1993.
26. ANON, *Textile Month*, 1982 July 14.
27. R SPECTOR, *Daily News Record*, 1987 **17**(31) February 22.
    Superior Technology Means Superior Performance, http://www.northern.com, July, 1997
28. J BRADSHAW, *Nonwovens Ind.*, 1981, 12, Aug., 18.
29. P A BAJAJ and A K SENGUPTA, *Protective Clothing, Textile Prog.*, 1992 **22**(2/3/4) 64.
30. M C MAGILL, *Report on Ski Glove Tests*, Outlast, Gateway Technologies, December, 1996.
31. P A BAJAJ and A K SENGUPTA, *Protective Clothing, Textile Prog.*, 1992 **22**(2/3/4) 9–43.
32. R E BOUCHILLON, *Performance of Protective Clothing*, eds. R L Barker and G C Coletta, ASTM Special Technical Publication 900, Philadelphia, 1986, p. 389.
33. L BENISEK and G K EDMONDSON, *Textile Res. J.*, 1981 **51**(3) March 182.
34. L BENISEK et al, *Performance of Protective Clothing*, eds. R L Barker and G C Coletta, ASTM Special Technical Publication 900, Philadelphia, 1986, p. 405.
35. M GSTEU, K OTT and H D EICHHORN, *Textile Horizon*, 1997 **17**(5) October/November 12.
36. H H FORSTEN, *Protective Clothing: An Update on Personal Protection Against Chemical, Thermal and Other Hazards*, Du Pont, May, 1989.
37. A R HORROCKS, *Rev. Prog. Colouration*, 1986 **16**(62).
38. J KRASNY, *Performance of Protective Clothing*, eds. R L Barker and G C Coletta, ASTM Special Technical Publication 900, Philadelphia, 1986, p. 463.
39. J H VEGHTE, *Performance of Protective Clothing*, eds. R L Barker and G C Coletta, ASTM Special Technical Publication 900, Philadelphia, 1986, p. 487.
40. M W SPENCE, *Performance of Protective Clothing*, eds. R L Barker and G C Coletta, ASTM Special Technical Publication 900, Philadelphia, 1986, p. 32.
41. S Z MANSDORF, *Performance of Protective Clothing*, eds. R L Barker and G C Coletta, ASTM Special Technical Publication 900, Philadelphia, 1986, p. 207.
42. N W HENRY, *Performance of Protective Clothing*, eds. R L Barker and G C Coletta, ASTM Special Technical Publication 900, Philadelphia, 1986, p. 51.
43. A P NIELSEN and R W MARASKI, *Performance of Protective Clothing*, eds. R L Barker and G C Coletta, ASTM Special Technical Publication 900, Philadelphia, 1986, p. 95.
44. P A BAJAJ and A K SENGUPTA, *Protective Clothing, Textile Prog.*, 1992 **22**(2/3/4) 79.
45. J M LAUGHLIN, *Performance of Protective Clothing*, eds. R L Barker and G C Coletta, ASTM Special Technical Publication 900, Philadelphia, 1986, p. 136.
46. H D BRANSON et al, *Performance of Protective Clothing*, eds. R L Barker and G C Coletta, ASTM Special Technical Publication 900, Philadelphia, 1986, p. 114.
47. G A LLOYD, *Performance of Protective Clothing*, eds. R L Barker and G C Coletta, ASTM Special Technical Publication 900, Philadelphia, 1986, p. 121.
48. N E HOBBS et al, *Performance of Protective Clothing*, eds. R L Barker and G C Coletta, ASTM Special Technical Publication 900, Philadelphia, 1986, p. 151.
49. P A BAJAJ and A K SENGUPTA, *Protective Clothing, Textile Prog.*, 1992 **22**(2/3/4) 75.
50. C E GARLAND et al, *Performance of Protective Clothing*, eds. R L Barker and G C Coletta, ASTM Special Technical Publication 900, Philadelphia, 1986, p. 276.

18

Textiles in transportation

Walter Fung

Collins and Aikman, PO Box 29, Manchester Road, Walkden,
Manchester M28 3WG, UK

18.1   Introduction

Transportation is the largest user of technical textiles. Textiles provide a means of decoration and a warm soft touch to surfaces that are necessary features for human well being and comfort, but textiles are also essential components of the more functional parts of all road vehicles, trains, aircraft and sea vessels.

Textiles in transportation are classed as technical because of the very high performance specifications and special properties required. Seat coverings, for example, are not easily removable for cleaning and indeed in automobiles they are fixed in place and must last the lifetime of the car without ever being put in a washing machine. In trains, aircraft and passenger vessels they are exposed to much more rigorous use than domestic furniture. In addition they have to withstand much higher exposure to daylight and damaging ultraviolet radiation (UV) and because they are for public use they must satisfy stringent safety requirements such as flame retardancy.

In more functional applications, textiles are used in articles as diverse as tyres, heater hoses, battery separators, brake and clutch linings, air filters, parts of the suspension, gears, drive belts, gaskets and crash helmets. They are present in all forms of transport and, apart from tyres, are in applications of which the non-technical person is not even aware. Fibre/plastic composites are replacing metallic components and more traditional materials with considerable benefits, especially savings in weight. The most significant growth area in transportation textiles is expected in composites that straddle the textile and plastics industries. Volumes of the more traditional applications of textiles such as clothing and furnishing are not expected to grow substantially in the developed countries of Western Europe, North America and Japan. The largest growth will be in technical applications and a 40% increase on 1996 figures is expected by the year 2000 for fibre composites.

The most familiar technical textile in transportation is car seat fabric which is amongst the largest in volume and is growing annually in the developing world of the Pacific rim, Eastern Europe and South America (see Table 18.1). Car seat fabric

Table 18.1   World car sales ($\times 10^3$ units)

	1996	1997	1998	1999	2000	2001
Western Europe	12829	13030	13366	13698	14087	14134
Germany	3497	3522	3583	3779	3954	3846
Italy	1744	1899	1917	1854	1950	2089
France	2132	1889	2064	2202	2243	2218
UK	2025	2117	2054	1973	1960	2013
Spain	914	1000	1023	1055	1111	1128
East Europe	1637	1695	1861	1964	2132	2277
North America	9366	9320	8853	8637	9383	9813
USA	8527	8339	7826	7563	8210	8652
Canada	661	715	700	694	738	758
Mexico	179	266	327	380	434	403
Latin America	1761	1874	1825	1811	1830	1918
Japan	4643	4471	4842	4879	4932	4988
Asia Pacific	2912	3149	3562	3947	4287	4518
TOTAL	37566	38061	39311	40023	42055	43715

Source: J D Power-LMC Automotive Services/Automotive International March, 1997.

requires considerable technical input to produce both the aesthetic and also the very demanding durability requirements. The processes developed for car seat fabric and the technical specifications provide some indication of the requirements for seat materials in other transport applications.

In all transportation applications certain important factors recur, like comfort, safety and weight saving. In public transportation situations as far as textiles are concerned safety means reduced flammability. Environmental factors have also become important and these have influenced the transportation textile industry in a number of ways including design, choice of materials and manufacturing methods. Conservation of world resources by using less fuel, not to mention reduced atmospheric pollution by reduced exhaust emissions, is now a concern for world governments.

Reduced flammability properties are understandable considering the restrictions on escape routes especially in the air and at sea. Flame-retardancy (FR) requirements of private cars are not especially high but are stringent for passenger trains and standards are increasing for passenger coaches. Transportation disasters which become headline news are frequently the impetus for increased FR standards and improvements in public safety, for example, the Salt Lake City air disaster of 1965, the Manchester Airport fire of 1985 and more recently the Channel Tunnel fire.

The whole area of transportation is growing with increasing trade between all the nations of the world generating higher volumes both in freight and also commercial passenger travel. Leisure travel is also increasing dramatically with larger disposable incomes, increased leisure time and increased interest in foreign cultures; the largest growths are expected in air travel.

The stresses of modern living require transportation interiors to be more pleasing and mentally relaxing, to ease travelling and to make journeys more enjoyable. Indeed the various forms of transport now compete with each other for passengers.

For many national internal journeys the travelling times and costs of, say, air and rail are very similar from city centre to city centre.

A further recurring requirement of textiles involving passenger transport is cleanability. The only opportunity for servicing is at the end of a journey and just before the start of the next one. Expensive items of equipment such as jumbo jets must earn their keep by being on the move as much as possible and indeed jumbo jets are in the air 22 hours a day. Being out of service for lengthy overhauls or cleaning is money wasted and so cleaning must be done as quickly as possible. Easy care and maintenance are very important requirements; dirty carpets or seats would deter passengers.

Technical textiles are important, especially for the developed countries of the world, because they generally rely less on labour costs and are more dependent on technical know-how. However, even with technical textiles the manufacturer is facing new challenges to reduce costs even further still. To a certain extent this is being achieved by rationalisation, consolidation of production, transfer of production to developing countries, company mergers and joint ventures; but at the end of the day, much depends on the innovative technologist to devise new and more effective materials and production methods at lower cost and with increased performance as well!

Technical textiles are relative newcomers to the textile industry, which is probably one of the world's oldest industries, but there are still opportunities to learn from more traditional methods and from 'synergies' with other industries; for example, the film industry developed a process for producing novel film properties on a piece of apparatus that is essentially a textile stenter.

Although textiles have been used in some car seats since the invention of the car, widespread use has only occurred since the mid-1970s. The technology and manufacturing methods are still on the 'learning curve' compared to other sectors of the textile industry. Fabric car seats could still benefit from certain developments and processes that have been available to the garment and finishing industry for many years, for example advanced finishing techniques providing softer handle and touch, antistatic finishes, antimicrobial finishes, encapsulated chemicals, specialist yarns and techniques for improved thermal comfort.

18.1.1 Fibre requirements

For seat coverings the main technical requirements are resistance to sunlight (both colour fading and fabric degradation by UV), abrasion resistance[1-5] and, for public transport vehicles, reduced flammability (see Table 18.2). Seats frequently get damp from contact with wet clothing and, in the case of seats in public transport, subject to abuse by vandals and other irresponsible individuals.[6] The fabrics need to be resistant to mildew, hard wearing and strong with high tear strength. Soil resistance and easy cleanability are also necessary. Other requirements will become evident later in the chapter.

18.1.1.1 Resistance to sunlight and UV degradation

Resistance to sunlight is perhaps the most important property a fabric must have. Choice of the wrong fabric can lead to breakdown of the seat cover within weeks, depending on the intensity and spectral distribution of the sunlight. Spectral distribution of sunlight varies with geographical location, cloud cover and even the

Table 18.2   Properties of fibres used in transportation[a]

	Density (g cm^{-3})	Melting point[b] (°C)	Tenacity (g den^{-1})	Stiffness (flexural rigidity) (g den^{-1})	LOI (% oxygen)	Abrasion resistance	Resistance to sunlight
Acrylic	1.12–1.19	150d[b]	2.0–5.0	5.0–8.0	18	Moderate	Excellent
Modacrylic	1.37	150d[b]	(HT)	3.8	27	Moderate	Excellent
Nylon 6	1.13	215	2.0–3.5 4.3–8.8	17–48	20	Very good	Poor–good (stabilised)
Nylon 6.6	1.14	260	(HT)	5.0–57	20	Very good	Poor–good (stabilised)
Polyester	1.40	260	4.3–8.8 (HT)	10–30	21	Very good	Good–excellent (stabilised)
Polypropylene	0.90	165	4.2–7.5 (HT) 4.0–8.5 (HT)	20–30	18	Good	Poor–good (stabilised)
Wool	1.15–1.30	132d[b]	1.0–1.7	4.5	25–30 (Zirpro)	Moderate	Moderate
Cotton	1.51	150d[b]	3.2	60–70	18	Moderate	Moderate
UHM							
Polyethylene	0.97	144	30	1400–2000	19		
Aramid	1.38–1.45	427–482d[b]	5.3–22	500–1000	29–33		
Carbon	1.79–1.86	3500d[b]	9.8–19.1+	350–1500	64+		
Glass	2.5–2.7	700	6.3–11.7	310–380	–		
PBI	1.30	450d[b]	–	9–12	41		
Inidex	1.50	–	1.2	–	40		
Panox	1.40	200–900d[b]	–	–	55		
Steel	7.90	1500	2.5–3.2	167–213	–	–	–
Aluminium	2.70	660	–	–	–	–	–

LOI, limiting oxygen index; HT, high tenacity; UHM, ultra high modulus; PBI, polybenzimidazole.
[a] Data compiled from several sources and intended only as a guide.
[b] d, does not melt but starts to degrade.

time of day. Because glass windows are being placed more at an angle, the temperature within a vehicle can exceed 100 °C and during the course of a day relative humidity can vary from 0–100%. These factors, combined with sunlight, contribute to breakdown of seat fabric. Glass filters out a section of the sunlight spectrum including part of the UV area, which is most damaging to most fibres and in particular polyester. Hence polyester exposed behind glass exhibits much better performance compared to polyester exposed directly to sunlight. This factor is a major reason why polyester has emerged as the most used fibre for car upholstery (see Table 18.3).

Actual degradation by UV radiation is influenced by the thickness of the yarn, the thicker the better because less radiation will penetrate into the centre. This is particularly the case for nylon yarns. Matt or delustred yarn often breaks down the fastest because the titanium dioxide delustrant may photosensitise degradation and the lower specific surface area reduces the rate of photo-oxidative attack. UV degradation will therefore also be influenced by cross-section, the poorest again being those presenting the greater surface area for a given linear density.

Table 18.3 Light durability (in Florida) of some natural and synthetic fibres exposed simultaneously (months required to reach loss in strength indicated)

	Initial tenacity (g denier⁻¹)	Outdoors (direct sunlight)		Behind glass	
		50% Loss	80% Loss	50% Loss	80% Loss
Acrylic semidull	2.1	13.6	36 (72%)[a]	19	36 (63%)[a]
Polyester bright	4.2	3.7	7.9	24	36 (75%)[a]
Polyester semidull	3.1 (spun)	4.0	9.1	36	36 (49%)[a]
Polyester dull	4.2	3.6	8.0	20	36 (79%)[a]
Nylon bright	5.3	9.5	17.0	10.3	20.7
Nylon semidull	5.4	3.2	6.5	4.5	8.2
Nylon dull	5.1	3.1	5.1	4.1	7.7
Rayon bright	1.6	2.6	6.3	3.0	14.2
Acetate bright	1.0	5.1	11.8	8.1	27
Cotton deltapine	1.8	2.9	5.8	4.9	14.0
Flax Irish	3.5	0.9	2.5	4.5	5.0
Wool worsted	0.7	2.3	3.2	4.5	7.6
Silk	4.2	–	–	0.8	3.9

[a] Loss per cent indicated after 36 months.
Source: B F Faris (DuPont), in *Automotive Textiles*, ed. M Ravnitzky, SAE PT-51 1995, p. 23. Copyright held by Society of Automotive Engineers, Inc. Warrendale PA. Reprinted with permission.

Significant improvements in UV resistance can be obtained by addition of certain chemicals that are UV absorbers[7,8] and these are used extensively with polyester, nylon and polypropylene for transportation applications. UV absorbers in nylon are usually added to delustred yarns which deactivate the sensitising effects of the titanium dioxide present.

18.1.1.2 Abrasion resistance
Seating fabric needs to be of the highest standard of abrasion resistance. Only polyester, nylon and polypropylene are generally acceptable, although wool is used in some more expensive vehicles because of its aesthetics and comfort. Wool has other specialist properties such as non-melting and reduced flammability which, as will be seen, make it suitable for aircraft seats. Fabric abrasion is influenced by yarn thickness, texture, cross-section and whether spun or continuous filament. Those factors that result in larger surface area or provide points of frictional stress reduce abrasion resistance. Fabric construction and weight have an effect on abrasion, not to mention fabric finishes and processing variables. Reproducing damage by accelerated testing is well known to cause problems. The simulation of UV degradation,[9–12] abrasion damage and the associated problem of pilling that occurs in actual use with accelerated laboratory tests is therefore not straightforward and has been the subject of much research.[13–15] All fabric property requirements are demanding and need to be of general 'contract' standard[16] or higher.

18.1.1.3 Reduced flammability
Reduced flammability testing[17–20] has become much more sophisticated as the mechanisms of fire disasters and the causes of fatalities are analysed. Thus it is now important to test for toxicity of smoke generated (see Tables 18.4–18.6) and its effect on

Table 18.4   Composition of off-gases of Kevlar and other fibres under poor combustion conditions[a]

	Combustion products of sample $(mg\,g^{-1})$									
	CO_2	CO	C_2H_4	C_2H_2	CH_4	N_2O	HCN	NH_3	HCl	SO_2
Kevlar	1850	50	–	1	–	10	14	0.5	–	–
Acrylic	1300	170	5	2	17	45	40	3	–	–
Acrylic/modacrylic (70/30)	1100	110	10	1	18	17	50	5	20	–
6.6 Nylon	1200	250	50	5	25	20	30	–	–	–
Wool	1100	120	7	1	10	30	17	–	–	3
Polyester	1000	300	6	5	10	–	–	–	–	–

[a] The sample is placed in a quartz tube through which air is drawn at a controlled flow and heated externally with a hand-held gas–oxygen torch. Air flow and heating are varied to give a condition of poor combustion (i.e. deficiency of oxygen). Combustion products are collected in an evacuated tube and analysed by infrared.
Source: KEVLAR Technical Guide (H-46267) 12/92 Table II-8 (DuPont), December 1992.

Table 18.5   Combustion products and their physiological effects

Product	Sources	Physiological effects		
Oxygen depletion	All fires	21%	=	Normal concentration
		12–15%	=	Headache, dizziness, fatigue, loss of coordination
		<6%	=	Death in 6–8 min
Carbon monoxide	All fires (incomplete combustion)	1000 ppm	=	Death after 2 hours
		5000 ppm	=	Death within 5 min
Carbon dioxide	All fires	250 ppm	=	Normal concentration
		5%	=	Headache, dizziness, nausea, sweating
		12%	=	Death within 5 min
Hydrogen cyanide	Nitrogen-containing polymers (nylon, wool, modacrylics etc.)	50 ppm	=	Death in up to 1 hour
		180 ppm	=	Death after 10 min
Hydrogen chloride	PVC, PVDC fibres, neoprene coatings	10 ppm	=	Irritation
		100 ppm	=	Death within 5 min
Oxyfluoro compounds	PTFE membranes	50 ppm	=	Irritation
		100 ppm	=	Death within 1 hour
Acrolein	Polyolefins, Cellulosics (cotton)	1 ppm	=	Severe irritation
		150 ppm	=	Death in 10 min
Antimony compounds	Some modacrylics some rubber coatings, tentage	>0.5 mg m^{-2}	=	Pulmonary and gastrointestinal problems

PVC = polyvinyl chloride, PVDC = polyvinylidene chloride, PTFE = polytetrafluoroethylene.
Source: see footnote to Table 18.6.

Table 18.6   Heat release rate (HRR), total heat release rate (THRR) and time to peak of heat release (T_p) for a variety of fabrics

Fabric	HRR ($kW\,m^{-2}$)	THRR ($kW\,min\,m^{-2}$)	T_p (s)
Cotton/polyester	170	53	33
Wool	117	39	24
Modacrylic	83	28	27
Zirpro wool	64	24	25
Panox	27	15	30
Meta-aramid	13	6	40

Source: M Masri, 'Survival under extreme conditions', in *Technical Textiles Int.*, 1992 June.

visibility as well as for ignitability and rate of propagation. Heat generated has also been identified as important and tests have been developed to measure this. Testing of whole assemblies such as seats is now carried out in addition to testing of the individual components.

For comfort, foam materials are used beneath the covering fabric. Despite much development, which has significantly reduced the foam flammability,[21,22] 'fireblocker' materials have been introduced between the face fabric and the foam. Fireblockers, first used on aircraft seats, are textile fabrics made from fibres with a very high level of inherent flame retardancy and heat stability,[23,24] for example Panox (Lantor Universal Carbon Fibres), Inidex (Courtaulds) and aramid. They are being used increasingly on trains, buses and coaches.

18.1.2   Fibre/plastic composites

Composites can be regarded as a macroscopic combination of two or more materials to produce special properties that are not present in the separate components. How composites function can be explained by the analogy of the use of straw in clay bricks by the ancient Egyptians. A strong brick was obtained because the straw reduced and controlled the occurrence of cracks in the hard but brittle clay. Glass fibres have a very high tensile strength but are brittle because of their extreme sensitivity to cracks and surface defects. When incorporated in a plastic matrix, the tensile properties of the fibres define that of the composite to which the plastic is added. The plastic protects their surface thus preventing crack developments; this results in a strong composite material.[25]

Glass reinforced plastics (GRP) date from the 1920s and combine high strength and stiffness with light weight. From the early 1960s more advanced fibres became available (e.g. carbon, aramid, boron and ceramic fibres), which are all very strong and many times stiffer than glass. Carbon fibre properties vary enormously depending on the conditions of manufacture. Nomex and Kevlar (DuPont), the first aramid fibres,[26] offer high thermal stability and very high strength at relatively low weight. Their density is about $1.45\,g\,cm^{-3}$ compared to about $2.5\,g\,cm^{-3}$ for glass and $7.9\,g\,cm^{-3}$ for steel.

18.1.2.1   Advantage of composites

Composites have made very significant advances because of their high strength and stiffness combined with low weight and, in many cases, less bulk. These properties are especially suited to transport applications offering fuel savings and more useful space within the aircraft, vessel or vehicle. A man-powered aircraft, 'Gossamer Albatross' weighing 32 kg, flew across the English Channel.[27] This would have been impossible without Kevlar as a structural reinforcement. In many cases several metal parts joined together can be replaced by a single composite component.[28] Life-cycle analyses on composites show that, although they can be more expensive to produce, they are generally more environmentally friendly because they consume less fuel during their lifetime compared to the heavier metallic items they replace. Yet composite technology is still in its infancy compared to the use of metals, which has centuries of accumulated know-how. More significant innovations can be expected, however; carbon fibres are already used extensively in aircraft, trains, some buses and racing cars. The use of carbon fibre is predicted eventually in volume passenger cars. Carbon fibre growth is estimated to be at about 10% annually to the year 2001, although not all of it will be in transportation.[29]

More advanced fabric structures are being developed (e.g. three-dimensional knitting and weaving, multiaxial knitting, cartesian braiding, 'noobing' and so on). These are extending the scope of composites for all industrial applications, including transportation.[30-32] A UK Civil Aviation Authority project is developing composite material involving 20 layers of fabric capable of containing terrorist bomb explosions on board aircraft.

18.1.2.2   Tyres

The tyre, a rubber/textile composite (approximately 10% w/w of textile) dates from 1888 when canvas was the reinforcement used in Dunlop's first tyre. Continuous filament rayon began to be used in the 1930s. Research subsequently led to high tenacity variants and much needed improvements in fibre/rubber adhesion. Nylon, offering toughness at a lighter weight, was introduced into aircraft tyres during World War II. Once again, fibre/rubber adhesion limitations needed to be overcome and nylon 6 was claimed to have better adhesion than nylon 6.6. 'Flat spotting', probably caused by low nylon elasticity when hot, restricted the growth of nylon in tyres for cars but nylon tyres were used extensively on trucks and farm vehicles where heat generation during use is less of a problem. Use of rayon declined from the mid-1950s owing to competition from polyester, which like its predecessors required improvement in its adhesion to rubber[33] and from steel cords, which were cheaper to produce. Tyre cord development continues with new yarn variants, for example high modulus low shrink 'DSP'™ (dimensionally stable polyester) (Allied Signal) and novel means of improving fibre adhesion to rubber.[34] Aramids, offering high strength to weight ratios and high temperature resistance, find use in high performance car and aircraft tyres.

18.2   Textiles in passenger cars

Car interiors have become increasingly important for a variety of reasons. People spend more time in cars generally with increased daily commuter distances to and from work, increased traffic density and because more people are working away from home and travelling long distances by car at weekends. The car has become a place of work for sales representatives, sales engineers and businessmen who are now also able to communicate with the office and customers by mobile telephone. The general public has more leisure time and larger disposable incomes and so travel more on days out, on holidays, and make more journeys to the supermarket and out of town shopping centres. The car has become not just an office and a living room on wheels but also a 'shopping bag' on wheels.

This increased time spent inside the confined space of a car, plus the stresses of modern living, not to mention the frustrations of gridlocks and roadworks, has produced a need for more pleasant, comfortable and relaxing interiors. Comfort in all its forms reduces stress and fatigue and, therefore, contributes to road safety. The buying public has become more discerning and demands more value for money and this is coupled with more choice from an increasing number of competing vehicle manufacturers. The increasing numbers of female car buyers means that they have more influence in the area of family car purchase than before.

With the importance of fuel economy all cars must be aerodynamic and because exterior shapes of models from different manufacturers can look generally similar, greater emphasis must be put upon interiors to compete effectively. From the manufacturers' point of view, changing the car interior via a 'facelift', is an economical way of revamping and relaunching a model that is not selling well quickly or prolonging the life of a model (see Fig. 18.1).

18.2.1   Interior design

The colour, texture and overall appearance of the car interior, especially the seat, has become extremely important in attracting a potential buyer's attention. There is only one opportunity to make a first impression. The textile designer must be able

18.1   Computer-assisted design plays an important role in modern automobile seat cover design.

to produce innovative interior appearances which reflect or even set current fashion trends, social and economic moods and customer lifestyles,[35] whilst at the same time being compatible with the exterior colour and car shape. The design must be aimed at the market sector at which the particular model is aimed and also be in the particular car company's chosen overall style.

New designs originate from the fabric producers who submit several ideas for each car model to the vehicle producer. These are developed and fine tuned until eventually a single selection is decided upon. European designers are considered world leaders. Lead times for new car models are currently about two years, so the designer must be able to forecast future trends. The logistics of availability of materials, attainable colours and fabric performance limitations also have to be considered carefully.

Modern designers have access to computers that allow designs to be created and viewed in three-dimensional simulation on actual car seats within a car without any actual fabric being made. Colours and other variants can be varied at the touch of a keyboard and usually national preferences exist for interior colours and texture. Flat woven fabrics account for about half of car seat fabric in Europe but only about a quarter in the USA and Japan. Woven velours are the most important type in the USA, and tricot knit the most important in Japan.

Circular knits are increasing in Europe, and woven velours increasing further in the USA. Worldwide, designs are becoming more sophisticated with increased use of more colours and Jacquard-type designs.[36] In Japan printing on to tricot knit has been attributed to a need for a more economical alternative to Jacquard equipment, which is in short supply in Japan. Printers in the USA and Europe are developing the technology, but some manufacturers hold the view that printing may not be necessary because adequate Jacquard equipment is readily available in these areas. Car models increasingly must have global appeal which cuts across national boundaries and is suited to different cultures on different continents. Lead times are also shortening in an increasingly competitive marketplace.

18.2.2 General requirements

Interior textiles must generally last the life of the vehicle and show no significant signs of wear for at least 2–3 years to maintain a good resale value of the car. If the seats look worn, buyers will assume the engine is worn. Car interiors are subjected to the full range of temperatures (−20 to 100+ °C) and relative humidities (0–100%). In addition, the car seat occupants' clothes may be wet and rough in texture and so the seat fabric must always appear uncreased and without noticeable colour fading or soiling.

18.2.2.1 Fibre selection

The two most important factors governing the selection of fabrics for car seat covers are resistance to light (UV radiation) and abrasion. Several fibres were initially used as replacements for the widely used PVC (polyvinyl chloride) in the late 1960s and early 1970s, namely nylon 6 and nylon 6.6, acrylic, wool and polyester. Nylon, especially nylon 6, suffers rapid sunlight degradation, which has made car seat makers still cautious about its use. Abrasion of acrylic is limited and wool is relatively expensive. These factors have resulted in polyester emerging as the predominant fibre now accounting for about 90% of all textile seat covers worldwide.

Manufacturers of polypropylene strive to have their fibre more widely accepted as seat material[37] but disadvantages like low melting point, low yarn extensibility, and limitations in colour spun dyeable at present outweigh the advantages, which are significantly lower density $0.90\,g\,cm^{-3}$ compared with $1.38\,g\,cm^{-3}$ for polyester, cheaper cost, and easier recyclability. Polypropylene is, however, used in nonwovens in headliners, floor coverings and parcel shelves. Addition of light stabilisers has improved the stability of polypropylene fibre to light and thermal degradation.[8]

18.2.2.2   Yarn types

Fabrics are generally produced from bulked continuous filament (BCF) textured polyester yarns; false twist, knit de knit and air texturising are common, although the latter method is the most used. Courtaulds Textiles Automotive Products[38] were the first company in Europe to use 100% air textured for car seats in the late 1970s. Staple spun yarns are less common because of their limited abrasion resistance in flat woven constructions. They are used, however, in woven velvets where the abrasion or wear is on the tips of the yarn rather than across its width.

The multinational fibre producers such as Hoechst, DuPont and Rhone Poulenc supply flat continuous filament yarns, and sometimes partially orientated yarn to yarn (POY) texturisers, for example Neckelmann, Autofil. These 'feedstock' yarns are doubled, tripled or even quadrupled during the texturising process. Special effects can be produced by mixing the yarns or by applying significant overfeed of one component to produce core and effect yarns. Here the overfeed yarn (the 'effect' yarn) may hang loosely around the core yarn. In flat woven fabrics, these yarns produce a certain amount of surface texture and give a more pleasant touch.

Typical yarns for weaving are 167 dtex/48 filaments primary feedstock yarn which when quadrupled produces 668 dtex/192 filaments and 835 dtex/240 filaments yarn made from five ends of a primary yarn. Heavy duty yarns over 3000 dtex with 550 filaments are used for heavy goods vehicles (HGVs) or for special effects. Knitting yarns are lighter up to say 300 dtex.

18.2.2.3   Fabric structure

The main fabric types with typical weight ranges are: flat woven fabric (200–400 g m⁻²), flat woven velvet (360–450 g m⁻²), warp knit tricot (generally pile surface, 160–340 g m⁻²), raschel double needle bar knitted (pile surface, 280–370 g m⁻²) and circular knits (generally pile surface, 160–230 g m⁻²). Fabrics in nylon tend to be towards the lower weight range. Woven fabrics have been produced for many years using mechanical Jacquard systems and while they once offered the greatest design potential, knits have now caught up with them with the introduction of computer-controlled knitting machines. The growth of automotive fabrics, especially knitted, has been the subject of several papers.[39–48]

Woven fabrics have limited stretch, which sometimes restricts their use in deep drawer moulding applications for door casings. Some polybutylterephthalate (PBT) yarns, which have increased stretch, are being used in certain instances, but they are significantly more expensive than regular polyester even though they are more easily dyeable. However, woven fabrics, even with PBT yarns, cannot compare with the stretch capabilities of knitted fabrics.

Flat woven velvet fabrics are the most expensive to produce but are considered top of the range in quality. Knitted fabrics with raised surfaces are more softer to the touch than flat wovens. Modern weft knitting machines equipped with advanced electronics that allow intricate design patterns offer considerable potential. These utilise yarn packages, compared with warp beams where yarn preparation is necessary in both weaving and warp knitting. Thus machine setup is much quicker and production volumes are much more flexible. Both of these factors are ideally suited to the shorter lead times and sometimes unpredictable production programmes required by car companies.

18.2.3 Composition of car seat covers

Whatever the construction of the fabric, it is usually made into a trilaminate consisting of face fabric and polyurethane foam with a scrim lining on the back.[49] The foam backing ensures that the fabric never creases or bags during the life of the car. It also produces attractive deep contoured sew lines in the seat cover, imparts a soft touch to the seat surface and contributes to seat comfort. The polyurethane foam density is usually between 26–45 $kg\,m^{-3}$, anything from 2–20 mm thick and can be either polyester polyurethane or polyether polyurethane, the latter being more hydrolysis resistant and is necessary for humid climates. Foam can be either standard or flame retardant, the test method being FMVSS302, although car companies each have their own performance specifications.

The scrim fabric lining is usually warp knitted nylon or polyester 30–90 $g\,m^{-2}$ in weight. This fabric acts as a 'slide aid' when the seat cover is sewn and when it is pulled over the foam seat body. It also contributes to seam strength, helps prevent strike through of liquid foam in 'pour in' techniques and can, by choice of construction, assist laminate dimensional stability by controlling excessive stretch. Nonwovens are sometimes used when only a slide aid property is required. Exact specification of cover components depends on where in the car the laminate will be used, that is, seat centre panel, bolster or back. Matching door panels are frequently made from the same material as the seats.

18.2.4 Manufacturing processes

18.2.4.1 Dyeing and finishing

Polyester yarn is package dyed usually with a UV light absorbing agent such as Fadex F (Clariant). This improves the lightfastness of disperse dyes and also contributes to yarn UV degradation resistance. Similar products are available for polyamide yarn, which is dyed mainly with 1:2 premetallised dyes. Higher lightfastness of certain shades is possible compared to aqueous dyeing by melt (also referred to as dope, solution or spun) dyeing during yarn manufacture. The drawback here is that it is not economical to produce small amounts of yarn in this way, which is a serious restriction in a commercial world where flexibility is an important advantage. Dyeing and finishing procedures have been published.[50-53] Flocked yarns and space dyed yarns are also used.

Processing routes for the production of woven and knitted fabric can be summarised as follows:

- Yarn, texturise, package dye, warp/beam, **weave**, scour, stenter/finish, laminate, cut/sew, fit to seat;

- yarn, texturise, warp/beam, **warp/knit**, brush/crop, stenter preset, scour/dye, stenter, brush, stenter finish, laminate, cut/sew, fit to seat;
- yarn, texturise, package dye, cone, **weft knit**, shear, scour, stenter/finish, laminate, cut/sew, fit to seat;
- yarn, texturise, package dye, cone, **three-dimensional knit**, heat stabilise, fit so seat.

When spun dyed yarns are used, the dyeing stage is, of course, omitted. The sequence of brushing, stentering and dyeing can be varied according to the particular warp-knit product. Piece dyeing has the advantage of colour flexibility because the fabric can be dyed to the required colour at the last minute. Some woven fabrics are coated with acrylic or polyurethane resin to confer reduced flammability properties and to improve abrasion. Woven velvets, however, must be coated to improve pile pull-out properties. Compared to the apparel and domestic furnishings industry, relatively few finishes are used on automotive fabrics. Antistatic and antisoil finish are applied either by padding or by foam processing on to the face of the fabric. Any finish must be carefully tested not only for effectiveness but also for any harmful side effects such as catalytic fading or discolouration of the dye, fogging, unpleasant odours or waxy or white deposits developing on the car seat during use. Some finishes, especially silicone-based ones, have a harmful effect on adhesion during lamination and should be avoided. The fabric must be stentered to provide a stable, flat, tension free substrate for lamination and eventual seat fabrication.

18.2.4.2 Printing
This is a relatively new development in Europe and the USA for automotive fabrics, although it has been carried out in Japan since the late 1980s.[54-56] The initial problems of dye penetration and sublimation of UV absorbers during the printing process are now being overcome. Printing offers a means of producing almost unlimited design options on both wovens and warp knits and quicker setup than either piece dyeing or the yarn dyed route. The constraints of weaving or knitting are absent and the design decision can be made closer to the launch date of the car, thus enabling the design to be right up-to-date. Another significant advantage is the cost saving gained by not having to put a new printed fabric through the full testing and acceptance procedure if a different print is put on to an existing and already approved base fabric.

18.2.4.3 Lamination
Most automotive fabric worldwide is flame laminated to foam. This is a quick economical process whereby all three components are fed into the laminator and the triple laminate emerges at speeds up to and exceeding $25–40\,\mathrm{m\,min^{-1}}$. The process depends on a gas flame licking moving foam to melt the surface of the foam, which acts as the adhesive. The flame lamination process, which produces potentially toxic emissions, has come under environmental scrutiny[57,58] and alternative methods of lamination have been developed using hot melt adhesives.[59-62] However, very few major laminations of automotive fabric have changed to the hot melt route, but some have installed improved methods of controlling emissions, such as carbon filter absorption.

Also on environmental grounds, alternatives to polyurethane foam have been sought. The triple laminate comprising polyester face fabric, polyurethane foam and nylon or polyester scrim is made from two or three dissimilar materials which are joined together and cannot be easily separated for recycling. Polyester nonwovens have been evaluated as alternatives to the foam,[63–64] as have three-dimensional knitted structures such as spacer fabric (Karl Mayer) and fleece materials such as 'Kunit' and 'Multiknit'.[65,66] All, however, lose thickness during use, especially at the higher temperatures that sometimes prevail inside the car. Nevertheless some car companies have replaced the laminate foam with wool or wool/polyester non-woven material.

Some nonwoven fabrics made from recycled polyester have been used.[67–72] Attempts are being made to replace the foam in the seat squab (back) and cushion (bottom) both with nonwoven material and also rubberised natural products such as coconut hair, horse hair and pigs' hair which are considered recyclable. 'BREATH AIR'™ (Toyobo), random continuous loops of a thermoplastic elastomer are claimed to offer improved thermal comfort and recylability, amongst other benefits.

18.2.4.4   Quality control and testing

Fabric laminate testing is carried out for two main reasons, first to determine suitability for further processing so the processes will be right first time every time and second, to simulate actual wear conditions during the life of the car. Accelerated ageing and wear tests are applied.

The actual tests and performance required depend where in the car the fabric laminate will be used, for example, seat centre panel, bolster, seat back, door casing, headliner, parcel shelf. Each car maker has its own test methods and performance specifications.

For seat make-up, panels must first be cut accurately and invariably several layers are cut together. It is important that the material lies flat and does not change its dimensions due to residual fabric shrinkage or tensions introduced during lamination. Fabrics dyed in different dyebaths or even different dye works may come together in the same car seat set and there must not be any noticeable differences in shade.

Lightfastness testing of textiles is not easy. The objective is to reproduce several years of simulated exposure in conditions of actual use within a short test period. This is especially difficult for automotive fabrics, where the damaging effect of sunlight inside a car means the combined effect of light and UV radiation, heat and varying relative humidity (and dampness). In addition, the test substrate itself, chemical finishes, processing conditions and possibly industrial and traffic fumes can all influence results. There are different types of test equipment, and although most car makers now use xenon arc, some Japanese require the carbon arc as a source. In Europe DIN 75202, the FAKRA method of continuous exposure is widely used. The American method of SAE J1885 includes an intermittent exposure, that is, a period with the light switched off. Performance requirements vary, but values around blue wool scale 6 is the required lightfastness rating necessary for seating fabric. Some US manufacturers were considering raising their present lightfastness standards substantially by the year 2000.

Abrasion is carried out using the Martindale, Taber or Schopper machines, again each car manufacturer specifying their own method and performance requirements. The related fabric wear phenomena of pilling and snagging are also assessed. Typical requirements are around 50 000 Martindale rubs for seating fabric.

Other tests carried out on fabric laminates include the following:

- Peel bond adhesion (face fabric-to-foam and scrim-to-foam) including testing wet, after heat ageing tests and treatment with solvents and cleaning fluids
- Cleanability after soiling with items such as chocolate, hair lacquer, lipstick, coffee, ball point pen ink and engine oil. Some vehicle manufacturers also include tests for 'minking' (loose hair deposits on the seat from fur coats) and 'linting' (white fibrous deposits)
- Dye fastness to perspiration
- Dye fastness to crocking (both dry and wet)
- Dimensional stability
- Flammability
- Tear/tensile strength
- Sewing seam strength
- Bursting strength
- Stretch and set
- Fogging (see below).

Although vehicle manufacturers detail their own preferred test method, many are based on national and international specifications. Table 18.7 lists many of these tests. Details of actual test methods are generally confidential between customer and supplier and performance specifications are stringent, and typically of 'contract' standard.[16]

'Fogging' is a mist-like deposit on car windscreens which reduces visibility.[73-79] It is caused by volatile materials vaporising from interior trim components and probably is aggravated by other materials introduced from outside the car or even from other areas of the car, for example ducting in the ventilation system.

Attempts are being made by the car companies and their suppliers to harmonise test methods.[80] The Transportation Division of the Industrial Fabrics Association International in the USA is especially active in this exercise. A 'data bank' has been proposed by one of the American technical journals and contributors have been requested.[81] Thus testing could become simpler and require fewer items of expensive test equipment, thereby releasing time and effort that could be directed towards development of new products, increasing efficiency and reducing costs.

18.2.5 Other parts of the car interior

Headliners used to be simple items in warp-knitted nylon, or PVC, sometimes 'slung' that is, held in place only at certain points. Modern headliners are multilayer materials[82] that have become a structural part of the car roof supporting accessories, such as sun visors, interior lights, assist handles, electrical components and some even contain brake lights. They are engineered to give sound insulation and sound absorbing properties. The majority of headliner face fabrics in Japan and Europe are nonwovens, but in the USA most are still warp-knitted nylon or polyester.

Warp knits have better abrasion and pilling resistance and mould better because of their superior stretch properties. However, nonwovens have the advantage of non-recovery after moulding. Nonwoven headliners are typically made from fine denier polyester or polypropylene fibre for maximum cover at low weight, about $200 \, g \, m^{-2}$, with an antiabrasion finish. Rotary screen-printed needle-punched polyester and malifleece headliners are used in some European cars.

Table 18.7 Summary of test methods applied to automotive seating fabrics

	British Standard test methods	Selected related test methods

Colour fastness	BS 1006: 1990 (1996)
	Methods for determining
	  colour fastness to about 70
	  different agencies
	BS 1006 Grey scales for
	  assessing changes in colour
	  AO2
	BS 1006 Grey scales for
	  assessing staining A03
	BA 1006 B01 Blue wool
	  standards
		ASTM methods
		ASTM evaluation procedures
		DIN 54022 (fastness to hot pressing)
		DIN 54020 (rub fastness)

Crocking (wet and dry)	BS 1006: 1990 (1996)	SAE J861 Jan 94 AATCC TM8 DIN 54021

Light fastness	BS 1006: 1990 (1996)	SAE J1885 Mar 92 Water cooled
		  Xenon-arc
		SAE J2212 Nov 93 Air cooled
		  Xenon-arc
		DIN 75202
		FAKRA 7/91

Abrasion	BS 5690: 1991 (Martindale)
	NB: Sometimes tested after
	  UV exposure
		SAE J365 Aug 1994 Scuff resistance
		  (Taber)
		ASTM 3884 (Taber)
		DIN 53 863 3/4 (Martindale)
		DIN 53 863/2 (Schopper)
		DIN 53528 (Frank Hauser, loss in
		  mass for coated fabrics)
		DIN 53 754 (Taber)

Pilling	BS 5811: 1986 pill box	ASTM D3511-82 (Brush) ASTM D3512-82 (Tumble) DIN 53863/3 (Modified Martindale) DIN 53865 (Modified Martindale)
Snagging		SAE J748 ASTM D5362-93 (Bean bag) ASTM-D3939-93 (Mace Test)

Tear strength	BS 4303: 1968 (1995) Wing
	  tear
	BS 3424 pt5: 1982 (for coated
	  fabrics)
	BS 4443 pt6 Method 15 (for
	  foam laminates)
		ASTM D2261: 96 (Tongue tear)
		ASTM D1117-95 (Trapezoidal tear)
		DIN 1424-96 Elmdorf tear apparatus
		DIN 53 356 (Tear propagation)

Tensile strength/ breaking and elongation	BS 3424: 1982 Method 6
	  (coated fabric)
	BS 2576: 1986 (Woven fabric/
	  strip method)
	BS 4443 pt6 1980 Method 15
	  cellular foam (laminates)
		ASTM D5034:95 (Grab method)
		ASTM D1682 (Grab method)
		DIN 53857 (Nonwovens)
		DIN 53571 (Tensile and elongation)
		ASTM D-751 (Test for coated
		  fabrics)

Stretch and set	BS 3424 pt 21: 1987 (for
	  coated fabrics) but BS 3424
	  pt 24 1973 (still in use)
		SAE J855 Jan 94
		DIN 53853
		DIN 53857

Table 18.7 *Continued*

	British Standard test methods	Selected related test methods
Stretch and recovery	BS 4952: 1992 (for elastic fabrics–replaces BS 4294: 1968)	
Bursting strength	BS 4768: 1972 (1997) Bursting strength and distension	DIN 53861
Dimensional stability	BS 4736: 1996 (cold water)	SAE J883 Jan 94 Cold water SAE J315A DIN 53894
Stiffness	BS 3356: 1990 bending length and flexural rigidity	D1338-96
Drape	BS 5058 1973 (1997)	DIN 53350 (bendability)
Crease recovery	BS EN 22313: 1992	
Steam strength		ASTM D1683 for woven fabrics SAE J1531
Peel bond	BS 3424 Pt 7 1982 Method 9 (coating adhesion)	ASTM D-751 ASTM D-903 DIN 53357
Compression (For foam/ laminates)	BS 4443 Pt 1 Method 5A stress strain characteristics BS 4443 Pt 1 Method 6A compression set	ASTM D2406-73 Method B DIN 53 572 Compression set DIN 53 577 Stress strain characteristics
Air permeability	BS 5636: 1978 for fabrics now BS EN ISO 9237: 1995 BS 4443 Pt 6 1980 Method 16 (For foam laminates) BS 6538 Pt 3 1987 (Gurley method)	ASTM D737 DIN 53 887
Surface resistivity (antistatic)	BS 6524: 1984 (Surface resistivity)	DIN 53282 (Surface resistivity) ASTM F365-73 Charge decay Federal Method 101C – 4046 (Charge decay) BTTG Body voltage chair test
Cleanability		AATCC Method 118 – 1983
Stain repellency	BS 4948: 1994 Soiling by body contact	
Fogging		SAE J1756: 1994 ASTM D5393 DIN 75201
Flammability resistance		FMVSS302 DIN 75200 SAE J369 ASTM D2859-70
Water wicking		SAE J913

Table 18.7 *Continued*

	British Standard test methods	Selected related test methods
Accelerated ageing methods	BS 3424: 1996 Pt 12 for coated fabrics BS 4443 Pt 4 Method 11 for cellular materials (foam) humidity and elevated temperatures BS 4443 Pt 6 Method 12 (heat ageing)	ASTM D2406-73 DIN 53 378 'Environmental cycles' of individual manufacturers as pretreatments for further testing, e.g. peel bond, dimensional stability and effect on appearance. Sometimes includes cooling to as low as −40 °C and heating to as high as 120 °C
Resistance to microorganisms		AATCC Method 30 resistance to mildew and rot AATCC Method 100 resistance to bacteria AATCC Method 174 bacteria resistance for carpets Federal test method standard 191 Method 5750 mildew resistance, mixed culture method

Parcel trays, being just beneath the slanting rear window, demand the highest resistance to sunlight, for example 450 kJ compared to 150 kJ for a headliner. They are made by a press lamination technique or by direct pouring of the polymer on to the back of a polyester or polypropylene nonwoven. Many of the decorative interior nonwovens are needle punched in the usual manner, but increasing numbers are formed using the Dilour technique to produce a velour-type material with deep draw mouldability.

18.2.5.1 Nonwoven applications
Nonwovens are used under the bonnet to reinforce acoustic insulation material. A novel material, Colback (Akzo), made from a bicomponent yarn/nylon 6 sheath over a polyester core/provides very high strength with light weight.

With increasing awareness of air quality within cars, air filters are being installed as standard equipment. Nonwovens with an effective surface area of about $0.3 \, m^2$ are required to filter out solid particles with diameters as small as $3 \, \mu m$. Development is being carried out to improve performance and up to $0.5 \, m^2$ of nonwoven material may be required with activated carbon and antibacterial chemicals to remove malodours.[83,84]

Textile battery separators experienced substantial growth in the late 1970s to early 1980s when woven and nonwoven polyesters, impregnated with specially selected acrylic resins, replaced many of the PVC type. Batteries are used in all transportation vehicles, especially in electric powered vehicles such as golf carts, road sweepers, milk floats and forklift trucks and represent an increasing market.

Needlefelt nonwoven carpets are increasingly popular in Europe especially those comprising polypropylene, which offers savings both in weight and cost and is claimed to be better for recycling. More expensive cars generally use tufted carpets in nylon, however recyclability of automobile carpets is becoming an important issue and both nonwoven polypropylene and tufted nylon carpets are ideal in principle. In practice, the presence of dirt (up to 1 kg m^{-2}) is a major problem.

18.2.5.2 Flocked fibre
Surfaces such as window seals and dashboard components have textile flocked surfaces. The flock is usually polyester or nylon 6.6, but viscose and acrylic fibre are also used. Flock is useful in eliminating rattles and squeaks in the car as well as contributing to the overall aesthetics.[85,86] Flocked yarns are sometimes used for seating and door panel fabric. Novartis (Rhone Poulenc) are targeting automotives with their improved flock technology and they claim enhanced seat thermal comfort together with a velour-type appearance at an economical price.

18.2.5.3 Seat belts
Seat belts are multiple layer woven narrow fabrics in twill or satin construction from high tenacity polyester yarns, typically 320 ends of 1100 dtex or 260 ends of 1670 dtex yarn. These constructions allow maximum yarn packing within a given area for maximum strength and the trend is to use coarser yarns for better abrasion resistance. For comfort they need to be softer and more flexible along the length, but rigidity is required across the width to enable them to slide easily between buckles and to retract smoothly into housings. Edges need to be scuff resistant but not unpleasantly hard and the fabric must be resistant to microorganisms. Nylon was used in some early seat belts but because of its better UV degradation resistance, polyester is now used almost exclusively worldwide.

Melt-dyed yarns are used, but other colours are obtainable by pad thermosol dyeing with dyes selected for the highest resistance to light and UV degradation and excellent wet rub and perspiration fastness. Fabric is about 50 g/linear metre (about 5 cm wide) loomstate but about 60 g/linear metre after finishing. This is because shrinkage is induced in the finishing process to improve the energy absorption properties. Controlled, limited non-recoverable (i.e. not elastic) stretch reduces deceleration forces on the body in a collision.[87,88]

Performance standards (e.g. BS 3254) typically require a belt to restrain a passenger weighing 90 kg involved in a collision at 50 km h^{-1} (about 30 mph) into a fixed object. Straight pull tensile strength should be at least 30 kN/50 mm. Other tests include accelerated ageing and, in the made up form, resistance to fastening and unfastening 10 000 times. The seat belt must last the lifetime of the car without significant deterioration.

Studies in the 1970s concluded that seat belts could reduce fatal and serious injuries by 50%, consequently front seat belts became compulsory in the UK in January 1983 and are now compulsory in many countries of the world. Many US drivers, however, still refuse to wear them and rely only on the air bag. Because of this the US front seat air bag is larger than those used in Europe where air bags are used in conjunction with seat belts.

About 14 m of seat belt fabric weighing about 0.8 kg are present in every new car and total usage is about 32 000 tonnes per annum. Recycling of seat belts is feasible because they are very easily removed and are of uniform composition.[89] Belts are mainly black in Europe, light grey in the USA and Japan, but this is changing to coordinate more with interior colours.

18.2.5.4   The airbag

Airbags were first introduced in the late 1960s, but it is only in the 1990s that their use has grown spectacularly and is set to grow even further. This justifies the considerable research and development still being conducted on design, deployment and base fabric material.

A triggering device sets off explosive chemicals when it senses an accident above 35 km h^{-1} is about to occur. These chemicals inflate the bag to restrain and cushion the car occupant from impact with harder objects.[90] The fabric from which the bag is made must be capable of withstanding the force of the propellant chemicals. More important, the hot gases must not penetrate the fabric and burn the skin of the car occupant. The earliest airbags were Neoprene (DuPont)-coated, woven nylon 6.6, but lighter and thinner silicone-coated versions soon followed.[91] Later, however, uncoated fabrics have appeared.

There are advantages and disadvantages for each type; coated fabrics are easier to cut and sew with edges less likely to fray and air porosity can be better controlled, whilst uncoated bags are lighter, softer, less bulky and easier to recycle.[92] Airbags vary in size and configuration depending on in which car they are to be used. In addition driver side airbags (from 35 litre capacity upwards) are smaller than for the front passengers, from about 65 litres capacity upwards.

Airbags are typically made from high tenacity multifilament nylon 6.6 in yarn quality finenesses from 210, 420 to 840 denier although some polyester and even some nylon 6 is used.[93] Nylon 6 is said to minimise skin abrasion because it is softer. Airbag fabric is not dyed but has to be scoured to remove impurities which could encourage mildew or cause other problems. It needs to have high tear strength, high antiseam slippage, controlled air permeability[94] (about 10 l m^{-2} min^{-1}) and be capable of being folded up into a confined space for over 10 years without deterioration. Some tests require 75% property retention after 4000 hours at 90–120 °C, the equivalent of 10 years UV exposure and also cold crack resistance down to –40 °C. A new fibre nylon 4.6 (Akzo) with a melting point of 285 °C has been introduced especially for airbags.

In the USA, FMVSS 208 requires all passenger cars sold during 1997 to have airbags both for the driver and front seat passenger.[95] Worldwide production of airbags was about 43 million units in 1997 and this was expected to grow to 120–200 million units (up to 50 000 kg of mainly woven nylon) by the year 2000.[96] Production could be much more with the development of head protection via inflatable tubular structures (ITS), side airbags, rear seat airbags, even knee and foot well airbags have been mentioned.[97-99] Indeed, in the USA, FMVSS 201 has required that by May 1999 10% of cars must be fitted with some kind of head protection. This requirement will rise to 100% by May 2003. BMW has introduced a side impact airbag and an ITS for head protection in their 1997 '7 series' to be extended later to their '5 and 3 series'. A BMW 'concept car' features 12 airbags![100] Future airbags are likely to be smaller, lighter and more compactable.

Whilst airbags undoubtedly save lives they can also cause serious injury. Following the deaths of children in accidents under 20 mph, the design of airbags in the USA has come under considerable scrutiny. The search is on for a reliable 'smart' airbag which can sense the size of the passenger or even if the seat is unoccupied and react accordingly. Furthermore, integrated safety systems combining the seat belt and other safety items are under development especially for child passengers.

18.2.6   Seating developments

18.2.6.1   Three-dimensional knitting of car seats
This novel development originated from garment research at Courtaulds, Research laboratories in Derby. The objective was to knit garments in one piece, thus eliminating panel cutting and making up along with the associated cutting waste of up to 30%. The potential benefits to the automotive industry were soon recognised and General Motors became involved.

Initial progress was hampered by the mechanical flat bed weft-knitting machine controls and its Jacquard card needle selection mechanisms, but these limitations were overcome with the appearance of the computer.[101–103] With computer assistance, each needle can be individually controlled enabling almost infinite colour combinations and design patterns and car seat covers can now be knitted in just one piece with accurate placement of logos if necessary. This single item includes all flaps, tubes and tie downs necessary for direct fitting. The stages of panel cutting and sewing together of up to seventeen pieces of fabric are reduced to just one or two with no cutting waste.

Other benefits include rapid setup and dramatically reduced stock holding especially end of model surplus.[104,105] A new design can be produced simply by changing the yarns and inserting a new floppy disk, so that within minutes a different seat cover becomes available for fitting. Design themes can cover two or more seats and can even include the door panel.

Three-dimensional seat covers made their European debut in the Vauxhall Rascal van and in the 1993 Chevrolet Indy Pace car in the USA. They are being used in the General Motors electric car EVI (see Fig. 18.2) and in the GM car GEO Prizm.[106] Research work however continues with a variety of different yarns to develop the aesthetics and handle further. Although the 25 or so worldwide patents relevant to automobile application are held by GM of the USA, the original inventors are mainly British and continue their work at Derby.

18.2 New generation of environmentally friendly vehicles. The EVI contains many novel features including three-dimensional knitted car seat covers.

18.2.6.2   New seat technology

The traditional way of seat making, involving cutting and sewing panels into a cover that is then pulled over the squab (seat back) and cushion (seat bottom), is time consuming and cumbersome in the modern age. Efforts have been made to develop alternative quicker and more efficient methods. Three-dimensional knitting is an option but this has had only limited usage so far.

'Pour in foam' techniques for small items such as head rests and arm rests are carried out, but the preparation of whole seats by this method on a large scale by one car manufacturer in the early 1990s was discontinued. A barrier film was needed to allow vacuum to be applied and to prevent liquid foam penetration before solidification. The barrier films, however, affected seat comfort. For small items, the laminate foam itself is sufficient to prevent this 'weeping through' of liquid foam.

The latest techniques involve bonding the seat cover laminate to the squab/cushion using a vacuum technique and a hot-melt adhesive film. The cover and film are held together by vacuum, the moulded squab/cushion is placed on top and the adhesive film activated by hot air or by steam in some variations of the process.[107] Many seats are made in the USA by this method. It is especially suited to increasingly curvaceous seat contours and could reduce the need for thick cover laminate foam. Because of the adhesive layer, seat breathability and comfort may be affected, but the seat makers are aware of this.

DuPont have developed polyester nonwoven fibre 'clusters', which can be formed into seat cushions and squabs in place of polyurethane foam. Benefits claimed include 20% reduced weight and increased comfort through improved breathability and recyclability. With polyester-based material used as webbing and a polyester fabric covering and scrim, the whole seat in just one material becomes easily recycled.[108,109]

In an entirely different approach, cushion foam or springs are replaced by woven fabric with the correct stretchability. Two systems Sisiara (Pirelli) and Dymetrol (DuPont)[110] have appeared in production cars. Benefits include savings in both weight and space. Developments of the same concept using elastomeric materials have appeared.[107]

18.2.6.3   Artificial leather and suede

A leather shortage is forecast in the near future and at the same time designers anticipate an increased demand in cars, especially for leather/textile combinations. Leather is universally regarded as the ultimate in seat luxury. The shortfall is likely to be filled by artificial products and the manufacturers, almost entirely Japanese, have expansion plans. Toray estimate an increased production of artificial suede from 16 million m^2 in 1995 to 25 million m^2 by 2005 with a significant part going into European cars.[111] The base material is polyester microfibres approximately 68% by weight (32% polyurethane). Automotive grades are backed by polyester/cotton or knitted nylon scrim fabric. The best known product in Europe is Alcantara, made in Italy since 1975 in a Toray/Enichen joint venture. At present about 1 million metres are used in cars, mainly of Italian manufacture.[112]

A new artificial leather, Lorica, which was launched in Italy in 1994 by Enichen,[113] has several advantages over natural leather, including increased elongation and tear strength, mouldability and high frequency microwave weldability. It is produced from polyamide microfibres and polyurethane and is available in a variety of colours. Artificial leathers and suedes have the advantage over natural products in uniformity of quality and thickness and availability in roll form which facilitates production planning and minimises waste. Significant improvements in the quality have been achieved by microfibres and the latest Japanese products use ultra fine filaments of 0.001–0.003 dtex. While Alcantara and other successful suedes are produced by a solvent coagulation process, attempts have been made to develop more environmentally friendly aqueous methods.[114]

18.2.7 Aspects of management

A revolution is taking place within the car industry. Car makers, 'original equipment manufacturers' (OEMs) have become assemblers of components, making few individual items themselves. As competition intensifies, OEMs have to manufacture and sell worldwide and their immediate suppliers, the Tier 1 level suppliers have had to follow suit. In turn, suppliers to Tier 1, that is Tier 2, also need to have a global presence. Global strategies are now vital for global products[115] and the QS-9000 quality standard (based on ISO 9001 and specific to the automotive industry), is becoming required increasingly.

OEMs are continually applying pressure to cut costs, indeed some required 3–6% annual cuts to the year 2000, that is, 20% compounded.[116] To meet the challenge 'Tier 1s' are integrating interior components and assuming responsibility for design and specification and because of this in some ways the design and specification control of interiors is moving away from the OEMs. Lear Corporation is buying up interior component companies with the declared intention of eventually being able to offer entire interiors of cars at an agreed price.[117,118]

The trend for OEMs and others is to reduce the number of suppliers to simplify administration and reduce cost. They expect more from the select few, requiring supply 'just-in-time' at the required quality. These companies are in almost minute-by-minute contact with their customers. In such a fast moving environment, the successful textile technologist must know his/her subject and all the downstream processes inside out to be able to identify and solve problems quickly. However, all involved must provide all relevant information without hesitation. The cooperative culture must be created and fostered. OEMs rarely rely on a single supplier and a supplier who has developed a new product may be required to hand over all details to a competitor for them to become a second supplier.

18.2.8 Recycling

The car seat laminate comprising up to three different chemical types, polyester face fabric, polyurethane foam and nylon scrim cannot be easily recycled, although use of polyester scrim reduces the number to two. Chemical hydrolysis can break the three polymers down into simpler chemicals, which can be repolymerised but this is not commercially feasible at present. Use of nonwoven polyester, both in the seat cover laminate and in the seat cushion/squab, has already been mentioned. Replacing the laminate foam with a polyester nonwoven may not solve the problem if the adhesive joining the two polyester components (i.e. face fabric and non woven) is chemically dissimilar.

Fibre manufacturers Hoechst and EMS have shown the possibility of recycling polyester face fabric into nonwoven material. Shredded face fabric mixed with 30% of virgin polyester polymer has been re-extruded into polyester nonwoven fibre.

Nonwovens, notably from Wellman produced from recycled polyester bottles, are already being used commercially.

A key factor is the time taken for disassembly. The 'dirt factor' for carpets has already been mentioned and this must also be a factor for car seat fabric and adds to the case for better cleanability. The impetus for recycling arose within the last six years and it is only relatively recently that cars are actually being designed for disassembly. Possibly in the future a second (recycled) use for a material could be decided in advance and influence the initial choice. Recycling has been and is the subject of much research[119–124] some of it funded by the European Union (through Brite Euram programmes), which has also funded efforts to make the car lighter. Choice of materials are now influenced by 'cradle to grave' (life cycle) analyses of the various options and commonisation of materials.[125]

18.2.8.1   Voluntary action by the industry
About eight million 'end of life vehicles' (ELVs) are scrapped annually in the European Union. Early in the 1990s voluntary accords were set up. Amongst them are the Automotive Consortium on Recycling and Disposal (ACORD)[126] in the UK and PRAVDA[127] in Germany to provide national frameworks for economic break-even for recovery systems, to reduce waste disposal from ELVs and to ensure by 1998 that all were properly collected. CARE (Consortium for Automotive Recycling)[128] was set up in the UK in 1996. Composed of ten car manufacturers and a number of car dismantlers, it works with government bodies and others to obtain specific results from practical work by helping individual companies. Since 1993 Recytex (a subsidiary of Viktor Achter, the car upholstery manufacturer) has processed textile waste from its parent company and others.[129]

18.2.8.2   European legislation
Proposed European Directive DGX1 (Environment) targets 80% of the car to be recycled by the year 2002 with not more than 15% by weight going to landfill. This will be decreased to 5% by 2015 with 90% recycled.[130] Already landfill fees in the UK and elsewhere have risen substantially. The draft directive also requires a system of collection and treatment to be created with the responsibility for reuse, recovery and recycling of the ELVs to rest with the automotive sector's 'economic operators'. This can be interpreted as meaning everyone involved with the vehicle. The European Car Industry considers these measures unreasonable because the car is already recycled 75% by weight and actual waste, the industry claims, represents only 0.2% of all European industrial waste. The EU however considers ELVs a priority.[131] These issues are very likely to affect the textile industry eventually, because fabric and fabric laminates are major components of vehicle interiors. In addition, the EC directive also proposes to scrutinize the use of PVC – if replacement is recommended, textiles are likely to replace at least some of it.

18.2.9   Future development in automotive textiles
Car production is expected to remain generally static up to 2005 in the developed world, but is likely to expand considerably in the developing nations. Globally there are excellent opportunities for the multinational OEMs and their suppliers, especially those with the imagination and will to innovate new products and design fea-

tures that will make car journeys more comfortable, safe and pleasant. The following paragraphs discuss the possibilities that are believed to exist.

The largest growth area in automotive textiles will be in air bags as they become standard equipment in more cars. Development is needed to improve their safe functioning, however, legislation may spur on such developments in a similar way to the USA. The latest system is an air bag that deploys outwards from the occupant's seat belt. Possible new applications for textiles within the car include the dashboard, sunvisor and seat pockets and circular knitted fabrics may be especially suited to these outlets.

Increased hygiene awareness, already present in Japan, may lead to greater use of antimicrobial finishes on car seats and carpets and perhaps the use of fibres with built in permanent protection, such as Bactekiller (Kanebo) and Diolen (Akzo), both polyesters, and Amicor (Courtaulds' acrylic fibre). This may be hastened by more food being consumed in cars, particularly in recreational vehicles (RVs) and multipurpose vehicles (MPVs). Improved cleanability or more durable soil-resistant finishes may also be necessary, especially with increased Jacquard designs and brighter colours. The quest for cleaner air is leading to more air filters being installed as standard features and existing ones being improved. One novel development, which does not appear to have been widely adopted in the USA or Europe, is the Japanese invention of odour-absorbing backcoatings on seat fabric.[132]

Engineered fabrics can contribute to comfort, making journeys less stressful and therefore safer. Fabrics can be designed to improve seat thermal comfort[133-137] by absorbing and transporting moisture away from the body. Softer touch and more pleasing handle would make car seats more comfortable as many car seat covers are rough compared to domestic furniture and of particular note is that clothing today is significantly softer than it was say in the mid 1980s. Much research, possibly assisted by the Kawabata objective measurement of fabric handle and touch,[138,139] has resulted in the soft and smooth surface to touch of modern clothing. Some attempt has been made to apply the Kawabata technique to automotive fabric.[140] The very high abrasion resistance requirements, however, restrict progress. Additionally, textiles could possibly contribute more to the control of noise in the car,[141-143] which after alcohol is believed to be one of the main causes of accidents, because it contributes to driver fatigue.

The layout of car seats is becoming more flexible with, for example, some individual seats able to be turned through 180° for family meals or for conferences. They can even be removed altogether in the latest vehicles and so increase load carrying capacity. Could this lead to renewable seats? Will car seats be changed in a similar way to renewing easy chairs in the living room? Renewable items may not need the extremely high durability requirements and this could open the door to many new developments not possible at present, such as use and introduction of significantly softer fabrics. Another possibility is that people will keep their vehicles for longer, which will push performance specifications even higher. DuPont have developed 'Xtra-life' nylon in anticipation of this and US OEMs are believed to favour significantly increased light-resistance properties. Generally higher durability requirements are anticipated in the USA, because of the growing market in pick-up trucks (a 'robust' product, treated accordingly!) and increased vehicle leasing. In the latter case, the vehicle's 'private life' begins when it is already 2–3 years old. The developed countries will see a larger retired population. People who have retired younger and are living longer, with reduced incomes, may not wish to change their cars as frequently as before.

'Mass customization' is expected in the garment industry. Garment scientists working with retailers are developing a 'body scan' which instantly measures a person's dimensions. The information will enable a garment to be made quickly in any colour and design of the customer's own choosing. Could this concept spread to car interiors? Younger persons and others seek designs to suit their individual lifestyle and so it is possible, if not probable, that this will occur.

Environmental issues already influence car design and manufacture. Chlorofluorocarbon (CFC)-expanded foams have disappeared, but the antimony/halogen synergy combinations are still used in flame retardants in textile coatings and these are currently giving concerns on ecotoxological grounds in Europe. Will the environmental laws of the future prohibit their use? Alternative FR chemicals are being developed, but so far are less effective weight for weight. Will it become necessary to recycle car seat covers? At present they consist of dissimilar materials joined together. However, the new technique of joining the seat cover directly to the cushion and squab would appear to make recycling more difficult.

Will a 'green car' sell better? Consumer research in the USA suggests that the public may well be prepared to pay more for environmentally friendly items. One of the simplest ways to make the car greener is to reduce fuel consumption by building it lighter. Already several parts have been replaced with lighter plastic composites and eventually carbon fibres are likely to appear in mass-produced cars. Some attempts are being made to make seat cover fabrics lighter by using finer yarns and reduced construction densities but this lowers abrasion resistance and other properties. Fewer fabrics are now backcoated to save both weight and cost. In Germany, the objective is to produce the 'three-litre car'[144] in other words a car which will cover 100 km on 3 litres of petrol, equivalent to 94 mpg.

New generations of specialist yarns based on established fibres are appearing, for example high tenacity low shrink polyester (for tyres) and nylon variants exclusively for air bags. Teijin have developed 'soft feeling' polyester yarns specifically for velvet pile fabric and a type that is resistant against pile crush.[145] The Japanese fibre companies have developed many yarns with novel cross-sections and ever finer microfibres to produce new generations of apparel fabrics (e.g. 'Shin-Goshen' fabrics). Courtaulds (working with Gateway Inc.) and DuPont are developing smart acrylic yarns for clothing which respond to body temperature, that is, provide warming when cold and cooling when hot.[146,147] In addition there are several modified yarns designed to improve thermal comfort of clothing.[148,149] These yarns may be adaptable for automotive seating, but they are likely to cost more and have lower abrasion resistance. As global volumes increase it may become economically feasible to develop speciality yarns for automotive seating, perhaps offering some novel feature of texture or appearance or with enhanced abrasion and light fastness durability.

The control of static electricity in cars in being studied carefully[150] because, whilst static shocks may be unpleasant, concern is being expressed that they present a safety hazard by interfering with the electronic management systems within the car. Fabric finishes wear off eventually and may not be fully effective if the car seat is not earthed. New solutions may be possible using conductive yarns.

Opportunities for textile innovation exist, but projects developing new products which are likely to add to the cost of a car must be scrutinized very carefully because

OEMs, at present, are mainly concerned with cost reduction. This is understandable in an intensely competitive industry, but it can be discouraging to their supplier's research and development staff who strive to produce the innovative features the industry needs.

Factors other than those directly concerned with textiles could significantly influence fabric development in the not too distant future, like better UV and other radiation screening glass or new lighting systems via fibre optics. The former factor may well lead to cooler interiors in sunlight and possibly eventually remove some of the restrictions on fibre type and attainable shades.

18.3   Textiles in other road vehicles

18.3.1   Heavy goods vehicles (HGV)

More use of textiles is even being made in HGV interiors, which are becoming more comfortable with livelier colouring, softer more rounder shapes and surfaces. In the USA there is a reported shortage of drivers and comfort and appearance are important factors in attracting and retaining employees. In addition there are a growing number of husband and wife teams in the industry.[151,152]

Composite materials are being used to replace bulky space dividers and doors to create more cab storage space. More cabs have sleeping quarters with beds, curtaining, carpets and textile wall coverings. Seating fabric requirements are very similar to automobiles except heavier fabrics about $430 \, g \, m^{-2}$ using yarns up to 3000 dtex are sometimes used and the performance requirements of the flame retardant test, FMVSS 302, are generally higher.

18.3.1.1   Tarpaulins

HGVs are a major user of tarpaulins, which are made from PVC plasticol-coated nylon and polyester, usually Panama and plain woven from high tenacity yarns.[153,154] Base fabrics vary from about $100 \, g \, m^{-2}$ to over $250 \, g \, m^{-2}$ and are coated with up to $600 \, g \, m^{-2}$ or more of PVC plastisol applied in several layers. The more up-market products, for example Complan–Trevira (Isoplan–Trevira in the USA) are coated on both sides with the face side lacquered with an acrylic or polyurethane resin to improve UV degradation resistance, abrasion resistance and antisoil properties and also to reduce plastisol migration. Tarpaulins must also pass flexing resistance, cold cracking, reduced flammability, coating adhesion, waterproofness and tear and tensile tests. They must be dimensionally stable over a wide range of temperatures and relative humidities and be resistant to common chemicals, oils and engine fuels. However, if the coating is damaged, microorganisms can migrate via moisture into the material. Both Hoechst (Trevira HT Type 711) and Akzo (Diolen 174 SLC) have developed special variants of polyester to prevent this.

Environmental pressure groups, especially Greenpeace, maintain that PVC is harmful to the environment, especially during manufacture and disposal. The PVC industry has countered such claims, but tarpaulins using other coatings such as polyethylene have appeared. Tarpaulins are secured with high tenacity polyester narrow fabric which must also be tested carefully for strength and UV resistance.

18.3.1.2 Spray guards

A European Community directive requires HGVs to be fitted with guards to reduce road spray. Suitable products have been produced from polyester monofilament yarn knitted in a spacer fabric construction about 12 mm thick. Textile guards are substantially lighter than ones produced in plastic and about six guards are required for the average HGV.[155]

18.3.1.3 Flexible intermediate bulk container

Flexible intermediate bulk containers are used for transporting materials such as powders and so may be considered as an adjunct to HGVs. They are woven from polypropylene tape yarn with a specially formulated coating. Because of the danger of static explosions when being filled or emptied they need to be carefully earthed with metal wire in the fabric. It has been possible to replace the wire with Negastat, DuPont's antistatic yarn that functions without the need to earth.

18.3.2 Buses and coaches

These vehicles cater for the general public and therefore require the highest standards of safety and durability. Seating fabric is typically 780 g m^{-2} after coating with acrylic latex and is generally in conservative designs. Wool or wool/nylon woven moquettes are very common with high standards of abrasion (80 000 Martindale rubs), light fastness (to at least wool blue scale 6), fastness to perspiration, crocking, tear strength, soil resistance and cleanability by shampooing all being important requirements. The life of seating fabric varies from less than 6 years in some commuter public transport vehicles to 10 years or more in luxury coaches. High flammability performance requirements are becoming more stringent. BS5852 ignition source 5 is necessary and 'fireblocker' materials similar to those used in aircraft (see below) are being used increasingly. Consideration is also being given to smoke opacity and toxicity and heat flux. Following a number of coach disasters, both in the UK and Europe, safety standards are being studied and the use of seat belts for passengers may become compulsory in the near future. In some cities of the world, vandalism and graffiti is a serious problem in public transport. Fabrics have been designed to minimise these effects, like fabrics with pile that stands up so that if slashed with a knife it will not show readily.[155,156]

Textile-reinforced rigid composites are being used increasingly in buses and coaches to reduce weight and therefore conserve fuel. A prototype bus was unveiled in California that included three structural composites to replace 250 parts in a conventional bus. The 'stealth bus', nicknamed after the stealth bomber, has an expected life of 25 years compared to 8–12 years for ordinary buses and, being over 4000 kg lighter, will allow very considerable savings on fuel.[157]

18.4 Rail applications

The railways went through a period of contraction in the early 1960s in the UK and, more recently, reorganisation into different companies. This may lead to more varied textile designs in seating fabric. Railways in Europe have a key role to play in an integrated transport system[158] and several primary routes using high speed trains are being developed, some of which include the Channel Tunnel. A plan envisaging

30 000 km of new and improved lines was presented to the European Union in January 1989. The railway is probably the most environmentally friendly mode of passenger travel, both for long distance and commuter traffic. Traffic congestion, both on the road and in the air, plus concerns over the environment, strengthens the case for rail travel. The development of high speed trains allows rail to compete effectively with airlines. Interior decor and comfort are key factors in winning passengers away from other forms of transport. The decor of wall panels, seats and carpets cannot be changed every year and so designs must be neutral and not driven by fashion or fad.[159] The major technical issue concerning textiles is reduced flammability. Seat upholstery, loose coverings, carpets curtains and bedding must all pass stringent tests.[160] The materials must also have the correct aesthetics and durability must be in line with planned maintenance schedules.

18.4.1   Seating

Woven moquette weighing about 800 g m^{-2} in 85% wool/15% nylon has been the standard fabric for many years.[161] To withstand high volumes of passengers, the fabric must satisfy high burst strength and breaking load tests and abrasion resistance must be in the order of 80 000 Martindale rubs. It must meet light fastness standard 6 (blue wool scale), be dimensionally stable and not change in appearance after shampooing. Polyesters, especially FR grades such as Trevira CS and FR have gained acceptance especially in Europe. Designs are generally conservative in dark colours to mask soiling but this may change with the privatisation of British Rail and the formation of several companies who may develop corporate colours.

The overall FR standard aspired to is BS6853 1987 '*Code of Practice for Fire Precautions in Design and Construction of Railway Passenger and Rolling Stock*'. This document includes requirements for smoke and toxic fumes assessment. Materials are not only tested singly but complete seats are evaluated according to BS5852 ignition source 7.[162] Fire blocker materials are being used increasingly for rail seats. Control of toxic fumes and smoke, which reduces visibility, is of especial importance for trains that pass through tunnels or are used in underground railways. Halogen-containing materials, such as PVC, and any other materials that have high toxicity indices (modacrylic fibre) are excluded from passenger coaches.[163] The building of the Channel Tunnel has influenced regulations and all international passenger trains must comply with the International Union of Railways specification UIC 574-2 DR. The French standard NF F 16-101, which contains a very structured procedure for the testing of individual materials for flammability and both smoke opacity and toxicity, is also sometimes required. Generally, however, the FR standards across Europe are quite diverse. In addition some local authority vehicles have graffiti-proof fabric and metal wire beneath face fabrics to minimise vandal damage by knife slashing.

18.4.2   Other textiles uses

Sleeping car textiles, such as bed sheets and blankets, generally require high standards of performance and durability and some FR properties. Carpets are important in helping to create an attractive relaxing appearance, but they must be extremely hard wearing to cope with the volume of foot traffic sometimes for up to 20 hours a day.[164] Wool and nylon are the fibres most used in FR qualities with smoke emission, toxicity of fumes and heat release carefully assessed. BS476 Part 7 *Spread of flame* is used, but ASTM E 648 *Critical heat flux* is applied in addition for certain cases. Nylon carpet is reported to be better than wool for cleanability and stain resistance but wool is preferred for flammability resistance. Durable antistatic properties are sometimes accomplished with the use of small amounts of conductive fibres in the carpet, such as Resistat (BASF) or Antron P140 (DuPont).

With the advent of the high speed train to compete with aircraft, weight is becoming more sensitive and metal and other traditional material parts of trains have been replaced with composite materials to reduce weight. The French TGV train capable of above $300 \, \text{km h}^{-1}$ contains significant amounts of carbon fibre/epoxy composites.

18.5   Textiles in aircraft

As the 21st century dawns, more effort is being put into design of aircraft interiors to make them more passenger 'friendly'. This means more head room and rounder and softer surfaces to give the impression of spaciousness.[165-167] Increased safety is also being researched to make seats stronger and bulk head airbags may soon appear in passenger aircraft.

The main technical challenges for the textile technologist are safety (mainly with respect to flame retardancy) and weight saving. It is estimated that for every 1 kg of weight saved in an aircraft, £150 a year is saved in fuel costs, whilst a 100 kg lighter load can increase the range by 100 km. Reduced flammability is vital and statistics show that fire accounts for over 25% of deaths in aircraft accidents.

About 500–600 large passenger airliners are constructed each year, almost entirely in the USA (75%) and Europe (25%). A further 250 smaller passenger aircraft, 1300 light aircraft and about 1500 helicopters for civil use are built every year.[167] Aircraft have a lifespan of about 30 years, so the numbers in service are increasing steadily. Of all the forms of transport, air, especially air freight, is the biggest growth area.[168] Furnishings and equipment are refurbished regularly or when signs of wear are evident.

18.5.1   Furnishing fabrics

Furnishing fabrics include seat covers, curtaining, carpets and on long distance flights, blankets and pillows. Designs are generally in the livery colours of the particular airline, sometimes with company logos appearing in prominent positions. The article requiring most technical attention is the seat cover assembly on top of polyurethane foam. The fabric itself is generally made from woven wool or wool/nylon blends (nylon in the warp) of $350–450 \, \text{g m}^{-2}$ weight. Before acceptance by the airline it is tested for colour fastness, crocking, cleanability, pilling, snag resistance and dimensional stability. Some woven polyester covers are now being used, giving saving in weight and improved easy care. The wool is generally Zirpro (IWS) treated for the highest FR performance and the polyester must have FR properties that are durable in washing. Seat covers are usually fully laundered every three months and life expectancy is about three years. Cleaning must be accomplished during the restricted 'turn around' times in between flights. The materials must therefore have soil-release properties and cleanability is evaluated by test staining with items such as lipstick, coffee, ball point ink, mayonnaise and other oils. After cleaning, the effect of residual stains or colour change is assessed, sometimes with the use of grey scales. There is an increasing interest in antistatic properties for all aircraft textiles both for comfort and also for non-interference with electronic equipment.

All textile furnishings must pass stringent vertical burn tests such as BS3119 or DIN 53906. An internationally accepted test is (USA) FAR 25.853b which limits burning time to 15s after removal of the source and a char length no greater than 20 cm. However vertical fabric strip tests on individual items are not sufficient for seating that contains foam. The whole seat assembly must satisfy the FAR 25.853c test procedure in which a paraffin burner delivering flame at 1038 °C is applied to the seat cushion for 2 min. The average weight loss must not exceed 10% and the char length must not exceed 17 inches (43 cm). All seats in all passenger aircraft have had to satisfy this test[169] from 1st July 1987. To pass this test, 'fireblockers' were introduced under the face fabric to encapsulate the polyurethane foam in the seat and shield it from flame.[23,24] Fireblockers are made from preoxidised acrylic fibre, for example Panox (Lantor Universal Carbon Fibres), aramid fibres (Nomex, Kevlar – DuPont) Zirpro (IWS) treated wool, Inidex (Courtaulds), PBI (Celanese – Hoechst) fibre or combinations of these materials. Fabric weight needs to be considered; amongst the heaviest fireblockers is 60% wool/40% Panox, and the lightest but amongst the most expensive is 100% aramid.

FR grades of foam are used, but the higher FR grades can compromise comfort and tend to crumble. Research to improve both foams and fireblockers continues to produce improved effectiveness at lower weight and cost.[21,22] Amongst the materials being developed is Visil (formerly Kemira, now Sateri Fibres, Finland) a modified viscose fibre[170] that may offer benefits in comfort.

Study of aviation fires showed that victims were overcome by smoke or toxic fumes rather than direct contact with flame. Disorientation or incapacitation caused by the products of combustion prevented them evacuating the aircraft. Airbus Industrie standard ATS 1000.001 controls smoke opacity and the concentration of toxic gases such as CO, HCl and HCN. Wool and aramids have especially low toxic emissions from combustion (see Tables 18.4 and 18.5), but PVC and modified acrylic fibres have high levels.

Tests to measure heat release have come into force, for example the Ohio State University test, OSU 65/65. This test requires that interior components larger than 10×10 inches square (25.4×25.4 cm) on new aircraft (i.e. those built from 1990 onwards) must not release more than $65 \, kW \, m^{-2} \, min^{-1}$ heat during 5 min of flame exposure and the heat must not peak at more than $65 \, kW \, m^{-2}$ at any time during the test.[171] Table 18.6 provides data on various fabrics. Fabrics must pass all the tests that are designed in anticipation that passengers can evacuate the aircraft in 1.5 min should a small fire start.

18.5.2 Fibre-reinforced composites

Fibre-reinforced composites are used extensively in all parts of the aircraft resulting in very significant savings in weight, for instance, about 1350 kg of composites are used in the Airbus A310 and approximately 690 kg are used in the Boeing 737-300 representing about 6% of the entire weight of the planes.[168] Actual weight savings in the parts replaced by composite materials are between 20–30%. Internal dividing structures are mainly glass fibre in phenolic resin. They must comply with the flammability standards and also those regulations governing smoke and toxic gases. Knitted fabrics are used for compound contours, woven fabrics for flat parts. Phenolic resin is widely used because of its fire-resistant properties, low smoke and low toxic gas emissions.[172]

Advanced composites such as boron, silicon carbide and ceramic fibres are used in miliary aircraft for functional reasons not restricted by cost. The NASA space vehicle, Challenger, uses composites in many areas. Composites can only be used for radomes, (radio and radar equipment protective covers) because radiowaves would be shielded by metal. Composites are used extensively in helicopters and the rotor blades are made from carbon fibres.

18.5.3 Technical fabrics

Each passenger seat has a safety belt and in emergencies there is a life jacket under every seat. The life jacket is generally coated nylon of total weight about $265\,g\,m^{-2}$. A hot-melt polyurethane layer is applied to woven nylon usually primed first with either a solvent or a water-based base layer of polyurethane. The seat belt is a lap-type only, usually in woven polyester. Some consideration is being given to making these more like car seat belts with three anchoring points that would more than double the present volume of material required.

Carpets are usually woven loop pile wool with a polypropylene backing to save weight and coated with FR neoprene foam. Aisle carpets which need to withstand food and drink trolleys are changed around every 3 months whilst carpet in other areas is changed much less frequently.[173] Needless to say, the carpet has to meet high FR standards (e.g. FAR 25 8536) and smoke emission tests and has to be easily cleaned and maintained. Typically, they must be under $2000\,g\,m^{-2}$ in weight and contain conductive fibres together with a conductive backcoating for permanent antistatic properties. Some airlines require a maximum generated voltage of $1000\,V$ in a static body voltage generation test. Carpets or any items aboard the aircraft must not contain any material that could give rise to corrosive chemicals, which could cause deterioration of the aircraft's metal structure.

In large aircraft there are life rafts and escape chutes generally made from coated woven nylon or polyester fabric. The coatings are usually polyurethane or a synthetic rubber; PVC is avoided because of toxic emissions if it is set alight.

18.6 Marine applications

As in other areas of transportation, fibres are used in functional applications and more overtly in decorative applications. Again safety, like flame retardancy, is crucial and weight savings are also important requirements, especially in racing craft. Many safety requirements for furnishings and standards are set by the International Maritime Organisation such as the IMO Resolution A471 (XII) for fire resistance.[174] As in other forms of transport, comfort, design and appearance are important in providing passengers with a relaxing atmosphere.

18.6.1 Furnishing fabrics

Cruise ships can be regarded as 'floating hotels'[175] and, therefore, textile properties requirements must be of 'contract standard'.[16] Flame retardancy standards need to be high because of escape restrictions at sea and also because narrow corridors and low ceilings in many vessels make panic more likely in the event of a fire.[176] Fires in hotels and cruise ships are frequently caused by carelessness on the part of smokers. Furnishing fabrics must have durable high standards of flame retardancy and more use is therefore being made of inherently FR textiles. Standards required include DIN 4102 class B and BS476 paragraph 6.[177]

Carpets are especially important on passenger vessels because of their noise and vibration absorbing properties. They are more pleasant to walk upon than a hard surface and help to reduce physical stress and to provide a calmer and quieter atmosphere. Dyes used must be fast to light, rubbing and salt water. Durability is important because some areas of vessels are in use 24 hours a day and cleaning is done to rigorous schedules. Some ferries in Scandinavia have a million passengers a year; the heavy duty carpet is expected to last over 7 years.[178] Flame retardancy is important and wool carpets are generally Zirpro (IWS) treated. Durable antistatic properties are also generally required, imparted sometimes by the use of conductive fibres which are more durable than chemical finishes.

18.6.2 Functional applications

Fibre composites of glass reinforced plastic are used extensively in small vessels, patrol boats and pleasure craft.[179,180] Polyester fibre is being used to replace some of the heavier and more costly glass fibre in the composite. The advantages are easy handling, corrosion resistance and low maintenance. Kevlar (DuPont) is also used, sometimes in combination with glass fibre. Examples of specific cases where metal cannot be used are minesweepers, sonar domes and in corrosive-cargo carriers. Composites are being increasingly used for navigational aids such as buoys so that no damage results to the craft in the event of an accidental collision.[181]

Coated fabrics are used for life rafts buoyancy tubes, canopies and life jackets.[153] The base fabric for life rafts is generally woven polyamide with butyl or natural rubber, polychloroprene or thermoplastic polyurethane coatings. The total weight of the material varies from $230\,g\,m^{-2}$ up to $685\,g\,m^{-2}$. Quality tests include air porosity, coating adhesion and breaking and tear strength both in the warp and weft direction, flexing and waterproofness measured by hydrostatic head test methods. The canopies used on life rafts are made from much lighter coated fabrics. Natural rubber, polyurethane or SBR (styrene butadiene rubber) coated on to woven polyamide fabric give a total weight of between $145–175\,g\,m^{-2}$. Life jackets are generally made from woven polyamide coated with butyl or polychloroprene rubber to give total weights of about $230–290\,g\,m^{-2}$. Performance specifications include polymer adhesion, tensile strength, flex cracking and elongation-at-break, including testing after immersion in water for 24 hours. Performance standards for life jackets and life rafts are usually subject to government departmental controls and specifications.

18.6.3 Sails

Natural fibres in sails were first replaced by nylon and polyester, which are lighter, more rot resistant, have lower water absorption and, in the case of polyester, higher sunlight resistance. Sail development has progressed to some lighter laminated types where film is bonded to the fabric. Thus the fabric does not form the surface of the sail, only the reinforcing structure. For racing yachts, where weight is crucial, aramids which provide high strength with light weight, began to be used for the reinforcing structure. However, aramids degrade in sunlight, so that the ultra high modulus polyethylene yarns Spectra (Allied Signal) and Dyneema (DSM) and carbon fibres are now used.[182] The new polyethylene yarn has also found application in heavy duty ropes.

18.7   Future prospects for transportation textiles

Probably the most important challenge facing the transportation industry is its effect on the environment. Greenpeace warn that transportation at present accounts for 30% of carbon dioxide production, the main global warming gas, from burning fossil fuels. Yet the industry continues to grow and growth is inevitable for the forseeable future as the economies of the developing nations and those of the Third World progress. International trade is essential for a prosperous world society. Tourism, now the largest single industry in the world, is an important leisure pursuit but it also builds bridges and spreads understanding between the nations of the world.

As transportation volumes increase, the environmental issues are likely to be addressed with more efficient and lighter aircraft, trains, road vehicles and sea vessels. The 'three-litre' car is likely to become a reality and greater use of fibre-reinforced composites will result in savings of large quantities of fuel. The 'stealth' bus will become common place and it is probably only a matter of time before we see 'stealth' trains and private cars as well. Spurred on by government legislation, high levels of recycling of vehicles will occur assisted by design for recycling from the outset and by 'commonisation' of materials.

At the same time, more welcoming and pleasing transportation interiors, as well as enhanced comfort and safety can be expected as competition intensifies, not only between individual car companies, train companies, airlines and passenger shipping lines, but also between the different modes of transport.

All these improvements and advances are likely to be assisted by the invention of novel processes and more specialist materials, both as new fibres and in the form of composites. Within the last thirty five years up to 2000, an extremely short period in the history of humanity, significant discoveries bringing tremendous benefits have been made with the introduction of carbon, aramid and other specialist fibres, the most recent being the ultra high modulus polyolefin fibres. More 'breakthroughs' can be expected in future years.

Society is now taking a broader view of issues and a whole generation is growing up educated in environmental affairs. Research and development teams are much better equipped than ever before, benefiting from synergy enhancing instant communication via fax machines, e-mail and the Internet. There are exciting times ahead. For every problem there is an opportunity for those with the energy, imagination and determination to build a better world.

Acknowledgements

Thanks are due to colleagues past and present for their help in compiling this chapter and to the directors of Courtaulds Textiles Automotive Products for their permission to publish it. Special thanks are due to the following: Ian Leigh (Mydrin),

Walter Duncan (Synthomer), David Dykes (British Vita), Neil Saville (LUCF), Mike Smith (Duflot), Jim Rowan (Courtaulds Aerospace), Tony Morris (Courtaulds Chemicals), Gerald Day (Delphi), Simon Beeley (Holdsworth), Roy Kettlewell (IWS), Lara Creasey (Automotive & Transportation Interiors), Mick Dyer (DuPont), and the staff of Hoechst (Frankfurt), Boeing (Seattle), Autoliv (Birmingham), DuPont (Wilmington and Geneva) and to John Gannon (for translation from French and German) and to Melanie Wray who typed the manuscript.

Thanks are also due to General Motors for the photograph of the Electric car, EV1 and to the following for permission to reproduce tables: LMC Automotive Services (Table 18.1), The Society of Automotive Engineers; SAE (Table 18.3), DuPont (Table 18.4), and International Newsletters (Tables 18.5 and 18.6).

References

1. J PARK, 'The technology and production of fabrics for the automotive industry', *Rev. Prog. Coloration*, 1981 **11** 19–24.
2. B MILLIGAN, 'The degradation of automotive upholstery fabrics by light and heat', *Rev. Prog. Coloration*, 1986 **16** 1–7.
3. A HORSFALL, 'One big technical headache on wheels', *J. Soc. Dyers Colorists*, May/June 1992 **108** 243–246.
4. M A PARSONS (Rover), 'Fabric requirements for automotive use', *Autotech, Seminar 9*, NEC Birmingham 1991.
5. T L SMITH, 'Taking the heat', *Auto Transp. Interiors*, June 1996 32–35.
6. C DOIREAU, 'Specific requirements for urban transport and future trends', *Techtextil*, lecture 314, Frankfurt 1991.
7. G REINERT (CIBA), 'Coloration of automotive textiles using stabilisers', *IMMFC*, Dornbirn, 22–24 September 1993.
8. R V TODESCO, R DIEMUNSCH and T FRANZ (CIBA), 'New developments in the stabilisation of polypropylene fibres for automotive applications', *IMMFC*, Dornbirn, 20–23 September 1993.
9. R D WAGNER, R C LESLIE and F SCHLAEPPI, 'Test methods for determining colourfastness to light', *Textile Chemist Colourist*, February 1985 **17**(2) 17–27.
10. J PARK, 'Assessment of fastness properties', *Rev. Prog. Coloration*, 1979 **10** 20–24.
11. J PARK, 'Colour fastness assessment of textile materials', *Rev. Prog. Coloration*, 1975 **6** 71–78.
12. Midwest AATCC Section Committee, 'Accelerated lightfastness testing of disperse dyes on polyester automotive fabrics', *Textile Chemist Colourist*, December 1993 **25**(12) 25–32.
13. G FRANCKE and A HENKEL, 'Friction wear on car upholstery materials causes and possibilities of avoiding it', *Textiles in Automobiles*, VDI Congress, Dusseldorf, 14–15 October 1992.
14. J HURTEN, 'How pilling in polyester weaves can be controlled in the finishing process', *Textil Praxis Int.*, 1978 **33** 823–836.
15. M BOSCH, 'Pilling on textiles, fundamentals, extent of influence and test procedures', *Textiles in Automobiles*, VDI Congress, Dusseldorf, 14–15 October 1992.
16. M O'SHEA, 'Interior Furnishings', *Textile Prog.*, 1981 **11**(1) 1–68.
17. A R HORROCKS, 'Flame retardant finishing of textiles', *Rev. Prog. Coloration*, 1986 **16** 62–101.
18. D L ROBERTS, M E HALL and A R HORROCKS, 'Environmental aspects of flame retardant textiles – an overview', *Rev. Prog. Coloration*, 1992 **22** 48–57.
19. A J G SAGER, 'Protection against flame and heat using man-made fibres', *Textiles*, 1986 **15**(1) 2–9.
20. J BAGNALL, 'Testing the reaction of textiles to fire', *Textiles Magazine*, 1995 (4) 12–17.
21. K T PAUL, 'Flame retardant polyurethane foam furniture testing and specification', *Rev. Prog. Coloration*, 1990 **20** 53–69.
22. R HURD, 'Flame retardant foams', *J Cellular Polym.*, 1989 (4) 277–295.
23. N SAVILLE and M SQUIRES, 'Latest developments in fire resistant textiles', *Textile Month*, 1990 May 47–52.
24. G KEIL, 'Fire-blockers – a protection for passengers', *Tech. Usage Textiles*, 1991 **4**(2) 46–47.
25. KIRK OTHMER, 'Composite Materials–Survey', *Encyclopaedia of Chemical Technology*, 4th edition, John Wiley, New York, 1993.
26. P M EATON, 'Aramid fibres', *Textiles*, 1983 **12**(3) 58–65.
27. P M EATON, 'Fibre reinforced composites', *Textiles*, 1986 **15**(2) 35–38.
28. Modern Plastics Magazine, *Encyclopaedia Handbook*, McGraw Hill, USA 1994, pp. 124–136.

29. ANON, 'Toray expanding PAN-based carbon fibres', *Japan Textile News (Japanese)* 1996 September 107.
30. ANON, 'Technical textiles', *Knitting Int.*, June 1996 **103**(1227) 38–39.
31. J W S HEARLE, 'Textiles for composites', *Textile Horizons*, December 1994 **14**(6) 12–15 and February 1995 **15**(1) 11–15.
32. N KHOKAR, 'An experimental uniaxial "Noobing" device', *Textiles Magazine*, 1996 **3** 12–14.
33. KIRK OTHMER, 'Tire Cords', *Encyclopaedia of Chemical Technology*, 4th edition, Vol 24, John Wiley, New York, 1997.
34. M O WEBER and D SCHILO, 'Surface activation of polyester and aramid to improve adhesion', *J. Coated Fabrics*, October 1996 **26** 131–136.
35. J WATERSON, 'Global trends in colour and trim', *Auto Interior Int. Directory*, 1996 16–18.
36. B TEN HOEVAL, 'Worldwide trends in automotive textiles', *Auto Interior Int. Directory*, 1996 16–18.
37. C GARNER, 'Polyolefin and the 10 year automobile', *America's Textiles Int.*, 1996 February 75–78.
38. D WARD, 'Car textiles – Courtaulds makes up for lost time', *Textile Month* 1980 December 30–32.
39. D WARD, 'Car makers realising merits of textiles', *Textile Month*, 1981 January 21–23.
40. G REINERT, 'Textiles as interior furnishing fabrics in vehicles', *IMMFC*, Dornbirn, 22–24 September 1993.
41. C WILKENS, 'Automotive Fabrics', *Knitting Int.*, August 1995 **102**(1218) 50–55.
42. J MILLINGTON, 'The rise, rise and prospective further growth of knitted fabrics in automotive applications', *Knitting International*, December 1992 **99**(1188) 13–17.
43. J MILLINGTON, 'Automotive fabrics-the role of warp knitting', *Knitting Int.*, January 1993 **100**(1189) 13–18.
44. J MILLINGTON, 'Prospective growth of knitted fabrics in automotive applications', *Knitting Int.*, February 1993 **100**(1190) 13–18.
45. W R SCHMIDT (Eybl), 'The application of circular knitted fabrics in the motor industry', *IMMFC*, Dornbirn, 22–24 September 1993.
46. S ANAND, 'Knitted fabrics take the lead in automotive market', *Textile Month*, 1993 September 41–42.
47. ANON, 'Guildford Automotive knows the business', *America's Textiles Int.*, 1995 June 44–52.
48. W C SMITH, 'Automotives – a major textiles market', *Textile World*, 1993 September 68–73.
49. M KAWALSKI, 'Automotive textile presentation', *Autotech* NEC Birmingham 1991, Seminar 9.
50. T D FULMER, 'Dyeing textiles for automotive interiors', *America's Textiles Int.*, 1993 December 88–90; B KRAMRISCH (Oldham S-Ciba), 'Dyeing of automotive fabrics', *Int. Dyer Printer*, 1986 December 28; P OSMAN (Ford), 'Dyestuff selection and colour control for the car industry', *Int. Dyer Printer*, June 1985 **170** 7–8.
51. J R ASPLAND, 'Disperse dyes and their application to polyester', *Textile Chemist Colourist*, December 1992 **24**(12) 18–25.
52. D J PEARSON (Guildford), 'The finishing of automotive textiles', *J. Soc. Dyers Colorists*, December 1993 388–390.
53. M KOWALSKI, 'Automotive textiles update', *Textiles*, 1991 (2) 10–12.
54. J KATH, H P HARRI, P JOHNSON, L MCGARRIE and R ROMMEL (CIBA), 'Printing on automotive textiles', *IMMFC*, 20–22 September 1995 Dornbirn.
55. T L SMITH, 'Printed fabrics make their mark on US interiors', *Auto Transp. Interiors*, 1994 October 34–35.
56. T L SMITH, 'Rotary screen printing offers possibility for colour variety, speed of production', *Auto Transp. Interiors*, 1996 May 74–76.
57. C GARNER, 'The low down on laminating', *Inside Automotives*, 1995 May/June 23–25, 47.
58. R GARNER, 'Flame or dry? The debate is on', *Auto Transp. Interiors*, 1994 May 52–54.
59. D C MILES (Dermil), 'Dry powder bonding adhesives in automotive trim laminates', *J. Coated Fabrics*, April 1991 **20** 229–239.
60. J HOPKINS (Nordson), 'A comparative analysis of laminating automotive textiles to foam', *J. Coated Fabrics*, 1995 January 250–267.
61. F WOODRUFF (Web Processing), 'Environmentally friendly coating and laminating; new processes and techniques', *J. Coated Fabrics*, April 1992 **21** 240–259.
62. J HALBMAIER (Bostik), 'Overview of hot melt adhesives application equipment for coating and laminating full-width fabric', *J. Coated Fabrics*, April 1992 **21** 301–310.
63. S KMITTA (Fehrer), 'Use of polyester fibre padding in shaped upholstery for car seating', *Textiles in Automobiles*, VDI Conference, Dusseldorf, 14–15 October 1992.
64. S KMITTA (Fehrer), 'Polyester nonwovens – an alternative material for car seats', *IMMFC*, Dornbirn, 20–22 September 1995.
65. C WILKENS, 'Raschel knitted spacer fabrics', *Melliand English*, 1993 (10) E348–E349.
66. KARL MEYER/MALIMO Technical information leaflet, 'Manufacture of fabrics for automotive interiors using warp knitting and stitch bonding', We 75/1/8/93.
67. G SCHMIDT and P BOTTCHER, 'Laminating non woven fabrics made from or containing secondary or recycled fibres for use in automotive manufacture', Index Conference, 1993.

68. H FUCHS and P BOTTCHER, 'Textile waste materials in motor cars-potential and limitations', *Textil Praxis Int.*, 1994 April (4) II–IV.
69. H HIRSCHEK, 'Recycling of automotive textiles', *IMMFC*, Dornbirn, 22–24 September 1993.
70. M COSTES (Rhone-Poulenc), 'Use of textiles in vehicles and recycling; state of the art and outlook', *IMMFC*, Dornbirn, 22–24 September 1993.
71. B KIEFER, A BORNHOFF, P EHRLERN, H KINGENBERGER and H SCHREIBER, 'Assessing second hand automotive textiles for use in new vehicles', *IMMFC*, Dornbirn, 20–22 September 1995.
72. E EISSLER, H NONNER and G SCHMIDT, 'Use of textiles in the Mercedes-Benz S Class', *Textiles in Automobiles*, VDI Conference, Dusseldorf, 30–31 October 1991.
73. E BAETENS and E ALBRECHT, 'Fogging characteristics of automotive textiles' *Techtextil*, Frankfurt, May 13–16 1991, lecture 336.
74. E BAETENS and E ALBRECHT, 'Reducing the fogging effect in cars', *Technical Usage Textiles*, 1992 **4** 42–44.
75. P HARDT, 'Estimation of the amounts of volatile substances as applicable to stenter frames for automotive fogging,' *IMMFC*, Dornbirn, 22–24 September 1993.
76. W BEHRENS and T LAMPE (VW), 'Fogging behaviour of textile materials', *Textiles in Automobiles*, VDI Congress, Dusseldorf 30–31 October 1991.
77. P EHLER, H SCHREIBER and S HALLER, 'Emissions from textiles in vehicle interior trim causes and assessment of short term and long term fogging', *IMMFC*, Dornbirn, 22–24 September 1993.
78. J B MCCALLUM, 'Ford Motor Co develops its own test method for predicting light-scattering window film', *Textile Chemist Colourist*, December 1989, **21**(12) 13–15.
79. W BEHRENS, 'Fogging behaviour of car interiors', *Technische Textilen*, March 1993 **36** E25–E27.
80. G S RINK (Opel), 'Harmonisation of textile testing in the German automotive industry', *IMMFC*, Dornbirn, 20–22 September 1995.
81. T L SMITH, 'There's no time to re-invent the wheel-or the seat cover', *Auto Transp. Interiors*, 1996 November 40–41.
82. C GARDNER, 'Headliners rising to the occasion', *Inside Automobiles*, 1994 December 19–23.
83. ANON, *Edana Automotive Non-wovens Newsletter*, (4) 1993 and (5) 1996.
84. A CASHIN, 'Preserving the environment', *Auto Interior Int.*, Autumn 1995 **4**(4) 16–21.
85. A WEHLOW, 'Flocked parts for car interiors', *IMMFC*, Dornbirn, 22–24 September 1993; J P BIANCHI (Novalis), 'Flock technology for car interiors', *IMMFC*, Dornbirn, 17–19 September 1997.
86. S W BOLGEN, 'Flocking technology', *J. Coated Fabrics*, October 1991 **21** 123–131.
87. W J MORRIS, 'Seat Belts', *Textiles*, 1988 **17**(1) 15–21.
88. C ROCHE, 'The seat belt remains essential', *Technical Usage Textiles*, 1992 (3) 63–64.
89. W KRUMMHEUR (Akzo), 'Recycling of used automotive seat belts,' *IMMFC*, Dornbirn, 22–24 September 1993.
90. A DAVIDSON, 'Growing opportunities for airbags', *Tech. Textiles Int.*, 1992 May 10–12.
91. F BOHIN and M LADREYT, 'Silicone elastomers for airbag coatings', *Auto Interior Int.*, Winter 1996/7 **5**(4) 66–71.
92. T L SMITH, 'Tough stuff', *Auto Transp. Interiors*, 1996 August 30–32; H R ROSS (AlliedSignal), 'New future trends in airbag fabrics', *IMMFC*, Dornbirn, 17–19 September 1997.
93. DUPONT Automotive TI leaflets H-48030 and H-48032 (USA); V SIEJAK (Akzo), 'New yarns for lighterweight airbag fabrics', *IMMFC*, Dornbirn, 17–19 September 1997.
94. J A BARNES, J PARTRIDGE and S MUKHOPADHYAY, 'Air permeability of nylon 66 airbag fabrics', *Yarn & Fibre Conference*, Textile Institute, Manchester 2–3 December 1996.
95. T L SMITH, 'Airbags and seat belts; fabric's role in safety restraint systems', *Auto Transp. Interiors*, 1995 December 53–54.
96. J BRAUNSTEIN, '2001 and beyond, a safety odessey', *Auto Transp. Interiors*, 1996 April 27–29.
97. K N CRIGHTON, 'Tubular side airbag technology takes another shape', *Auto Transp. Interiors*, 1995 April 16.
98. ANON, 'New head impact rules spur research', *Auto Transp. Interiors*, 1996 April 28–29.
99. P SONDERSTROM, 'Side impact airbags, the next step', *Inside Automotives*, 1996 May/June 12–15.
100. H MOUND, 'BMW gives a boost to hyperinflation', *The Times*, 12 April 1997.
101. ANON, 'Knitted car upholstery', *Knitting Int.*, June 1993 **100**(1194) 34.
102. F ROBINSON and S ASHTON, 'Knitting in the third dimension', *Textile Horizons*, 1994 December 22–24.
103. C GARDNER, 'CAD and CAE, balancing new technology with traditional design', *Inside Automotives*, 1994 June 17–20.
104. T L SMITH, '3-D knitting adds new dimension to interiors', *Auto Transp. Interiors*, 1994 November 42–45.
105. ANON, 'Technical textiles', *Knitting Int.*, June 1996 **103**(1227) 38–39.
106. M JONES and W GIRARD, '3-D knitting', *Auto Interior Int.*, 1995 Autumn 24–30.
107. J BRAUNSTEIN, 'Whole lot goin' on underneath', *Auto Transp. Interiors*, 1997 April 22–32; R BORROFF, 'Rubber soul', *Automotive Seating Review 1997 (Auto Interior International)*, 1997 22–26.
108. I ADCOCK, 'The inside track', *European Plastic News*, 1996 December 37.

109. C GARDNER, 'Interior industry's one stop shop-Dupont automotive', *Inside Automotives*, 1995 March/April 40–45.
110. J GRETZINGER, 'Dymetrol seating supports for automotive applications', *Automotive Textiles* Symposium, Greenville, S. Carolina, October 1991. Also J Gretzinger and R L Rackley US Patent No 4,469,739.
111. C BORRI, 'Comfort for car interiors-Alcantara-features and advantages of a unique product', *IMMFC*, Dornbirn, 22–24 September 1993.
112. ANON, 'Man-made leather grows in product and demand', *JTN (Japanese)*, 1997 June 66–73.
113. G BAGNOLI and G POLETTE, 'The revolutionary high-tech leather, Lorica, for automotive interiors', *IMMFC*, Dornbirn, 22–24 September 1993.
114. J HEMMRICH, J FIKKERT and M VAN DER BERG (Stahl), 'Porous structured forms resulting from aggregate modification in polyurethane dispersions by means of isothermic foam coagulation', *J. Coated Fabrics*, 1993 April 268–278.
115. H MILLER SIR, 'Global strategies imperative to supplier success', *Auto Transp. Interiors*, 1995 August 52.
116. A LORENZ, 'Ford slam the brake on parts prices', *Sunday Times Business*, 22 October 1995.
117. L E SULLIVAN, 'System integration; the race is on', *Auto Transp. Interiors*, 1997 June 20–23.
118. ANON, 'Life in the fast lane', *Non-wovens Rep Int.*, 1997 (311) March 23.
119. A WEBER, 'Potential for recycling plastics from scrap cars', *Plastics Rubber Weekly*, 1990 14 April 13.
120. 'Recyclable moulded parts made from Grilene needled mats', *Techtextil telegramm*, 26E 10 December 1992.
121. D WARD, 'There's nothing new about recycling', *Textile Month*, 1995 June 30–32.
122. ANON, 'Opel undertakes comprehensive review of its automotive plastics usage', *Plastics Rubber Weekly*, 9 May 1992, p 10.
123. ANON. 'Recycling becomes a key', *Plastics Rubber Weekly*, 1993 23 October 18–19.
124. R ROWAND, 'Polyester may replace PU', *Urethanes Tech.*, 1995 October/November 34–36.
125. ANON, 'Common materials to cut costs from manufacture', *Automotive Int.*, 1997 March 20.
126. ANON, 'ACORD deal agreed', *Materials Recycling Weekly*, 1997 18 July 3.
127. ANON, 'PRAVDA-the moment of truth?', *European Plastic News*, 1991 September 58–59.
128. B JAMES, 'Rover takes CARE to meet European goals', *Plastics Rubber Weekly*, 1997 7 February 7.
129. ANON, 'Automotive textiles', *Kettenwirk-Praxis*, 1995 (3) E30–E32.
130. S EMINTO, 'Vehicle recycling-UK to take voluntary route', *Materials Recycling Weekly*, 1997 31 January 12–14.
131. ANON, 'ELV recovery', *Automotive Textiles Newsletter*, Rhone Poulenc Setila/Mavel, 1997 April 4.
132. Y YAMADA, 'Holding the odours', *Auto Interior Int.*, 1992/3 Winter 52–58.
133. H BOLLINGER and K R DUWEL, 'New concepts in seating', *Textiles in Automobiles*, VDI Congress, Dusseldorf, 30–31 October 1991.
134. G T KNOZINGER, H THEYSOHN and H VOGT, 'Physiology of seat comfort', *Textiles in Automobiles*, VDI Congress, Dusseldorf, 30–31 October 1991.
135. M JARRIGEON, 'Sensorial and thermal comfort', *Techtextil*, Frankfurt, May 13–16 1991, lecture 335; S E HANEL, T DARTMAN and R SHISHOO, 'A new method for measuring mechanical and physiological comfort in car seats', *IMMFC*, Dornbirn, 20–22 September 1995.
136. W FUNG and K C PARSONS, 'Some investigations into the relationship between car seat cover materials and thermal comfort using human subjects', *J. Coated Fabrics*, October 1996 **26** 147–176.
137. W FUNG, 'How to improve thermal comfort of the car seat', *Vehicle Comfort and Ergonomics*, ATA (Italy), Bologna 6–7 October 1997; V T BARTELS and K H UMBACH, 'Laboratory tests of thermophysiological seat comfort', *Vehicle Comfort and Ergonomics*, ATA (Italy), Bologna 6–7 October 1997.
138. S KAWABATA and M NIWA, 'Fabric performance in clothing and clothing manufacture', *J. Textile Inst.*, 1989 **80**(1) 19–50.
139. J W S HEARLE, 'Can fabric hand enter the dataspace?' Part 1, *Textile Horizons*, April 1993 **13**(2) 14–16 and Part 2, *Textile Horizons*, June 1993 **13**(3) 16–20.
140. C J KIM, 'The Kawabata system use in the fabric hand evaluation of automotive textiles', *Automotive Test Procedures*, IFAI Transportation Division, Atlanta, 7–8 October 1992.
141. R WHITE, 'Fabrics for acoustic control', *Tech. Textiles Int.*, 1993 April 26–29; Y SHOSHANI, 'Studies of textile assemblies used for accoustic control', *Tech. Textiles Int.*, 1993 April 32–34.
142. B RIEDEL, 'Sound-insulating materials for cars based on chemical recycling of PUR flexible foams from car seats', *IMMFC*, Dornbirn, 20–22 September 1995.
143. J DIJKEMA 'Acoustic protection for cars', *Technical Usage Textiles*, 1992 **4**(6) 47–49; J LASER, 'Moulded automotive carpets-influence on the interior noise level', *IMMFC*, Dornbirn, 17–19 September 1997.
144. K GRACE, 'Polymers are crucial for the motor industry to meet its aspirations', *British Plastics Rubber*, 1996 November 26–30.
145. T MATSUI (Teijin), 'Development of polyester fibres for car seats', *IMMFC*, Dornbirn, 22–24 September 1993.

146. A P ANEJA (DuPont), 'New fibre technology for thermal comfort in performance apparel', *Textiles Sports Sportswear*, Huddersfield, 1995 11 April.
147. ANON, 'Coating produces dynamic insulation', *America's Textiles Int.*, 1995 June 6–8.
148. A DOCKERY, 'New process lets polyester breath', *America's Textiles Int.*, 1993 October 12–13.
149. T I DUPONT sheet, *CoolMax*, H-53717 December 1993.
150. P HALL, 'Triboelectronics', *Saturday Telegraph*, 14 May 1994; J CHUBB, 'Testing how well materials dissipate static electricity', *British Plastics Rubber*, 1995 May 8–10.
151. M HENRICKS, 'Truck interiors become home from home', *Auto Transp. Interiors*, 1994 October 16–19.
152. M HENRICKS, 'Works like a truck, rides like a car', *Auto Transp. Interiors*, 1996 March 22–25.
153. P BAJAJ and A K SENGUPTA, 'Industrial applications of textiles for filtration and coated fabrics', *Textile Prog.*, 1982 **14**(1) 15–17 and 25–26; M WILKINSON, 'A review of industrial coated fabric substrates', *J. Coated Fabrics*, October 1996 **26** 87–106.
154. T I HOECHST sheet, *Trevira in focus – Complan*, 12687/98e.
155. C WILKENS, 'Raschel knitted spacer fabrics', *Melliand English*, 1993 (10) E348–E349.
156. L MOORE, 'Bus interior components softer, more flexible', *Auto Transp. Interiors*, 1994 April 26–29.
157. J A GRAND, 'High tech bus draws heavily on aerospace technologies', *Modern Plastics Int.*, February 1997 **27**(2) 30.
158. H-J FRANK (Deutsche Bank), 'The development of transport until the year 2000', *Techtextil*, Frankfurt, May 13–16 1991, lecture 311.
159. C B SUMMERS, 'Transit agencies position for the inside track', *Auto Trans. Interiors*, 1995 April 36.
160. J TROITZSCH, *International Plastics Flammability Handbook*, 2nd edition, Hanser, New York, 1989, 299–310.
161. E J LOWE, 'Textiles in railways', *Textiles*, February 1972 **1**(1) 8–11.
162. H R JONES, 'Textiles in the railway passenger environment', *Flammability*, BTTG, Britannia Hotel, London, 1–2 December 1993.
163. J BAKER-COUNSELL, 'Testing for fire safety on London underground', *Plastics Rubber Weekly*, 1988 30 January 8–9.
164. IWS TI leaflet, *Wool contract carpets for rail passenger vehicles.*
165. R GARNER, 'Aircraft interiors feel the pressure', *Auto Transp. Interiors*, 1996 February 12–22.
166. T L SMITH, 'Fabrics and fibres review; aircraft designers go corporate', *Auto Transp. Interiors*, 1995 March 28–32.
167. M HENRICKS, 'Return of the pampered passenger?', *Auto Transp. Interiors*, 1996 April 48–50.
168. H LAURENT, 'Fabrics used in transport vehicles', *Techtextil*, Frankfurt May 13–16 1991, lecture 312.
169. C C BARROW, 'Standards for textiles used in commercial aircraft', *Textile Horizons*, 1992 April/May 30–34.
170. S GARVEY, 'Visil-the hybrid viscose fibre', *Textiles Magazine*, 1996 (3) 21–24.
171. J BUCHER, 'Regulations, economics limit plastics choice in aircraft', *Auto Transp. Interiors*, 1995 September 48–50.
172. J M ANGLIN, 'Aircraft applications (of composites)', *Engineered Materials Handbook*, Vol 1, ASM Metal Park, Ohio, 1987, 801–809.
173. IWS TI leaflet, '*Wool contract carpets for passenger aircraft*'.
174. International Maritime Organisation, '*Fire Test Procedures*', 2nd edition IMO 844E, London 1993.
175. U GIRRBACH (Hoechst), 'Decorating ship interiors with flame retardant fabrics', *Textile Month*, 1995 April 25.
176. J TROITZSCH, '*International Plastics Flammability Handbook*, 2nd edition, Hanser Publications, New York, 1990, 322–323.
177. P LEWIS, 'Polyester "safety fibres" for a fast-growing market', *Textile Month*, 1997 April 39–40.
178. IWS TI leaflet, *Wool contract carpets for passenger ships*'.
179. J G KAREGEANNES, 'Discovering and exploiting new markets for a fibre', *Tech. Textiles Int.*, 1992 May 14–17.
180. R F PINZELLI, 'Use of composites in maritime structures', *Techtextil*, Frankfurt May 13–16 1991, lecture 323.
181. J SUMMERSCALES, 'Marine applications (of composites), *Engineered Materials Handbook*, Vol 1, ASM Metal Park Ohio, 1987, 837–844.
182. G BELGRANO and C O'CONNELL, 'Carbon sails to the front', *Tech. Textiles Int.*, 1992 May 28–31.
NB. *IMMFC* = International Man-Made Fibres Congress held annually at Dornbirn, Austria organised by Osterreichisches Chemiefaser-Institut Tagungsburo Dornbirn.

19

Textiles and the environment

Keith Slater

School of Engineering, University of Guelph, Guelph, Ontario NIG 2WI, Canada

19.1 Introduction

The existence of human beings on earth is the result of a fortuitous set of circumstances in which conditions for development of the species were present so that evolution could take place allowing us to reach our present state of being. Our tenuous continuation could be jeopardized at any time by changes in these conditions, and this far-reaching effect could result from shifts which might be totally insignificant by cosmic standards. They could bring about, for example, our inability to breathe, or stay warm or cool enough, or grow the food we need. Thus, we are only able to survive because our planet provides all the sustenance we need without major effort on our part. We can broadly define this set of conditions to which we are exposed as our environment.

One of the minor ways by which we reduce the risk of premature extinction is to guard our bodies from excessive temperature fluctuation by the use of textiles. Textiles are also used to make life more comfortable or convenient for us. Without them, we would find life harsher, and probably not survive with the same life expectancy as we do now. Specifically, technical textiles can provide direct protection in the form of architectural structures, tentage or sleeping bags to provide protection against a cold climate, and geotextiles to guard against swamping by rough water in a harbour or against the escape of harmful chemicals from confinement.

The interaction between textile materials and the environment is a complex one taking two distinct forms. There is, first, the effect of a change in properties that the environment can bring about in the textile, generally classed as degradation. Second, there is the manner in which the production or use of textiles can impinge on the environment, generally classed under the term' pollution for the negative impact', but also including environmental protection by pollution reduction where, say, a landfill liner is used to prevent leaching. A further need is for the production of textiles to be possible by using the resources available on the earth without depleting them irreplaceably. Each of these factors is important and should be considered separately in order to build up a complete view of how textiles and the environment can impinge on one another.

Because this text is intended to be aimed primarily at the area of technical textiles, it is important to consider, at the outset, where such materials fit into the environmental field. If technical textiles are defined broadly as all those not intended for personal clothing or household use, then such end-uses as industrial, automotive, sporting, or architectural ones must be included, as also must geotextiles, tarpaulins, ropes and reinforcement fibres for composites. With such a diversity of products, it is impossible to be specific about fibre content, because all fibres are applicable to one or more members of this range and it is important to take a widely inclusive view in a consideration of the environmental effects of textile production or use.

19.2   Degradation

19.2.1   General factors

Textile materials are exposed to a wide variety of environmental factors during manufacture and use, many of them caused by the very conditions necessary for a proper development of the final textile product. The successful growth of crops, such as wool or cotton, is dependent to a considerable extent on environmental factors. If the temperature is too warm crops can die, so that cotton plants are unable to blossom and sheep cannot be fed. Too much rain can result in plants rotting or in floods that carry away the sheep, again lowering the yield of fibres. The retting of flax or other bast fibres is controlled by environmental conditions, while the presence of too high a relative humidity during storage of some types of fibre, particularly cellulosic ones, can cause damaging mould to develop, rendering them unusable.

In the processing of fibres, artificially developed environments may need to be established. Acid conditions are often essential for various scouring, carbonizing, bleaching, dyeing, printing and finishing operations. Alkaline conditions may be needed, conversely, for mercerizing and other bleaching or dyeing treatments. Heat is needed in some of these processes too, as well as in solvent spinning, drying or tentering. Chemical reagents must be used in sizing or desizing, in some dyeing or printing methods, and in applying finishes such as soil release, antistatic, durable press, flame-retardant, optical modification and shrinkproofing treatments. Moisture and mechanical force are needed in washing, fulling, crabbing, and decating, while dry force is used in beetling, raising and shearing. There may also be specific needs involved with the production of technical textiles, especially if unfamiliar fibres or finishes are to be incorporated.

In this chapter, discussion will be restricted to the changes brought about by exposure of the end-product, as put into service, to expected environmental factors. The effects of any artificially imposed environments present in maintenance of the textile products after they have been put into service are omitted, such as high mechanical force, unnaturally high or low temperatures, chemical agents or organic solvents. These types of stress are important issues for the durability of the textile materials, but word limitation and the nature of the chapter's content do not allow either space or reason for their inclusion.

19.2.2   The degradative process

Degradation may thus be manifested as either a visual effect, which is merely unsightly, or as a physical one in which the material disintegrates in some way under a load. In either case, the net result is that the end-product becomes unacceptable to the consumer for useful service. In the case of technical textiles, there may also be a risk of danger if, say, a sling or rope breaks, an architectural flexible structure or a fibre-reinforced composite structural element collapses, or a landfill liner or tarpaulin bursts.

19.2.3   Importance of degradation

The degradation of textiles in use is universal and inevitable. As soon as the product is put into service, it is subjected to the action of all the agents that can bring about the molecular changes responsible for degradation. Air alone apparently does not cause any noticeable change in textile materials in the absence of other stress factors, but it is rare for all other potentially harmful sources of degradation, such as moisture, light or microbiological agents, to be absent.

There are many ways in which changes occur. Heat and light, separately or in combination, can bring about problems. Each can damage materials, and the combination can cause enough tendering, brittleness or discoloration to render fabrics unusable. A fabric can be exposed to both conditions simultaneously in, for example, the process of drying or ironing.

The contact of fabrics with swimming pool chemicals can also bring about degradation. The chlorine or other disinfectant present can make fibres weak or brittle, shortening the useful life of the article, whether a garment, a liner or a cover.

Fabrics left on or near moist ground quickly experience microbiological growth, a change that it is crucial to counteract for succesful use of such technical fabrics as geotextiles, tentage or horticultural plant covering cloth. Natural fibres, less resistant to microbial agents, tend to rot, visible changes being discernible within days. Synthetic ones, lacking substituent groups used by these agents as food, resist weakening, but may be discoloured. Changes in pH, during perspiration or anaerobic decomposition, for instance, can cause weakening, again usually in natural rather than synthetic fibres. Architectural fabrics, flags, tents and tarpaulins exposed to sun and movement arising from wind force can be torn or tendered. Thus, degradation is all-pervasive and can have drastic effects that render fabrics useless.

19.3   Resource depletion and pollution

The textile industry is one of those that are blamed, often unfairly, for the general decline in the planet's health, and before considering the effects that textile production and use have on ecology, the following terms are defined:

1   The environment: In the context of global harm, the term 'environment' will be restricted to the physical, rather than social, one. The two often impinge on each other, and damage of a physical kind can affect our social environment.
2   Resource depletion: This term is defined as meaning the use of any material present on or within the earth in such a way that it is difficult, or impossible, for it to be recovered without inflicting harm on the planet.

3   Pollution: this is to be understood as meaning the production of any substance or condition harmful to any species, plant or animal on the earth in a manner which is difficult or impossible to reverse without the use of added resources.

19.4   Textile sources of environmental harm

With these definitions in mind, it is possible, in a consideration of textile subjects, to focus on ways in which relevant factors affect the environment. It should be recognised that the production and use of textiles are no more harmful than are those of any other material, and may be less so. There are, nevertheless, ways in which they influence the planet, and these should be considered.

Several papers in the literature deal with this matter. Horstmann[1] notes that environmental hazards in textile production are growing in the finishing sector. He discusses ways of reducing pollution by eliminating or lowering resource consumption, mainly by reusing or recycling goods. Perenich[2] also examines environmental issues in the industry, identifying water quality and use, air quality and emissions and waste minimization as major areas needing attention. Kramar[3] takes a similar view, but focuses on aqueous effluents. He feels that the best way of being environmentally responsible is to reduce the use of water, suggesting that this will cut down both pollution and cost.

Blum[4] takes a slightly more pessimistic view, questioning whether the textile/environment interaction constitutes a threat or a challenge. He feels that it is the industry's responsibility to rise to the situation, discussing political, economic and legal aspects of meeting it. He foresees conflicts between environmental legislation and competition policies, putting his finger on an important issue. Even though the eventual benefits of environmental responsibility can be financially rewarding, there will be an initial outlay of capital to meet higher standards, and this may well force a manufacturer out of business if his competitors do not have to meet the costs at the same time.

It is possible, then, to find harmful effects in the manufacture of fibres, in their subsequent treatment during processing, in their marketing, distribution or use, and in their disposal. The narrow criteria defining harmful practices, normally considered, should not be allowed to restrict our opinion on how textiles bring about ecological damage. Norgaard[5] analyzes the environmental impact of cotton production, recommending the Ecomanagement and Audit System (EMAS) to give quantitative meaning to the process. In addition, Kralik[6] points out that the auxiliary activities associated with the manufacture and use of textile products are often more harmful than the specific textile factors usually considered. Each of the factors relating to environmental degradation should therefore be considered in turn with particular reference to textile applications.

19.4.1   Resource depletion

It is disconcerting to find that the textile industry as a whole does not seem to regard the problem of resource depletion as a serious one. Only one paper in the more recent literature was found to mention the need to use resources effectively, and the reasons are biased more towards economic rather than environmental factors.[4] However, renewable and non-renewable resources alike are at risk from the needs of textile manufacture and marketing. Oil for the manufacture of plastics, water for the manufacturing processes, iron and other metals for machinery or dyes, are all resources which either are not infinite or are finite. Trees are heavily harvested for fibre and paper production and the resulting loss of forest cover must be replaced if the earth's ability to regenerate oxygen is to continue.

19.4.2   Energy consumption

The manufacture, distribution and use of energy are also responsible for resource depletion in the form of the fuels (or oxygen) in combustion. The apparent plenitude of seemingly clean energy can lull us into a false sense of security that encourages waste, though there is increased awareness on the part of many people that shortages are looming.

The often-voiced assumption that new technology, and especially new energy-generating techniques, will resolve the situation is a vain one. There are serious (usually unrecognized) flaws with all the new methods proposed. Even the 'ideal' and 'clean' generation technique of nuclear energy is receiving less enthusiastic support as plant faults compel closure or redesign. Solar energy falling on the earth is so great that a three-day influx, harnessed completely, would provide enough electricity to meet all needs for a decade. Unfortunately, problems involving the means of harnessing it are not taken into account.

Wind, tidal, or geothermal energy production processes all need bulky, unsightly equipment for the capture, conversion and distribution of energy. In addition, they tend to need frequent replacement because equipment is subjected to severe stresses from weather, changes in temperature or tidal action.

19.4.3   Recycling

In many applications, especially where metals, glass or polymers (including synthetic textile materials) are involved, the recycling process can only slow down, not reverse, damage to the planet. Virtually all recycling, in the sense of making a new product from the waste of an old one (rather than just reusing an old product) needs heat, thus using energy. This brings about resource depletion and produces pollution, even if (a highly unlikely presupposition) that pollution is only carbon dioxide, a greenhouse gas. One form of recycling which falls between the two types and is environmentally friendly is practised within the textile industry. This is the process by which surplus fibres or fibre assemblies are returned to the production train for reprocessing instead of merely being discarded; unfortunately, there may be a diminution of properties (such as fibre length, yarn evenness or fabric strength) as a result, which can lower the quality of goods that can be produced from this recycled material. The effects of using reworkable waste on yarn quality, for instance, include a significant reduction in yarn tenacity.[7]

19.5   Textile sources of pollution

19.5.1   Introduction

It is now time to discuss how discarded substances find their way into the environment as contaminants, together with the reasons behind the concern for their presence.

The textile industry is a complex one. The production, coloration, finishing and distribution of fibres, yarns or fabrics are carried out with the aid of large, complicated, expensive machines and a range of chemical substances. The difficulties inherent in manipulating the textiles mean that there are many opportunities for materials, either textile components or reagents added to them, to escape from the equipment. It is this difficulty in maintaining control over movement of materials that is responsible for pollution. Inevitably, the effort to produce all the goods needed leads to the dispersion of impurities into the air, water or land, as well as to undesirable noise levels or visual ugliness. These can be considered in turn with a view to establishing how much responsibility the textile industry must bear in each case.

19.5.2 Air pollution

Air pollution within the textile industry affects people, machinery and products. There is an increased incidence of health problems, especially byssinosis, tuberculosis and asthma,

Air pollution can also arise from use of textiles after manufacture. For indoor furnishings, many pollutants are related to building materials,[8] but furniture, carpets, draperies and wood or fabric furnishings probably give rise to more consumer complaints. This may be because of the presence of formaldehyde or volatile organic substances from wood and office furniture. Guidelines on chemical levels are available (though they do not cover non-industrial buildings), but research to investigate biological pollutants (potentially of more interest in the textile context) is much less extensive.

Secondary emissions from floorcoverings include harmful substances (especially formaldehyde) given off, for example from back coatings. Tests which can be carried out in order to provide a 'green' certification for carpets using suitable chemicals, processes, dyes and colorants are slowly becoming available.

Textiles can, though, play a valuable rôle in contributing to the reduction of air pollution. Many types of filter fabrics are produced, with an ability to remove particles with a range of sizes. The fine pores in a fabric are ideal for preventing the transmission of impurites while allowing air flow to take place. Filter fabrics, indeed, form a major class of technical fabrics and are used throughout the world in all kinds of situations.

19.5.3 Water pollution

Water pollution is more apt than any other type of pollution to be associated with the textile industry by the general public, mainly because, when it occurs, evidence of its existence in the form of coloured dyestuffs from dyeing and printing or detergent foam from scouring or washing is clearly visible. There are, though, other sources of water pollution generated by textile production.

Pollution in wet processing has reached alarming levels, and measures are being developed to reduce water consumption by changing or modifying processes, by lowering the concentration of waste products in water, by using only the optimum quantities of dyes or chemicals of an ecofriendly nature, and by carrying out appropriate restorative treatment. Using less water in manufacturing, reducing the number of steps in bleaching, and recovering chemicals from waste streams reduces both costs and pollution.

Sizing agents and starch are frequently considered to be the most serious sources of pollution in the textile industry, primarily from the volume of emissions present. The total cost of desizing is about 2.1 times the cost of sizing and is responsible for 3–4% of the cost of a loom state fabric. Starch/PVA (polyvinyl alcohol) size discharge can easily exceed legally permitted levels.

In the public mind, though, it is almost always the dyeing process which is associated with textile pollution, particularly with metals such as chromium, cobalt, nickel and copper. The dyeing process itself, possible substitutes, alternative reactive or acid dyes, and ways of minimizing residual dyestuff content may all be able to make some contribution to the overall aim of ecological improvement.

Natural dyes are not without their environmental problems and colour removal does not necessarily mean toxic substances have also been removed; the carcinogenic and toxic effects of dyes are of crucial importance. Mutagenic changes caused by ingesting textile effluents may mean that toxic effects will still be present after biological treatment, a fact which may form a basis for selecting dyes and chemicals for textile plants.

Many finishes can produce pollutant byproducts in the water stream. The use of oils, resins or other chemicals in finishing treatments is so diverse and so widespread that it is impossible to consider them all in a brief survey of this kind, but the application of flame-retardant, softening, durable-press, antistatic, soil-release, stain-resistant, waterproofing or oil-repellent finishes invariably uses materials which are harmful to the environment if discarded. Loss of lubricating or spinning oil from machinery can result in the accidental release of harmful substances, and spillage of diesel or other fuels from vehicles can occur. All of these products can bring about harmful side effects in either or both of the two ways mentioned earlier, by the poisoning of aquatic life or the enhancement of species such as algae, which remove oxygen from water and deprive aquatic creatures of this vital element. Waste discard of other kinds, from floor sweepings to excess chemical leachate from containers, can find its way into streams, either during a storm or by careless handling in cleaning or tidying. As mentioned earlier, an entire industry involving technical textiles has developed with the sole purpose of constraining, as far as possible, damage brought about by such unfortunate polluting events.

Again, textile effects on water pollution are not entirely negative. Indeed, one of the most important developments of the past few decades has been the use of geotextiles to contain pollution. Industrial waste in harbours and oil spills near sensitive coastal regions have been prevented from causing irreparable damage by the effective use of geotextile membranes to prevent widespread dissipation of the polluting substances, while ditch liners, landfill liners and stabilization fabrics for banks of vegetation have prevented the loss of valuable topsoil and the movement of soil containing pesticides or other harmful reagents into water supplies.

19.5.4   Land pollution

Land pollution can arise when a textile, or a substance used during its production, is thrown away on a landfill site. Fibres or chemicals can be harmful if their decomposition (as mentioned earlier) under the influence of air, water or sunlight produces a toxic agent. It is surprising and sad to see that there is virtually no attention paid to this problem in the textile literature, perhaps because it is so obvious and yet so easily accepted that it seems not to be interesting. Examples illustrating the omnipresence of the problem include a range of toxic breakdown products from materials such as polyester, nylon, or other polymers which have been discarded into the waste stream and find their way into a landfill site. Steps taken to render them 'biodegradable' include the use of starch as a source of bacterial nutrition or the incorporation of a substance decomposed by ultraviolet radiation, both of which facilitate disappearance of the waste material. Unfortunately, ultraviolet decomposition is only effective until the polymer is buried, and breakdown products are not attacked by either biodegradation technique, but aided in entering the soil more rapidly. From there, they can find their way into the water supply, acting as contaminants in the same way as if they had been discarded into a stream initially. Again, the valuable contribution of technical textiles in the form of barriers to this transfer cannot be forgotten, and may well prevent serious escape of pollutants from taking place.

19.5.5 Noise pollution

Noise pollution, ignored as an annoying but essentially harmless nuisance until recently, is becoming of more concern in the general population, though not, apparently, in the textile industry. The impression received from a review of the literature is that all that can be done to reduce noise has already been accomplished, and the residual problem is one that has to be tolerated. High noise levels are still generated in, for instance, twisting, spinning and weaving processes. Unpleasantly loud noise can also arise from the use of vehicles or other equipment in loading, shipping or handling raw materials or finished goods.

There is evidence to indicate that the effects of noise pollution are numerous, the most obvious one being hearing loss. Exposure to high intensity noise leads to deafness, at a rate which increases rapidly as the decibel level of exposure increases. The usual assumption made in legislation is that continuous exposure to a sound pressure level of 90 dB(A) is permissible throughout an eight-hour working day, but that exposure to higher noise levels, or to that level for a longer continuous period of time, must be restricted. This is accomplished by using some kind of derating curve or equation, with permitted exposure time continuously reduced as sound level increases, until no exposure at all is legally permitted at levels of 125 dB(A) or higher.

Other effects of noise exposure are less easily identifiable. There is a growing body of evidence to indicate that high noise levels bring about psychological changes, which may include frustration, carelessness, withdrawal or sullenness. Noise exposure over long periods of time has also been associated with increased absenteeism and even wilful destruction.

As before, textile products (and especially technical textiles) can be of service in controlling the effects of noise pollution. They enjoy widespread use as acoustic absorbent materials to reduce the annoyance of high sound levels for human beings.

19.5.6 Visual pollution

Visual pollution, often not considered as a problem, is in fact all pervasive. Not only are textile materials evident as waste strewn around the countryside, but advertisement hoardings frequently include material intended to increase the sale of these goods. Paper documentation and packaging, or plastic sheets used to wrap textiles displayed for sale, often find their way into landfill sites or are scattered around, from such sites or haphazardly, to offend the viewer by ruining the pristine sight of a peaceful natural vista. Landfill sites themselves, even where they manage to contain the waste goods, are gradually encroaching on more and more of the beauty of the earth and, once they are full, can often not be safely capped for fear of toxic breakdown substances being leached into the water table, as mentioned already. Once again, there appears to be little or no interest in developing research to alleviate the problem.

Textiles, once more, are useful in alleviating the problem by exploiting their aesthetic qualities to beautify our domestic surroundings.

19.6   Effects on the environment

As with all modern industries, substances released into the environment by textile producers are generally not harmless and are likely to have far-reaching effects if their release is at concentrations above safe levels. This limit is difficult to decide, as we do not know that a substance will be sufficiently diluted before it is absorbed by an organism, nor can we be certain that that organism will not concentrate toxic materials before any harm is caused. Hence, the most serious concern should be felt (and is often demonstrated) for the discharge of chemical pollutants into the natural environment via air or water or land.

Special consideration needs to be given to water contamination levels from dyeing, printing and finishing, and there is a need to make available adequate information on the ecological impact of chemical products. Some of the better-recognized problems include abnormal pH levels, suspended or settleable solids, oxygen demand, toxicity, colour, persistent bioaccumulative organic substances, mutagenic chemicals and a fish-flesh tainting propensity.

It is usually accepted that many substances that are discarded in textile production can bring about untold harm to nature. Fish can ingest toxins, can even be killed by them, and birds can be rendered sterile. Even if the lower creatures survive in spite of all these hazards, the potentially harmful contents of their bodies can be transmitted up the food chain to affect human beings. Because of concentration at each stage in this process, the end-consumer is exposed to a relatively high level of toxin that may have similar effects on human beings, in terms of sterility, altered genetic structure, or deformed births, as it could have had on other species.

Animals (including human ones) are also subjected to additional pollution-related risks in the form of long-term exposure to harmful substances, such as carcinogenic or other disease-causing agents, as a result of breathing contaminated air or being exposed to harmful agents. Toxic emissions from solvent spinning or tentering operations, and increased incidence of liver cancer among dyehouse personnel, have been noted and are examples of this type of hazard in the textile industry.

Plants can also suffer from the use of insecticides, herbicides or fertilizers which can cause stunted growth, sterility, or death from disease. Once again, even if the plant does not die, the presence of a harmful material can be passed up the food chain until it eventually reaches human beings and causes them harm as before.

19.7   Environmental harm reduction

In view of the damage which our careless attitude to planetary welfare has brought about in the past when our awareness was much lower, it is hardly surprising that a great deal of effort is being exerted by textile companies to reduce the harm that production is doing to the earth.

The ideas put forward for ecological conservation can be divided essentially into four classes. These are, respectively, the adoption of recycling as a means to cut down resource depletion, the use of ecologically friendly fibres or other materials, a reduction in the amount of pollution produced and improvement in methods of removing pollution after it has been generated.

Technical textiles may provide particular difficulties with respect to recycling. They are frequently used in conjunction with other materials, such as coatings and hardened oils, or as components of fibre-reinforced composites. These end-products may be difficult or impossible to break down satisfactorily into their original constituents, so that the textile polymer would be decomposed by any attempt at recovery. In additon, even where some degree of recovery is possible, the environmental cost of the extensive energy and reagent use needed may make the process prohibitively unattractive. Some use of the undestroyed materials may occur, though; if they are suitable for incorporation into, say, road-bed construction or concrete reinforcement, their chemical inertness and ability to resist mechanical stress may be of considerable advantage.

Ecofriendly textile processing is presented as a global challenge,[9] as ecological criteria are increasingly accepted in all parts of the world in selecting consumer goods and 'green' products can command a higher price. Two aspects of textile production, the limitation of harmful products and the reduction in air or water pollution, should be tackled in particular. Public interest in environmentally friendly processes is increasing, and examples of this trend can be found (in addition to the use of recycled bottles in making polyester) by a tendency to use organically grown cotton with naturally coloured dyes. Caution must be used, though, in order to prevent the real nature of so-called green products being overlooked when the green designation is a marketing ploy rather than a genuine benefit to the environment. Perkins[10] notes the transition towards green management, stating that good environmental management can lead to a quick return on investment. He discusses the new ISO standards in this context. Tyagi[11] discusses the environmental audit process, dividing it into the three sections of preaudit, at-site audit and postaudit. Arcangeli[12] feels that environmental management is here to stay, giving details of British developments and information on how companies can obtain accreditation.

The textile industry's resistance to environmental regulations was initially high[13] because of cost, but hard work has nevertheless been carried out to reduce waste and decrease resource use. Manufacturers have now discovered that waste minimization, as well as providing environmental benefits, can be financially beneficial to the industry. Standards, though, unfortunately suffer from the usual defect of ignoring all but the process for which they are established. Thus, no real account is taken in any of the work mentioned of the environmental cost of supplying the energy to run a plant, or of the way in which the planet is harmed by the extraction of ores to manufacture the steel or other special-purpose elements, like manganese or chromium, needed to improve its properties. The environmental costs of transportation are also ignored.

It is mainly in the area of water improvement that most efforts to reduce pollution are taking place. There is need for a reduction in textile waste water[14] to make manufacturers less dependent on changes in government regulations. Recycling cuts down waste, water use, energy and chemical costs, and Kramar[15] proposes practical solutions for this, suggesting not using water as a substrate, or developing measures to reduce consumption and pollution such as the use of plasma under vacuum, ink-jet printing, or dyeing with supercritical carbon dioxide.

Biotechnology has been recommended to reduce pollution after it has been produced[16] and to provide a cleaner industry, while Frey and Meyer[17] describe a new oxidation reactor that can treat heavily polluted water flows and reduce the consumption of both water and chemicals in textile finishing. They outline problems relating to current in situ disposal installations and give details of the benefits of the new system, which can convert up to 90% of the organic sewage present to carbon dioxide and water, as well as saving 80% of water and 20–30% of chemicals used.

However, most attention is being paid to the area of dyeing, presumably because the problem is seen to be greatest in that process. The recommendations of the Paris Commission[18] on environmental protection introduce sweeping new restrictions with regard to dyeing and finishing processes and the growing conditions for natural fibres. Health, safety and environmental regulations for dyeing have undergone significant changes in the last 25 years[19] and are now a major force in shaping the industrial workplace. Hohn[20] summarises the methods available for treating dyehouse effluent for the purposes of purification, including reductive, oxidative, filtration, dispersion, evaporation and condensation techniques.

A cost/benefit analysis of decolorisation by means of coagulation and flocculation treatment processes[21] indicates that the technique is very effective, but controversial because of sludge problems and ecological objections to the flocculation methods used. Activated carbon prepared from coir pith[22] is used to decolorise waste water from reactive dyes, although other work[23] shows that adsorption of dyes on peat has a similar extraction ability, presumably because surface changes and solution pH are important.

In the finishing area, some work is in progress. Regulations pertaining to exhaust air emissions are so difficult to meet that a compromise may have to be made if finishing plants are to stay operative. Textile manufacturers have to be persuaded that the benefits of complying with existing legislation for clean air are that finishing costs can be reduced significantly if the legal position can be met by identifying processes running optimally from the energy point of view.

19.8   Future prospects

Environmental damage is still occurring despite all efforts to reduce it. Harm cannot be prevented or ameliorated by the development of new energy sources, so we are compelled to face up to the fact that we can only reduce the impact of our activities on the planet by reducing consumption or by reusing goods. In the context of personal textiles, this means that clothing will have to last longer, and will only be discarded when it is worn out, not when it no longer appeals to the aesthetic or fashion sense of the wearer. There is the possibility of using second-hand garments, but it is conceivable that the supply of such articles will diminish as the ability to produce new ones falls, or as the range of styles is reduced. Where technical textiles are concerned, attempts to achieve extended life may well result in the continuation of use past the point where it is safe to do so. The collapse of buildings constructed from architectural textiles, the breaking of ropes in a climbing accident, and the tearing of sails or of a tent in a storm all provide further examples of the type of potential disaster that could occur.

There is current concern regarding textile chemicals, environmental protection and aspects of finishing, and the use of ecolabelling, mentioned already, is seen by various workers as a key to future control of pollution. However, even though environmental demands and ecolabelling may well bring about problems in developing countries,[24] environmental constraints deserve to be paramount in new textile development.

Printing may also require special attention. It seems that ozonation has the best potential for success and will best meet future directives. These are likely to be more restrictive, since conventional treatment times and discharge via municipal systems will no longer be tolerated. It is thus critical to reduce pollution and know exactly the nature and appropriate treatment that any product will need to meet colour removal under more stringent future laws. A word of caution should be sounded. Simultaneous compliance with customers' needs and the requirements of present or future legislation may be impossible and the effort to satisfy both criteria will provide major problems for textile printers and finishers. The total print management (TPM) system is suggested as a means of giving the greatest possible reduction in waste (or its total prevention) in water and printing paste use.

An issue that arises is the effect of environmental demands on fashion creativity. Strong, brilliant colours, modified surfaces and unusual material combinations may be needed at times for fashion, but may be unacceptable ecologically. Designers therefore need help, especially regarding ecological information, from suppliers.

One final approach needs to be mentioned, that of meeting the challenge by using 'waste' textiles in unexpected ways. Suggestions include[25] adopting wool batts in facades of buildings for durable heat and sound insulation, serving also as pollution-binding elements. Textile waste is used in concrete aggregate, where polyamide warp-knit fabric waste gives excellent strength and can be used to provide elasticity in on-site cast concrete, as in road building. Polluted land can be used to grow non-edible crops, such as flax or hemp, and to detoxify land at the same time since metal concentration in the soil is lower after harvest, indicating that toxins must be absorbed. The crop is excellent for the textile industry and as a cellulose source in paper making. There must, of course, be some concern about the final fate of the metals once the fabric is discarded.

The future of textile production capability is also likely to be dependent to a considerable extent on environmental factors. Current predictions are that population levels will approximately double every 35 years or so. While this might be taken as an indication of an increased need for textile goods, the large number of people to feed will also require the use of more land for growing foodstuffs. This will, in turn, mean less available space for growing textile-related crops, especially cotton. Some fibres, such as linen or hemp, may still be produced without any competition from food crops, since they grow on marginal land or in ditches, while wool, or other animal hair fibres, may enjoy the advantage of being able to grow on marginal land, and of providing a ready source of food as well as fibres.

In addition, people consume other commodities besides food. Their needs will strain the world's manufacturing resources (especially oil) to the utmost. In consequence, oil will become a scarce commodity and textile fibres derived from it may be given lower priority than that accorded to other, more easily recognised uses, such as transportation or aerospace, thus preventing or impeding the manufacture of artificial fibres to compound the overall fibre shortage.

Yet the demand for fibres will still exist. Textile goods are essential to human presence in most parts of the world, because we have lost the ability as a species to survive the rigours of climate without some form of protection in the form of body covering. The other reasons for clothing use, adornment, the display of wealth or status, physical or psychological comfort and modesty, will presumably still be important, at least in the foreseeable future. Thus the greatly increased demand stemming from a much larger population base will be extremely difficult to meet. This means that textile materials will have to be much more durable, since they cannot be replaced as often, and that the fashion industry (which depends for its very existence on a rapid replacement schedule of garments) can expect to be drastically curtailed.

New uses are also appearing on a continuing basis for technical textiles. There is an added need for industrial goods based on textile structures. Sporting and outdoor activities are becoming more popular as leisure time increases, and new means of satisfying demands are constantly being sought. The tremendous value of geotextiles is becoming more obvious as industrial expansion occurs and the increased need for environmental protection is recognized.

Still other environmental concerns will affect textile use. As population and consumption rise, so also do resource depletion and pollution production. The planet is already stretched to the limit in coping with the demands placed on it; the continuing onslaught on the tropical forest cover that is so crucial to the earth's ability to replace lost oxygen, for instance, is of grave concern to thinking people everywhere but shows little sign of diminishing. The parallel loss of fresh water and pure air will, it is to be assumed, also continue at an accelerated rate as the population rises, at least until some shock event (such as mass asphyxiation or poisoning) brings the human race to its senses. All industry, not just the textile one, will have to share the responsibility for this increased planetary load, driven essentially by the human greed to consume, so industrial activity will at some point need to be curbed. The exact time at which this will have to take place is unpredictable, but as long as the population rises and no steps are taken to reduce consumption drastically, the end-result of a choice between survival or reduction in lifestyle is inevitable.

Steps that can be taken by the textile industry to reduce global damage include the modification of processes to make them more simple and more labour intensive, because these changes tend to lower energy or resource use and pollution production. The results of this trend, though, will include lower quality products, price increases and a smaller selection of goods, together with an increased importance for durability properties in comparison with aesthetic ones. Marketing will be adversely affected; a reduction in exports will result from the lower availability of goods and the need to reduce fuel consumption, and longer replacement times for existing textile products will have to be accepted.

The future for textile production, and especially for the use of technical textiles, appears to be assured, but the need for preservation of the fragile environment may well become a major cause of concern that tempers the possibility of unfettered expansion in response to the ever-rising demand.

References

1. G HORSTMANN, 'Environmental trends in textile production', *Pakistan Textile J.*, 1996 **45**(2) 58–62.
2. T A PERENICH, 'Environmental issues in the textile industry', *Colourage*, 1996 **43**(8) 19–22.
3. L KRAMAR, 'Environmental opportunities in textiles', 35th International Man-made Fibres Congress, Proceedings Conference, Dornbirn 1996 (Österreichisches Chemiefaser-Institut), Dornbirn, 1996.
4. C BLUM, 'The textile industry and the environment – a threat or a positive challenge? *Asian Textile J.*, 1995 **4**(2) 41–45.
5. L NORGAARD, '"Green" cotton – lifecycle assessment in textiles', *Niches in the World of Textiles*, Proceedings Textile Institute 77th World Conference, Vol 1, Institute of Fiber, Textile and Clothing Science, Tampere University, 1996, 169–175.
6. M KRALIK, 'Environmental influences decisive for manufacture and application of textile auxiliaries', *Vlakna a Textil* 1995, **2**(2) 55–60.
7. P STRAUSE, C ROGERS and WEI TIAN, 'Influence of reworkable waste on yarn quality', Beltwide Cotton Conferences. Proceedings Conference Nashville, 1996, Vol. 2, ed. P Dugger and D Richter, National Cotton Council, Memphis, 1996, 1334–1335.
8. D FRANKE, C NORTHEIM and M BLACK, 'Furnishings and the indoor environment', *J. Textile Inst.*, 1994 **85**(4) 496–504.
9. R PAUL, J THAMPI and M JAYESH, 'Eco-friendly textile processing – a global challenge', *Textile Dyer and Printer*, 1996 **29**(16) 17–20.
10. W S PERKINS, 'The transition toward "green" management', *Amer. Textiles Int.*, 1996 **25**(5) 62–63.
11. P C TYAGI, 'Environmental audit', *Asian Textile J.*, 1995 **3**(11) 49–53.
12. L ARCANGELI, 'Environmental management – here to stay', *Laundry and Cleaning News*, 1996 **18**(9) 24–25.
13. E J ELLIOTT, 'Textiles' role in the environment', *Textile World*, 1995 **145**(9) 221–222.
14. J J PORTER, 'Reduction of textile wastewater using automatic process control, recycle and filtration', in *Environmental Chemistry of Dyes and Pigments*, eds. A Reife and H S Freeman, Wiley, 1996, 193–214.
15. L KRAMAR, 'To keep clear of water pollution: state of the art and solutions', *Industrie Textile*, 1995 **1266**(6) 44–48.
16. A WILSON, 'Biotechnology for a cleaner textile industry', *Int. Dyer*, 1996 **181**(12) 10.
17. F FREY and M MEYER, 'Cracking industry's wastewater challenge', *America's Textiles International*, 1996 **25**(11) 56–58.
18. ANON, 'Parcom recommendations 94/95 III', *Tinctoria*, 1996 **93**(9) 64–68.
19. H M SMITH, 'U.S. safety, health and environmental regulatory affairs for dyes and pigments', in *Environmental Chemistry of Dyes and Pigments*, eds. A Reife and H S Freeman, Wiley, 1996, 295–306.
20. W HOHN, 'Effluent purification methods for indirect discharge', *Int. Textile Bull.*, Dyeing/Printing/Finishing, 1995 **41**(4) 62–67.
21. J JANITZA and S KOSCIELSKI, 'Cost/benefit ratio of chemical coagulation/flocculation/flotation methods for clearing and decolouring coloured textile effluents', *Melliand Textilberichte*, 1996 **77**(10) 684–687.
22. H GURUMALLESH-PRABU and K RAVICHANDRAN, 'Studies in colour removal by coir pith', *Asian Textile J.*, 1995 **3**(8) 53–54.
23. K RAMAKRISHNA and K VIRARAGHAVAN, 'Dye removal using peat', *Amer. Dyestuff Reporter*, 1996 **85**(10) 28–34.
24. A HYVARINEN, 'Environmental demands and ecolabelling', *Textile Asia*, 1996 **27**(4) 90–91.
25. K DOMSTIFT BRANDENBURG, STEGLICH, W LEUCHT and W THUMANN, 'Use of natural fibres for facade elements and process for producing the same', EP 0700477, 13 March 1996.

Index

Lightning Source UK Ltd.
Milton Keynes UK

177612UK00001B/1/P